DATE	

DISCARD

DIFFERENTIAL EQUATIONS FOR ENGINEERS

DIFFERENTIAL EQUATIONS FOR ENGINEERS

Thomas M. Creese

Assistant Professor of Mathematics
The University of Kansas

Robert M. Haralick

Professor of Electrical Engineering
The University of Kansas

McGraw-Hill Book Company

New York / St. Louis / San Francisco
Auckland / Bogotá / Düsseldorf / Johannesburg
London / Madrid / Mexico / Montreal / New Delhi
Panama / Paris / São Paulo / Singapore
Sydney / Tokyo / Toronto

DIFFERENTIAL EQUATIONS FOR ENGINEERS

1234567890 DODO 783210987

This book was set in Times Roman. The editors were A. Anthony Arthur and James W. Bradley; the cover was designed by John Hite; the production supervisor was Leroy A. Young.
The drawings were done by Oxford Illustrators Limited.
R. R. Donnelley & Sons Company was printer and binder.

Library of Congress Cataloging in Publication Data

Creese, Thomas M
    Differential equations for engineers.

    Includes index.
    1. Differential equations. 2. Engineering mathematics. I. Haralick, Robert M., joint author. II. Title.
TA347.D45C73   515'.35   77-22340
ISBN 0-07-013510-X

CONTENTS

PREFACE

We are marching in the footsteps
Of those who've gone before

Old song

This book is dedicated to the footsteps of those who will come after—the students and colleagues who will use it.

TO THE STUDENT

This text was written for you—to read, use, and understand. It emphasizes the calculations and techniques which are really used to set up and solve real differential-equations problems. It contains many engineering applications and other examples which you should work through, calculating and writing out for yourself, using the text to guide you. Such a detailed guide is for your first time through the material. The sample problems are the next step in mastering the material. They are sprinkled liberally through the book, and the answers are given to help you check your work. Finally, there are problems for which no answers are given, just as real-world engineering problems do not come with answers listed in answer sections.

Chapter I shows some of the diversity of engineering problems. In each, we are asked to predict the behavior of an engineering system. Each leads to a differential equation whose solution provides the required prediction. The solution techniques are calculus techniques.

In contrast to Chapter I, Chapter II shows the systematic and routine treatment of whole classes of electrical and mechanical systems. You will see that mechanical systems and electrical systems lead to the same differential equations. The chapter illustrates the fact that different engineering disciplines solve their problems with the same mathematical tools, but the emphasis is on the technique of "modeling," beginning from an engineering problem and progressing systematically to a differential equation describing the system.

Chapter III is concerned with two transform methods for solving those equations. The first, the Laplace transform, is sufficiently versatile to solve any equation derived by the methods of Chapter II.

In any particular problem, some other method may be more efficient or give further insight into the nature of the problem. Therefore, the remainder of the book is devoted to other techniques and useful unifying ideas, particularly the concepts of linear equation and linear operator.

It is very important that you read and study the book sufficiently to develop the confidence that you are *learning from the printed page*. This is important to your career, for professional engineers do not have an instructor around to explain. They must go to whatever written materials are available and learn what they can from those materials, which will be rather mathematical, condensed, and uncompromising by comparison with the book you now hold.

Our book is designed for you to read; it is designed to be an early step on the way to obtaining mathematical information from the printed page. Therefore, practice with it. Practice reading, studying, and learning at the same time and in the same way that you practice solution techniques for engineering differential-equations problems.

TO THE INSTRUCTOR

We cannot expect that most engineering students will have enjoyed calculus or achieved the level of proficiency to which they must now aspire. However, every student should benefit from seeing how the work undertaken leads toward the goal of solving real engineering problems. Therefore, we have taken as a principle that the applications should come first, that the techniques should be selected for their practical use and practiced until they are used proficiently, and that the unifying ideas should be introduced only when there are things already familiar to the student which justify the effort required to unify them.

As a concomitant, we have avoided partial differentiation and multiple integration as prerequisites. We teach differential equations to first-semester sophomore engineering students. The multivariable calculus course comes the following semester and uses topics from differential equations for motivation. In particular, the derivation of the Laplace transform of the convolution integral appears in this book, but in class we use the formula without deriving it. The derivation becomes one of the first examples in the multivariable calculus course which follows.

Since power-series solution methods have largely disappeared from under-graduate engineering curricula, we have made power series prerequisite only to one appendix. Elsewhere it can be avoided.

Our organization is also unusual in that Laplace transforms (Chapter III) come before the characteristic polynomial (Chapter IV), because we find that our students definitely prefer it that way. Consequently, Chapter III is the core of the book, and systems analysis is its turning point. Thereafter, the techniques are the motivation for the unifying ideas; but earlier, the techniques require motivation from the engineering examples.

Chapter I, the introduction, has a theme, but it is essentially a collection. Chapter II, by contrast, shows that widely differing practical systems lead systematically to a single class of equations. It reveals the similarities of basic engineering problems. It is sufficiently axiomatic so that it can be taught confidently by a nonengineer. Section A should be covered. The beginning of Section B should be followed by either the latter part of Section B or by Section C. Systems of linear differential equations, introduced here, become a theme worth repeated attention thereafter. Elimination of variables from such a system need not be the obstacle which students have found it in the past, because Section D is really effective. It deserves painstaking coverage. The matter is brought to fruition in Chapter IV, where the parallels with Gaussian elimination are made more apparent.

We build gradually toward the idea of a linear operator, and in Chapter IV we give it a workout. The treatment of the method of undetermined coefficients is unusually operator-oriented, and in it full use is made of the characteristic polynomial.

For the sake of those who wish to teach power-series solution methods, an appendix on the subject is included. It is more condensed than the rest of the book, however, and we take the occasion to point out to the student that most of the useful properties of special functions are not contingent on deriving a power series or evaluating the function.

In one 5-hour semester (72 class meetings), we do Sections A and B, and the essentials of Sections D, E, and F of Chapter I. The material from Sections H and I we work in gradually, because we consider efficient graphing to be important. From Chapter II we omit Section C, and from Chapter III we omit Sections F and H. In Chapter IV we usually treat linear independence with only portions of Section E. This leaves time for some additional applications early in the semester, or a few lectures on vector- and matrix-valued functions at the end.

A more traditional 3-hour course would be based on:

Chapter I: Sections A, B, D, E, F, H, J
Chapter II: Section A or B and part of D
Chapter III: Sections A to E and Part 7 of Section G
Chapter IV: Nearly all

In any section, the choice of those numbered parts which may be omitted without loss of continuity depends on plans for the rest of the semester, but in general the more readily expendable parts are entitled "Remarks on . . ." or come at the end of the section.

Either or both of the appendixes to Chapter I, on the treatment of nonlinear equations, can be inserted after Section D or Section B. With a little care they could even come before Section B.

ACKNOWLEDGMENTS

We have each team-taught this material with many others, and we are deeply indebted to our former teammates. In particular, Section C of Chapter I is based on a lecture given by E. J. McBride; the aerospace examples were among those offered by Jan Roskam and C. T. Lan; and the draining-tank problem was presented by George Forman. Most of the drawings for Chapter I and many of those for Chapter II were originally prepared for class by Judith Vago and Patricia Moore under a grant from the University of Kansas Office of Instructional Resources.

Most of the typing was done by Jean Reinfrank, Shirley Roberts, Rockne Grauberger, Sharon Amundson, and Lynn Ertebati, whose skill and forbearance have made this their project as well as ours.

The text has benefited from the comments of our students. James Hague and Steven Esch have looked for errors and solved problems. The mistakes that may remain are ours, and we invite your comments, for the sake of those whose footsteps will come after yours.

Thomas M. Creese
Robert M. Haralick

DIFFERENTIAL
EQUATIONS
FOR ENGINEERS

INTRODUCTION, FEATURING FIRST-ORDER LINEAR DIFFERENTIAL EQUATIONS

We begin our study by deriving an equation for the concentration of a chemical in a lake. The equation is a differential equation which we solve in order to predict the concentration at times in the future. With the equation and its solution, we illustrate the uses of "mathematical models" of practical situations and suggest their importance to the engineer.

The solution of the equation, in Section A, is accomplished by reduction to the form

$$\frac{dy}{dt}(t) = g(t)$$

Solutions of such equations are easy to find, by integration. Sections B and C are therefore devoted to using such equations to illustrate the "general solution" and "initial values," on which predictions can later be made concerning the behavior of engineering systems.

In Section D we finish our discussion of the differential equation of Section A. Together with Sections A and B, this section constitutes the core of the chapter.

Many ideas and formulas from calculus appear in Sections A through D. The reader should review differentiation and integration generally, with particular attention to:

1. The logarithm
2. The exponential function, including the law of exponents and the fact that $e^x > 0$ for all x
3. The Fundamental Theorem of Calculus

Additional models are presented in Section E. Some of these lead to the same equation as that of Sections A and D, but others lead to "nonhomogeneous" equations, for which a solution method is given in Section F. After this section, one can reasonably proceed to Chapter II.

Section G contains a collection of models from many parts of engineering. Some lead to equations of "order" 2, 3, and 4, and one leads to a "nonlinear" equation, illustrating ways in which the equations of Chapter I are only a beginning for our study.

The chapter closes with a synopsis of the properties of the first-order linear differential equation, together with appendices on graphing and nonlinear equations.

The wide variety of examples in this chapter illustrates the "problem-oriented" use of mathematics in engineering. The differences between the examples will be obvious; the similarities require some pointing out. In Chapter II the similarities will be treated systematically.

A. A MIXING PROBLEM: THE POLLUTED LAKE

1. Deriving the Equation

A lake with a constant volume, V cubic meters, has an inlet admitting a constant flow, I cubic meters/second, and an outlet through which the flow is the same I. At time t the lake contains water and $P(t)$ cubic meters of pollutant; the concentration is $y(t) = P(t)/V$. Water flows into the lake, but there is no pollutant flowing into the lake. If the concentration of pollutant in the outflow is the same as the concentration in the lake (for instance, if the concentration is uniform throughout the lake), then the rate of decrease of the volume of pollutant in the lake is

$$-\frac{dP}{dt} = Iy \tag{1}$$

This is why: dP/dt is the rate of change of P. Its values are positive when P is increasing, negative when P is decreasing. Thus dP/dt may be called the *rate of increase* of P. Observe that a negative increase is a decrease. Now it is easy to see that $-dP/dt$ should be called the *rate of decrease* of P. The rate of decrease of pollutant in the lake is the rate at which pollutant flows through the outlet. Specifically, it is the concentration y (the fraction of the volume which is pollutant) times the rate of flow I (in cubic meters per second). Equation (1) states that these two expressions for the rate of decrease must give the same value. (For more details see Part 2, Verifying the Role of the Derivative, on page 4.)

Equation (1) is a *differential equation*, in which P and y are unknown functions (but I is known). Our task is to find a formula for one or the other of them, in order to predict the volume of pollutant and the concentration in the lake at a future time.

Equation (1) has a drawback, in that both P and y are unknown. On the other hand, $P(t) = Vy(t)$ as remarked above, and V is constant, so that

$$\frac{dP}{dt} = V\frac{dy}{dt}$$

This quantity may be substituted into the left-hand side of (1), giving

$$-V\frac{dy}{dt} = Iy$$

Look at this equation closely to see what it contains. Both V and I are determined by the shape of the lake and its outlet. They are known and constant, even though their values have not been given. On the other hand, y is unknown (and so is its derivative, dy/dt or y'), and this unknown function is the one we seek. Division by $-V$ in the last equation gives

$$y'(t) = -\frac{I}{V}y(t) \tag{2}$$

If the lake has been correctly described, then the function y, the concentration of pollutant in the lake, must satisfy Equation (2).

What did the description of the lake consist of?

1. The lake has constant volume V.
2. It has constant inflow I, containing no pollutant.
3. The concentration of pollutant in the lake $y(t)$ (a function of time) is the same as the concentration in the outflow, for each time t.
4. It has constant outflow, also I.

Under these conditions, Equation (2) is said to be a *model* of the lake, for if $y(t)$ is the concentration of pollutant in the lake, then the function y must satisfy Equation (2).

If I is not taken to be constant, but if the inflow and the outflow are still equal so that V is constant, then the same analysis holds. That is, the constancy of I is not required. At time t the rate of decrease is still $-\frac{dP}{dt}(t)$. The rate of flow of pollutant through the outlet is still $y(t)I(t)$, and so the rate of decrease of $P(t)$ is still $y(t)I(t)$ also. Thus,

$$-\frac{dP}{dt}(t) = y(t)I(t)$$

Division by $-V$, as before, gives

$$\frac{dy}{dt}(t) = \frac{-I(t)}{V}y(t) \tag{3}$$

If one considers $y(t)$ to be the value at t of the function y, and likewise $I(t)$ is the value at t of I, and $(dy/dt)(t)$ is the value at t of $y' = dy/dt$, then (3) yields an equation relating the functions y', y, and I:

$$y' = -\frac{I}{V}y \tag{4}$$

2. Verifying the Role of the Derivative

Engineers should check their derivations very carefully, and when developing a model from first principles, as we did, they will often check dimensions and the role of the derivative. We shall do both:

$P(t) =$ volume of pollutant in the lake, cubic meters
$V =$ volume of lake, cubic meters
$y(t) =$ concentration of pollutant in the lake (volume of pollutant divided by total volume of the lake), dimensionless
$I =$ volume rate of flow of lake water into the inlet and out of the outlet, cubic meters per second
$\Delta t =$ time interval, seconds

If the concentration of pollutant in the outflow is the same as the concentration in the lake, then the amount of pollutant flowing out of the lake in a time interval of Δt seconds is the concentration of pollutant in the lake times the volume rate of flow times the time interval Δt. Thus the volume of pollutant flowing out of the lake between time t and time $t + \Delta t$ is (approximately)

$$y(t)I\,\Delta t \quad \frac{\text{cubic meters}}{\text{second}} \times \text{seconds} = \text{cubic meters} \tag{5}$$

Since the volume of pollutant in the lake at $t + \Delta t$ must equal the volume of pollutant in the lake at t less the amount which flows out in Δt seconds, the

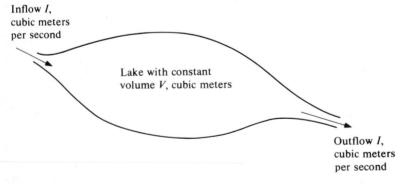

Inflow I,
cubic meters
per second

Lake with constant
volume V, cubic meters

Outflow I,
cubic meters
per second

Figure 1 Flow in the lake.

principle of conservation of mass gives

$$P(t + \Delta t) - P(t) = -y(t)I \; \Delta t \qquad \text{cubic meters} \tag{6}$$

This equation contains a pair of expressions for the increase in volume of pollutant in the lake, the negative of (5), which was the decrease. Division by Δt gives
the average rate of increase,

$$\frac{P(t + \Delta t) - P(t)}{\Delta t} = \frac{-y(t)I \; \Delta t}{\Delta t}$$

$$= -y(t)I \qquad \text{cubic meters per second}$$

Taking the limit as Δt goes to zero gives the rate of increase at time t:

$$\frac{dP}{dt}(t) = -y(t)I \qquad \text{cubic meters per second}$$

Thus we have verified the role of the derivative in (1), and while we did it we
checked the dimensional consistency of our expressions, and of Equation (1).
    The approximation in Equation (5) deserves a word. The expression $y(t)I \; \Delta t$
is approximate because $y(t)$ depends on t. It is exactly because the concentration
varies with time that we need to make predictions of its values. However, since the
concentration clearly washes out continuously, it follows that $y(t)$ is a continuous
function of t, and therefore the approximation improves as Δt goes to zero.

3. Making the Prediction

In the lake of the example above, suppose that $I/V = 10^{-6}$, so that

$$y'(t) = -10^{-6}y(t) \tag{7}$$

and divide by $y(t)$, leaving for later any worries about the possibility that $y(t) = 0$:

$$\frac{y'(t)}{y(t)} = -10^{-6}$$

The left-hand side is the derivative of $\ln y(t)$:

$$\frac{d}{dt} \ln y(t) = \frac{1}{y(t)} y'(t)$$

by the chain rule. Thus,

$$\frac{d}{dt} \ln y(t) = -10^{-6} \tag{8}$$

or $$\qquad \ln y(t) = -10^{-6}t + C \qquad \text{for some } C \tag{9}$$

Note here that the lake *is* polluted, there *is* a function y, it *does* satisfy
Equation (7), it *does* therefore have property (8), and hence there *does* exist some
number C such that (9) is also satisfied at each time t. Imagine, now, that we go to

the lake just once, at time $t = 0$, and measure the concentration $y(0)$. We measure it at that one time, and on the basis of that one measurement and Equation (9) predict the future concentrations. Certainly,

$$\ln y(0) = -10^{-6} \cdot 0 + C$$
$$= C$$

and so C is determined by the one measurement. Now replace C in (9), the one place it appears, by its measured value $\ln y(0)$, so that (9) becomes

$$\ln y(t) = -10^{-6}t + \ln y(0)$$

Exponentiation now gives $y(t)$, the desired function:

$$y(t) = e^{\ln y(t)}$$
$$= e^{-10^{-6}t + \ln y(0)}$$
$$= e^{-10^{-6}t} e^{\ln y(0)}$$
$$= y(0)e^{-10^{-6}t}$$

If our lone measured value was $y(0) = 3 \cdot 10^{-7}$, then

$$y(t) = 3 \cdot 10^{-7} e^{-10^{-6}t} \tag{10}$$

That is, the concentration diminishes exponentially at a rate dependent upon the initial value $y(0)$ and upon I/V.

In Part 2 on page 4, we verified that Equation (1) describes the lake [you may want to check that this is so even if the volume of the lake varies with time, so long as $y(t) = P(t)/V(t)$ is taken as the concentration]. As long as the volume is assumed constant, Equation (1) is equivalent to (4). It remains to check that the function in Equation (10) satisfies (4). That is, we must differentiate (10), then substitute into (4), and see if (4) is satisfied at every value of t. Differentiating (10)

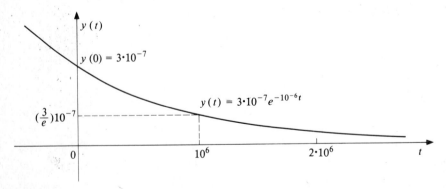

Figure 2 Graph of (10), predicting pollution concentration.

gives

$$y'(t) = 3 \cdot 10^{-7} e^{-10^{-6}t}(-10^{-6})$$
$$= y(t)(-10^{-6})$$
$$= -\frac{I}{V} y(t)$$

for every value of t, as required, since I/V was taken to be 10^{-6}. Notice also that the value of $y(0)$ is $3 \cdot 10^{-7}$, as required.

4. Summary and Definition

We have worked a practical example, first deriving Equation (4) and checking that its description of the lake satisfied the assumptions about the lake, and then solving the equation given the values of I, V, and $y(0)$. The process of deriving differential equations to model practical situations will reappear in this chapter and the next, and occasionally thereafter. The solution method we used will be explained in more detail in Section D of this chapter. The way in which a solution such as (10) is determined by the *initial condition*, the value of y at some single point, is the subject of Section B. There remains, however, a question: What does it mean for a function to "satisfy" a differential equation such as that considered here? The answer to that may be almost obvious: If y has a derivative at each point t, and if for each t the values of $y(t)$ and $y'(t)$ are substituted into both sides of the equation, and if the two sides are then equal for all values of t, then y *satisfies* the equation. The important points are that y must have a derivative and that the equation must be satisfied at each t, for instance,

$$y'(t) = -\frac{I}{V} y(t) \qquad \text{for each } t$$

If y has these properties, then y is a *solution* of the equation.

B. INTRODUCTION TO INITIAL VALUE PROBLEMS AND GENERAL SOLUTIONS

1. Problems

Use integration to solve the following *initial value problems*.

1. $y'(t) = \sin t$
   $y(0) = 1$

2. $y'(t) = \sin 5t$
   $y(0) = 0$

3. $y'(t) = \sin t$
   $y(0) = -3$

4. $y'(t) = 2e^{3t}$
   $y(0) = -5$

5. $y'(t) = 2e^{3t}$
   $y(1) = 0$

6. $y'(t) = 2e^{3t}$
   $y(-2) = 1$

7. $y'(t) = t^{-1}$
   $y(1) = 0$

8. $y'(t) = 3t^{-1}$
   $y(1) = -7$

9. $y'(t) = 3t^{-1}$
   $y(-1) = 0$

The problems given above may all be solved by integration, in the following way: find the indefinite integral of the right-hand side, then evaluate the constant of integration so as to satisfy the initial condition.

2. Example

$$\begin{aligned} y'(t) &= 5t^2 \\ y(1) &= 4 \end{aligned}\Bigg\} \tag{1}$$

The problem is solved by finding the indefinite integral of y',

$$y(t) = \frac{5t^3}{3} + C \qquad \text{for some } C \tag{2}$$

and then evaluating C:

$$4 = y(1) = \frac{5 \cdot 1^3}{3} + C$$

$$4 = \tfrac{5}{3} + C$$

$$\tfrac{7}{3} = C$$

This value may be substituted back into the indefinite integral (2), giving the following antiderivative of $5t^2$:

$$y(t) = \frac{5t^3}{3} + \frac{7}{3} \tag{3}$$

Differentiation shows that y satisfies the differential equation of (1) at each point t, and so y is a solution. Also

$$y(1) = \tfrac{5}{3} \cdot 1^3 + \tfrac{7}{3} = \tfrac{12}{3} = 4$$

and so y satisfies the condition $y(1) = 4$ as well. Since the constant C in (2) is arbitrary, it is clear that we could have solved *any* initial value problem of the form

$$\begin{aligned} y'(t) &= 5t^2 \\ y(1) &= y_0 \end{aligned}\Bigg\} \tag{4}$$

in which y_0 is any arbitrary number.

If $y_0 = -5$, then

$$-5 = y(1) = \frac{5}{3} \cdot 1^3 + C$$

$$-5 - \frac{5}{3} = C$$

$$-\frac{20}{3} = C$$

$$y(t) = \frac{5t^3}{3} - \frac{20}{3} \tag{5}$$

Is it possible that some other function solves the problem? Any function which satisfies (4) has $5t^2$ as its derivative. Functions with the same derivative can differ only by a constant and hence are of the form (2):

$$y(t) = \frac{5t^3}{3} + C \qquad \text{for some } C$$

Since the value $y(1) = y_0$ is specified at $t = 1$, it follows that C is determined. In case $y_0 = -5$ we have $C = -\frac{20}{3}$, and in case $y_0 = 4$ we have $C = \frac{7}{3}$, as we saw previously. In fact, for any value of y_0,

$$y_0 = y(1) = \tfrac{5}{3} \cdot 1^3 + C$$

and so
$$C = y_0 - \tfrac{5}{3}$$

Substitution into (2) gives

$$y(t) = \tfrac{5}{3}t^3 + y_0 - \tfrac{5}{3}$$
$$= \tfrac{5}{3}(t^3 - 1) + y_0 \qquad (6)$$

For any given value of y_0, (6) is the unique solution of (4).

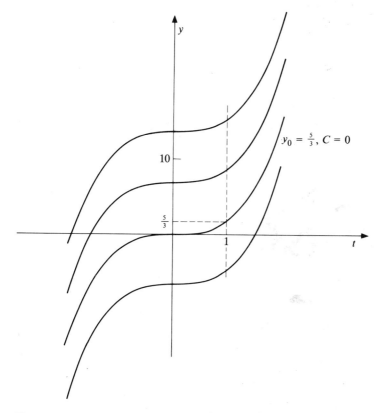

Figure 1 Representative graphs of $y(t) = \tfrac{5}{3}t^3 + C$, the general solution of $y'(t) = 5t^2$.

3. General Solution and Particular Solutions

The graphs in Figure 1 show only a few of the solutions of the equation

$$y'(t) = 5t^2 \tag{7}$$

but those which are not shown differ from those that are by some constant, and their graphs are vertical translates of those given.

Since all solutions of that equation are of the form

$$y(t) = \tfrac{5}{3}t^3 + C \tag{8}$$

and since (as we have seen) all functions of this form are solutions of Equation (7), we call (8) the *general solution* of (7). The term is actually used in two ways. On the one hand, it refers to the actual formula (8). On the other hand it also refers to the collection of all functions of the form of (8), the collection of all solutions of (7). This collection can be written two ways in the " set builder " notation you may have seen previously:

$$\{y \,|\, y'(t) = 5t^2 \text{ for all } t\} = \{y \,|\, \text{for some } C, y(t) = \tfrac{5}{3}t^3 + C, \text{ for all } t\}$$

Any solution of (7) is a member of this collection and is referred to as a *particular solution*. In the problems given at the beginning of the section, you were required to find particular solutions satisfying specified initial conditions.

4. Initial Conditions

In the example studied earlier,

$$\left. \begin{aligned} y'(t) &= 5t^2 \\ y(1) &= 4 \end{aligned} \right\}$$

the condition

$$y(1) = 4$$

is called the *initial condition*. The terminology seems natural if one thinks back to the situation of the polluted lake. There, the differential equation related the rate of change of concentration to the value of the concentration, and we made a prediction of concentration, as a function of time, after a single measurement $y(0)$. That is, $y(0)$ was the *initial* concentration, and $y(t)$ was the predicted value of the concentration for time $t > 0$. Now look at Figure 1. The differential equation determines the whole family of curves, but each curve is then uniquely determined by going through some given point. Among all the solutions of the equation, any particular solution may be uniquely determined by a single value.

Do not be confused by the word *initial*. In engineering problems it *usually* refers to a value measured (or a condition set) at the beginning of a time interval, but it could as well be measured at the termination, for the particular solution is determined to the left as well as to the right by its value at the initial point. This is clear from Figure 1, for the initial conditions were set at $t = 1$, an arbitrary point.

5. Sample Problems with Answers

Find the general solution of each of the following equations.

A. $y'(t) = \sin 5t$ *Answer:* $y(t) = -\frac{1}{5}\cos 5t + C$

B. $y'(t) = 4e^{-7t}$ *Answer:* $y(t) = -\frac{4}{7}e^{-7t} + C$

Solve the following initial value problems.

C. $\left.\begin{array}{l} y'(t) = \sin 5t \\ y(0) = 0 \end{array}\right\}$ *Answer:* $y(t) = -\frac{1}{5}\cos 5t + \frac{1}{5} = \frac{1}{5}(1 - \cos 5t)$

D. $\left.\begin{array}{l} y'(t) = 4e^{-7t} \\ y(1) = -2 \end{array}\right\}$ *Answer:* $y(t) = -\dfrac{4e^{-7t}}{7} + \dfrac{4e^{-7}}{7} - 2$

6. Problems

Find the general solution.

10. $y'(t) = \sin t$ **11.** $y'(t) = \cos 5t + 2$

12. $y'(t) = 2e^{3t}$ **13.** $y'(t) = 4t + 6$

7. Integrating a Discontinuous Function

It is possible to find the integral of a function which is continuous but given by different formulas on adjacent intervals, or of some discontinuous functions. However, we shall start by considering the corresponding differentiation problem, for instance that of the function

$$y(t) = \begin{cases} t^2 & t < 0 \\ t & t \geq 0 \end{cases}$$

Since the function y is given by the formula t^2 on all of the interval $t < 0$, it can be differentiated there, giving

$$y'(t) = 2t \qquad t < 0 \tag{9}$$

(See Figure 2.) On the right half line $t > 0$, we have $y(t) = t$ so that

$$y'(t) = 1 \qquad t > 0 \tag{10}$$

At the point $t = 0$, where the intervals are adjacent, the function y is not differentiable. Putting together formulas (9) and (10) gives the derivative of y:

$$y'(t) = \begin{cases} 2t & t < 0 \\ 1 & t > 0 \end{cases}$$

Functions such as y occur very frequently in applications, where machines are turned off and on, valves are opened and shut, or circuits are closed and opened. Each such operation causes a change in the formulas for the functions describing the behavior of the system. We shall see many examples later on.

Now consider the reverse process, which is to take us from y' to a continuous function y. Let us take a different example to illustrate the integration, letting

$$f'(t) = \begin{cases} -t & t < 2 \\ t & t > 2 \end{cases}$$

and finding that continuous function f which takes the value 3 at $t = 1$.

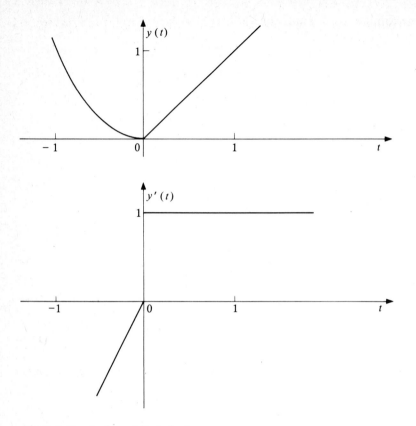

Figure 2 Graph of y and its derivative.

Since the initial point $t = 1$ occurs in the interval $t < 2$, we shall start by solving the initial value problem on that interval. The problem may be stated

$$\left.\begin{array}{l} f'(t) = -t \\ f(1) = 3 \end{array}\right\} \quad t < 2$$

By integration, the general solution of the differential equation is

$$f(t) = -\frac{t^2}{2} + C \qquad t < 2$$

and evaluation at the initial point gives

$$3 = -\tfrac{1}{2} + C$$

$$3\tfrac{1}{2} = C$$

$$f(t) = -\frac{t^2}{2} + 3\frac{1}{2} \qquad t < 2 \tag{11}$$

Since f is to be continuous, the function value $f(2)$ at the end of the interval must be given by formula (11) also, so that

$$f(t) = -\frac{t^2}{2} + 3\frac{1}{2} \qquad t \le 2 \tag{12}$$

In fact, $\qquad f(2) = -\frac{4}{2} + 3\frac{1}{2} = 1\frac{1}{2}$

Now let us find f on the interval $t > 2$, where

$$f'(t) = t \qquad t > 2$$

The general solution is

$$f(t) = \frac{t^2}{2} + D \qquad t > 2$$

Since f is to be continuous, the limit at $t = 2$ must be $1\frac{1}{2}$. Thus,

$$1\frac{1}{2} = \frac{4}{2} + D$$

$$D = -\frac{1}{2}$$

$$f(t) = \frac{t^2}{2} - \frac{1}{2} \qquad t > 2 \tag{13}$$

Putting (12) and (13) together gives a formula for the function value of f at t, for each t, namely

$$f(t) = \begin{cases} -\dfrac{t^2}{2} + 3\dfrac{1}{2} & t \le 2 \\[2mm] \dfrac{t^2}{2} - \dfrac{1}{2} & t > 2 \end{cases}$$

(See Figure 3.)

The calculation needs to be checked for three things. First, the initial value must be correct:

$$f(1) = -\frac{1^2}{2} + 3\frac{1}{2} = 3$$

using the formula for the interval $t < 2$ since $1 < 2$. Second, the function f must be continuous. Since it is continuous on the interval $t < 2$ and on the interval $t > 2$, where it is a polynomial, there remains the verification of continuity at $t = 2$. Here the formulas must give the same limit.

$$\lim_{t \to 2} \left(-\frac{t^2}{2} + 3\frac{1}{2} \right) = 1\frac{1}{2}$$

$$\lim_{t \to 2} \left(\frac{t^2}{2} - \frac{1}{2} \right) = 1\frac{1}{2}$$

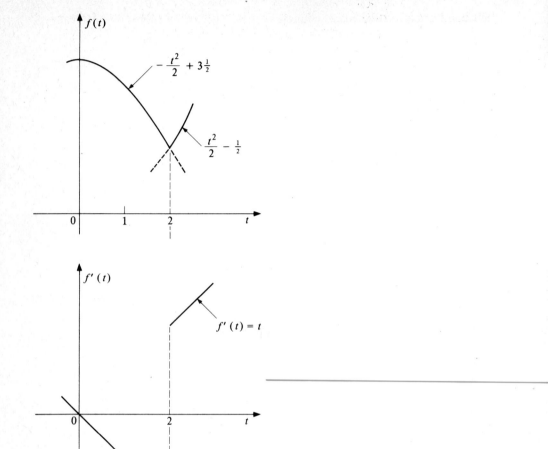

Figure 3 The function f and its derivative.

That limit must give the function value there, $f(2)$. Finally, the derivative must be correct on each interval. It is correct for $t < 2$ because

$$\frac{d}{dt}\left[-\frac{t^2}{2} + 3\frac{1}{2}\right] = -t$$

and for $t > 2$ because

$$\frac{d}{dt}\left[\frac{t^2}{2} - \frac{1}{2}\right] = t$$

Here is another example, illustrating the fact that the first integration should be carried out on the interval where the initial value is given, whether that is the left-hand interval or the right. Let

$$y'(t) = \begin{cases} 2 & t < 2\pi \\ \sin t & t > 2\pi \end{cases}$$

and let $y(3\pi) = -1$. We solve first the initial value problem on the interval containing the initial point $t = 3\pi$:

$$\left.\begin{array}{l} y'(t) = \sin t \\ y(3\pi) = -1 \end{array}\right\} \quad t > 2\pi$$

Solution

$$y(t) = -\cos t + C \qquad t > 2\pi$$

$$-1 = y(3\pi) = -\cos 3\pi + C$$

$$= 1 + C$$

$$C = -2$$

$$y(t) = -\cos t - 2 \qquad t > 2\pi$$

That gives us the formula for y on the right-hand interval and at $t = 2\pi$, because y is to be continuous. Thus,

$$y(2\pi) = -3 \tag{14}$$

and

$$y(t) = -\cos t - 2 \qquad t \geq 2\pi \tag{15}$$

Now we can solve the differential equation on the other interval.

$$y'(t) = 2 \qquad t < 2\pi$$

$$y(t) = 2t + D \qquad t < 2\pi$$

Since y is to be continuous, the limit of $2t + D$ at $t = 2\pi$ must be $y(2\pi)$ or -3, as calculated in (14). Thus,

$$-3 = 2(2\pi) + D$$

$$D = -4\pi - 3$$

$$y(t) = 2t - 4\pi - 3 \qquad t < 2\pi \tag{16}$$

Putting together formulas (15) and (16) for the values of y on the two intervals, we get

$$y(t) = \begin{cases} 2t - 4\pi - 3 & t < 2\pi \\ -\cos t - 2 & t \geq 2\pi \end{cases}$$

8. Sample Problems with Answers

Solve the following initial value problems. Be sure that your solutions are continuous.

E. $y'(t) = \begin{cases} 0 & t < 1 \\ 1 & t > 1 \end{cases}$ $\quad$ *Answer:* $y(t) = \begin{cases} 2 & t \leq 1 \\ t + 1 & t > 1 \end{cases}$

$y(-1) = 2$

F. $y'(t) = \begin{cases} 0 & t < 1 \\ 1 & t > 1 \end{cases}$ *Answer:* $y(t) = \begin{cases} -2 & t < 1 \\ t - 3 & t \geq 1 \end{cases}$

$y(2) = -1$

G. $y'(t) = \begin{cases} e^{-2t} & t \leq 0 \\ e^{2t} & t > 0 \end{cases}$ *Answer:* $y(t) = \begin{cases} \frac{1}{2}(e^{-2t} - 1) & t \leq 0 \\ \frac{1}{2}(e^{2t} - 1) & t > 0 \end{cases}$

$y(0) = 0$

9. Problems

Solve the following initial value problems. Be sure that your solutions are continuous.

14. $y'(t) = \begin{cases} 0 & t < 1 \\ 1 & t > 1 \end{cases}$ **15.** $y'(t) = \begin{cases} 1 & t < 1 \\ 0 & t > 1 \end{cases}$

$\quad\;\; y(-1) = 1$ $\qquad\qquad\qquad\qquad y(0) = 1$

16. $y'(t) = \begin{cases} e^{3t} & t > 0 \\ 1 & t < 0 \end{cases}$ **17.** $y'(t) = \begin{cases} e^{3t} & t > 0 \\ 1 & t < 0 \end{cases}$

$\quad\;\; y(-1) = -1$ $\qquad\qquad\qquad\quad\; y(1) = -1$

18. $y'(t) = \begin{cases} t^{-1} & t > 1 \\ 1 & t < 1 \end{cases}$ **19.** $y'(t) = \begin{cases} \cos 3t & t > 0 \\ 0 & t < 0 \end{cases}$

$\quad\;\; y(0) = 0$ $\qquad\qquad\qquad\qquad y(-1) = -1$

C. OBTAINING DIFFERENTIAL EQUATIONS FROM SIMPLE GRAPHS

In Figure 1 a linear relation between the independent variable t and the dependent variable y is illustrated. This can be expressed by the algebraic equation

$$y(t) = mt + b \qquad (1)$$

If the derivative of y with respect to t is taken, it is seen that

$$\frac{dy}{dt}(t) = m \qquad (2)$$

This is a differential equation, one of whose solutions is given by Equation (1) and is shown in Figure 1. Note that in the differential equation the constant b was "lost." Equation (2) is solved by integrating both sides:

$$\int \frac{dy}{dt}(t)\, dt = \int m\, dt$$

$$y(t) = mt + C \qquad (3)$$

The constant of integration C is arbitrary and can only be fixed when the specific initial condition is known. If the initial condition on y is specified when $t = 0$, then the statement of the initial condition is $y(0) = b$, meaning that at $t = 0$, the value of the function y is the number b. Without the initial condition, Equation (2) has an infinite number of solutions, one solution for each value of C in Equation (3). This is shown in Figure 2, where every member of the family of straight lines of slope m is a solution of Equation (2).

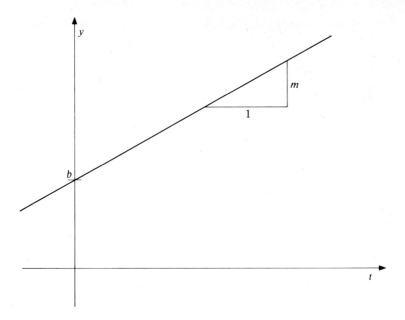

Figure 1 The line $y(t) = mt + b$.

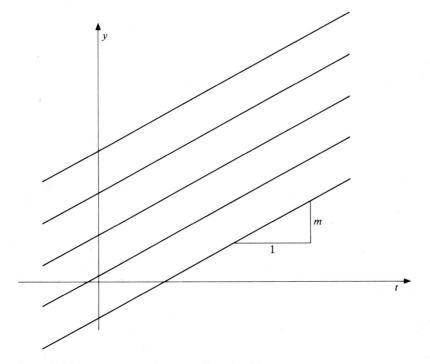

Figure 2 Lines of slope m, solutions of Equation (2).

Another simple relation between the independent variable t and the dependent variable y is given by a polynomial containing both the first and second powers of t,

$$y(t) = a_0 + a_1 t + a_2 t^2 \tag{4}$$

The graph of this relation is shown in Figure 3.

Equation (4) can be differentiated once, giving

$$\frac{dy}{dt}(t) = a_1 + 2a_2 t \tag{5}$$

Again it should be noted that the constant a_0 has been lost.

A second differentiation, differentiation of Equation (5), yields

$$\frac{d}{dt}\left[\frac{dy}{dt}(t)\right] = 2a_2$$

This can, of course, be written

$$\frac{d^2 y}{dt^2}(t) = 2a_2 \tag{6}$$

Now the constants a_0 and a_1 have been lost.

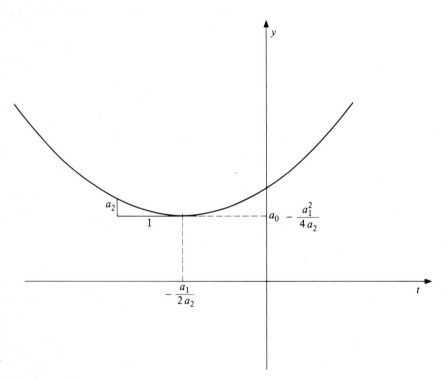

Figure 3 The parabola $y(t) = a_2 t^2 + a_1 t + a_0$.

This last differential equation contains a second derivative. It can be solved by two integrations. The first gives:

$$\int \frac{d}{dt}\left[\frac{dy}{dt}(t)\right] dt = \int 2a_2 \, dt$$

$$\frac{dy}{dt}(t) = 2a_2 t + C_1$$

The second then follows:

$$\int \frac{dy}{dt}(t) \, dt = \int 2a_2 t \, dt + \int C_1 \, dt$$

$$y(t) = a_2 t^2 + C_1 t + C_2 \tag{7}$$

The solution has two arbitrary constants C_1 and C_2. This certainly means that the solution of Equation (6) may be any parabola of the shape of that shown in Figure 3. The constants C_1 and C_2 merely determine the vertex of the parabola. Why? Can you show it by completing the square in Equation (7)?

The exponential relation is a familiar one:

$$y(t) = y_0 e^{at} \tag{8}$$

Graphs of this function are shown in Figure 4. One curve is for $a > 0$, the other for $a < 0$; the horizontal line is for $a = 0$.

Differentiation of Equation (8) gives

$$\frac{dy}{dt}(t) = a y_0 e^{at} \tag{9}$$

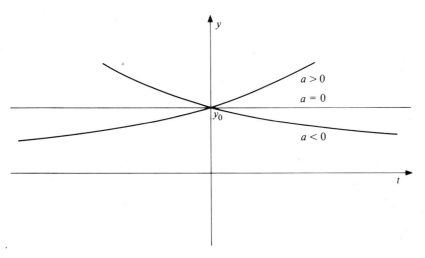

Figure 4 The graph of $y(t) = y_0 e^{at}$, for three values of a.

Because of (8), the expression $y_0\, e^{at}$ may be replaced in (9) by the expression $y(t)$, giving the differential equation satisfied by the exponential:

$$\frac{dy}{dt}(t) = ay(t) \tag{10}$$

Equation (10) is solved by the separation of variables technique, as illustrated in Section A.

$$\frac{1}{y(t)}\frac{dy}{dt}(t) = a$$

$$\frac{d}{dt}\ln y(t) = a$$

$$\ln y(t) = at + C$$

$$y(t) = e^C e^{at} \tag{11}$$

Without an initial condition, Equation (10) has an infinite number of solutions, one solution for each value of C appearing in Equation (11). This is shown in Figure 5, illustrating the graphs of the family of exponential functions. Every member of this family is a solution of Equation (10).

A technique which is adaptable to many situations consists of solving for y_0 and then differentiating. Since y_0 is constant, the calculation in the present example gives the differential equation satisfied by y, as follows:

$$y(t) = y_0 e^{at}$$

$$y_0 = y(t)e^{-at}$$

$$0 = y'(t)e^{-at} - ay(t)e^{-at} = e^{-at}[y'(t) - ay(t)]$$

$$0 = y'(t) - ay(t)$$

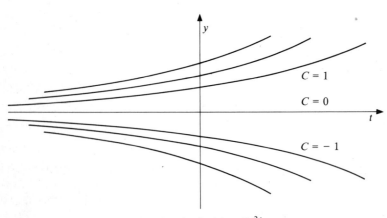

(a) The graph of $y(t) = Ce^{2t}$,
for several values of C

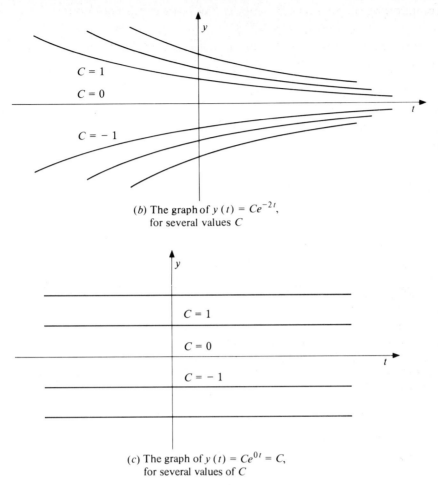

(b) The graph of $y(t) = Ce^{-2t}$,
for several values C

(c) The graph of $y(t) = Ce^{0t} = C$,
for several values of C

Figure 5 Graphs of Ce^{at} for $a = 2, -2,$ and 0.

D. A SEPARATION OF VARIABLES METHOD TO SOLVE THE EQUATION $y'(t) + a(t)y(t) = 0$

1. The Method of Separation of Variables

The method of separation of variables has already been used in the polluted-lake example in Section A. The method must now be explained in greater detail. The general form of the equation is

$$y'(t) + a(t)y(t) = 0 \qquad (1)$$

where $a(t)$, the coefficient of $y(t)$, is a known function; we seek a function $y(t)$ which "satisfies" this equation on some interval. That is, y should be differentiable on that interval (y' must exist there), and at each t in the interval the equation must hold. The equation of the example in Section A,

$$y' = -10^{-6}y \qquad \text{or} \qquad y' + 10^{-6}y = 0$$

21

has a solution valid on the whole real line, for we checked at the end of the example that

$$y(t) = 3 \cdot 10^{-7} e^{-10^{-6}t}$$

actually satisfies the differential equation for all t.

Now the task is to find a function y satisfying

$$y'(t) + a(t)y(t) = 0$$

for functions $a(t)$. Proceeding as before, transpose the second term

$$y'(t) = -a(t)y(t)$$

and divide by $y(t)$:

$$\frac{y'(t)}{y(t)} = -a(t) \qquad \text{if } y(t) \neq 0 \tag{2}$$

If we are interested only in the function $y(t) \equiv 0$ then we are done, for that function surely satisfies the equation everywhere. Let us look for some other solution, for instance one which takes a nonzero value $y(t_0)$ at some point t_0.

Let us suppose first that y has a positive value at t_0, that is, $y(t_0) > 0$. Since y is to be differentiable it must certainly be continuous, and so $y(t)$ only has values near $y(t_0)$ if t is near t_0. That is, there is an interval about t_0 on which $y(t) > 0$. (See Figure 1.) From the calculus formula $(d/dt)[\ln y] = y'/y$, and from (2), we have

$$\frac{d}{dt}[\ln y(t)] = -a(t) \tag{3}$$

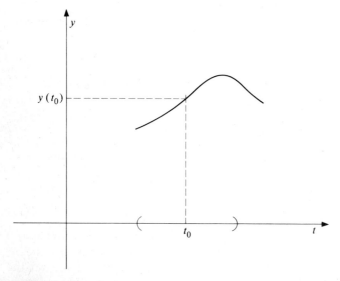

Figure 1 A differentiable function which is positive at t_0 is also positive on some interval about t_0.

[We need to know that $y(t) > 0$ before we can take the natural logarithm. We do not take logarithms of negative numbers. This is a point to which we shall return later, but we shall do the calculation on the interval where $y(t) > 0$.]

If A is any function whose derivative is the coefficient function, that is, if $A'(t) = a(t)$ for all t in that interval, then integration of (3) gives

$$\ln y(t) = -A(t) + C \tag{4}$$

where C is the arbitrary constant of integration, and exponentiation gives

$$y(t) = e^{\ln y(t)} = e^{-A(t) + C}$$
$$= e^C e^{-A(t)} \tag{5}$$

The constant C may be arbitrary as far as Equation (3) is concerned, but if $y(t_0)$ is fixed then C is determined by $y(t_0)$. In (4), for instance,

$$\ln y(t_0) = -A(t_0) + C$$

and therefore
$$C = \ln y(t_0) + A(t_0)$$

Substitution into (5) gives

$$y(t) = e^{\ln y(t_0) + A(t_0)} e^{-A(t)}$$
$$= y(t_0) e^{-A(t) + A(t_0)} \tag{6}$$

Recall that $e^x > 0$ for each x, and so $e^{-A(t) + A(t_0)}$ is positive. Since $y(t_0) > 0$, it follows that the product (6) is also positive, neither negative nor zero, wherever it is defined.

Although the calculation above was performed assuming $y(t) > 0$, the function (6) will satisfy the differential equation (1) even if $y(t_0)$ is zero or negative.

Here are the main points in cookbook form. We are given $a(t)$ on some interval. Take any antiderivative $A(t)$, and write the solution

$$y(t) = y(t_0) e^{-A(t) + A(t_0)} \tag{7}$$

Differentiation and substitution into the left-hand side of (1) show that

$$y'(t) + a(t)y(t) = y(t_0) e^{-A(t) + A(t_0)} \left[-\frac{d}{dt} A(t) \right]$$
$$+ a(t) y(t_0) e^{-A(t) + A(t_0)}$$
$$= y(t_0) [-a(t) + a(t)] e^{-A(t) + A(t_0)}$$
$$= 0 \tag{8}$$

Thus (7) represents a differentiable function which satisfies Equation (1). It is defined on the interval where the coefficient function a has an antiderivative, and it never changes sign. It is always positive if $y(t_0) > 0$, always negative if $y(t_0) < 0$, and always zero if $y(t_0) = 0$, and it is a solution of (1) because it is differentiable and satisfies (1) at each point, as (8) shows.

2. The Uniqueness of the Solution

It is important to know whether there are any functions other than (7) which satisfy the equation and take value $y(t_0)$ at t_0, for our ability to predict the behavior of the solution, as exemplified by our prediction of the concentration of pollution in the lake, is dependent on being able to specify some *one* function. Suppose that there were a second function, also satisfying Equation (1) and with the same initial value. Specifically, suppose that

$$w'(t) + a(t)w(t) = 0 \qquad \text{and} \qquad w(t_0) = y(t_0) \qquad (9)$$

Can $w(t)$ differ from $y(t)$? Let us see. Suppose first, as before, that $y(t_0) > 0$. By the same analysis as before,

$$\frac{w'(t)}{w(t)} = -a(t)$$

Thus,
$$\frac{d}{dt}[\ln w(t)] = -a(t)$$

and
$$\ln w(t) = -B(t)$$

where B is some antiderivative of a. By the same analysis as before,

$$w(t) = w(t_0)e^{-B(t) + B(t_0)} \qquad (10)$$

However, it is one of the important facts of calculus that two functions with the same derivative differ by some constant function whose value is the value of the difference at t_0. This must be true of A and B:

$$A(t) - B(t) = A(t_0) - B(t_0)$$

or
$$A(t) - A(t_0) = B(t) - B(t_0)$$

Substituting the left-hand side into (10) gives

$$w(t) = w(t_0)e^{-B(t) + B(t_0)}$$
$$= w(t_0)e^{-A(t) + A(t_0)}$$
$$= y(t_0)e^{-A(t) + A(t_0)}$$
$$= y(t)$$

since $w(t_0) = y(t_0)$ by (9).

That is, if $y(t_0) > 0$, then w and y are the same solution of the differential equation (1) taking that value at t_0. That solution is defined on the largest interval where $a(t)$ is continuous, and it is given by

$$y(t_0)e^{-A(t) + A(t_0)}$$

where A may be any antiderivative of the coefficient function a.

But what if $y(t_0) < 0$? The computation (8) is correct even if $y(t_0) \leq 0$ (check it and see), and so $y(t) = y(t_0)e^{-A(t) + A(t_0)}$ is a solution of (1). We must now see that it is uniquely determined by its value $y(t_0)$. If $w(t)$ is another such function, then $-w(t)$ and $-y(t)$ both satisfy (1), both take the value $-y(t_0)$ at t_0, and both are positive. By what we have seen about positive solutions, it follows that

$$-w(t) = -y(t) \qquad \text{and hence} \qquad w(t) = y(t)$$

as required.

Finally, there is the possibility that $y(t_0) = 0$. If $w(t)$ is any function which satisfies (1) and if $w(t)$ takes a nonzero value anywhere, then $w(t)$ is *never* zero but either always positive or always negative, as we have seen. Therefore, only the function which is zero at every point can take the value zero at t_0.

Thus, a solution of (1) is either always positive, always negative, or identically zero, and it is entirely determined by any one of its values, that is, by its value at any one point.

3. Recapitulation

If A is any antiderivative of a given function a on an interval, if t_0 is a point of that interval, and if y_0 is any number, then on that interval

$$y'(t) + a(t)y(t) = 0$$

has solution
$$y(t) = y_0 e^{-A(t) + A(t_0)} \tag{11}$$

and that function is the only solution whose value at t_0 is y_0. That is, the problem

$$\left. \begin{array}{l} y'(t) + a(t)y(t) = 0 \\ y(t_0) = y_0 \end{array} \right\}$$

has a unique solution, which is

$$y(t) = y_0 e^{-A(t) + A(t_0)}$$

This is a fact which should be memorized.

4. Example

Solve the equation $y'(t) + ty(t) = 0$ for a function y such that $y(2) = 3$.

Note that $t_0 = 2$ and $a(t) = t$. We can find $A(t)$ at once, for any antiderivative of $a(t) = t$ will do:

$$A(t) = \frac{t^2}{2}$$

$$-A(t) + A(t_0) = -\frac{t^2}{2} + \frac{2^2}{2}$$

$$= -\frac{t^2}{2} + 2$$

Thus,
$$y(t) = y(t_0)e^{-A(t)+A(t_0)}$$
$$= y(t_0)e^{-t^2/2+2}$$
$$= 3e^{-t^2/2+2}$$
$$= 3e^2e^{-t^2/2} \tag{12}$$

Check that the function satisfies the differential equation:
$$y'(t) = 3e^2e^{-t^2/2}\frac{d}{dt}\left[-\frac{t^2}{2}\right] = 3e^2(-t)e^{-t^2/2}$$

and so
$$y'(t) + ty(t) = 3e^2(-t)e^{-t^2/2} + t3e^2e^{-t^2/2}$$
$$= -3e^2te^{-t^2/2} + 3e^2te^{-t^2/2}$$
$$= 0 \qquad \text{for all } t$$

Check that the function also has the required value at $t_0 = 2$:
$$y(2) = 3e^2e^{-2^2/2} = 3e^2e^{-2} = 3$$

And if, instead, we needed a solution to the same equation, but such that $y(2) = -7\sqrt{2}$?
$$y(t) = -7\sqrt{2}e^2e^{-t^2/2}$$

Check for yourself, by substitution in the given equation, that y satisfies the equation, and check also that it satisfies the condition $y(2) = -7\sqrt{2}$. And if, instead, we needed a solution to the same equation, but satisfying the condition $y(2) = 0$?
$$y(t) = 0e^2e^{-t^2/2} = 0$$

Notice that

If $y(2) > 0$	then $y(t) > 0$	for all t
If $y(2) < 0$	then $y(t) < 0$	for all t
If $y(2) = 0$	then $y(t) = 0$	for all t

For each number C there is exactly one function y with both of the following properties:
$$\left.\begin{array}{c} y'(t) + ty(t) = 0 \\ y(2) = C \end{array}\right\}$$

Thus every solution of the equation is one of the functions $y(t) = Ce^2e^{-t^2/2}$, and the set of all such functions, the set of all functions of the form
$$Ce^2e^{-t^2/2} \qquad C \text{ some number}$$
is the general solution of the equation $y'(t) + ty(t) = 0$.

5. Definitions

We have used the words "solution" and "general solution" previously. For reference, here is what is meant, precisely: If a is a continuous function on an interval,

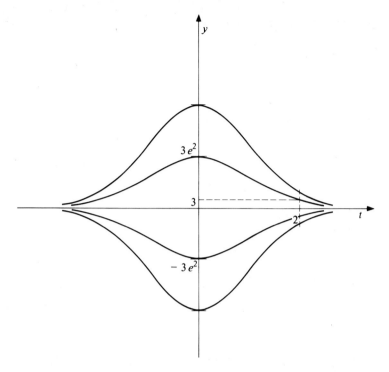

Figure 2 Particular solutions of $y'(t) + ty(t) = 0$.

then a function y is a *solution* of the equation

$$y' + ay = 0 \tag{13}$$

on that interval if y has a derivative $y'(t)$ at each point t of the interval and if

$$y'(t) + a(t)y(t) = 0 \tag{14}$$

at each point t of that interval. The *general solution* of (13) is the collection of all solutions of Equation (13).

Thus, a solution is a function (on an interval), but the general solution is a collection of functions. A member of the collection is often called a *particular solution*. (See also Section B.)

The difference between Equations (13) and (14) is worth attention. Very often it is said, "$\sin t$ is a function of t" or "e^t is a function." Many functions have names, in these instances "sine" and "exponential." In (13), y', a, y, ay, $y' + ay$, and 0 are names of functions, quite as much as "sine" or "log." If we want to refer to the value of such a function at, say, 3, then we write $\sin 3$, $\log 3$, or $y'(3) + a(3)y(3)$, and there is no disagreement about the meaning of these expressions. What, then, is the value of y at a point called t_0? It is $y(t_0)$, even though a value for t_0 has not been specified in advance. No exception needs to be made for t: if a point is called t, then the value of y there is $y(t)$. That is, y is a function, and $y(t)$ is the value of y at t. Notice that $y(t)$ is a number, while y is a function. We all agree that the value of $y(t)$ depends on the value of t, and in this sense we may continue to say, " $y(t)$ is a function of t," but please recognize now and hereafter that y is the name of the function in the expression $y(t)$.

Thus, (14) means "the value of y' at t plus the product of the values of a and y at t is the number zero."

Attempts to make a universally applicable consistent notation for the distinction between a function and its values necessarily run afoul of the traditional practice of letting e^t be the name of the exponential function and letting t^n be the name of the function which assigns to each number its nth power. Thus, we shall take $e^{-t^2/2}$ not only to be a number, a function value at t, but also to be the name of the function itself. Where possible, though, we shall distinguish between the function and its value, between a and $a(t)$.

6. Remarks on the Use of the Definite Integral

The quantity $A(t) - A(t_0)$, whose negative is the exponent appearing in (11), is in reality the integral of the function a, from t_0 to t:

$$A(t) - A(t_0) = \int_{t_0}^{t} a(z)\, dz \tag{15}$$

That is just the Fundamental Theorem of Calculus, which you should now review, but a question is invariably asked: "Where did the z come from?" We must settle that question completely.

Suppose first that we integrate from α to β:

$$A(\beta) - A(\alpha) = \int_{\alpha}^{\beta} a(t)\, dt$$

That is no trouble. In our examples,

$$\int_{\alpha}^{\beta} t\, dt = \frac{t^2}{2}\Big|_{\alpha}^{\beta} = \frac{\beta^2}{2} - \frac{\alpha^2}{2}$$

Note that the right-hand side, $\beta^2/2 - \alpha^2/2$, does not depend upon t. That is, $\int_{\alpha}^{\beta} t\, dt$ does not depend upon t:

$$\int_{\alpha}^{\beta} u\, du = \frac{u^2}{2}\Big|_{\alpha}^{\beta} = \frac{\beta^2}{2} - \frac{\alpha^2}{2} \qquad \int_{\alpha}^{\beta} v\, dv = \frac{v^2}{2}\Big|_{\alpha}^{\beta} = \frac{\beta^2}{2} - \frac{\alpha^2}{2}$$

In other words, in a definite integration, the integral depends not on the variable of integration, but only on the integrand and the end points of the interval. However, we *want the integral to depend on t*, as in (15). After all, we are just seeking to write $A(t) - A(t_0)$ as an integral, and so we want t_0 and t for the end points of the interval of integration:

$$A(t) - A(t_0) = \int_{t_0}^{t} (?)$$

The integrand, obviously, must be the function a. For the variable of integration we may choose any letter which has no other use in the problem, for instance z, and that is where the z in (15) comes from. Note that differentiation of the integral now gives no problem, for t appears only in one place:

$$\frac{d}{dt}[A(t) - A(t_0)] = \frac{d}{dt}\left[\int_{t_0}^{t} a(z)\, dz\right] = a(t)$$

and in our example

$$\frac{d}{dt}\left(\frac{t^2}{2} - \frac{t_0^2}{2}\right) = \frac{d}{dt}\left(\int_{t_0}^{t} z\, dz\right) = t$$

Thus, the solution to $y'(t) + ty(t) = 0$ may be written

$$y(t) = y(t_0)e^{-\int_{t_0}^{t} z\, dz}$$

and in general the solution of $y'(t) + a(t)y(t) = 0$ may be written

$$y(t) = y(t_0)e^{-\int_{t_0}^{t} a(z)\, dz} \tag{16}$$

This fact is a companion to (11), and it is worth remembering, too.

The variable z is called the *variable of integration* or *dummy variable*. In older usage, "dumb" meant "silent," and a "dummy" was a silent person or thing. In this case the variable is silent in that it does not appear again after the integration is performed.

7. Example: The Integral in the Lake Problem

In the lake problem discussed previously, the integral

$$-\int_{t_0}^{t} a(z)\, dz = -A(t) + A(t_0)$$

just derived has a very particular meaning, an interpretation in this problem. Look again at the model.

$$y'(t) = \frac{-I(t)}{V} y(t)$$

and so $$y(t) = y(t_0)e^{-(1/V)\int_{t_0}^{t} I(z)\, dz}$$

Now I is the rate of flow into the lake, and so $\int_{t_0}^{t} I(z)\, dz$ is the volume of water which flows into the lake during the period from t_0 to t.

Question: If $V = 10^6$, and if $y(t_0) = 3 \cdot 10^{-7}$, then what is the concentration at the time t_5 when $5 \cdot 10^6$ cubic meters of water will have flowed into the lake?

$$\int_{t_0}^{t_5} I(z)\, dz = 5 \cdot 10^6$$

$$y(t_5) = y(t_0)e^{-(1/V)\int_{t_0}^{t_5} I(z)\, dz}$$

$$= 3 \cdot 10^{-7}e^{-5 \cdot 10^6/10^6}$$

$$= 3 \cdot 10^{-7}e^{-5}$$

Note that we did not have to know the value of t_5 in seconds, nor did we need to know what formula gave $I(t)$. We needed to know only the volume of water which flowed in: when $5 \cdot 10^6$ cubic meters have flowed in, the concentration is reduced by a factor $e^{-5 \cdot 10^6/V} = e^{-5}$.

All this we can tell by solving the differential equation. Do you see now why a differential equation may be called a *model* of a system? A little work with the equation allows us to make a forecast of how a real system will work.

8. Sample Problems with Answers

These are suitable for any method discussed in the section. Try each method as you learn it.

A. $\begin{aligned} y'(t) + 2y(t) &= 0 \\ y(0) &= 3 \end{aligned}$ Answer: $y(t) = 3e^{-2t}$

B. $\begin{aligned} y' - 3y &= 0 \\ y(2) &= 5 \end{aligned}$ Answer: $y(t) = 5e^{-6}e^{3t} = 5e^{3(t-2)}$

C. $\begin{aligned} y' + 2y &= 0 \\ y(-4) &= 6 \end{aligned}$ Answer: $y(t) = 6e^{-8}e^{-2t} = 6e^{-2(t+4)}$

D. $\begin{aligned} y' - 3y &= 0 \\ y(2) &= -5 \end{aligned}$ Answer: $y(t) = -5e^{-6}e^{3t} = -5e^{3(t-2)}$

E. $\begin{aligned} y' + 2y &= 0 \\ y(5) &= 0 \end{aligned}$ Answer: $y(t) \equiv 0$

F. $\begin{aligned} y'(t) + 2ty(t) &= 0 \\ y(0) &= 4 \end{aligned}$ Answer: $y(t) = 4e^{-t^2}$

G. $\begin{aligned} y'(t) + 2ty(t) &= 0 \\ y(2) &= 3 \end{aligned}$ Answer: $y(t) = 3e^4e^{-t^2} = 3e^{-(t^2-4)}$

H. $\begin{aligned} y'(t) + 2ty(t) &= 0 \\ y(2) &= -3 \end{aligned}$ Answer: $y(t) = -3e^4e^{-t^2} = -3e^{-(t^2-4)}$

I. $\begin{aligned} y'(t) + 2ty(t) &= 0 \\ y(2) &= 0 \end{aligned}$ Answer: $y(t) \equiv 0$

Find the general solution.

J. $y' + 2y = 0$ Answer: $y(t) = Ce^{-2t}$
K. $y' - 3y = 0$ Answer: $y(t) = Ce^{3t}$
L. $y'(t) + 2ty(t) = 0$ Answer: $y(t) = Ce^{-t^2}$

9. Problems

Use the method of separation of variables to solve the following initial value problems.

1. (a) $\begin{aligned} y'(t) + 3y(t) &= 0 \\ y(0) &= 5 \end{aligned}$ (b) $\begin{aligned} y'(t) + 3y(t) &= 0 \\ y(0) &= 2 \end{aligned}$

(c) $\begin{aligned} y'(t) + 3y(t) &= 0 \\ y(4) &= 2 \end{aligned}$ (d) $\begin{aligned} y'(t) - 7y(t) &= 0 \\ y(4) &= 2 \end{aligned}$

2. (a) $\begin{aligned} y' - 4y &= 0 \\ y(-3) &= 2 \end{aligned}$ (b) $\begin{aligned} y' + 5y &= 0 \\ y(4) &= 7 \end{aligned}$ (c) $\begin{aligned} y' + \pi y &= 0 \\ y(100) &= \sqrt{2} \end{aligned}$

3. $\begin{aligned} y'(t) + (\cos 2t)y(t) &= 0 \\ y(0) &= 3 \end{aligned}$ 4. $\begin{aligned} y'(t) + (\cos 2t)y(t) &= 0 \\ y(\pi/2) &= 1 \end{aligned}$

5. (a) $\begin{aligned} y'(t) + e^{3t}y(t) &= 0 \\ y(-2) &= 10 \end{aligned}$ (b) $\begin{aligned} y'(t) &= e^t y(t) \\ y(4) &= 3 \end{aligned}$ 6. $\begin{aligned} y'(t) + (1/t)y(t) &= 0 \\ y(2) &= 3 \end{aligned}$

Write the solution to each of the following problems using the definite integral as in Part 6.

7. $\begin{aligned} y'(t) + 3y(t) &= 0 \\ y(0) &= -5 \end{aligned}$

8. $\begin{aligned} y' + 14y &= 0 \\ y(1) &= 10 \end{aligned}$

9. $\begin{aligned} y' - 7y &= 0 \\ y(0) &= 0 \end{aligned}$

10. $\begin{aligned} y'(t) + 6y(t) &= 0 \\ y(-3) &= -2 \end{aligned}$

11. $\begin{aligned} y'(t) + [1/(t + 2)]y(t) &= 0 \\ y(1) &= 5 \end{aligned}$

12. $\begin{aligned} y'(t) + (\sin 3t)y(t) &= 0 \\ y(0) &= -1/e \end{aligned}$

13. Suppose the lake of the examples is actually a tank with constant volume 2 cubic meters and water intake rate $I(t) = 1/(t + 10)$ cubic meters/second for $t \geq 0$.

    (a) Set up the corresponding differential equation for concentration.

    (b) Solve it to find the concentration of pollutant if the tank contains $\frac{1}{100}$ cubic meter of pollutant at $t = 0$.

    (c) Graph your solution to part b.

    (d) What is the volume of pollutant lost from the tank between $t = 0$ and $t = 30$, under the circumstances of part b.

    (e) From $t = 0$, how many seconds must elapse before 2 cubic meters of new water will have flowed into the tank? What will be the concentration at that time?

    (f) How many seconds must elapse before the concentration will have fallen by 20 percent?

14. Find the general solution:

    (a) $y' + \pi y = 0$ (b) $y'(t) + (\cos 2t)y(t) = 0$

    (c) $y'(t) + e^{3t}y(t) = 0$ (d) $y'(t) + (1/t)y(t) = 0$

10. Remarks on Direct Integration after Separation of Variables

If we start from (1) and separate variables as before,

$$\frac{y'}{y} = -a \tag{17}$$

then we may integrate both sides of the equation directly. This is very convenient and easy to remember, and many people prefer to do the problem this way. There is, however, a point that must be watched for, namely the absolute value in the formula

$$\int \frac{y'(t)}{y(t)} \, dt = \ln |y(t)| + C \tag{18}$$

where C is an arbitrary constant. The formula is from calculus and is checked by differentiation:

If $y(t) > 0$, then $|y(t)| = y(t)$, and so

$$\frac{d}{dt} \ln |y(t)| = \frac{d}{dt} \ln y(t) = \frac{1}{y(t)} y'(t) \qquad \text{if } y(t) > 0$$

If $y(t) < 0$, then $|y(t)| = -y(t)$, and so

$$\frac{d}{dt} \ln |y(t)| = \frac{d}{dt} \ln [-y(t)] = \frac{1}{-y(t)}[-y'(t)]$$

$$= \frac{y'(t)}{y(t)} \qquad \text{if } y(t) < 0$$

The real difficulty underlying this is that negative numbers do not have real logarithms. We have faced this difficulty once already, earlier in this section, where we performed the integration if $y(t) > 0$ but changed sign in order to draw conclusions if $y(t) < 0$. Thus, in either case we were considering $|y(t)|$.

Equation (18) gives the integration of the left-hand side of (17). Integration of both sides gives

$$\ln |y(t)| = -\int a(t)\,dt + C$$

from which exponentiation yields

$$|y(t)| = e^C e^{-\int a(t)\,dt}$$

A continuous function such as y cannot change sign without taking the value zero, which y does not do. Therefore,

$$y(t) = Be^{-\int a(t)\,dt}$$

where B is a new arbitrary constant. The sign of B is the sign of $y(t)$, for the exponential is positive.

For initial value problems, a definite integration may be preferred. In this case, start by

1. Changing to the dummy variable
2. Taking as the limits of integration the initial point t_0 and the variable point t

Thus,
$$\int_{t_0}^{t} \frac{y'(z)}{y(z)}\,dz = \ln|y(t)| - \ln|y(t_0)| = \ln\left|\frac{y(t)}{y(t_0)}\right|$$

$$= \ln \frac{y(t)}{y(t_0)}$$

and the whole computation would look like the following examples.

11. Example

Solve the initial value problem

$$\left. \begin{aligned} y'(t) + ty(t) &= 0 \\ y(2) &= -3 \end{aligned} \right\}$$

Solution

$$y'(t) = -ty(t)$$

$$\frac{y'(t)}{y(t)} = -t$$

$$\frac{y'(z)}{y(z)} = -z$$

$$\int_2^t \frac{y'(z)}{y(z)} \, dz = \int_2^t -z \, dz$$

$$\ln |y(z)| \Big|_2^t = -\frac{z^2}{2} \Big|_2^t$$

$$\ln |y(t)| - \ln |y(2)| = 2 - \frac{t^2}{2}$$

$$\ln \left| \frac{y(t)}{-3} \right| = 2 - \frac{t^2}{2}$$

Since $y(t)$ never changes sign,

$$\ln \frac{y(t)}{-3} = 2 - \frac{t^2}{2}$$

$$\frac{y(t)}{-3} = e^{2 - t^2/2}$$

$$y(t) = -3e^{2 - t^2/2}$$

12. Example

Solve the initial value problem

$$\left. \begin{array}{l} y'(t) + (1/t)y(t) = 0 \\ y(-4) = -2 \end{array} \right\}$$

Solution

$$\frac{y'(t)}{y(t)} = -\frac{1}{t}$$

$$\frac{y'(z)}{y(z)} = -\frac{1}{z}$$

$$\int_{-4}^{t} \frac{y'(z)}{y(z)} dz = \int_{-4}^{t} -\frac{1}{z} dz$$

$$\ln|y(z)|\Big|_{-4}^{t} = -\ln|z|\Big|_{-4}^{t}$$

$$\ln|y(t)|.- \ln|y(-4)| = -\ln|t| + \ln|-4| \qquad t < 0 \tag{19}$$

$$\ln\left|\frac{y(t)}{-2}\right| = -\ln\left|\frac{t}{-4}\right| = \ln\left|-\frac{4}{t}\right|$$

$$\ln\left(\frac{y(t)}{-2}\right) = \ln\left(-\frac{4}{t}\right)$$

$$\frac{y(t)}{-2} = -\frac{4}{t}$$

$$y(t) = \frac{8}{t} \qquad t < 0 \tag{20}$$

Check: $y'(t) = -8/t^2$, and so

$$y'(t) + \frac{1}{t}y(t) = -\frac{8}{t^2} + \frac{1}{t}\frac{8}{t} = 0$$

and

$$y(-4) = \frac{8}{-4} = -2$$

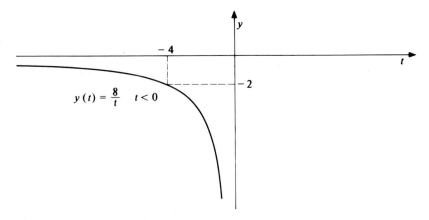

$$y(t) = \frac{8}{t} \qquad t < 0$$

Figure 3 The solution of Example 12.

13. Remarks on the Interval of Validity of the Solution

Notice that in the previous example it is not possible to integrate from -4 to any positive t, because the integrand "blows up" at the origin, $z = 0$. This is why the restriction $t < 0$ must appear in Equation (19). Hence the solution is valid and uniquely determined on $(-\infty, 0)$ by the value at -4.

On the other hand, formula (20), though restricted to $t < 0$, suggests that one solution of the problem is a function

$$v(t) = \frac{8}{t} \qquad t \neq 0$$

As the check shows, v is a solution of the differential equation, but its values on the interval $(0, \infty)$ are not determined by the initial condition at -4. Compare v, for instance, with the function

$$w(t) = -\frac{8}{|t|} \qquad t \neq 0$$

as shown in Figure 4. If $t < 0$, then

$$w(t) = -\frac{8}{|t|} = \frac{-8}{-t} = \frac{8}{t} = v(t) \qquad t < 0$$

but if $t > 0$, then

$$w(t) = -\frac{8}{|t|} = -\frac{8}{t} = -v(t) \qquad t > 0$$

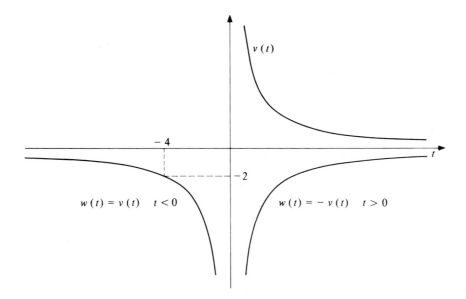

Figure 4 Comparison of $v(t) = 8/t$ and $w(t) = -8/|t|$.

Thus w also satisfies the differential equation, and it satisfies the initial condition too. On the other hand, v and w agree exactly on the largest interval which contains the initial point and on which $1/t$, the coefficient in the equation, is continuous.

Now reread Recapitulation, Part 3 in this section.

14. More Problems

Solve each of the following problems by direct integration after separation of variables. Use the indefinite integral for the odd-numbered problems, and the definite integral for the even-numbered problems. Graph each solution, and find the largest interval on which the solution is uniquely determined by the initial values. (Leave plenty of time for graphing.)

15. $\begin{aligned} y' - 3y &= 0 \\ y(2) &= 5 \end{aligned}$ 16. $\begin{aligned} y' + \pi y &= 0 \\ y(-2) &= 7 \end{aligned}$

17. $\begin{aligned} y'(t) + e^{2t}y(t) &= 0 \\ y(-3) &= 4 \end{aligned}$ 18. $\begin{aligned} y'(t) + 8t^2 y(t) &= 0 \\ y(0) &= 6 \end{aligned}$

19. $\begin{aligned} y'(t) + (\sin t)y(t) &= 0 \\ y(1) &= -2 \end{aligned}$ 20. $\begin{aligned} y'(t) + (\cos t)y(t) &= 0 \\ y(-4) &= -1 \end{aligned}$

21. $\begin{aligned} y'(t) &= 2y(t)/t \\ y(1) &= 1 \end{aligned}$ 22. $\begin{aligned} y'(t) + (3/t^2)y(t) &= 0 \\ y(-1) &= 2 \end{aligned}$

23. $\begin{aligned} y'(t) + (\tan t)y(t) &= 0 \\ y(\pi/4) &= 0 \end{aligned}$ 24. $\begin{aligned} y'(t) + e^t y(t) &= 0 \\ y(-3) &= 0 \end{aligned}$

E. ADDITIONAL MODELS AND NONHOMOGENEOUS EQUATIONS

1. Cooling by Dilution

Consider again the lake of Section A, but now do not suppose there is a foreign substance (pollutant) in the lake. Suppose instead that the lake has been heated above the inlet temperature U, the normal temperature of the lake. That is,

$V =$ constant volume of lake, cubic meters
$I(t) =$ rate of flow into (and out of) the lake at time t,
     cubic meters per second
$U =$ temperature of the inflow, degrees Celsius (°C)
$T(t) =$ temperature at time t at all points of the lake,
     degrees Celsius

No heat is lost except by being carried off in the outlet water.

In this situation we can define two useful functions, analogous to those in Section A. These are $y(t)$, the amount of excess heat stored in a cubic meter of lake water at time t, and $P(t) = Vy(t)$, the amount of excess heat stored in the lake as a whole at the same time.

At time t, $T(t) - U$ is the excess of the lake temperature T over the inlet temperature U. It follows that each gram of lake water stores

$$c[T(t) - U] \qquad \text{calories}$$

of excess heat, where c is the specific heat of water, 1 calorie/gram-degree. Each cubic meter contains $10^6 \rho$ grams of water, where ρ is the density of water (1 gram/cubic centimeter), and 10^6 is the number of cubic centimeters per cubic meter. Thus,

$$y(t) = 10^6[T(t) - U]$$

is the number of calories of excess heat stored in a cubic meter of lake water, and

$$P(t) = 10^6[T(t) - U]V$$

is the amount of excess heat stored in the lake.

The same analysis as in Section A now leads to the same differential equation. Thus, $-dP/dt$ is the rate of loss of heat in calories per second, and so is $(T - U)10^6 I = yI$, for $T - U$ is the outlet temperature less the inlet temperature in degrees, 10^6 is the heat capacity of water in calories per degree per cubic meter, and I is the flow rate in cubic meters. That is,

$$-\frac{dP}{dt} = yI \qquad \text{or} \qquad \frac{dy}{dt} = -\frac{I}{V}y \tag{1}$$

We have also seen the solution to this equation before:

$$y(t) = y(t_0)e^{-\int_{t_0}^{t} [I(z)/V] \, dz}$$

If $t_0 = 0$ and if I is constant, then the integral is tI/V. If, further, we use the definition of y as given above, then

$$y(t) = 10^6[T(t) - U] = 10^6[T(0) - U]e^{-(I/V)t}$$

The latter equation may be solved for $T(t)$:

$$T(t) - U = [T(0) - U]e^{-(I/V)t}$$
$$T(t) = U + [T(0) - U]e^{-(I/V)t} \tag{2}$$

Actually, Equation (1) is not a very good model of a heated lake, because the lake loses heat by conduction and radiation as well as by mixing, to say nothing of evaporation. However, an insulated vessel in a manufacturing process or a laboratory could very well be described by Equation (1), and (2) would be a reasonable model for predicting future temperatures.

2. Cooling by Conduction: Newton's Law of Cooling

The equation describing cooling by conduction is closely related to (1). Here is how we observe it in the laboratory.

We arrange a constant-temperature water bath at temperature U, and immerse in the water a small vessel containing a hot substance. (See Figure 1.) We

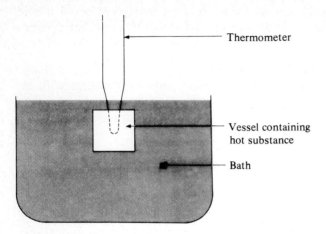

Thermometer

Vessel containing
hot substance

Bath

Figure 1 Laboratory equipment for measuring cooling rate.

read the temperature $T(t)$ from the thermometer at various times, plot the results, and get the graph in Figure 2.

With care, we can observe that $T - U$ is an exponential function with decay constant λ, which we can measure. That is,

$$T(t) - U = [T(0) - U]e^{-\lambda t}$$

If we define $P = T - U$, then

$$P(t) = P(0)e^{-\lambda t}$$
$$P'(t) = -\lambda P(0)e^{-\lambda t}$$
$$= -\lambda P(t)$$

or
$$P' + \lambda P = 0 \tag{3}$$

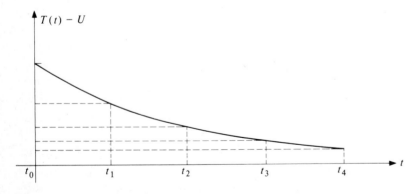

Figure 2 Sample observations for measuring the cooling rate.

This is one version of *Newton's law of cooling*. It is the same kind of differential equation as that which appeared in the mixing problems. Of course, the differential equation satisfied by the function T is a bit different:

$$T(t) = U + [T(0) - U]e^{-\lambda t}$$
$$T'(t) = -\lambda[T(0) - U]e^{-\lambda t}$$
$$= -\lambda[T(t) - U]$$
$$= -\lambda T(t) + \lambda U$$

or
$$T' + \lambda T = \lambda U \qquad (4)$$

where U is the temperature of the bath, and λ is determined empirically, once the material is immersed in the bath, by taking a few readings of $T(t)$. Now the initial value problem for predicting $T(t)$ is

$$\left.\begin{array}{c} T' + \lambda T = \lambda U \\ T(0) \text{ given} \end{array}\right\} \qquad (5)$$

and it must lead to a function

$$T(t) = U + [T(0) - U]e^{-\lambda t} \qquad (6)$$

3. Example

A constant-temperature bath is fixed at 50°C, and a small quantity of iron is immersed in it. At $t = 0$, the temperature of the iron is 90°C; at $t = 100$, the temperature of the iron is 80°C. Predict the temperature of the iron at later times t.

The first problem is to find λ, for clearly

$$U = 50 \qquad T(0) = 90$$

Actually, the figure $T(100) = 80$ together with this information determines λ. Here is the computation: From (6),

$$T(t) = U + [T(0) - U]e^{-\lambda t}$$
$$= 50 + 40e^{-\lambda t}$$

At $t = 100$,

$$80 = T(100) = 50 + 40e^{-100\lambda}$$

or
$$30 = 40e^{-100\lambda}$$

$$\tfrac{3}{4} = e^{-100\lambda}$$

$$\ln \tfrac{3}{4} = -100\lambda$$

$$\lambda = -\tfrac{1}{100} \ln \tfrac{3}{4}$$

Hence,
$$T(t) = 50 + e^{(1/100)[\ln (3/4)]t}$$

$$= 50 + (\tfrac{3}{4})^{t/100} \qquad \text{for all } t \geq 0$$

4. Comparison of Homogeneous and Nonhomogeneous Equations

Newton's law of cooling and the equations for the mixing of lake water, (3) and (1), both have the form

$$y' + ay = 0 \qquad a \text{ known, } y \text{ unknown} \tag{7}$$

while $\qquad y' + ay = f \qquad a \text{ and } f \text{ known, } y \text{ unknown} \tag{8}$

is the form of Equation (4) for T. The former is called a "homogeneous equation," the latter a "nonhomogeneous equation," on the basis of the following distinction: if y is any solution of (7) and k is any constant, then ky is also a solution of (7), but (8) and its solutions do not have the same property.

To see that this is so, substitute ky for y in (7). Since

$$\frac{d}{dt}[ky] = k\frac{dy}{dt}$$

the substitution into the left-hand side gives

$$\frac{d}{dt}[ky] + aky = k(y' + ay)$$

which is $k \cdot 0$ for all k if y is a solution of (7). That is, ky is a solution if y is. Equations with this property are *homogeneous*; equations lacking it are *nonhomogeneous*. Equation (8) is an example of a nonhomogeneous equation if f is a nonzero function, for if y satisfies (8) then

$$\frac{d}{dt}[ky] + aky = k(y' + ay)$$

$$= kf \neq f$$

In this situation, f, the term which is not a multiple of the unknown or its derivative, is the *nonhomogeneous term*.

We have already seen a nonhomogeneous equation [Equation (4)] and an initial value problem in (5). This section contains examples leading to nonhomogeneous equations. The next section contains a solution method for nonhomogeneous equations.

5. Example

In the lake problem in Section A, let us introduce a nonhomogeneous term into the equation by supposing that some pollutant flows into the lake through the inlet. Specifically,

$$V = \text{volume of the lake, cubic meters (constant)}$$
$$I = \text{rate of inflow (also the rate of outflow),}$$
$$\text{cubic meters per second (constant)}$$
$$P(t) = \text{volume of pollutant in the lake at time } t,$$
$$\text{cubic meters}$$
$$y(t) = P(t)/V = \text{volume concentration}$$

and $y(0)$ is fixed for an initial condition. It might be any number between 0 and 1, for that is where concentrations lie. Let us, this once, take $y(0) = 0$. If there were no pollutant flowing into the lake, then $y(t) \equiv 0$ would be the solution of the problem. Instead, let us suppose that pollutant flows into the lake in the inlet at a constant rate, with concentration r. That is, rI cubic meters/second of pollutant flows into the lake. Now our analysis must be a little different. The rate of increase of P is

$$\frac{dP}{dt} = rI - yI$$

for the right-hand side is the rate of inflow less the rate of outflow. Since $P = Vy$, it follows that

$$V\frac{dy}{dt} = rI - yI$$

or $\qquad\qquad y' = r\dfrac{I}{V} - y\dfrac{I}{V} \qquad$ or $\qquad y'(t) + \dfrac{I}{V}y(t) = r\dfrac{I}{V} \qquad\qquad$ (9)

Note that the equation we derived in Section A is the special case of (9) for which $r = 0$. The equation is nonhomogeneous, of the form (8), with $a = I/V$ and $f = rI/V$ both constant. We shall solve it by the method given previously, beginning with a transposition and division:

$$y' = -\frac{I}{V}(y - r)$$

$$\frac{y'}{y - r} = -\frac{I}{V} \qquad\qquad (10)$$

The point to notice is that

$$\frac{d}{dt}[y - r] = y'$$

since r is constant. Thus, the difference of the concentrations in the lake and the inlet may become a new variable

$$T(t) = y(t) - r \qquad\qquad (11)$$

with $\qquad\qquad T'(t) = y'(t)$

and (10) becomes

$$\frac{T'}{T} = -\frac{I}{V} \qquad\qquad (12)$$

where the right-hand side is known. By integrating both sides (as in Section D), we obtain

$$T(t) = T(0)e^{-(I/V)t}$$

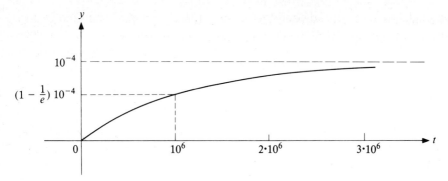

Figure 3 Example 5, pollutant concentration in the lake, $y(t) = 10^{-4}(1 - e^{-10^{-6}t})$.

Substitution from (11) gives

$$y(t) - r = [y(0) - r]e^{-(I/V)t} = -re^{-(I/V)t}$$
$$y(t) = r + [y(0) - r]e^{-(I/V)t} \qquad (13)$$
$$= r(1 - e^{-(I/V)t})$$

Checking the initial value, we have $y(0) = r(1 - e^0) = 0$. Checking that y satisfies the differential equation gives

$$y'(t) = -re^{-(I/V)t}\left(-\frac{I}{V}\right)$$

and $\qquad y'(t) + \dfrac{I}{V} y(t) = -re^{-(I/V)t}\left(-\dfrac{I}{V}\right) + \dfrac{I}{V}(r - re^{-(I/V)t}) = r\dfrac{I}{V}$

If $V = 10^6$, $I = 1$, and $r = 10^{-4}$, then $a = I/V = 10^{-6}$, $f = r(I/V) = 10^{-4} \cdot 10^{-6} = 10^{-10}$, and

$$y(t) = 10^{-4}(1 - e^{-10^{-6}t})$$

As before, the *general solution* of the equation is the formula for all solutions, which is (13). That is, for each value of $y(0)$, (13) is a solution of (10), and each solution can be written in the form (13) for some value of $y(0)$.

Alternatively, the definite integration of (10) runs this way:

$$\int_0^t \frac{y'(z)}{y(z) - r}\,dz = -\int_0^t \frac{I}{V}\,dz$$

$$\ln\left|\frac{y(t) - r}{y(0) - r}\right| = -\frac{I}{V}t$$

$$\ln \frac{y(t) - r}{y(0) - r} = -\frac{I}{V}t$$

from which exponentiation gives (13) and the solution obtained previously.

Of course, if $y(0)$ had been exactly r, then we would not have been able to perform the integration. Our conclusion would have been

$$|y(t) - r| \equiv 0$$

as in Section D, and thus

$$y(t) \equiv r$$

In the case of the lake, if the concentration in the inflow is the same as the concentration in the lake at time $t = t_0$, then the concentration is constant, for water enters and leaves at the same concentration.

6. Example: A Piecewise-Constant Nonhomogeneous Term

Reconsider the lake problem of the previous part, with $V = 10^6$ and $I = 1$. Suppose now that no pollutant flows into the lake during time $t \leq 0$, and that no pollutant flows into the lake during time $t > 10^6$, but that

$$r(t) = 10^{-4} \qquad 0 < t \leq 10^6$$

That is, the concentration of pollutant in the inflow is 1/10,000 during the time interval between zero and 10^6, but zero at other times:

$$r(t) = \begin{cases} 0 & t \leq 0 \\ 10^{-4} & 0 < t \leq 10^6 \\ 0 & 10^6 < t \end{cases}$$

(It might be that a factory upstream had a temporary failure of its filtering system.) If the concentration of pollutant in the lake is y, and if it happens that $y(0) = 0$, then (9) becomes

$$y'(t) + \frac{I(t)}{V} y(t) = \frac{I(t)r(t)}{V} = 10^{-4} \frac{I}{V} = 10^{-10} \qquad 0 < t \leq 10^6$$

and

$$y'(t) + 10^{-6} y(t) = \begin{cases} 0 & t \leq 0 \\ 10^{-10} & 0 < t \leq 10^6 \\ 0 & 10^6 < t \end{cases} \tag{14}$$

$$y(0) = 0$$

The function r is constant on each of three intervals and is said to be *piecewise-constant*.

The solution method is like that of Part 7 of Section B. For $t \leq 0$, the differential equation in (14) is

$$y'(t) + 10^{-6} y(t) = 0 \qquad t \leq 0$$

whose general solution is

$$y(t) = y(0)e^{-10^{-6}t} \qquad t \leq 0$$

Since $y(0) = 0$, it follows that

$$y(t) = 0 \qquad t \le 0 \tag{15}$$

For $0 < t \le 10^6$, the differential equation in (14) is

$$y'(t) + 10^{-6}y(t) = 10^{-10} \qquad 0 < t \le 10^6$$

whose general solution may be found as in Part 5 above:

$$y(t) = r + Ce^{-(I/V)t} = 10^{-4} + Ce^{-10^{-6}t} \qquad 0 < t \le 10^6$$

To have $y(0) = 0$, we must have $C = -r = -10^{-4}$, and so

$$y(t) = 10^{-4}(1 - e^{-10^{-6}t}) \qquad 0 < t \le 10^6$$

as we found before.

So far we have found that

$$y(t) = \begin{cases} 0 & t \le 0 \\ 10^{-4}(1 - e^{-10^{-6}t}) & 0 < t \le 10^6 \end{cases} \tag{16}$$

and we can make predictions of concentration as far into the future as $t = 10^6$. In particular, $y(10^6) = 10^{-4}(1 - 1/e)$. (See Figure 3.)

Now consider the time interval $10^6 < t$. We know that the differential equation satisfied there is

$$y'(t) + 10^{-6}y(t) = r(t)$$

$$= 0 \qquad 10^6 < t$$

and that $y(10^6) = 10^{-4}(1 - 1/e)$. Thus, we can predict $y(t)$ on the interval $10^6 < t$, using (for instance) a definite integral:

$$y(t) = y(10^6)e^{-10^{-6}(t - 10^6)}$$

$$= 10^{-4}\left(1 - \frac{1}{e}\right)e^{-10^{-6}(t - 10^6)} \qquad 10^6 < t \tag{17}$$

Combining this information with (16) gives the prediction of concentration $y(t)$ for all values of t:

$$y(t) = \begin{cases} 0 & t \le 0 \\ 10^{-4}(1 - e^{-10^{-6}t}) & 0 < t \le 10^6 \\ 10^{-4}\left(1 - \frac{1}{e}\right)e^{-10^{-6}(t - 10^6)} & 10^6 < t \end{cases}$$

The function y is graphed in Figure 4, which should be compared with Figure 3.

There is one last thing to point out. In (17), the expression $t - 10^6$ which appears in the exponent may be viewed as a new variable; call it τ:

$$\tau = t - 10^6$$

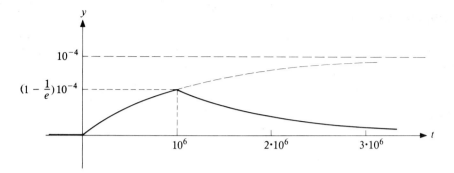

Figure 4 Example 6, pollutant concentration in the lake.

The behavior of the function y at some point $t = t_0$ is like the behavior at $\tau = t_0 - 10^6$ of a new function x, defined by

$$x(\tau) = 10^{-4}\left(1 - \frac{1}{e}\right)e^{-10^{-6}\tau}$$

or
$$x(\tau) = y(t) \qquad 10^6 < t \qquad (18)$$

If one imagines t as being measured on a clock, then one can imagine τ being measured on a clock running at the same speed but 10^6 seconds behind. In other words, the τ clock reads $\tau = 0$ when the t clock reads $t = 10^6$, and it reads $\tau = t - 10^6$ when the t clock reads t. When the τ clock reads τ, the t clock reads $\tau + 10^6$. Thus, the graph of $x(\tau)$ for $\tau \geq 0$, as given in Figure 5, is the same shape as the graph of $y(t)$ for $t \geq 10^6$ (as given in Figure 5), but the scale on the horizontal axis has been shifted. (See also *Translation to the right a distance t_0*, in Part 1 of Section I.)

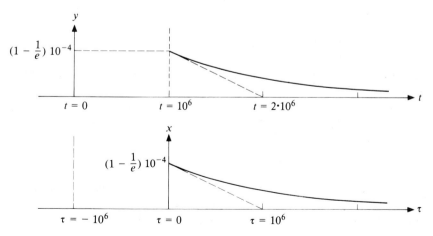

Figure 5 Graph of $y(t)$ for $10^6 < t$ and $x(\tau)$ for $0 < \tau$, showing that $y(t) = x(\tau)$ there.

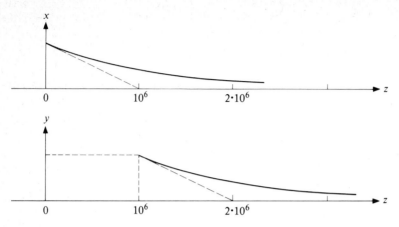

Figure 6 $y \neq x$, as shown by the corresponding parts of the graphs of y and x, plotted on the same axis, in this case called the z axis.

Notice that (18) does *not* say that x and y are the same function. If x and y were the same function, then $x(z)$ would equal $y(z)$ for every z; for instance, $x(3)$ would equal $y(3)$, which is not so. Instead, the functions x and y are different but related by the following equations, each of which tells the whole story:

$$y(t) = x(\tau) \qquad \text{where } \tau = t - 10^6 > 0$$

$$y(t) = x(\tau) \qquad \text{where } t = \tau + 10^6 > 10^6$$

$$y(t) = x(t - 10^6) \qquad \text{where } t > 10^6$$

$$y(\tau + 10^6) = x(\tau) \qquad \text{where } \tau > 0$$

See Figure 6.

7. Sample Problems with Answers

Try these using the definite integral, and again using the indefinite integral.

A. $\left.\begin{array}{l} y' + 2y = 5 \\ y(0) = 0 \end{array}\right|$ *Answer:* $y(t) = \frac{5}{2}(1 - e^{-2t})$

B. $\left.\begin{array}{l} y' + 2y = 5 \\ y(0) = 3 \end{array}\right|$ *Answer:* $y(t) = \frac{5}{2} + \frac{1}{2}e^{-2t}$

C. $\left.\begin{array}{l} y' + 2y = 5 \\ y(0) = \frac{5}{2} \end{array}\right|$ *Answer:* $y(t) \equiv \frac{5}{2}$

D. $y'(t) - 2y(t) = \left.\begin{array}{ll} 5 & t \leq 0 \\ 1 & t > 0 \end{array}\right|$ *Answer:* $y(t) = \left|\begin{array}{ll} -\frac{5}{2}(1 - e^{2t}) & t \leq 0 \\ -\frac{1}{2}(1 - e^{2t}) & t > 0 \end{array}\right.$

  $y(0) = 0$

E. $y'(t) + 2y(t) = \left.\begin{array}{ll} 5 & t \leq 0 \\ 1 & t > 0 \end{array}\right|$ *Answer:* $y(t) = \left|\begin{array}{ll} \frac{1}{2}(5 - 9e^{-2(t+1)}) & t \leq 0 \\ \frac{1}{2} + \frac{1}{2}(4 - 9e^{-2})e^{-2t} & t > 0 \end{array}\right.$

  $y(-1) = -2$

Find the general solution of each of the following.

F. $y' + 2y = 5$ *Answer:* $y(t) = \frac{5}{2} + Ce^{-2t}$

G. $y' - 7y = 4$ *Answer:* $y(t) = -\frac{4}{7} + Ce^{7t}$

8. Problems

Solve the following problems twice, using the definite integral the first time and the indefinite integral the second time.

1. $\left.\begin{array}{l} y' + 5y = 10 \\ y(0) = 3 \end{array}\right|$ 2. $\left.\begin{array}{l} y' - 5y = 10 \\ y(0) = 3 \end{array}\right|$ 3. $\left.\begin{array}{l} y' + 5y = 10 \\ y(0) = 1 \end{array}\right|$

4. $\left.\begin{array}{l} y' - 5y = 10 \\ y(2) = 1 \end{array}\right|$ 5. $\left.\begin{array}{l} y' + 5y = 10 \\ y(3) = 2 \end{array}\right|$

What makes Problem 5 different from all the others?

Find the general solution of each of the following.

6. $y' + 5y = 10$ **7.** $y' - 5y = 10$ **8.** $y' - 5y = 0$

Solve the following initial value problems, and graph the solutions.

9. (a) $y'(t) + 5y(t) = \left\{\begin{array}{ll} 0 & t \le 0 \\ 4 & t > 0 \end{array}\right\}$ (b) $y'(t) - 5y(t) = \left\{\begin{array}{ll} 4 & t \le 0 \\ 0 & t > 0 \end{array}\right\}$

$\qquad\qquad y(0) = 0$ $y(0) = 0$

(c) $y'(t) + 5y(t) = \left\{\begin{array}{ll} 4 & t \le 0 \\ 0 & t > 0 \end{array}\right\}$ (d) $y'(t) + 5y(t) = \left\{\begin{array}{ll} 4 & t \le 0 \\ 0 & t > 0 \end{array}\right\}$

$\qquad\qquad y(1) = 0$ $y(-1) = 0$

9. A Dynamics Problem: A Rocket Ascending under Power

You are familiar with Newton's second law of motion, which states that the time rate of change of the momentum of a body is equal to the applied force. In many problems, the mass is constant and $F = Mv'$. The problem of predicting the velocity of a rocket falls in the more general case $F = (Mv)'$, in which the mass is not constant.

Let us assume, as shown in Figure 7, that the rocket is rising vertically, its velocity is $v(t)$, and its mass is $M(t)$ at time t. Its momentum, then, is $M(t)v(t)$. As time passes, it burns fuel (propellant) and ejects it as the exhaust. After Δt seconds it has ejected mass of propellant Δp. Its new mass is $M(t + \Delta t) = M(t) - \Delta p$, and its new velocity is $v(t + \Delta t) = v(t) + \Delta v$. The new momentum of the rocket is

$$(M - \Delta p)(v + \Delta v)$$

In order to avoid computing the force on the rocket due to the fuel burning, we consider our "body" to be the rocket plus its fuel; its momentum after Δt is

$$(M - \Delta p)(v + \Delta v) + \Delta p(v - u) \tag{19}$$

where u is the velocity of the exhaust relative to the rocket, and $v - u$ is the velocity of the exhaust relative to a fixed point, for the exhaust moves rearward from the rocket. The first term of (19) is the momentum of the rocket itself; the second is the momentum of the gas which was exhausted between t and $t + \Delta t$.

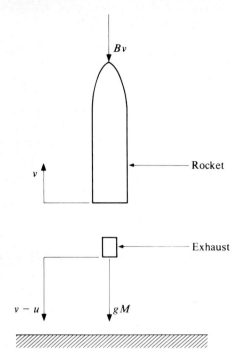

Figure 7 The rocket problem: forces and speeds.

Notice that the exhaust mass is Δp, but the total mass of the system is constant:

$$(M - \Delta p) + \Delta p = M$$

The net change in the momentum of the system, then, is (19) less Mv, which simplifies to

$$M \, \Delta v - \Delta p \, \Delta v - u \, \Delta p$$

Dividing by Δt and letting Δt tend to zero gives the derivative of the momentum of the system, that is, the net applied force on the system at time t, according to Newton's second law:

$$F(t) = \lim_{\Delta t \to 0} \frac{M(t) \, \Delta v - \Delta p \, \Delta v - u(t) \, \Delta p}{\Delta t}$$

$$= M(t) \frac{dv}{dt}(t) - u(t) \frac{dp}{dt}(t) \tag{20}$$

where $(dp/dt)(t)$ is the rate of propellant consumption at time t.

F in this case is the sum of the weight of the system, which at time t is $gM(t)$ acting downward, and a drag force due to atmospheric friction, which we shall take to be $Bv(t)$, proportional to $v(t)$ but acting downward. Let us assume as well that the exhaust-gas speed u is constant, and that the rate of fuel consumption

$p' = -dM/dt$ is also constant. With these assumptions, (20) becomes

$$M(t)\frac{dv}{dt}(t) - up' = -gM(t) - Bv(t)$$

or
$$Mv' + Bv = up' - gM \tag{21}$$

Our problem is to determine the velocity v_f when the entire fuel supply is exhausted, subject to the conditions that the initial velocity is v_0 and the initial mass is M_0. This velocity v_f is called the *burnout velocity*. The final mass of the rocket is that of its structure and payload.

In Equation (21) there are two functions M and v, but $M = M_0 - tp'$ since $dM/dt = -p'$ is constant, and $M(0) = M_0$. Thus, Equation (21) becomes

$$(M_0 - tp')\frac{dv}{dt} + Bv = up' - g(M_0 - tp')$$

Division by the coefficient of dv/dt gives

$$v'(t) + \frac{B}{M_0 - tp'}v(t) = \frac{up'}{M_0 - tp'} - g \tag{22}$$

This is a nonhomogeneous equation of the form

$$y'(t) + a(t)y(t) = f(t)$$

where
$$a(t) = \frac{B}{M_0 - tp'} \quad \text{and} \quad f(t) = \frac{up'}{M_0 - tp'} - g$$

Equation (22) will be solved in Section F.

10. Problems

10. Bacteria divide (reproduce) in proportion to their own number. That is, in the same circumstance two million bacteria of a given strain will have twice as many divisions per minute as one million, so that if $N(t)$ is the number of bacteria present on day t then

$$\alpha N(t) = \frac{dN(t)}{dt}$$

where α is the constant of proportionality. Solve the equation as before to find $N(t)$ if $N(0) = 1,000,000$. Compute α from the additional information that $N(1) = 2,000,000$. If there were 1,000,000 bacteria on Saturday at noon and 2,000,000 on Sunday at noon, how many are there at noon on Monday? How many were there on the previous Friday? Check your solution and graph it.

11. At any given moment, the rate of decay of a radioactive substance is proportional to the amount $s(t)$ of the substance present. If the constant of proportionality is called λ, one may write a differential equation

$$\lambda s = -s'$$

to describe the system. In this situation it is usually the "radioactivity" $-s'$ which is measured in the laboratory. Imagine you have on hand a device (such as a Geiger counter) which will measure $-s'$ in grams per second, and describe how you would use it to calculate λ and $s(t)$ for a given sample.

12. Dr. Mad is a noted experimentalist who performed the experiment suggested in Problem 11. This is the table of his data:

t	0	1	2	3
$-\dfrac{ds}{dt}(t)$	1	$\frac{1}{3}$	$\frac{1}{9}$	$\frac{1}{27}$

Find $s(t)$ for $t = 0$, 1, 2, and 3.

13. H and S are sitting together. They have identical cups of hot coffee and identical containers of cream. The cream is at room temperature. At time $t = 0$, H pours his cream into his coffee and leaves it to cool until $t = 5$. S leaves her coffee to cool first, and then adds the cream at $t = 5$. Then both drink. Who drinks the cooler coffee?

[Here is a fact you will want to know: If amount x of cream at temperature X is added to amount y of coffee at temperature Y, the mixture will have $(xX + yY)/(x + y)$ for its temperature. This kind of problem will be elaborated upon in Section G.]

14. A "first-order chemical reaction" is one which may be described as follows: The rate of reaction is proportional to the concentration of a single reagent. In other terms, the rate of change of the concentration of one of the reagents is proportional to (the negative of) the concentration of that reagent. Write the corresponding differential equation, and graph its solution.

15. A soft drink is mixed at a soda bar. One ounce of syrup is put in a glass, and then soda water is added at the rate of 2 ounces/second. The glass holds 12 ounces and is supposed to be filled to the top, but the attendant is thinking about other things and so the glass actually overflows for 1 second before the soda water is shut off. Find the concentration of syrup (as a function of time) during the period before the glass overflows.

16. In Problem 15, find the concentration (as a function of time) during the period when the glass is overflowing.

17. In Problem 15, find how much syrup was lost in the overflow.

F. AN INTEGRATING FACTOR FOR SOLVING NONHOMOGENEOUS EQUATIONS

1. Method

This is a four-step algorithm for solving initial value problems

$$\left. \begin{aligned} y' + ay &= f \\ y(t_0) &= y_0 \end{aligned} \right\} \tag{1}$$

applicable if a and f are continuous functions on the interval of interest and sometimes even if f is discontinuous. It is a special case of the method of the *integrating factor*, found in most textbooks.

Step 1. Write and solve the *associated homogeneous equation*

$$y'_h + ay_h = 0 \tag{2}$$

The associated homogeneous equation differs from the given equation (1) by having no nonhomogeneous term, no term independent of y. You should solve (2) at once, select any nonzero solution and label it plainly as y_h. The function y_h is called a *homogeneous solution*, but it is not a solution of the given equation (1).

Step 2. (Notice that y_h is never zero.) Divide both sides of the given equation (1) by y_h, leaving it in the form

$$\frac{y'}{y_h} + a\frac{y}{y_h} = \frac{f}{y_h} \tag{3}$$

and identify the left side of (3) as

$$\frac{d}{dt}\left[\frac{y}{y_h}\right] = y'\frac{1}{y_h} + y\left(-\frac{y_h'}{y_h^2}\right)$$

$$= \frac{y'}{y_h} + y\left(-\frac{-ay_h}{y_h^2}\right)$$

$$= \frac{y'}{y_h} + a\frac{y}{y_h}$$

where the first equality is the product rule for derivatives, applied to

$$y\frac{1}{y_h}$$

the second follows from (2), which allows the substitution of $-ay_h$ for y_h', and the third is an algebraic simplification. Equating to the right-hand side of (3) gives

$$\left.\begin{array}{c} \dfrac{d}{dt}\left[\dfrac{y}{y_h}\right] = \dfrac{f}{y_h} \\[2ex] \dfrac{y(t_0)}{y_h(t_0)} \text{ is known} \end{array}\right\} \tag{4}$$

where y_h is the function already found in step 1.

Step 3. Solve the initial value problem (4) for y/y_h by integration:

$$\frac{y(t)}{y_h(t)} - \frac{y(t_0)}{y_h(t_0)} = \int_{t_0}^{t} \frac{f(z)}{y_h(z)}\,dz$$

It is because (4) can be solved by integration that $1/y_h$ is called an *integrating factor* for (2).

Step 4. Solve for $y(t)$:

$$y(t) = \frac{y(t_0)}{y_h(t_0)}y_h(t) + y_h(t)\int_{t_0}^{t} \frac{f(z)}{y_h(z)}\,dz \tag{5}$$

Although the solution to the homogeneous equation may be remembered as a formula [Equation (11) or (16) from Section D], the solution of the nonhomogeneous equation should be remembered as a procedure: solve the homogeneous equation, divide, integrate, and solve.

2. Example: The Lake Problem in Case Some Pollutant Is Flowing into the Lake

As in Part 5 of Section E, let

V = volume of lake, cubic meters (constant)
$y(t)$ = (volume) concentration in the lake at time t
$P(t) = Vy(t)$ = volume of pollutant in the lake, cubic meters
$I(t)$ = rate of inflow (and rate of outflow),
      cubic meters per second
$r(t)$ = concentration of pollutant in the inflow

The rate of increase of the volume of pollutant is

$$rI - yI = P' = Vy'$$

so that
$$y' + \frac{I}{V}y = \frac{I}{V}r \tag{6}$$

Thus, in the terminology of (1),

$$a(t) = \frac{I(t)}{V} \qquad f(t) = \frac{I(t)r(t)}{V}$$

Let us do the example for $V = 10^6$ and $I(t) = 1$, but leave r and $y(0)$ unspecified, to see how the solution depends on them.

Step 1. The associated homogeneous equation is

$$y_h' + \frac{I}{V}y_h = 0 \qquad \text{or} \qquad y_h' + 10^{-6}y_h = 0$$

and its solution is

$$y_h(t) = Ce^{-\int (I/V)(t)\, dt} \qquad \text{or} \qquad y_h(t) = y_h(0)e^{-10^{-6}t}$$

As remarked in step 1 of the method, it does not matter which nonzero homogeneous solution one uses, so let $y_h(0) = 1$:

$$y_h(t) = e^{-10^{-6}t}$$

(Note that y_h is not zero at any point.)
Step 2. Division of (6) by y_h gives

$$\frac{y'(t)}{e^{-10^{-6}t}} + \frac{1}{e^{-10^{-6}t}}\frac{y(t)}{10^6} = \frac{r(t)}{e^{-10^{-6}t} \cdot 10^6} \tag{7}$$

The left side of (7) is the derivative of $y/y_h = e^{10^{-6}t}y$:

$$\frac{d}{dt}\left[\frac{y(t)}{y_h(t)}\right] = \frac{d}{dt}[e^{10^{-6}t}y(t)]$$

$$= e^{10^{-6}t}y'(t) + 10^{-6}e^{10^{-6}t}y(t)$$

$$= \frac{y'(t)}{e^{-10^{-6}t}} + \frac{y(t)}{10^6 e^{-10^{-6}t}}$$

Hence, (7) is the same as the equation

$$\frac{d}{dt}[e^{10^{-6}t}y(t)] = \frac{e^{10^{-6}t}r(t)}{10^6}$$

but it really is easier to think of this equation:

$$\frac{d}{dt}\left[\frac{y(t)}{y_h(t)}\right] = \frac{Ir(t)}{Vy_h(t)}$$

Step 3. The value of $y(t)/y_h(t)$ at $t = 0$ is $y(0)$, and so the initial value problem for y/y_h is solved as follows:

$$\frac{y(t)}{y_h(t)} - y(0) = \frac{I}{V}\int_0^t \frac{r(z)}{y_h(z)}\,dz$$

Step 4. Solving for $y(t)$ gives

$$y(t) = y(0)y_h(t) + y_h(t)\frac{I}{V}\int_0^t \frac{r(z)}{y_h(z)}\,dz$$

$$= y_h(t)\left[y(0) + \frac{I}{V}\int_0^t \frac{r(z)}{y_h(z)}\,dz\right]$$

$$= e^{-10^{-6}t}\left[y(0) + \int_0^t \frac{e^{10^{-6}z}}{10^6}r(z)\,dz\right]$$

If, for example, $y(0) = 10^{-5}$ and $r(t) = 10^{-7}e^{-10^{-6}t}$, then the integrand is 10^{-13} and

$$y(t) = e^{-10^{-6}t}(10^{-5} + 10^{-13}t)$$

3. Example: Solution of the Rocket Problem from Section E

The rocket is ascending vertically. At time t, it has mass $M(t)$ and velocity $v(t)$. Air friction is $Bv(t)$, a constant multiple of the velocity, and the weight is $gM(t)$, where g is the acceleration due to gravity. The exhaust leaves the rocket at the constant velocity u relative to the rocket, going downward, and propellant (fuel) is consumed at the constant rate

$$p' = -\frac{dM}{dt}$$

As shown in Part 9 of Section E, under these conditions

$$Mv' + Bv = up' - gM$$

or
$$v'(t) + \frac{B}{M(t)}v(t) = \frac{up'}{M(t)} - g \tag{8}$$

If the rocket starts from rest with $v(0) = 0$ and $M(0) = M_0$, then $M(t) = M_0 - tp'$ and

$$
\left.
\begin{aligned}
v'(t) + \frac{B}{M_0 - tp'}v(t) &= \frac{up'}{M_0 - tp'} - g \\
v(0) &= 0
\end{aligned}
\right\}
$$

The solution algorithm may now be employed.
    The associated homogeneous equation is

$$v_h'(t) + \frac{B}{M_0 - tp'}v_h(t) = 0$$

and since
$$\int \frac{B}{M_0 - tp'}\, dt = -\frac{B}{p'}\ln\,(M_0 - tp') + C$$

$$= -\frac{B}{p'}\ln\,M(t) + C$$

it follows that

$$
\begin{aligned}
y_h(t) &= (M_0 - tp')^{B/p'} \tag{9}\\
&= [M(t)]^{B/p'}
\end{aligned}
$$

or any constant multiple of that. Step 2 requires division of the given equation by v_h and identification of the left-hand side of the resulting equation as

$$\frac{d}{dt}\left[\frac{v}{v_h}\right]$$

Step 3 requires solution for v/v_h. Step 4 yields the following formula, as in (5):

$$v(t) = \frac{v(0)}{v_h(0)}v_h(t) + v_h(t)\int_0^t \frac{1}{v_h(z)}\left(\frac{up'}{M_0 - zp'} - g\right)dz$$

Since $v(0) = 0$, the first term vanishes. Substitution of the formula for v_h from (9) yields

$$v(t) = (M_0 - tp')^{B/p'} \int_0^t [up'(M_0 - zp')^{-B/p'-1} - g(M_0 - zp')^{-B/p'}] \, dz$$

$$= (M_0 - tp')^{B/p'} \left[\frac{up'}{B} (M_0 - zp')^{-B/p'} + \frac{g}{p'-B} (M_0 - zp')^{-B/p'+1} \right]_0^t$$

$$= \frac{up'}{B} + \frac{g}{p'-B}(M_0 - tp') - \frac{up'}{B}\left(\frac{M_0 - tp'}{M_0}\right)^{B/p'} - \frac{gM_0}{p'-B}\left(\frac{M_0 - tp'}{M_0}\right)^{B/p'}$$

$$= \frac{up'}{B} + \frac{gM(t)}{p'-B} - \left(\frac{up'}{B} + \frac{gM_0}{p'-B}\right)\left[\frac{M(t)}{M_0}\right]^{B/p'} \tag{10}$$

The V-2 rocket made in Germany in the early 1940s was used for rocket and high-altitude research in the United States in the following decade. For historical reasons its specifications are still given in the English measurement system. The important data are:

$$\text{Initial weight} = 28{,}300 \text{ pounds}$$

$$\text{Fuel weight} = 19{,}400 \text{ pounds}$$

$$\text{Fuel consumption rate} = 275 \text{ pounds/second}$$

$$\text{Exhaust-gas velocity} = 6560 \text{ feet/second}$$

$$B = 0.1 \text{ slug/second}$$

From these data it follows that:

$$u = 6560 \text{ feet/second}$$

$$p' = \frac{275}{32.2} = 8.54 \text{ slugs/second}$$

$$M_0 = \frac{28{,}300}{32.2} = 878.88 \text{ slugs}$$

$$\text{Duration of fuel burning} = \frac{19{,}400}{275} = 70.54 \text{ seconds}$$

$$\frac{p'}{M_0} = \frac{8.54}{878.88} = 0.0097/\text{second}$$

$$\frac{up'}{B} = \frac{6560 \times 8.54}{0.1} = 560{,}224 \text{ feet/second}$$

$$\frac{gM_0}{p'-B} = \frac{32.2 \times 878.88}{8.54 - 0.1} = 3353.07 \text{ feet/second}$$

$$\frac{gp'}{p'-B} = \frac{32.2 \times 8.54}{8.54 - 0.1} = 32.58 \text{ feet/second/second}$$

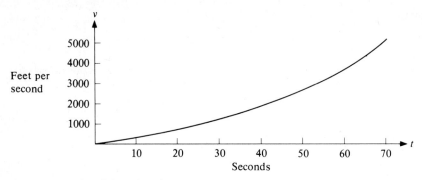

Figure 1 Velocity of the V-2 rocket.

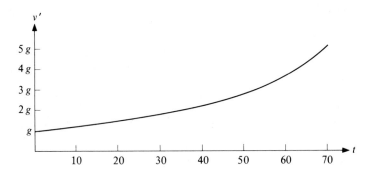

Figure 2 Acceleration of the V-2 rocket.

If $v(0) = 0$ as before, then (10) can be used in the form

$$v(t) = \frac{up'}{B} + \frac{gM_0}{p' - B} - \frac{gp't}{p' - B} - \left(\frac{up'}{B} + \frac{gM_0}{p' - B}\right)\left(1 - \frac{p'}{M_0}t\right)^{B/p'}$$

$$= 563{,}577[1 - (1 - 0.0097t)^{0.0097}] - 32.58t \qquad \text{feet/second}$$

if $0 \leq t \leq 70.54$ seconds. "Burnout" occurs at $t = 70.54$.

The graphs of $v(t)$ and the acceleration $v'(t)$ are given in Figures 1 and 2. The acceleration can be obtained from (8) and the formula for $v(t)$:

$$v'(t) = \frac{up'}{M(t)} - \frac{Bv(t)}{M(t)} - g$$

$$= \frac{6560 \times 8.54}{878.88 - 8.54t} - \frac{0.1}{878.88 - 8.54t}v(t) - 32.2$$

$$0 \leq t < 70.54 \text{ seconds}$$

Of particular interest to an aeronautical engineer are the velocity at burnout (5252 feet/second) and the peak acceleration (greater than $5g$).

4. Problems

1. $\left. \begin{array}{l} y'(t) + y(t) = t \\ y(0) = 0 \end{array} \right|$ **2.** $\left. \begin{array}{l} ty'(t) - 2y(t) = t^2 \\ y(1) = 1 \end{array} \right|$ **3.** $\left. \begin{array}{l} ty'(t) + (1 - t)y(t) = te^t \\ y(1) = e \end{array} \right|$

4. In rocketry, a *burning program* is a function p' specifying the propellant consumption rate. Its integral p is the mass lost by the rocket. That is, if M_0 is the mass of the rocket at time t_0, then

$$\left. \begin{array}{l} \dfrac{dM}{dt} = -p' \\[2mm] M(t_0) = M_0 \end{array} \right\} \quad \text{or} \quad \left. \begin{array}{l} \dfrac{dp}{dt} = p' \\[2mm] p(t_0) = 0 \end{array} \right\}$$

and $M(t) = M_0 - p(t)$. Putting these quantities into Equation (21) of Section E leads to a differential equation for p.

(a) Find a differential equation for p such that the burning program p' will give the rocket constant velocity $v(t) = v(t_0)$ for all $t \geq t_0$.

(b) Assuming B and u are constant, solve the equation and find the burning program of part a.

(c) Find a burning program such that $v(t) = gt$, where g is the acceleration due to gravity, and B and u are also constant.

(d) Find a burning program such that $v(t) = -gt$.

5. If the rocket engine of Section E shuts off at time t_1 with mass M_1 and velocity v_1, then, beginning at t_1, the propellant consumption rate p' is zero. Write and solve the initial value problem governing the subsequent flight.

6. In Figure 3, a body of mass M has been placed in a barrel containing thick oil, which is being heated continuously. While in the oil, the body settles to the bottom with velocity $v(t)$. The forces on the body are those due to its weight Mg and its buoyancy f, both of which are constant, and a viscous friction force $B_0(1 - \alpha t)^2 v(t)$. $B_0(1 - \alpha t)^2$ is the coefficient of viscous friction, a function of time because the oil is being heated. Set up the differential equation governing the velocity v, and solve it subject to the initial condition $v(0) = v_0$. Do not evaluate the final integral.

Either now or later you may wish to try to write equations modeling the situations described in the following problems.

7. Fluids A and B flow into vat X in a particular industrial process. (See Figure 4.) The mixture overflows at a fixed depth d, and so the vat is always full. In the vat the fluids are thoroughly mixed by a stirrer. A minor power failure suddenly shuts down the flow of fluid B into the vat, although fluid A and the stirrers continue as before. How long will it be until the concentration of fluid B is reduced by 1 percent? What if the flow of fluid B is not cut off completely but reduced by half?

8. Foundation Savings Bank compounds interest daily on its savings certificates. Its annual rate is 7.30 percent, so every day it adds to the account 7.30 percent/365 = 0.02 percent = 2/10,000 of what was already in the account, but otherwise nothing is added to or taken out of the account. On July 1, the

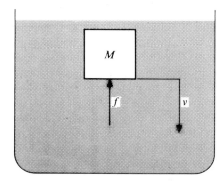

Figure 3 Problem 6: A mass sinking through thick oil.

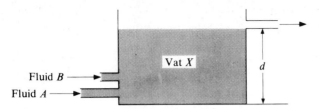

Vat X

Fluid B

Fluid A

d

Figure 4 The mixing vat of Problem 7.

certificate belonging to depositor X was worth \$3312. How much will it be worth on July 1 a year later? How much was it worth on July 1 a year earlier?

9. Over the space of 20 years of peace, the Army reserves are kept at 1,000,000 soldiers. Those who retire are replaced, but the strength is neither increased nor decreased. At the beginning of the period, three-quarters of the soldiers are combat veterans. Veterans and nonveterans are equally likely to retire, but replacements are nonveterans only. If X soldiers retire annually, what fraction is veterans at the end of T years? At the end of 20 years? What fraction is veterans after 100,000 soldiers have retired and been replaced? At the end of the twentieth year, the strength is allowed to decline by letting soldiers retire but making no additions. The retirement rate is one-tenth each year. What happens to the ratio of veterans to nonveterans during this decline?

G. MORE COMPLICATED MODELS

1. Example: Friction on a Cable Wound about a Post

Consider a post standing upright with a cable wound around it, as in Figure 1, with friction between the cable and the post. It is usually the case that $T_1 \neq T_2$; that is, the tension in the cable varies from point to point because of this friction. We shall compute how large T_2/T_1 may be without the cable slipping.

Each point of the cable determines an angle from a fixed direction. In Figure 1 we have measured the angle θ counterclockwise from the beginning of the winding. In fact, we shall take θ to be increasing in the direction of increasing tension, and so we are supposing that $T_2 \geq T_1$. For each value of θ there is a corresponding

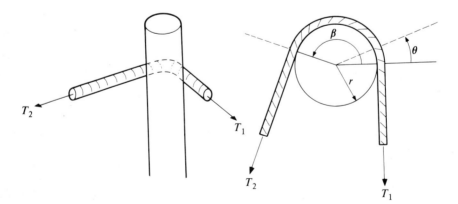

Figure 1 Cable wound about an upright post of radius r, oblique pictorial and plan.

tension $T(\theta)$ in the cable. As indicated, $T(0) = T_1$, and $T(\beta) = T_2$. We shall see that $T(\theta)$ satisfies a differential equation, and we shall find $T(\theta)$ by solving an initial value problem.

Consider the segment of cable from angle $\theta - \Delta\theta$ to $\theta + \Delta\theta$, a segment of length $2r\,\Delta\theta$, where r is the radius of the post. Since the sum of the radial forces must be zero, it follows that if N is the force of the post on the segment, then

$$N(\theta) = T(\theta + \Delta\theta)\sin\Delta\theta + T(\theta - \Delta\theta)\sin\Delta\theta$$

$$= [T(\theta + \Delta\theta) + T(\theta - \Delta\theta)]\sin\Delta\theta$$

Of course, N is distributed as pressure of the post on the cable, and so the average pressure may be defined as

$$n_{\mathrm{av}}(\theta) = \frac{N(\theta)}{2r\,\Delta\theta}$$

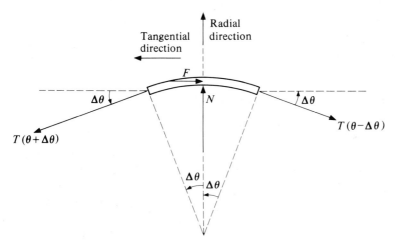

Figure 2 The forces on a short segment of the cable.

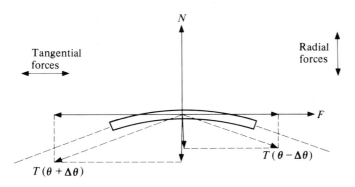

Figure 3 Radial and tangential forces on the segment.

When the formula for N is substituted into it, this formula for $n_{av}(\theta)$ becomes

$$n_{av}(\theta) = \frac{1}{r} \frac{T(\theta + \Delta\theta) + T(\theta - \Delta\theta)}{2} \frac{\sin \Delta\theta}{\Delta\theta}$$

The limit of $n_{av}(\theta)$ as $\Delta\theta$ goes to zero is the pressure of the post on the cable at the point θ, which we shall call $n(\theta)$. It follows from the formula for $n_{av}(\theta)$ that

$$n(\theta) = \lim_{\Delta\theta \to 0} n_{av}(\theta) = \frac{1}{r} \left[\lim_{\Delta\theta \to 0} \frac{T(\theta + \Delta\theta) + T(\theta - \Delta\theta)}{2} \right] \lim_{\Delta\theta \to 0} \left(\frac{\sin \Delta\theta}{\Delta\theta} \right) = \frac{1}{r} \frac{2T(\theta)}{2}$$

or

$$n(\theta) = \frac{T(\theta)}{r} \tag{1}$$

That is, the radial pressure $n(\theta)$, the force per unit cable length, is proportional to $T(\theta)$, the cable tension at θ; the proportionality constant is $1/r$, where r is the radius of the post. Notice that the cable bends into line with the tension force at the end of the segment. To assume such an alignment is to assume that the cable is quite flexible.

Now consider the tangential forces. The tension increases counterclockwise, as θ does, and the friction force F counteracts it. Hence,

$$T(\theta + \Delta\theta) \cos \Delta\theta - T(\theta - \Delta\theta) \cos \Delta\theta = F(\theta) \tag{2}$$

F is really dependent on the length of the segment $2r \, \Delta\theta$, and to analyze it we need an assumption based on empirical observation: The friction force is proportional to the product of the pressure $n_{av}(\theta)$ and the area of the region on which it acts $2r \, \Delta\theta$. That is,

$$f = \frac{F(\theta)}{n_{av}(\theta) 2r \, \Delta\theta} \text{ is constant, independent of } \Delta\theta \text{ and } \theta \tag{3}$$

Empirical observations have shown that independence of $\Delta\theta$ is realistic if $\Delta\theta$ is small enough. There is also an assumption required that the constant $F/n_{av}(\theta) 2r \, \Delta\theta = f$ must be the same at all θ. That is, the cable should not both slip in some places and be fixed at others. In its simplext form the assumption is that the cable should not stretch. In many applications this is quite reasonable. The value of the constant f is called the *coefficient of friction*, and with it one calculates the maximum friction force, $F(\theta) = fn_{av}(\theta) 2r \, \Delta\theta$. Substitution into (2) gives the equation

$$[T(\theta + \Delta\theta) - T(\theta - \Delta\theta)] \cos \Delta\theta = fn_{av}(\theta) 2r \, \Delta\theta$$

Dividing by $2\Delta\theta \cos \Delta\theta$ gives the equation

$$\frac{T(\theta + \Delta\theta) - T(\theta - \Delta\theta)}{2\,\Delta\theta} = \frac{1}{\cos \Delta\theta}\, fn_{av}(\theta)r \tag{4}$$

of which we now take the limit on each side as $\Delta\theta$ goes to zero. On the right, $\cos \Delta\theta$ has limit 1 and $n_{av}(\theta)$ has limit $n(\theta)$. On the left,

$$\lim_{\Delta\theta \to 0} \frac{T(\theta + \Delta\theta) - T(\theta - \Delta\theta)}{2\,\Delta\theta}$$

$$= \frac{1}{2} \lim_{\Delta\theta \to 0} \frac{T(\theta + \Delta\theta) - T(\theta) + T(\theta) - T(\theta - \Delta\theta)}{\Delta\theta}$$

$$= \frac{1}{2}\left[\lim_{\Delta\theta \to 0} \frac{T(\theta + \Delta\theta) - T(\theta)}{\Delta\theta} + \lim_{\Delta\theta \to 0} \frac{T(\theta) - T(\theta - \Delta\theta)}{\Delta\theta} \right]$$

$$= \tfrac{1}{2}[T'(\theta) + T'(\theta)] = T'(\theta)$$

Thus, taking the limits of the left- and right-hand sides of (4) gives

$$T'(\theta) = fn(\theta)r \tag{5}$$

Now we may substitute (1) into (5), obtaining

$$T'(\theta) = fT(\theta) \tag{6}$$

a differential equation whose general solution may be written

$$T(\theta) = T(0)e^{f\theta}$$

Since f is the coefficient giving the maximum tension, it follows that

$$T_2 = T(\beta) \leq T(0)e^{f\beta} = T_1 e^{f\beta}$$

or

$$\frac{T_2}{T_1} \leq e^{f\beta}$$

If T_1 were greater than T_2, the angles would have been measured in the opposite sense, clockwise.

2. Example: Cable Friction around a Horizontal Drum

In this case the weight of the cable increases the radial pressure at the top of the drum, decreases it at the bottom, and at the sides adds to one tangential force or the other (see Figure 4).

If θ is measured from the horizontal, and if W is the weight of the segment of length $2r\,\Delta\theta$, then the radial force at θ is increased by

$$W \cos\left(\theta - \frac{\pi}{2}\right) = \rho 2r\,\Delta\theta \sin \theta$$

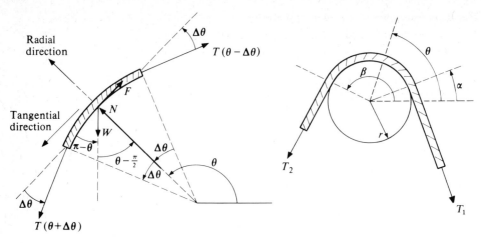

Figure 4 Cable wound about a horizontal drum, segment and plan.

where ρ is the weight of the cable per unit length. Thus the pressure at the point θ is increased by $\rho \cos (\theta - \pi/2) = \rho \sin \theta$:

$$n(\theta) = \frac{1}{r} T(\theta) + \rho \sin \theta \tag{7}$$

This relation, which replaces Equation (1), summarizes the condition of radial equilibrium.

In the weightless case, Equation (5) summarized the condition of tangential equilibrium. In the present case, W is a force with magnitude $\rho 2r \, \Delta\theta$, and we must account for the tangential component. Therefore the equation of tangential equilibrium is

$$T(\theta + \Delta\theta) \cos \Delta\theta - T(\theta - \Delta\theta) \cos \Delta\theta + W \cos (\pi - \theta) = F(\theta)$$

That is, $$\frac{T(\theta + \Delta\theta) - T(\theta - \Delta\theta)}{2 \, \Delta\theta} = \frac{1}{\cos \Delta\theta} \left(\frac{W}{2 \, \Delta\theta} \cos \theta + \frac{F(\theta)}{2 \, \Delta\theta} \right)$$

$$= \frac{1}{\cos \Delta\theta} \left(\rho r \cos \theta + \frac{F(\theta)}{n_{av}(\theta) 2r \, \Delta\theta} n_{av}(\theta) r \right)$$

Taking the limit as $\Delta\theta$ goes to zero on both sides gives

$$T'(\theta) = \rho r \cos \theta + f n(\theta) r \tag{8}$$

since we have assumed here, as we did in (3), that $F/n_{av} 2r \, \Delta\theta$ has limit f everywhere. Substitution of (7) into (8) gives

$$T'(\theta) - f T(\theta) = \rho r (\cos \theta + f \sin \theta) \tag{9}$$

If α is the angle where the cable meets the drum (see Figure 4), the angle at which the tension is least, then at α an initial condition may be set:

$$T(\alpha) = T_1$$

This, together with (9), makes an initial value problem which may be solved using the method of Section F. The solution is

$$T(\theta) = T_1 e^{f(\theta - \alpha)} + pr e^{f(\theta - \alpha)} \int_\alpha^\theta e^{-f(z - \alpha)}(\cos z + f \sin z)\, dz$$

$$= T_1 e^{f(\theta - \alpha)} + \frac{pr}{1 + f^2}[(1 + f^2) \sin \theta - 2f \cos \theta]$$

$$- \frac{pr e^{f(\theta - \alpha)}}{1 + f^2}[(1 + f^2) \sin \alpha - 2f \cos \alpha]$$

3. Example: Mutual Heating and Cooling

Recall Newton's law of cooling (Section E): If a body has temperature $T_1(t)$ and its surroundings have uniform constant temperature T_2, then the temperature difference $P(t) = T_1(t) - T_2$ satisfies a differential equation

$$P' = -\lambda P$$

Since $P(t)$ is proportional to the amount of excess heat stored in the body, this equation may be interpreted as describing the rate of heat flow. That is, if

$$E = cmP \qquad (10)$$

is the excess heat stored in the body, where c is the specific heat in calories per gram-degree and m is the mass of the body in grams, then $E' = cmP'$:

$$E'(t) = -\lambda cm[T_1(t) - T_2] \qquad (11)$$

This reformulation of Newton's law will lead presently to equations describing the transfer of heat by conduction between two bodies insulated from everything else.

Let two bodies have masses m_1 and m_2, specific heats c_1 and c_2, and temperatures $T_1(t)$ and $T_2(t)$. Let U be a convenient fixed temperature, let $P_1 = T_1 - U$ and $P_2 = T_2 - U$ be the excess temperature of the two bodies, and let $E_1(t) = c_1 m_1 P_1(t)$ and $E_2(t) = c_2 m_2 P_2(t)$ be the corresponding excess heats. Since the bodies, though touching each other, are insulated from everything else, their combined excess heat is constant:

$$0 = \frac{d}{dt}(E_1 + E_2) = E_1' + E_2'$$

That is, the heat given off by one is collected by the other:

$$E'_2 = -E'_1 \tag{12}$$

Newton's law, Equation (11), says the rate of heat transfer is proportional to the temperature difference:

$$E'_1(t) = -\alpha[T_1(t) - T_2(t)]$$
$$E'_2(t) = -E'_1(t) = -\alpha[T_2(t) - T_1(t)]$$

where α is a positive constant of proportionality depending on the amount of contact between the two bodies. These two equations may now be reinterpreted in terms of temperature rather than heat, using the definition of $E_1(t)$ and $E_2(t)$:

$$T'_1(t) = P'_1(t) = -\frac{\alpha}{c_1 m_1}[T_1(t) - T_2(t)]$$

$$T'_2(t) = P'_2(t) = -\frac{\alpha}{c_2 m_2}[T_2(t) - T_1(t)]$$

Subtracting the second equation from the first gives the differential equation

$$(T_1 - T_2)' = -\alpha \left(\frac{1}{c_1 m_1} + \frac{1}{c_2 m_2} \right)(T_1 - T_2) \tag{13}$$

which can be solved in order to predict temperature difference:

$$T_1(t) - T_2(t) = [T_1(0) - T_2(0)]e^{-\alpha(1/c_1 m_1 + 1/c_2 m_2)t}$$

The formula predicts the difference. To predict the functions T_1 and T_2 themselves, however, we need a relation between T_1 and T_2, such as (12):

$$T'_2 = P'_2 = \frac{E'_2}{c_2 m_2} = -\frac{E'_1}{c_2 m_2} \tag{14}$$

$$= -\frac{c_1 m_1}{c_2 m_2} P'_1 = -\frac{c_1 m_1}{c_2 m_2} T'_1$$

In one respect even (14) is not enough, for while it allows us to eliminate T'_2 from (13), it does not allow the elimination of T_2 itself. To get around this difficulty, we differentiate (13) all the way across:

$$T''_1 - T''_2 = -\alpha \left(\frac{1}{c_1 m_1} + \frac{1}{c_2 m_2} \right)(T'_1 - T'_2)$$

Now substitution of (14) eliminates T'_2 and T''_2:

$$T''_1 + \frac{c_1 m_1}{c_2 m_2} T''_1 = -\alpha \left(\frac{1}{c_1 m_1} + \frac{1}{c_2 m_2} \right) \left(T'_1 + \frac{c_1 m_1}{c_2 m_2} T'_1 \right)$$

or

$$T''_1 = -\alpha \left(\frac{1}{c_1 m_1} + \frac{1}{c_2 m_2} \right) T'_1 \tag{15}$$

This equation is a little different from those we have been studying, for it involves a second derivative. A differential equation which involves no higher derivatives than the second is called a *second-order equation*, and (15) is an example of one.

In Chapter III we shall introduce methods for solving second- and higher-order equations, but (15) can be solved for T_1' at once, and then for T_1 itself. Specifically, rewrite (15) as

$$\frac{d}{dt} T_1' = -\alpha \left(\frac{1}{c_1 m_1} + \frac{1}{c_2 m_2} \right) T_1'$$

Now solve for T_1' using separation of variables:

$$T_1'(t) = T_1'(0) e^{-\alpha(1/c_1 m_1 + 1/c_2 m_2)t}$$

Definite integration from 0 to t gives

$$T_1(t) = T_1(0) + T_1'(0) \frac{c_1 m_1 c_2 m_2}{\alpha(c_1 m_1 + c_2 m_2)} \left(1 - e^{-\alpha(1/c_1 m_1 + 1/c_2 m_2)t} \right) \tag{16}$$

Equation (16) shows very clearly that two pieces of initial data are required in solving an initial value problem involving a second-order differential equation

$$\left. \begin{array}{c} T_1'' = -\alpha \dfrac{c_1 m_1 + c_2 m_2}{c_1 m_1 c_2 m_2} T_1' \\[2mm] T_1'(0) \text{ given} \\[2mm] T_1(0) \text{ given} \end{array} \right\}$$

one arbitrary constant being required for each of the two integrations. In the model described above, $T_1(0)$ is a natural piece of information to measure or be given, but $T_2(0)$ is much more likely to be measurable than $T_1'(0)$ is. Hence, to finish our prediction, we must calculate $T_1'(0)$ from $T_1(0)$ and $T_2(0)$. It follows from (14) that

$$(T_1 - T_2)' = T_1' - T_2'$$

$$= T_1' + \frac{c_1 m_1}{c_2 m_2} T_1'$$

$$= \frac{c_1 m_1 + c_2 m_2}{c_2 m_2} T_1'$$

Substituting this expression for $(T_1 - T_2)'$ into (13) gives

$$\frac{c_1 m_1 + c_2 m_2}{c_2 m_2} T_1' = -\alpha \frac{c_1 m_1 + c_2 m_2}{c_1 m_1 c_2 m_2} (T_1 - T_2)$$

or

$$T_1' = -\frac{\alpha}{c_1 m_1} (T_1 - T_2)$$

Substitution of this into (16), where $T_1'(0)$ is required, gives

$$T_1(t) = T_1(0) + [T_2(0) - T_1(0)]\frac{c_2 m_2}{c_1 m_1 + c_2 m_2}(1 - e^{-\alpha(1/c_1 m_1 + 1/c_2 m_2)t}) \qquad (17)$$

The specific heat of a substance is the number of calories required to raise the temperature of 1 gram of the substance through 1°C. It varies somewhat with temperature, but nearly any substance has temperature ranges on which the specific heat is nearly constant. The heat capacity of a body of a given substance is the product of the specific heat of the substance and the mass of the body. When that much heat is provided, the temperature of the body will rise 1 degree. When cooled 1 degree, the body will give off that much heat. In the example, $c_1 m_1$ and $c_2 m_2$ are the heat capacities of the bodies, and $c_1 m_1 + c_2 m_2$ is the heat capacity of the system.

With this in mind, let us find the steady-state temperature of the system, the limiting temperature. From (17),

$$\lim_{t \to \infty} T_1(t) = T_1(0) + [T_2(0) - T_1(0)]\frac{c_2 m_2}{c_1 m_1 + c_2 m_2}$$

$$= \frac{c_1 m_1 T_1(0) + c_2 m_2 T_2(0)}{c_1 m_1 + c_2 m_2}$$

$$= \frac{c_1 m_1[T_1(0) - U] + c_2 m_2[T_2(0) - U] + (c_1 m_1 + c_2 m_2)U}{c_1 m_1 + c_2 m_2}$$

$$= \frac{E_1(0) + E_2(0)}{c_1 m_1 + c_2 m_2} + U$$

In other words, the steady-state excess temperature (above U) is the excess heat of the system divided by the total heat capacity of the system. The limiting value of T_2 will be the same.

4. Example: Vehicle Response to Fluid Dynamic Control

The example has analogs in ship and submarine control. Figure 5 shows an airplane in steady, level flight. The lift of the airflow over all surfaces is symmetric with respect to Y. If the pilot raises the right aileron and lowers the left, then the lift on the left wing is increased, that on the right wing is reduced, and a torque or moment about the X axis is produced. It is the *aileron-induced rolling moment* L_A.

Bodies in rotation about an axis have dynamic properties analogous to those of bodies in straight-line motion. Angular displacement ϕ, measured in radians, has as its derivative the angular velocity ϕ', measured in radians per second. The angular momentum is computed as $I\phi'$, where I is the moment of inertia of the body about the axis of rotation. It is a consequence of Newton's second law that the rate of change of angular momentum is the net moment about the axis.

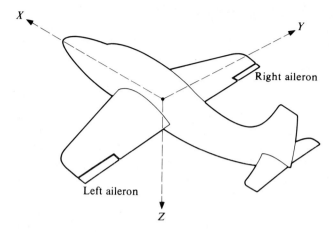

Figure 5 Airplane in steady, level flight.

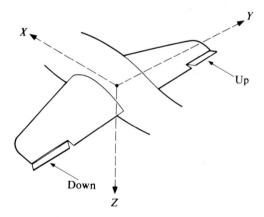

Figure 6 Deflection of ailerons.

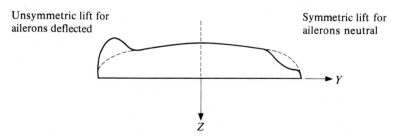

Figure 7 Comparison of wing lift distributions for ailerons neutral and ailerons deflected.

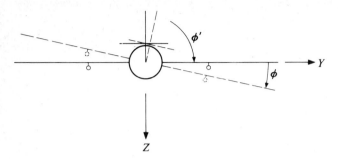

Figure 8 Airplane roll.

Therefore, the rolling moment L_A produces a rotation about the X axis, called *roll*. The *roll rate* is the angular velocity ϕ'; ϕ is called the *bank angle*.

The rotation itself produces an opposing antisymmetric lift distribution and hence another moment, opposite in direction to ϕ' and proportional to ϕ'. This is the *roll damping moment* $L_R = -B\phi'$.

Summing the moments and applying the appropriate form of Newton's second law gives

$$I\phi'' = L_A + L_R = L_A - B\phi'$$

or

$$\phi'' + \frac{B}{I}\phi' = \frac{L_A}{I}$$

The moment of inertia I is a constant determined by the mass distribution of the airplane, B is determined by its aerodynamic characteristics, and L_A by the extent to which the ailerons are deflected. If, at $t = 0$, the airplane is level and steady so that

$$\phi(0) = 0 \quad \text{and} \quad \phi'(0) = 0$$

then taking ϕ' as the unknown and solving for it (using the integrating factor as in Section F) gives

$$\phi'(t) = \frac{L_A}{B}(1 - e^{-(B/I)t})$$

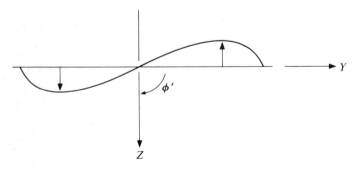

Figure 9 Lift distribution responsible for roll damping moment.

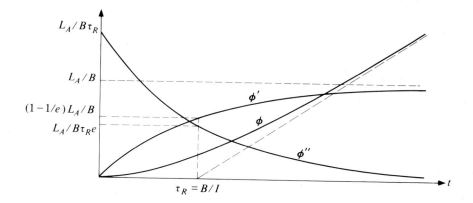

Figure 10 Bank angle ϕ and its derivatives.

Direct integration then gives

$$\phi(t) = \frac{L_A}{B} t - \frac{IL_A}{B^2} (1 - e^{-(B/I)t})$$

These formulas take the following simpler form if we introduce the parameter $\tau_R = I/B$:

$$\phi(t) = \frac{L_A}{B} t - \frac{L_A}{B} \tau_R (1 - e^{-t/\tau_R})$$

$$\phi'(t) = \frac{L_A}{B} (1 - e^{-t/\tau_R})$$

$$\phi''(t) = \frac{L_A}{B} \frac{1}{\tau_R} e^{-t/\tau_R}$$

Note that the graph of ϕ is asymptotic to the graph of $(L_A/B)(t - \tau_R)$.

5. Example: Machine with an Eccentric Rotor

A front-loading washing machine (as in Figure 11) is a machine with an eccentric rotor. Let us imagine it so constrained that it can move vertically but not horizontally.

Let M be the total mass of the machine including the load, let m be the unbalanced mass, and suppose that $\theta(t) = \omega t$ describes the internal motion of the machine, the spinning of the rotor. If $x(t)$ is the (vertical) displacement of the machine as measured at the axis, then the displacement of the unbalanced mass is $x(t) + r \sin \omega t$, and the center of mass of the system has displacement

$$\frac{(M - m)x(t) + m[x(t) + r \sin \omega t]}{M} = x(t) + \frac{m}{M} r \sin \omega t$$

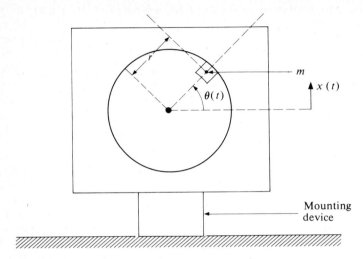

Figure 11 Front-loading washing machine and mounting (idealized).

The momentum of the machine is M times the derivative of this displacement:

$$Mx'(t) + mr\omega \cos \omega t \qquad (18)$$

Typically the mounting device consists of a combination of springs and dampers. The springs act to restore the machine to an equilibrium position, which we shall call $x = 0$; they provide a force proportional to the displacement but opposite in direction:

$$F_{spring}(t) = -Kx(t)$$

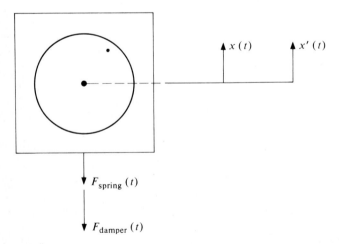

Figure 12 Washing machine: displacement, velocity, and external forces.

The dampers act to restrain the velocity. They provide a force proportional to the velocity but in the opposite direction:

$$F_{damper}(t) = -Bx'(t)$$

By Newton's second law, the net external force on the machine is the derivative of the momentum (18), and so

$$\frac{d}{dt}[Mx'(t) + mr\omega \cos \omega t] = -Kx(t) - Bx'(t)$$

Performing the differentiation gives

$$Mx''(t) - mr\omega^2 \sin \omega t = -Kx(t) - Bx'(t)$$

or
$$Mx''(t) + Bx'(t) + Kx(t) = mr\omega^2 \sin \omega t \tag{19}$$

It should be noted that the left-hand side of the equation contains the unknown function x and its derivatives, the gross mass M, and the mounting-device parameters B and K. The right-hand side contains all the information about the excitation of the system. Equations such as (19) will be solved in Chapter III.

6. Example: Hanging-Cable Problems

Let us consider a suspension bridge. The cable hangs from fixed supports, and the load is hung from the cable. We shall assume that the mass of the cable is negligible by comparison with the mass of the load, and that the load is evenly distributed along the x axis. We shall find the shape of the cable.

Consider the segment of cable over the interval of length Δx between x and $x + \Delta x$. Because the sum of the horizontal forces is zero, it follows that T_H, the horizontal component of tension, is constant:

$$T_H(x + \Delta x) = T_H(x) \qquad \text{for all } x \text{ and } \Delta x$$

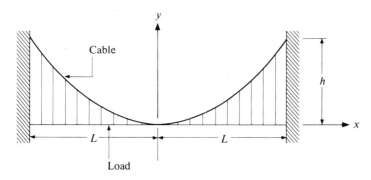

Figure 13 Simple suspension bridge.

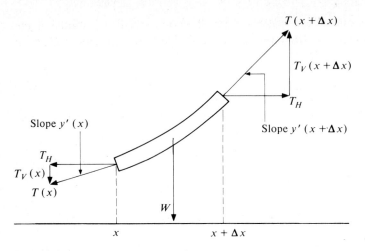

Figure 14 Forces on a segment of cable.

Since $y'(x)$ is the slope of the cable at x and therefore the tangent of the angle of inclination there, it follows that

$$T_V(x) = T_H y'(x)$$

and similarly

$$T_V(x + \Delta x) = T_H y'(x + \Delta x)$$

The remaining vertical force upon the segment is a portion W of the load. Since the load is uniformly distributed along the x axis, it follows that $W = \rho \, \Delta x$, where ρ is the weight of the load per unit length.

$$T_V(x + \Delta x) = T_V(x) + W$$

$$T_H y'(x + \Delta x) = T_H y'(x) + W$$

$$y'(x + \Delta x) - y'(x) = \frac{W}{T_H}$$

Thus,

$$y'(x + \Delta x) - y'(x) = \frac{\rho}{T_H} \Delta x \quad \text{or} \quad \frac{y'(x + \Delta x) - y'(x)}{\Delta x} = \frac{\rho}{T_H}$$

Passing to the limit as Δx goes to zero gives

$$y''(x) = \frac{\rho}{T_H} \tag{20}$$

The differential equation can be solved by direct integration, using the initial conditions $y(0) = 0$ and $y'(0) = 0$, which represent the fact that the height of the cable above the load is zero at the midpoint of the span and that the cable is horizontal there:

$$y(x) = \frac{\rho}{T_H} \frac{x^2}{2} \tag{21}$$

The cable thus describes a parabola between the supports. For instance, if the height of the supports above the load is h and the length of the span is $2L$, as in Figure 13, then (21) shows that

$$h = \frac{\rho}{T_H}\frac{L^2}{2} \tag{22}$$

from which the horizontal component of the tension, T_H, can be found directly:

$$T_H = \frac{\rho L^2}{2h}$$

The vertical component of the tension, T_V, is computed as follows:

$$T_V(x) = T_H y'(x) = T_H \frac{\rho x}{T_H} = \rho x$$

where the formula for $y'(x)$ is obtained by differentiating (21). The tension in the cable can also be computed as the resultant force:

$$T(x) = (T_H^2 + T_V^2)^{1/2}$$

$$= \left[\left(\frac{\rho L^2}{2h} \right)^2 + (\rho x)^2 \right]^{1/2}$$

$$= \rho \left(\frac{L^4}{4h^2} + x^2 \right)^{1/2}$$

This is maximized at the ends of the span, where

$$T(L) = \rho \left(\frac{L^4}{4h^2} + L^2 \right)^{1/2} = \frac{\rho L}{2h}(L^2 + 4h^2)^{1/2}$$

For a different hanging-cable problem, consider a catenary, typically a power cable, in which the load is the cable itself. We shall see that it hangs quite differently, and its shape requires a differential equation that is not at all like (20). As before,

$$y'(x + \Delta x) - y'(x) = \frac{W}{T_H} \tag{23}$$

but $W = \rho\,\Delta s$, where s is the length of the cable over the interval of length Δx, and ρ is the weight per unit length of the cable. A good approximation to Δs is given by the Pythagorean theorem:

$$(\Delta s)^2 = (\Delta x)^2 + (\Delta y)^2$$

Thus
$$W = \rho\,\Delta s = \rho[(\Delta x)^2 + (\Delta y)^2]^{1/2}$$

and if Δx is divided out of (23), it follows that

$$\frac{y'(x + \Delta x) - y'(x)}{\Delta x} = \frac{\rho}{T_H}\left[1 + \left(\frac{\Delta y}{\Delta x} \right)^2 \right]^{1/2}$$

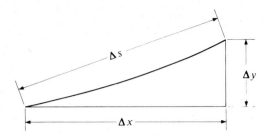

Figure 15 Estimating the length of a segment of a curve.

Taking the limit as Δx goes to zero gives

$$y''(x) = \frac{\rho}{T_H}\{1 + [y'(x)]^2\}^{1/2} \tag{24}$$

Equation (24) is not at all of the form $y' + ay = 0$, nor of the form $y'' + ay' = 0$, but trying the method of Section D (separation of variables) does produce a solution. After division, the left- and right-hand sides should be integrated with respect to x and an arbitrary constant of integration inserted:

$$\frac{\rho}{T_H} = y''(x)\{1 + [y'(x)]^2\}^{-1/2}$$

$$\frac{\rho}{T_H}x + C = \int \frac{y''(x)}{\{1 + [y'(x)]^2\}^{1/2}}\,dx$$

$$= \int \frac{\cosh z}{(1 + \sinh^2 z)^{1/2}}\,dz \qquad \text{if } y'(x) = \sinh z$$

$$= \int \frac{\cosh z}{\cosh z}\,dz$$

$$= \int dz = z$$

$$\sinh\left(\frac{\rho}{T_H}x + C\right) = \sinh z = y'(x)$$

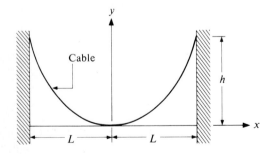

Figure 16 Catenary.

If $y'(0) = 0$ as in Figure 16, it follows that $C = 0$ and

$$y'(x) = \sinh\left(\frac{\rho}{T_H}x\right)$$

If $y(0) = 0$, then integration gives

$$y(x) = \frac{T_H}{\rho}\left[-1 + \cosh\left(\frac{\rho}{T_H}x\right)\right]$$

7. Example: The Deflection of a Prismatic Beam

A beam is *prismatic* if every section of the beam is the same. We shall assume as well that the section is symmetric about some plane and that all applied forces are either applied in this plane or are applied symmetrically about this *plane of symmetry*, so that the resulting deformation will not bend the beam toward either side of this plane. Such a deformation is called a *plane deflection*. (See Figure 17.)

It is observed in the laboratory that when the beam is deflected the inner extremity of the beam shortens, the outer extremity lengthens, and that there is an intermediate plane, the *neutral plane*, which is bent but in which distance remains unchanged. The intersection of the plane of symmetry and the neutral plane is the *neutral axis*. (See Figures 18 and 19.)

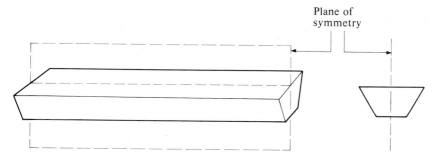

Figure 17 Prismatic beam with plane of symmetry, oblique pictorial and section.

Figure 18 Deflected prismatic beam, showing deformation of the neutral plane, oblique pictorial and section.

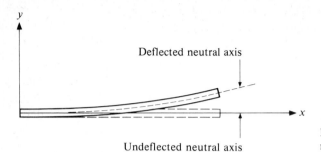

Figure 19 Coordinate system for measuring beam deflection.

Let x represent distance along the neutral axis of the undeflected beam; let $y(x)$ be the deflection of the beam at x, positive in the upward direction; and let $w(x)$ be the *load intensity* at x, the applied force per unit length along the beam, also positive in the upward direction. It may be shown that y satisfies the differential equation

$$y^{(4)}(x) = \frac{1}{EI} w(x) \tag{25}$$

This is a "fourth-order" differential equation. It relates the deflection y to a quantity which can be calculated externally, the loading on the beam. The equation is used to calculate the deflection. The parameter EI, the product of the modulus of elasticity and the moment of inertia of the section, is called the *stiffness coefficient*. If it is large, the deflection is small. Solutions are obtained by direct integration.

An initial value problem will require initial values for y''', y'', y', and y. Of course, $y(0)$ is the displacement at $x = 0$. If the beam is supported at $x = 0$, then the displacement there is zero and

$$y(0) = 0$$

The beam is said to be *cantilevered* at $x = 0$ if, in addition, its inclination there is fixed by its support. In Figure 19, the beam depicted may be considered to be supported only at $x = 0$, and its inclination at $x = 0$ is therefore necessarily fixed. In this case

$$y'(0) = 0$$

Since the beam of Figure 19 is supported only at $x = 0$, it is easy to compute the initial values $y''(0)$ and $y'''(0)$ directly from the load intensity w as follows:

$$y''(0) = -\frac{1}{EI} \int_0^L w(X)\, dX$$

and

$$y'''(0) = \frac{1}{EI} \int_0^L X w(X)\, dX \tag{27}$$

where L is the length of the beam. [Equations (25) to (27) are derived in texts on strength of materials. We shall not give the derivation.]

H. SYNOPSIS OF DEFINITIONS AND RESULTS ON LINEAR DIFFERENTIAL EQUATIONS

Differential equations are classified in several ways. Those in this book all depend on a single independent variable, usually t. Such equations are called *ordinary differential equations*, to distinguish them from those depending on several independent variables. If several independent variables are present, then the derivatives must be partial derivatives; the equations containing them are called *partial differential equations*, which will not concern us.

Ordinary differential equations may contain many derivatives of an unknown function. The *order of a derivative* is the number of differentiations required to achieve it, so that y, y', y'', y''', $y^{(4)}$, $y^{(5)}$, and $y^{(n)}$ are derivatives of order 0, 1, 2, 3, 4, 5, and n, respectively. *The order of a differential equation* is the maximum of the orders of the derivatives appearing. Thus,

$$y' + \frac{I}{V} y = 0 \qquad \text{is of order 1 (see Sections A and D)} \qquad (1)$$

$$y^{(4)} = \frac{1}{EI} w \qquad \text{is of order 4 (see Section G, Part 7)} \qquad (2)$$

$$y'' = \lambda \sqrt{1 + (y')^2} \qquad \text{is of order 2 (see Section G, Part 6)} \qquad (3)$$

Equations such as (2) or (3), in which one side of the equation gives a formula for the highest derivative, are sometimes said to be in *normal* form, but in fact there are many "normal" forms, each with its own use. For us, the most useful is the normal form for linear equations, to be defined.

A differential equation of order n is *linear* if it can be written in the following *normal form*:

$$y^{(n)}(t) + a_{n-1}(t)y^{(n-1)}(t) + a_{n-2}(t)y^{(n-2)}(t) + \cdots$$
$$+ a_2(t)y''(t) + a_1(t)y'(t) + a_0(t)y(t) = f(t) \qquad (4)$$

Each side is a sum of terms. On the left, each term is the unknown function y or one of its derivatives, multiplied by a given coefficient which is a function of the variable t alone. The coefficient of the highest derivative is the constant function 1. The term in which neither y nor any of its derivatives appears is on the right-hand side and is called the *nonhomogeneous term* (or *forcing function*, for reasons which may become apparent from the models in Chapter II). In a linear differential equation, if the nonhomogeneous term is identically zero, that is, if f is the zero function, then the equation is *homogeneous*. Otherwise it is *nonhomogeneous*. (Equations which are not linear are called *nonlinear*.)

In part, our purpose is to motivate and provide techniques for the use of systems of linear equations. Therefore, from this point on, we shall consider linear equations virtually exclusively. Equation (3) above is nonlinear; Sections J and K of this chapter treat a few nonlinear equations, and a few more appear in Chapter IV for comparative purposes. Nonlinear equations are important in applications but will be left to other sources.

The equations solved in Sections A through F are first-order linear equations, for they are of the form

$$y' + ay = f \tag{5}$$

[In Sections A and D we take $f(t) \equiv 0$, and in Section B we take $a(t) \equiv 0$.] A function y is said to *satisfy* (5) at the point t if y is differentiable there and if

$$y'(t) + a(t)y(t) = f(t)$$

The function y is a *solution* of (5) on an interval (α, β) if it satisfies (5) at every point of that interval, that is, if

$$y'(t) + a(t)y(t) = f(t) \qquad \text{for } \alpha < t < \beta$$

A solution, then, is a function which satisfies the differential equation on an interval (that is, at each point of an interval), even though the interval of interest may go unmentioned.

In Section B, solutions were found by integration, and integration is carried out on intervals. For instance, if

$$\left. \begin{aligned} y'(t) &= \frac{1}{t + 10} \\[4pt] y(0) &= 1 \end{aligned} \right\}$$

then
$$y(t) - 1 = y(t) - y(0) = \int_0^t y'(z)\, dz = \int_0^t \frac{1}{z + 10}\, dz$$

$$= \ln(t + 10) - \ln 10 = \ln\left(\frac{t}{10} + 1\right)$$

$$y(t) = 1 + \ln\left(\frac{t}{10} + 1\right)$$

Since the integral is divergent at $z = -10$, the solution is valid only on the interval $(-10, \infty)$, the set of points t such that $-10 < t$.

In the lake problem of Section A, the interval of interest was never specified, except that it had to contain the point $t = 0$ where the initial concentration in the lake was measured. If the coefficient I/V is constant, the solution is valid on the whole real line.

In Section D, it was shown that if the function a which appears as coefficient in (5) is continuous on the interval (α, β), then (5) has *homogeneous solutions* on (α, β):

$$y_h(t) = Ce^{-A(t)} \tag{6}$$

where A is any antiderivative of a, and C is an arbitrary constant. The functions y_h are not themselves solutions of (5), but of the *associated homogeneous equation*

$$y_h' + ay_h = 0 \tag{7}$$

which differs from (5) on the right-hand side. A homogeneous linear equation has the property that if y_1 is a solution then so is Cy_1, for every constant C. (We shall return to this in Chapter IV.) It was shown in Section D that every solution of (7) is of the form (6), and every function of the form (6) is a solution of (7). Thus, (6) is the *general solution* of (7) and the *general homogeneous solution* of (5). Any one function of the form (6) is a *particular homogeneous solution*.

In Section F, it was shown that division of both sides of (5) by any nonzero homogeneous solution y_h reduces (5) to the form studied in Section B, which can be integrated directly on the interval (α, β) provided f and a are continuous there. Specifically, if t_0 is an arbitrary point of the interval (α, β), usually called the *initial point*, and if A is any antiderivative of a, then integration performed from t_0 to t gives

$$y(t) = y(t_0)e^{-A(t) + A(t_0)} + e^{-A(t)} \int_{t_0}^{t} e^{A(z)}f(z)\,dz \qquad (8)$$

$$= \frac{y(t_0)}{y_h(t_0)} y_h(t) + y_h(t) \int_{t_0}^{t} e^{A(z)}f(z)\,dz \qquad (9)$$

for all t in (α, β). Functions of this form are solutions of (5), no matter what the value $y(t_0)$ may be. That is, (8) and (9) are formulas for the general solution of (5) on the interval (α, β).

For the special case in which a is constant, that is, $a(t) \equiv \lambda$, formulas (8) and (9) are useful to memorize in this form:

$$y(t) = y(t_0)e^{-\lambda(t - t_0)} + e^{-\lambda t} \int_{t_0}^{t} e^{\lambda z}f(z)\,dz$$

If $y(t_0) = 0$, then

$$y(t) = y_h(t) \int_{t_0}^{t} \frac{f(z)}{y_h(z)}\,dz \qquad (10)$$

$$= e^{-A(t)} \int_{t_0}^{t} e^{A(z)}f(z)\,dz$$

Thus (10) is itself a particular solution of (5). The right-hand side of equation (9) is just the right-hand side of (10) plus the right-hand side of (6).

That is, the general solution of a first-order linear equation is the sum of a particular solution of the equation plus the general solution of the associated homogeneous equation.

Here, as always with linear differential equations, the term *general solution* has two uses with the same mathematical content. On the one hand, it refers to formula (8), for instance, in which $y(t_0)$ is arbitrary. On the other hand, it refers to the collection of all functions satisfying (5) on the interval (α, β)

$$\{y \mid y'(t) + a(t)y(t) = f(t) \text{ for } \alpha < t < \beta\}$$

which is the same as the collection of all functions of the form (8). This collection is entirely determined by (5) and the interval (α, β), and every member of the collection is called a *particular solution*. Any one particular solution may then be determined by its value at any one point t_0 in (α, β). This, then, summarizes the results:

Theorem 1 *If a and f are continuous functions on an interval* (α, β) *and A is an antiderivative of the function a there, if* t_0 *is any point of the interval* (α, β) *so that* $\alpha < t_0 < \beta$, *and if* y_0 *is any number whatsoever, then the formula*

$$y(t) = y_0 \frac{e^{-A(t)}}{e^{-A(t_0)}} + e^{-A(t)} \int_{t_0}^{t} e^{A(z)} f(z)\, dz \qquad \alpha < t < \beta$$

defines the unique function on (α, β) *which both satisfies the differential equation*

$$y' + ay = f$$

there and satisfies the condition

$$y(t_0) = y_0$$

A problem of the type

$$\left.\begin{matrix} y' + ay = f \\ y(t_0) = y_0 \end{matrix}\right\}$$

is called an *initial value problem*. It consists of a linear differential equation of order 1 in normal form with coefficients continuous on some interval (α, β), a point t_0 of the interval, and a number y_0. The point t_0 is the *initial point*, y_0 is the *initial value*, and the equation

$$y(t_0) = y_0$$

is the *initial condition*. The theorem given above is simply a statement of the fact we have observed: *For first-order linear equations, initial value problems have unique solutions.*

Initial value problems for higher-order equations will be solved in Chapter III. In Part 3 of Section G we saw a second-order linear equation and solved an initial value problem. The solution was determined uniquely by two initial conditions, specifying $T(0)$ and $T'(0)$. Similarly, the fourth-order linear equation

$$y^{(4)}(t) = \frac{1}{EI} w(t) \tag{11}$$

appearing in Part 7 of Section G can be solved for y by definite integration. Four integrations are required, and an arbitrary constant is introduced each time. For instance, if $w(t)$ is constant with value ρ, then

$$y'''(t) = \frac{\rho}{EI} t + y'''(0)$$

$$y''(t) = \frac{\rho}{EI} \frac{t^2}{2} + y'''(0)t + y''(0)$$

$$y'(t) = \frac{\rho}{EI} \frac{t^3}{6} + y'''(0)\frac{t^2}{2} + y''(0)t + y'(0)$$

$$y(t) = \frac{\rho}{EI} \frac{t^4}{24} + y'''(0)\frac{t^3}{6} + y''(0)\frac{t^2}{2} + y'(0)t + y(0) \qquad (12)$$

For the case $w(t) \equiv \rho$, (12) is the *general solution* of (11): all functions of the form (12) are solutions, and all solutions are of the form (12). In this case it is easy to see the four arbitrary constants and their relation to initial values. The corresponding initial value problem would be written

$$\left. \begin{aligned} y^{(4)} &= \frac{\rho}{EI} \\ y'''(0) &= y_3 \\ y''(0) &= y_2 \\ y'(0) &= y_1 \\ y(0) &= y_0 \end{aligned} \right\}$$

and our derivation of (12) by definite integrations shows that the solution to this initial value problem is unique, whatever values the numbers y_0, y_1, y_2, and y_3 may have.

The solution of a first-order linear equation is given by (8). There is no known corresponding formula for the solution of an arbitrary higher-order linear equation. There is, instead, a collection of techniques for solving special cases. Some of these techniques are discussed in Chapters III and IV.

However, one very satisfactory fact is known, telling us at least how to write initial value problems so as to be sure that the solution, if ever it is found, will be unique:

Theorem 2 *If a_{n-1}, a_{n-2}, $\ldots$, a_2, a_1, a_0, and f are continuous functions on an interval (α, β), if t_0 is any point of the interval (α, β) so that $\alpha < t_0 < \beta$, and if y_0, y_1, y_2, $\ldots$, y_{n-2}, and y_{n-1} are any (arbitrary) n numbers, then there exists one* *and only one function y satisfying all the following properties:*

$$y^{(n)} + a_{n-1} y^{(n-1)} + a_{n-2} y^{(n-2)} + \cdots + a_2 y'' + a_1 y' + a_0 y = f \qquad \text{on } (\alpha, \beta)$$

$$y(t_0) = y_0$$
$$y'(t_0) = y_1$$
$$y''(t_0) = y_2 \qquad\qquad (13)$$
$$\cdots\cdots$$
$$y^{(n-2)}(t_0) = y_{n-2}$$
$$y^{(n-1)}(t_0) = y_{n-1}$$

Further, if the functions $a_{n-1}, \ldots, a_0$, and f have continuous kth derivatives on (α, β), then y will have a continuous derivative of order $n + k$ there.

A problem like (13) is clearly to be called an *initial value problem*, and the theorem says that if all the coefficients are continuous on the interval, then the initial value problem has a unique solution.

I. APPENDIX: GRAPHING

The graph of a function has several uses, including

1. Giving approximate function values
2. Showing the behavior of the function on some interval

Figure 1 is the graph of the solution of the lake-pollution problem from Section A. With this graph one can explain in a moment to expert and layman alike that the concentration will decline, but at a decreasing rate, and that there will be small amounts of pollutant in the lake for a very long time. That is, it is easy to see the

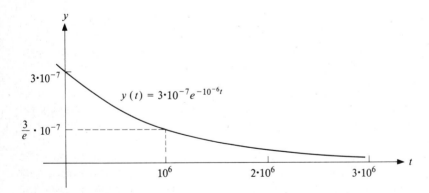

Figure 1 Concentration of pollutant in the lake of Section A.

general behavior of the function on the graph. Also, if the graph is accurately drawn, it is possible to make estimates of the concentration directly from the graph, over a fairly long time span. The graph, therefore, is an important medium for communicating information to others.

You will need to graph solutions of initial value problems. For this you should make a careful review of general graphing properties and techniques, and learn to sketch quickly and accurately the graphs of those common functions given below.

1. Several Important Features of Graphs

Several key concepts are listed below. As you study them here, you should also review them in your calculus book, so as to mobilize your calculus background for use now.

a. Slope The graph is rising (has rising tangent) where the derivative is positive, and falling (has falling tangent) where the derivative is negative. A local maximum occurs where the sign of the derivative passes from plus to minus, and a local minimum occurs where it passes from minus to plus. If you must plot several points in order to sketch a graph, draw in the tangents at those points as well, for it helps a lot. See, for instance, $t = -1$ and $t = -\frac{5}{2}$ in Figure 2 (also Figures 14 and 15).

b. Concavity The graph is concave upward where the second derivative is positive, and concave downward where the second derivative is negative.

c. Horizontal asymptotes The graph of y will have a horizontal asymptote on the right if and only if $y(t)$ has a limit as t increases unboundedly, and on the left if and only if $y(t)$ has a limit as t becomes unboundedly negative. (See Figure 3.)

d. Periodicity A function y is periodic with period T if $y(t + T) = y(t)$ for every value of t. (See Figure 4.)

e. Translation to the right a distance t_0 Any feature which $y(t)$ exhibits at a point T will be exhibited by $f(t) = y(t - t_0)$ wherever $t - t_0 = T$, that is, at the point $t = T + t_0$. For instance, if $y(t) = t^2$, then the graph of y is a parabola with vertex $T = 0$, while that of $f(t) = y(t - t_0) = (t - t_0)^2$ has vertex $T + t_0 = 0 + t_0 = t_0$. (See Figure 5, and also Part 6 of Section E.)

The purpose which a graph is to serve may determine the way in which it should be drawn. A graph smoothed in over a few experimentally determined values must necessarily begin with axis, scales, and data points. However, a graph of a well-known function may be much more efficiently sketched in the opposite order, beginning with the general shape of the curve and ending with the fixing of scales and axes. For two of the functions to be studied now, the latter graphing technique should be used.

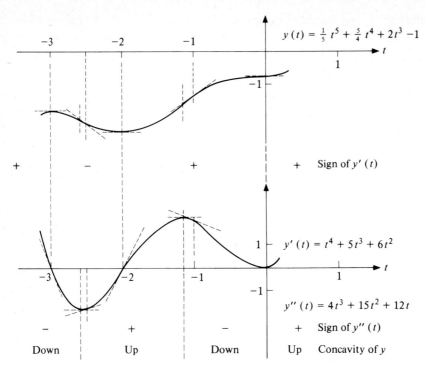

Figure 2 Sketching a graph using function values and tangent lines. A fairly good sketch of y' (lower graph) can help make a good sketch of y (upper graph).

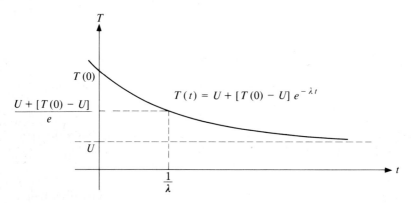

Figure 3 Graph with a horizontal asymptote on the right.

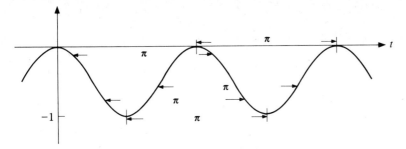

Figure 4 $-\sin^2 t = \frac{1}{2}(\cos 2t - 1)$ has period π.

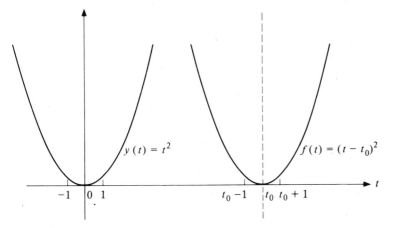

Figure 5 The graph of $y(t) = t^2$ is translated to the right by t_0 to give the graph of $f(t) = (t - t_0)^2$.

2. The Graphs of Several Important Functions

These functions are solutions of one homogeneous linear equation or another, and you should become familiar with them and their graphs.

a. Exponential The *exponential* e^{at} has three possibilities. It increases unboundedly if a is positive, it is constant if a is zero, and it decreases to zero if a is negative (see Section C). In the first case it has $y \equiv 0$ as a horizontal asymptote on the left; in the third case it has $y \equiv 0$ as an asymptote on the right. Very many engineering quantities "die out exponentially," because their energy dissipation rates are proportional to the amount of energy they store (see Part 2 of Section E and Part 3 of Section G).

When graphing an exponential, pay careful attention to two things: the function value at the first point of interest (the initial point, for instance) and the rate of decrease (or increase) near that point. Specifying these two features is entirely equivalent to specifying the parameters a and C in the formula $y(t) = Ce^{at}$.

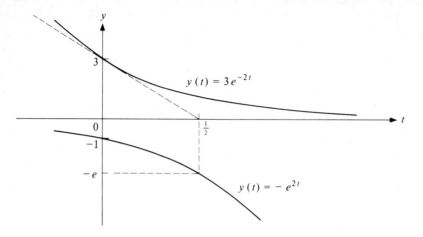

Figure 6 Graphs of exponentials.

In Figure 6 the function value at $t = 0$ is marked clearly on each graph. The rate of decrease of the upper graph is indicated by the tangent line at $t = 0$. This tangent is easily drawn when graphing the solution to initial value problems for linear equations of order 2 or more, where both $y(t_0)$ and $y'(t_0)$ are given. The rate of decrease of the lower graph is indicated by a different technique, which is very frequently used: the point t is marked where the value of y has changed by the factor e. (See Figure 10 of Section G, where the rate of decrease of ϕ'' is shown this way; also see Figure 1 of this section.)

The recommended steps for sketching the graph of an exponential will be illustrated with the function $y(t) = -4e^{-(t-1)/2}$. First, identify the fact that y has a limit as t increases, that is, it goes to zero. Then sketch any exponential with that property, *omitting* scales and the vertical axis (Figure 7a).

Next, identify the fact that y is negative everywhere, by reflecting the graph of Figure 7a across the x axis (Figure 7b). Select (and mark) any point on the horizontal axis as the point at which $t = 1$, and sketch in the tangent to the curve at $t = 1$. This determines both scales, though it remains to mark them in (Figure 7c). The intersection t_1 of the tangent in Figure 7c with the t axis can also be computed numerically, using the slope of the tangent:

$$\frac{\Delta y}{\Delta x} = y'(1) = -\frac{y(1)}{t_1 - 1}$$

$$\frac{4}{2} = \frac{4}{t_1 - 1}$$

$$t_1 - 1 = 2$$

$$t_1 = 3$$

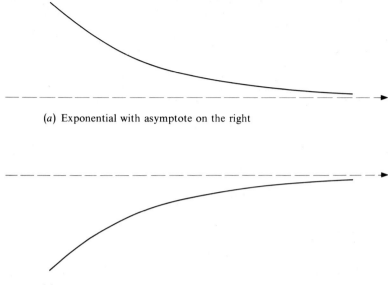

(*a*) Exponential with asymptote on the right

(*b*) Exponential with negative values and an asymptote on the right

(*c*) Exponential with tangent marked at $t = 1$

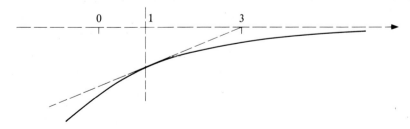

(*d*) Exponential with the horizontal scale computed

Figure 7 Graphing the exponential $-4e^{-(t-1)/2}$: the first steps.

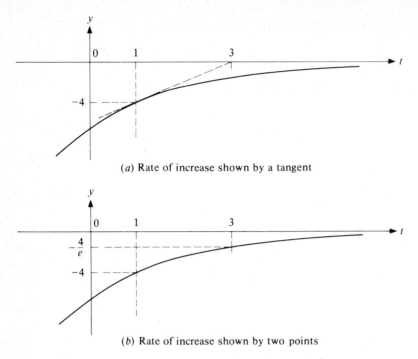

(*a*) Rate of increase shown by a tangent

(*b*) Rate of increase shown by two points

Figure 8 Exponential showing scales and rate of increase.

Mark this value at the appropriate point, and determine the location for $t = 0$ (Figure 7*d*).

Finally, draw in the vertical axis, and mark on it the value of y at $t = 1$ (see Figure 8*a*). Notice that the point $t = 3$ is also the point at which y declines to $(1/e)y(1)$. Thus, the graph could also be drawn as in Figure 8*b*.

If, however, the scales are already established, it is easy to sketch in an exponential. Plot a point in a region of interest, draw in the tangent at that point (see Figure 15*a* and *b*), sketch in the asymptote, and then sketch the curve.

In this example, $t = 1$ was selected as the point to plot because there the exponent took value zero.

b. Sinusoid The *sinusoids* are given by formulas such as

$$y(t) = A \sin (\omega t + \phi) \tag{1}$$

Sinus is the Latin name for the sine function, *oid* is a common taxonomic suffix meaning "of the same form as," and the sinusoids are the functions whose graphs are like the sines. The parameters in (1) are

$\omega = angular\ velocity$ ($\omega/2\pi$ is the *frequency*)

$A = amplitude$ (sometimes called *peak amplitude*)

$\phi = phase$

The sinusoids are solutions of the second-order equation

$$y'' + \omega^2 y = 0$$

because
$$y''(t) = \frac{d^2}{dt}[A \sin (\omega t + \phi)]$$

$$= A \frac{d}{dt}[\omega \cos (\omega t + \phi)] = -A\omega^2 \sin (\omega t + \phi)$$

$$= -\omega^2 y(t)$$

Typically, if an engineering system does not dissipate its energy, then it will exhibit a periodic behavior. The sinusoids are the simplest examples.

The graph of a sinusoid should exhibit clearly all three of the parameters. The amplitude A needs to be marked. It is often indicated with enveloping lines, such as the dashed lines in Figure 9b, but otherwise it is simply marked on the vertical scale, as in Figure 9a. It is usual to indicate on the graph not the frequency ω, but the *period* T:

$$T = \frac{2\pi}{\omega}$$

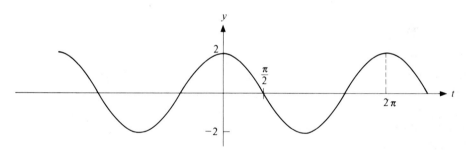

(a) The graph of $y(t) = 2 \cos t$

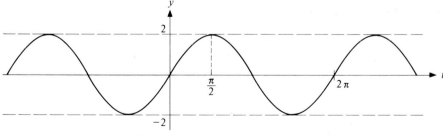

(b) The graph of $y(t) = 2 \sin t$

Figure 9 Graphs of sinusoids: $2 \sin t = 2 \cos (t - \pi/2)$.

which is essentially the reciprocal of the frequency. In Figure 9, the period is 2π. Enough information must be given on the horizontal scale so that the period is easy to read off. The period is indicated in another way in Figure 10.

The quantity $t = \frac{1}{12}$ which is marked on the horizontal axis of Figure 10 is called the *lag time*, the value of t by which features of the graph of $\sin 2\pi(t - \frac{1}{12})$ follow those of the graph of $\sin 2\pi t$. If one thinks of t as representing "time," then left is "earlier" on the t axis, and right is "later." Things which begin "later" appear translated to the right. The function sine "lags" the cosine by $\pi/2$, as illustrated in Figure 9. The lag time t_0 is related to the phase angle ϕ by the following formulas:

$$\sin(\omega t + \phi) = \sin \omega(t - t_0)$$

$$\omega t + \phi = \omega(t - t_0)$$

$$\omega t + \phi = \omega t - \omega t_0 \qquad \text{or} \qquad t - t_0 = t + \frac{\phi}{\omega}$$

$$\phi = -\omega t_0 \qquad \text{or} \qquad t_0 = -\frac{\phi}{\omega}$$

Thus, the phase angle $\phi = 2\pi(-\frac{1}{12}) = -\pi/6$ is not represented directly in Figure 10, but rather by means of the lag time $\frac{1}{12}$:

$$4 \sin 2\pi(t - \frac{1}{12}) = 4 \sin\left(2\pi t - \frac{\pi}{6}\right)$$

Among the sinusoids is the cosine, and very often it is preferable (or even conventional) to express a sinusoid in terms of a cosine

$$y(t) = A \cos(\omega t + \phi)$$

The recommended steps for sketching the graph of a sinusoid so represented will be illustrated with the function

$$y(t) = \tfrac{1}{2} \cos(3t - 1)$$

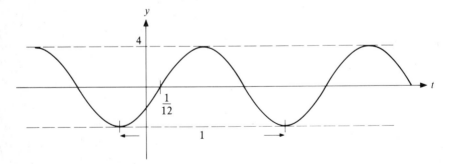

Figure 10 The graph of $y(t) = 4 \sin 2\pi(t - \frac{1}{12})$.

First, sketch an arbitrary cosine wave (which is the same as an arbitrary sine wave) omitting the vertical axis; then indicate the period $2\pi/\omega = 2\omega/3$, and add it to the sketch as in Figure 11a and b. Now compute the lag time, $t_0 = -\phi/\omega = -(-1)/3 = \frac{1}{3}$, and add it to the sketch, being careful to make use of the fact that y is expressed in terms of a cosine (Figure 11c).

From the lag time and the period, the whole horizontal scale can be marked out. What is needed is the length of the unit and the location of the origin (Figure 11d). Finally, the vertical scale is determined from the amplitude, a caption is added to identify the figure, and the sketch is finished (Figure 11e).

We have assumed in our discussion so far that ω and A are both positive, for in engineering applications it is conventional to take them so. However, if you have to graph a sinusoid with negative values for these parameters, then use a little trigonometry to change the function to the desired form. For instance,

$$6 \sin (-2t - 3) = -6 \sin (2t + 3)$$

$$= 6 \sin (2t + 3 + \pi)$$

$$= 6 \sin 2 \left(t + \frac{3 + \pi}{2} \right)$$

c. Damped sinusoid A *damped sinusoid* is of the form

$$y(t) = Ae^{-at} \sin (\omega t + \phi) = Ae^{-at} \sin \omega(t - t_0)$$

with A, a, and ω all positive. Such functions are exceedingly common in engineering applications where an oscillating system also dissipates energy. A sketch of the graph of a damped sinusoid should show clearly the *exponential envelopes* Ae^{-at} and $-Ae^{-at}$ and the properties associated with periodic behavior: the frequency ω (or the period $2\pi/\omega$) and the phase ϕ (or the lag time $-\phi/\omega$).

Look at Figure 13 and notice the following point about the function

$$y(t) = e^{-(1/2)t} \sin \left(t + \frac{\pi}{2} \right) = e^{-(1/2)t} \cos t$$

Its value at $t = 0$ is 1, and its derivative there is $-\frac{1}{2}$, since

$$y'(t) = -\tfrac{1}{2}e^{-(1/2)t} \cos t - e^{-(1/2)t} \sin t$$

Hence it is tangent to the exponential envelope $e^{-(1/2)t}$ at $t = 0$. Likewise, at $t = k\pi$ for k any integer, where the cosine is maximum or minimum,

$$\sin t = 0$$

and

$$y'(t) = -\tfrac{1}{2}e^{-(1/2)t} \cos t$$

$$= \pm\tfrac{1}{2}e^{-(1/2)t}$$

The maxima and minima of a damped sinusoid can be computed using the usual techniques of calculus, but one would do this calculation only if great accuracy were required.

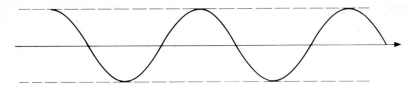

(a) Arbitrary sine wave with envelope

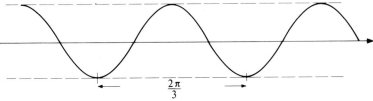

$$\frac{2\pi}{3}$$

(b) Sine wave with period $2\pi/3$, frequency 3

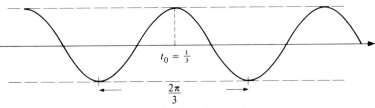

$$t_0 = \tfrac{1}{3}$$

$$\frac{2\pi}{3}$$

(c) Marking the lag time for a cosine

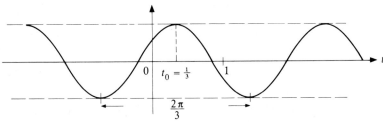

$$t_0 = \tfrac{1}{3} \qquad 1$$

$$\frac{2\pi}{3}$$

(d) Setting the horizontal scale

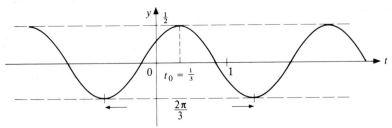

$$y \quad \tfrac{1}{2}$$

$$t_0 = \tfrac{1}{3} \qquad 1$$

$$\frac{2\pi}{3}$$

(e) Adding the vertical scale to finish the sketch

Figure 11 Steps in graphing a sinusoid: $y(t) = \tfrac{1}{2}\cos(3t - 1)$.

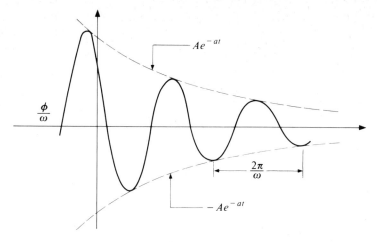

Figure 12 Damped sinusoid: $y(t) = Ae^{-at} \sin(\omega t + \phi)$.

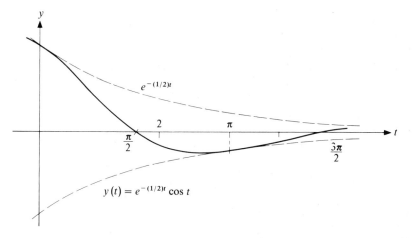

Figure 13 A damped sinusoid is tangent to its exponential envelope.

The recommended steps for sketching the graph of a damped sinusoid will be illustrated with the function

$$y(t) = 3e^{-(1/2)t} \sin\left(\tfrac{5}{2}t + 1\right) \qquad (2)$$

First, sketch the upper and lower envelopes, $3e^{-(1/2)t}$ and $-3e^{-(1/2)t}$, as described previously. Only one need be labeled, however, because of the symmetry. These curves establish the horizontal and vertical scales, which then should be marked in (see Figure 14a). Next compute the lag time and the period:

$$t_0 = -\frac{\phi}{\omega} = -\frac{2}{5} \qquad T = \frac{2\pi}{\omega} = \frac{4\pi}{5}$$

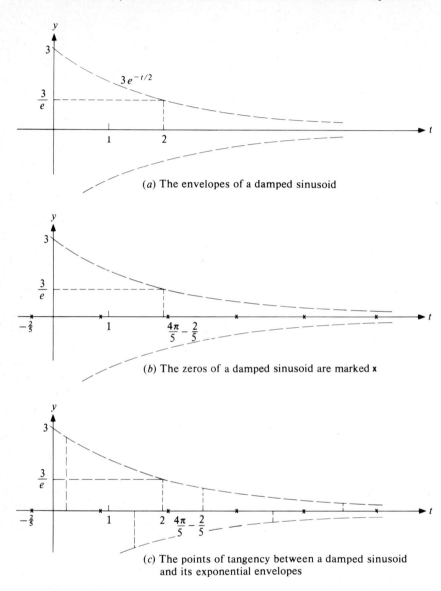

(a) The envelopes of a damped sinusoid

(b) The zeros of a damped sinusoid are marked x

(c) The points of tangency between a damped sinusoid
and its exponential envelopes

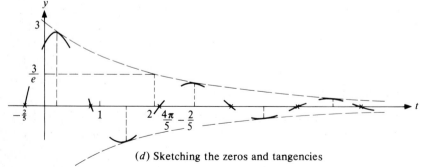

(d) Sketching the zeros and tangencies

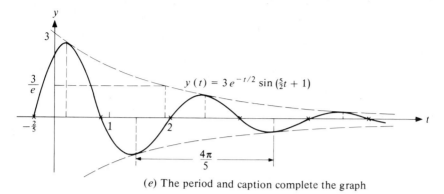

$$y(t) = 3e^{-t/2} \sin\left(\tfrac{5}{2}t + 1\right)$$

(e) The period and caption complete the graph

Figure 14 Steps in graphing a damped sinusoid.

Since (2) is written with the sine, t_0 will be a zero point of y:

$$y\left(-\tfrac{2}{5}\right) = 3e^{-(1/2)(-2/5)} \sin\left[\tfrac{5}{2}\left(-\tfrac{2}{5}\right) + 1\right] = 0$$

Mark in with small x's, therefore, t_0 and its periodic repetitions:

$$t_0 = -\frac{2}{5} \qquad t_0 + \frac{4\pi}{5} = -\frac{2}{5} + \frac{4\pi}{5} \qquad t_0 + 2\left(\frac{4\pi}{5}\right) = -\frac{2}{5} + \frac{8\pi}{5} \qquad \cdots$$

Also mark the points halfway between these, for the sine takes value zero twice in each period (Figure 14b).

Midway between each pair of x's, draw in light, dashed, vertical lines to the points at which the curve will be tangent to the envelope. These lines alternate up and down, beginning after t_0 (Figure 14c). Now sketch the curve, remembering that it must cross the axis at the x's and be tangent to the envelope where indicated (Figure 14d). Finally, label a "period" and add a caption (Figure 14e). (The function is not actually periodic if $a \neq 0$. The term "period" is used here only to indicate the behavior an undamped sinusoid would have.)

When graphing a sinusoid written with a cosine, such as

$$4e^{-t} \cos\left(\tfrac{5}{2}t + 1\right) = 4e^{-t} \cos \tfrac{5}{2}\left(t + \tfrac{2}{5}\right)$$

remember that the lag time $t_0 = -\tfrac{2}{5}$ identifies not a point where the function is zero, but a point of tangency. Hence, one would first draw the vertical lines to the points of tangency, upward at t_0 and alternately down and up thereafter at half-period intervals, and then draw the intermediate x's to mark the zero points (see Figure 13, in which $t_0 = 0$).

3. The Graph of a Sum of Two Functions

It is often quite easy to sketch the graph of a sum of two functions whose graphs are already drawn on the same axes. The steps will be illustrated using

$$y(t) = 2e^{-t} - 2e^{-3t}$$

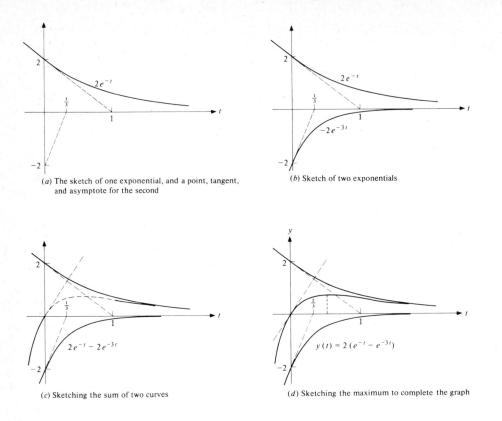

(a) The sketch of one exponential, and a point, tangent, and asymptote for the second

(b) Sketch of two exponentials

(c) Sketching the sum of two curves

(d) Sketching the maximum to complete the graph

Figure 15 Graphing the sum of two functions.

First, sketch in (lightly) either exponential, for instance, $2e^{-t}$. Then, plotting point, tangent, and asymptote (Figure 15a), sketch in (lightly) the other exponential (Figure 15b).

Note that each function value of $y(t)$ is a sum of function values, and therefore the height of each point of the graph above the x axis is the sum of two heights. For instance, the height of $2e^{-t}$ at 1 is $2e^{-1}$, that of $-2e^{-3t}$ is $-2e^{-3}$, and that of y is

$$y(1) = 2e^{-1} + (-2e^{-3})$$

These heights may be read directly off the two graphs previously sketched.

If $t \geq 1$, then $-2e^{-3t}$ is small by comparison with the other, and so the graph of $y(t)$ is very near that of $2e^{-t}$:

$$y(t) = 2e^{-t}(1 - e^{-2t}) \tag{3}$$

where e^{-2t} is small if $t \geq 1$.

On the other hand, if t is very much less than zero, then $-2e^{-3t}$ is large by comparison with $2e^{-t}$:

$$y(t) = -2e^{-3t}(-e^{2t} + 1)$$

where $-e^{2t}$ is small if t is much less than zero. Here the graph of the sum follows that of $-2e^{-3t}$. These parts of the graph may be sketched in now, but the region $0 \le t \le 1$ should be done only very lightly (Figure 15c), for there are several things to do before one can be sure of the graph in this region.

First, the behavior along the asymptote should be checked. Neither summand is large here, and so it is only the sign which is in question. Since (3) shows that $y(t) \ge 0$ for $t \ge 0$, it follows that the curve for y lies above the x axis, below but near to the graph of $2e^{-t}$. Also, $y(t) \le 0$ if $t \le 0$.

Second, check the value and tangent at $t = 0$. Clearly the value is zero, and the slope is the sum of the slopes of the summands:

$$y'(t) = \frac{d}{dt}(2e^{-t}) + \frac{d}{dt}(-2e^{-3t})$$

$$= -2e^{-t} + 6e^{-3t} \tag{4}$$

$$y'(0) = 4$$

The tangent at $t = 0$ may be sketched in (Figure 15c) as a check.

Now the graph indicates very clearly the presence of a maximum for y somewhere between $t = 0$ and $t = 1$. Its location may be found from the derivative (4),

$$y'(t) = 2e^{-3t}(-e^{2t} + 3)$$

Since $2e^{-3t}$ is positive, $y'(t)$ changes sign exactly once, from plus to minus, when e^{2t} becomes larger than 3. This fact verifies the presence of a single maximum t_{max} which must be located where $y'(t_{max}) = 0$:

$$0 = -e^{2t_{max}} + 3$$

$$e^{2t_{max}} = 3 \qquad \text{or} \qquad e^{t_{max}} = 3^{1/2} \tag{5}$$

$$t_{max} = \tfrac{1}{2} \ln 3 \tag{6}$$

From (6) the location can be sketched, and from (5) the function value can be computed:

$$y(t_{max}) = 2(e^{-t_{max}} - e^{-3t_{max}})$$

$$= 2e^{-t_{max}}(1 - e^{-2t_{max}})$$

$$= 2 \cdot 3^{-1/2}(1 - \tfrac{1}{3}) = \frac{4}{3\sqrt{3}}$$

Now the maximum can be sketched. The graph is completed with a caption (Figure 15d).

J. APPENDIX: SOLVING NONLINEAR EQUATIONS BY SEPARATION OF VARIABLES

The method of separation of variables, already introduced to start our study of linear equations, may be applied effectively to an important collection of nonlinear equations of the first order. This section contains examples of such equations and a short discussion of the method. To go further into the vast subject of nonlinear equations would involve us in a miscellany of techniques and prerequisites.

1. Examples and Comparison with Linear Equations

In Section G, the hanging-cable problems led to Equation (24),

$$y''(x) = \frac{\rho}{T_H}\{1 + [y'(x)]^2\}^{1/2}$$

a nonlinear equation. Since the unknown function y does not appear, but merely its derivatives y' and y'', let us define $v = y'$ and observe that the equation becomes first-order:

$$v' = k(1 + v^2)^{1/2} \qquad (1)$$

where $k = \rho/T_H$. Division by the radical "separates" the variables v and x, putting all references to v and v' on the left-hand side of the equation, along with *no* other nonconstant functions of x:

$$\frac{v'}{(1 + v^2)^{1/2}} = k \qquad (2)$$

The right-hand side is constant and can be integrated with respect to x:

$$\int k \, dx = kx + C$$

The left-hand side can also be integrated (for it is the derivative of $\sinh^{-1} v$) using the substitution

$$\left. \begin{array}{c} v = \sinh z \\[4pt] v' = \cosh z \, \dfrac{dz}{dx} \\[4pt] 1 + v^2 = 1 + \sinh^2 z = \cosh^2 z \end{array} \right\}$$

We obtain
$$kx + C = \int \frac{v'(x)}{\{1 + [v(x)]^2\}^{1/2}} \, dx = \int \frac{\cosh z}{\cosh z} \frac{dz}{dx} \, dx$$

$$= \int \frac{dz}{dx} \, dx = z$$

Taking the sinh of both sides gives

$$\sinh (kx + C) = \sinh z = v(x)$$
$$= y'(x)$$

Integrating again gives y itself:

$$y(x) = \frac{1}{k}\cosh (kx + C) + D$$

where C and D are arbitrary constants, one from each integration. Substitution shows that the result is correct.

In practice the computations are invariably simplified in one respect: the integration of the left side of (2) is carried out with respect to v rather than x. Beginning from (1) it would look like this:

$$\frac{dv}{dx} = k(1 + v^2)^{1/2}$$

$$\frac{1}{(1 + v^2)^{1/2}} \frac{dv}{dx} = k \tag{3}$$

$$\frac{dv}{(1 + v^2)^{1/2}} = k \, dx \tag{4}$$

$$\int \frac{dv}{(1 + v^2)^{1/2}} = \int k \, dx$$

$$\left. \begin{aligned} v &= \sinh z \\ dv &= \cosh z \, dz \\ 1 + \sinh^2 z &= \cosh^2 z \end{aligned} \right\}$$

$$\int \frac{\cosh z}{\cosh z} \, dz = \int k \, dx$$

$$z = kx + C$$

$$y' = v = \sinh z = \sinh (kx + C)$$

$$y(x) = \frac{1}{k} \cosh (kx + C) + D$$

Every solution to a nonlinear equation *must be checked* by substitution, for reasons which we shall see. This example checks by substitution.

Except for (4), which is just a handy substitute for (3), the steps of this solution are clearly the same as those of the previous one, but easier to remember. In particular, (4) shows very clearly why the variables are said to be "separated."

Every solution should be checked by substitution. In fact, the method can introduce extraneous functions which are not solutions, and can conceal functions which are solutions. The possibilities are illustrated in the following examples.

A nonlinear equation can be derived to describe the motion of a frictionless mass M attached to a spring. (Details of the model for such systems will be developed in Section B of Chapter II.) The displacement of the mass in Figure 1 will be positive to the right and negative to the left. If the spring constant is K and the mass is displaced from equilibrium a distance y, then the force of the spring on the mass is $-Ky$, and the potential energy stored in the spring is

$$\tfrac{1}{2}Ky^2 = \int_0^y Kz \, dz$$

If the mass of the spring is negligible by comparison with the mass M, then the kinetic energy stored in the system is $\tfrac{1}{2}M(y')^2$. Finally, if no energy is dissipated

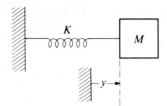

Figure 1 Frictionless mass attached to a spring.

from the system, then the energy E of the system, the sum of the potential energy and the kinetic energy, is constant:

$$E = \tfrac{1}{2}Ky^2 + \tfrac{1}{2}M(y')^2 \tag{5}$$

Equation (5) may be solved by separation of variables.

$$M(y')^2 + Ky^2 = 2E$$

$$(y')^2 = \frac{1}{M}(2E - Ky^2)$$

$$y' = \pm\sqrt{\frac{2E}{M}\left(1 - \frac{K}{2E}y^2\right)^{1/2}} \tag{6}$$

$$\frac{dy}{[1 - (K/2E)y^2]^{1/2}} = \pm\sqrt{\frac{2E}{M}}\,dt$$

$$\int \frac{dy}{[1 - (K/2E)y^2]^{1/2}} = \pm\sqrt{\frac{2E}{M}}\int dt$$

$$\sqrt{\frac{2E}{K}}\arcsin\left(\sqrt{\frac{K}{2E}}\,y\right) = \pm\sqrt{\frac{2E}{M}}\,t + C_0 \qquad C_0 \text{ arbitrary} \tag{7}$$

$$\arcsin\left(\sqrt{\frac{K}{2E}}\,y\right) = \pm\sqrt{\frac{K}{M}}\,t + C_0\sqrt{\frac{K}{2E}}$$

$$\sqrt{\frac{K}{2E}}\,y = \sin\left(\pm\sqrt{\frac{K}{M}}\,t + C_0\sqrt{\frac{K}{2E}}\right) \tag{8}$$

$$y(t) = \sqrt{\frac{2E}{K}}\sin\left(\sqrt{\frac{K}{M}}\,t + C\right) \qquad C \text{ arbitrary} \tag{9}$$

The necessary check is this:

$$y'(t) = \sqrt{\frac{2E}{M}}\cos\left(\sqrt{\frac{K}{M}}\,t + C\right)$$

$$\frac{1}{2}Ky^2 + \frac{1}{2}M(y')^2 = \frac{1}{2}K\frac{2E}{K}\sin^2\left(\sqrt{\frac{K}{M}}\,t + C\right)$$

$$+ \frac{1}{2}M\frac{2E}{M}\cos^2\left(\sqrt{\frac{K}{M}}\,t + C\right) = E$$

You may ask, "Where did the $\pm$ come from in Equation (6), and why does it not also appear in (9)?" It appears in (6) because $(y')^2$ has two square roots. Each leads to only one of the following differential equations:

$$y' = \sqrt{\frac{2E}{M}\left(1 - \frac{K}{2E}y^2\right)}^{1/2} \qquad y' = -\sqrt{\frac{2E}{M}\left(1 - \frac{K}{2E}y^2\right)}^{1/2} \qquad (10)$$

A solution of either may be a solution of (5), and in fact it will be. Therefore, we must solve *each* of the equations in (10) to see if either gives a solution overlooked in solving the other. For brevity, both sets of computations are done at one time, by carrying the symbol $\pm$ down through (8). However,

$$\sin(-x) = -\sin x = \sin(x + \pi)$$

and so the two collections of functions described by (8) are the same:

$$\left\{y \,\middle|\, y(t) = \sqrt{\frac{2E}{K}}\sin\left(-\sqrt{\frac{K}{M}}t + C_0\sqrt{\frac{K}{2E}}\right), \quad C_0 \text{ arbitrary}\right\}$$

$$= \left\{y \,\middle|\, y(t) = \sqrt{\frac{2E}{K}}\sin\left(\sqrt{\frac{K}{M}}t + C\right), \quad C \text{ arbitrary}\right\}$$

$$= \left\{y \,\middle|\, y(t) = \sqrt{\frac{2E}{K}}\sin\left(\sqrt{\frac{K}{M}}t + C_1\sqrt{\frac{K}{2E}}\right), \quad C_1 \text{ arbitrary}\right\}$$

Thus, (9) is the form of all solutions of (5) to be found by separation of variables. There are, however, two other solutions. They are $y \equiv \pm\sqrt{2E/K}$. It is clear that these two functions satisfy (5). (It is also clear that in the spring-mass system they are achieved by impressing a force of magnitude $\sqrt{2EK}$ in either direction and creating a new equilibrium.) They also appear in Figure 2 as the horizontal envelopes, above and below. They were "lost" from our computations above when we divided after Equation (6), for they are the solutions of the equation

$$0 = \pm\sqrt{\frac{2E}{M}\left(1 - \frac{K}{2E}y^2\right)}^{1/2}$$

or

$$0 = 1 - \frac{K}{2E}y^2$$

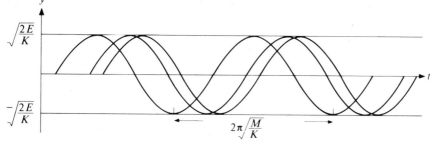

Figure 2 The solutions of $E = \frac{1}{2}Ky^2 + \frac{1}{2}M(y')^2$.

In Section D the need for caution in using the method was illustrated with the (linear) equation

$$y' + ty = 0 \tag{11}$$

Separation of variables gives the equations

$$\frac{1}{y}\frac{dy}{dt} = -t \tag{12}$$

$$\frac{1}{y}\,dy = -t\,dt$$

but these are valid only for $y \neq 0$. If $y > 0$, then

$$\ln y = -\frac{t^2}{2} + C_p$$

$$y = e^{C_p}e^{-t^2/2} > 0 \tag{13}$$

If $y < 0$, then

$$\ln(-y) = -\frac{t^2}{2} + C_n$$

$$-y = e^{C_n}e^{-t^2/2}$$

$$y = -e^{C_n}e^{-t^2/2} < 0 \tag{14}$$

Equations (13) and (14) can now be combined, giving

$$y = Ce^{-t^2/2} \qquad C \neq 0$$

Actually, there is yet another solution, as we know from Section D:

$$y \equiv 0 = 0e^{-t^2/2}$$

which is not a solution of (12), although it is a solution of (11). The general solution of (11), as we already know, is

$$y = Ce^{-t^2/2} \qquad C \text{ arbitrary}$$

The possibility of such special cases as occurred above requires that *every* computation be scrutinized very carefully against the possibility of omitting a solution.

For a final example, consider

$$y' = 2y^{1/2} \tag{15}$$

$$\int \frac{dy}{2y^{1/2}} = \int dt$$

$$y^{1/2} = t + C \tag{16}$$

$$y(t) = (t + C)^2 \qquad t + C \geq 0 \tag{17}$$

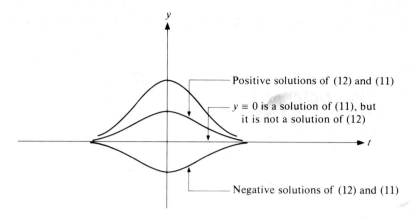

Positive solutions of (12) and (11)

$y \equiv 0$ is a solution of (11), but it is not a solution of (12)

Negative solutions of (12) and (11)

Figure 3 Not all solutions of an equation can be computed directly after separation of variables.

It is clear that functions of the form (17) satisfy (15), but it might be overlooked that in (15), $y^{1/2}$ is the positive square root of y, and so (16) must be nonnegative (see Figure 4). That is, the equations

$$(y')^2 = 4y \qquad \text{and} \qquad y' = 2y^{1/2}$$

have different solutions. The other solutions of $(y')^2 = 4y$ are solutions not of (15) but of

$$y' = -2y^{1/2} \tag{18}$$

as illustrated in Figure 5.

The examples given illustrate three important differences between the solutions of linear equations and those of nonlinear equations.

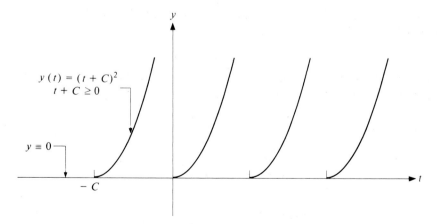

$y(t) = (t + C)^2$
$t + C \geq 0$

$y \equiv 0$

$- C$

Figure 4 Solutions of the equation $y' = 2y^{1/2}$.

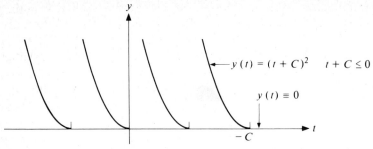

$$y(t) = (t + C)^2 \qquad t + C \leq 0$$

$$y(t) \equiv 0$$

$-C$

(a) The solutions of the equation $y' = -2y^{1/2}$

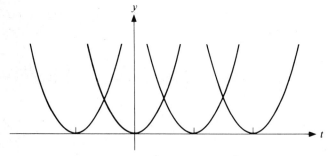

(b) The solutions of the equation $(y')^2 = 4y$

Figure 5 Related equations may have different solutions.

First, all the solutions of a linear equation exist on the interval where the coefficients (in normal form) are continuous, but among the solutions of a non-linear equation some may be defined on one interval, some on another. For instance, the function

$$y(t) = (t + 1)^2 \qquad t \geq -1$$

which satisfies (15) on the interval $t \geq -1$, is defined on a different interval from that on which the function

$$y(t) = (t + 2)^2 \qquad t \geq -2$$

is defined, even though the latter satisfies the same equation as the former.

Second, for the solution of a linear equation, an initial value may be any number at all, but nonlinear equations may lack solutions taking some particular value. For instance,

$$\tfrac{1}{2}Ky^2 + \tfrac{1}{2}M(y')^2 = E$$

does not have any solutions at all taking a value greater than $\sqrt{2E/K}$ or less than $-\sqrt{2E/K}$.

Finally, a linear equation has a unique solution satisfying suitable initial conditions, but both $y(t) \equiv 0$ and $y(t) = t^2$ are solutions of $(y')^2 = 4y$ with $y(0) = 0$.

This lack of uniqueness has practical illustrations, as in the model for draining a cylindrical water tank (Figure 6). It is a matter of empirical observation that the rate of flow out of a spigot at the bottom of such a tank is proportional to the square root of the pressure there and hence to the square root of the depth h of water in the tank:

$$\frac{dV}{dt}(t) = -K\sqrt{h(t)} \qquad \text{where } K > 0$$

The volume of water in the tank at time t is $V(t) = Ah(t)$, where A is the area of the water surface. Thus $V' = Ah'$, and

$$Ah'(t) = -K\sqrt{h(t)} \qquad \text{where } K > 0 \text{ and } A > 0 \tag{19}$$

Separation of variables gives the equation

$$h^{-1/2}h' = -\frac{K}{A} \qquad h \neq 0 \tag{20}$$

and integration yields the solution

$$2\sqrt{h(z)}\,\Big|_0^t = -\frac{K}{A}z\,\Big|_0^t$$

$$2\sqrt{h(t)} - 2\sqrt{h(0)} = -\frac{K}{A}t$$

$$2\sqrt{h(t)} = 2\sqrt{h(0)} - \frac{K}{A}t$$

$$h(t) = \left[\sqrt{h(0)} - \frac{K}{2A}t\right]^2 \tag{21}$$

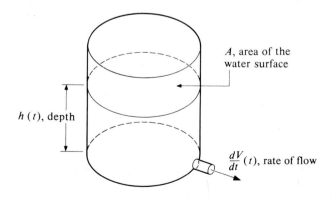

A, area of the water surface

$h(t)$, depth

$\frac{dV}{dt}(t)$, rate of flow

Figure 6 A cylindrical water tank being emptied through a spigot.

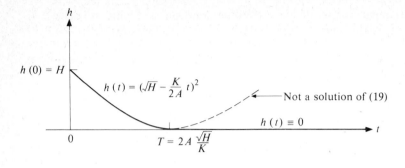

Figure 7 Depth in the tank and solutions of (19).

Now let us suppose that at $t = 0$ the depth in the tank was $h(0) = H$. According to (21), the tank would drain until the quantity in brackets was zero, at which time (T) the tank would be empty (see Figure 7).

$$\sqrt{H} - \frac{K}{2A} T = 0$$

$$\sqrt{H} = \frac{K}{2A} T$$

$$T = \frac{2A\sqrt{H}}{K}$$

The rising dashed curve in Figure 7 is the continuation of (21) to values of $t > T$, and it should be observed that here formula (21) does not give a solution of the differential equation (19), for at the very least the derivative of (21) is positive if $t > T$, whereas (19) shows that $h'(t) \le 0$ for all t. In fact, Equation (19) is of the same form as Equation (18), and its solutions are of the form illustrated in Figure 5a.

In our application, the interpretation of this mathematical fact is that the tank is being drained through the spigot. It does not immediately and automatically fill up again. Filling the tank requires a different model, no matter how it is done.

Is the tank described by Equation (19) for $t > T$? Yes, it is. The function which is identically zero,

$$h(t) \equiv 0$$

represents an empty tank, and it also satisfies Equation (19).

Thus, the fact that the tank is empty at $t = T$ determines uniquely the fact that

$$h(t) \equiv 0 \qquad \text{for } t \ge T$$

but it does not determine uniquely the function h on the interval $t < T$. All one can tell at $t = T$ is that at *some* moment prior to $t = T$ the tank became empty (and thereafter stayed empty). The fact that $h(0) = H$, however, determines h uniquely for $t \ge 0$.

(There are also practical illustrations of the case in which function values at a given time T do not determine the behavior of a system during the time interval $t > T$. Such systems are said to be *unstable*, and the removal of instabilities from a system may be of great importance to an engineer.)

2. Outline of the Method of Separation of Variables

The general form for a first-order differential equation is

$$F[t, y(t), y'(t)] = 0 \tag{22}$$

That is more important than you may think. Table 1 contains some examples to illustrate the equation. In each line of the table, an equation is given together with $F[t, y(t), y'(t)]$ and $F(t, y, z)$, to illustrate more clearly the way F depends on its three variables. The first step in solving (22) by separation of variables is to solve it for y', solving $F(t, y, z) = 0$ for z. In the examples of the previous part, this occurred in passing from (5) to (6). If the result is an equation of the form

$$\frac{dy}{dt} = h(y)g(t) \tag{23}$$

then the variables are separable, and the method will work. Notice the essential conditions of (23), that h be a function of y alone, that g be a function of t alone, and that y' be equal to their product. If the variables are separable, then the actual separation

$$\frac{dy}{h(y)} = g(t)\, dt \tag{24}$$

is either easy or impossible. It is impossible if

$$h(y) \equiv 0 \tag{25}$$

but in that case y is constant and may be found by solving (25). *This case must always be checked separately.*

Table 1 Illustrating the general first-order equation $F[t, y(t), y'(t)] = 0$, which relates t, $y(t)$, and $y'(t)$

Equation	$F[t, y(t), y'(t)] =$	$F(t, y, z) =$
1. $y'(t) + \dfrac{I}{V}y(t) = 0$	$y'(t) + \dfrac{I}{V}y(t)$	$z + \dfrac{I}{V}y$
2. $\frac{1}{2}K[y(t)]^2 + \frac{1}{2}M[y'(t)]^2 - E = 0$	$\frac{1}{2}K[y(t)]^2 + \frac{1}{2}M[y'(t)]^2 - E$	$\frac{1}{2}Ky^2 + \frac{1}{2}Mz^2 - E$
3. $y'(t) - 2y^{1/2}(t) = 0$	$y'(t) - 2y^{1/2}(t)$	$z - 2y^{1/2}$
4. $y'(t) + (t-1)^2 y(t) - 3t = 0$	$y'(t) + (t-1)^2 y(t) - 3t$	$z + (t-1)^2 y - 3t$

The third step is the integration of both sides of (24). If H is an antiderivative of $1/h$, and G is an antiderivative of g, then the result is

$$H(y) = G(t) + C \tag{26}$$

or
$$H[y(t)] - H[y(t_0)] = G(t) - G(t_0)$$

The last step is solving (26) for y or $y(t)$.

For those equations to which it may be applied, the method is often very efficient but, as the examples indicate, each of the steps must be carefully scrutinized for the possibility of omitting solutions or introducing extraneous functions which are not solutions of the given equation. In the end, every function produced by the method must be checked to show that it satisfies the given equation on the expected interval.

3. Sample Problems with Answers

Identify each of the following as " linear " or " nonlinear." Exactly one of the following is not separable. Find it. Solve the others by separation of variables. Check all solutions. Graph all solutions.

A. $y' + 3ty = 2t$ *Answer:* Linear; $y(t) = \frac{2}{3} + Ce^{-3t^2/2}$

B. $(y - 1)(y' + y^2 e^t) = 0$
     Answer: Nonlinear; $y \equiv 1$, $y \equiv 0$, $y(t) = 1/(e^t + C)$ for $e^t + C \neq 0$

C. $y' + 3y = t$ *Answer:* Linear; not separable

D. $yy' = t$ *Answer:* Nonlinear; $y(t) = \pm\sqrt{t^2 + C}$ for $t^2 + C \geq 0$

4. Problems

Identify each of the following as "linear" or "nonlinear." Exactly two of the following are not separable. Find them. Solve the others by separation of variables. Check all solutions. Graph all solutions to Problems 1, 3, and 6.

1. $ty' + y = 1$ **2.** $1 + y' = ty$ **3.** $y' = y^2$

4. $e^t y' + e^y = 0$ **5.** $\arccos (yy') = t$ **6.** $\ln y' = t$

7. $(y'')^2 = t + (y')^2$ **8.** $t(y'')^2 = 1 + (y')^2$

K. APPENDIX: NUMERICAL SOLUTION METHODS

When it is not possible (or perhaps not convenient) to obtain an exact formula for the solution, it is possible to use numerical methods, obtaining the solution as a table of values calculated for one value of t after another (iteratively). This kind of method is suitable for a digital computer. We illustrate with the simplest method and three very practical refinements, all of which are well suited to hand calculation. (We shall not discuss computer calculation.)

1. Example: Euler's Method

In Section J we considered a nonlinear equation of the form

$$y' = -a\sqrt{y}$$

and asked for a solution such that $y(0) = H$, and for the first number t_z such that $y(t_z) = 0$. Separation of variables gave the (exact) solution

$$y(t) = \left(\sqrt{H} - \frac{at}{2}\right)^2$$

and the value $t_z = 2\sqrt{H}/a$. To illustrate Euler's method, we shall take the initial value problem

$$y' = -2\sqrt{y} \qquad y(0) = 2 \tag{1}$$

and solve it numerically. We start the calculation at $t = 0$, where $y(0) = 2$ is known, calculating $y'(0)$ from the differential equation (1) and tabulating the numbers (Table 1).

Table 1 The initial point of the calculation

$t =$	$y(t) =$	$y'(t) =$
0	2	$-2\sqrt{y(0)} = -2.8284$

From this beginning we shall estimate the (numerical) value of y at $t = t_1 = \frac{1}{2}$ and then at $t = t_2 = 1$, each time increasing t by an amount $\Delta t = \frac{1}{2}$, the *step length*, which is traditionally written with the symbol h. When we have estimated the value of y at a point t_n, we may go on to estimate it at the next point $t_{n+1} = t_n + h$.

The graph of the solution may be unknown, but at t_0 the values of y and y' completely determine the line tangent to the graph of the solution there. (See Figure 1.) The tangent line lies close to the graph, and so its height

$$y(t_0) + (t - t_0)y'(t_0)$$

is used to estimate the height of the solution curve at t_1 (see Figure 2):

$$y(t_1) \cong y(t_0) + (t_1 - t_0)y'(t_0) = y(t_0) + hy'(t_0)$$

Thus we estimate

$$y(\tfrac{1}{2}) \cong y(0) + \tfrac{1}{2}y'(0) = 2 + \tfrac{1}{2}(-2\sqrt{2}) \cong 0.5858$$

We also estimate the value of the derivative from this, using the differential equation (1):

$$y'(\tfrac{1}{2}) = -2\sqrt{y(\tfrac{1}{2})} \cong -2\sqrt{0.5858} \cong -1.5307$$

Thus we produce the second line of Table 2.

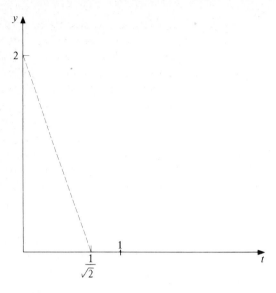

Figure 1 Graph of y: the initial point.

The remaining line of Table 2 is obtained by *iterating* (repeating) the procedure. (See Figure 2.) At each point t_n, the estimate for y is obtained from the estimates made of the value of the function and the value of its derivative at the previous point:

$$y(t_{n+1}) \cong y(t_n) + hy'(t_n) \tag{2}$$

where
$$y'(t_n) = -2\sqrt{y(t_n)} \tag{3}$$

and where h, the step length between t_n and t_{n+1}, is $\frac{1}{2}$ in this example, and (3) is obtained from the differential equation (1). The estimate for t_z is obtained from the last two lines of Table 2 by interpolation: $t_z \cong 0.77$.

Figure 2 shows that the errors in the numerical solution increase with distance from the initial point. It also shows how the gradual change in y' separates the true solution curve from even its own tangents. In order to make more efficient use of the differential equation, then, we must change the value of y' more often,

Table 2 Solutions of problem (1): Euler's method with $h = 0.5$, and the exact solution for comparison

		Numerical solution		Exact solution
$n =$	$t_n =$	$y(t_n) \cong$	$y'(t_n) \cong$	$y(t) = (\sqrt{2} - t)^2$
0	0	2	−2.8284	2
1	$\frac{1}{2}$	0.5858	−1.5307	0.8358
2	1	−0.1796		0.1716
			$t_z = 0.77$	$t_z = 1.414$

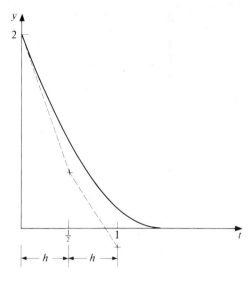

Figure 2 The numerical solution points compared with the exact solution.

reducing the distance $h = t_{n+1} - t_n$ between points where we evaluate. The result of reducing h to 0.1 is shown in Table 3.

In this example the procedure cannot be continued when the estimate for y is negative, because the impossibility of extracting the square root prevents our estimating y'. This situation illustrates how the steps of the procedure may be halted by what we shall call an *abnormal calculation*. It occurs at $n = 2$ in Table 2, and at $n = 13$ in Table 3.

Table 3 Numerical solution for (1) with $h = 0.1$

$n =$	$t_n =$	$y(t_n) \cong$	$y'(t_n) \cong$	$y(t) = (\sqrt{2} - t)^2$
0	0.0	2	−2.8284	2.0000
1	0.1	1.7172	−2.6208	1.7272
2	0.2	1.4551	−2.4125	1.4743
3	0.3	1.2138	−2.2035	1.2415
4	0.4	0.9935	−1.9935	1.0286
5	0.5	0.7941	−1.7823	0.8358
6	0.6	0.6159	−1.5697	0.6629
7	0.7	0.4589	−1.3549	0.5101
8	0.8	0.3235	−1.1375	0.3772
9	0.9	0.2097	−0.9159	0.2644
10	1.0	0.1181	−0.6874	0.1716
11	1.1	0.0494	−0.4444	0.0987
12	1.2	0.0049	−0.1405	0.0459
13	1.3	−0.0091		0.0130
		$t_z \cong 1.235$		$t_z = \sqrt{2} \cong 1.4142$

2. The Method

Euler's method for solving a first-order differential equation

$$y'(t) = f[t, y(t)]$$

consists of two steps, repeated alternately. From the initial value $y(t_0)$, and there-after from any estimated value for $y(t_n)$, estimate the corresponding value of the derivative, using

$$y'(t_n) = f[t_n, y(t_n)]$$

by evaluating f. Then, from the estimate for $y'(t_n)$, find the next function value

$$y(t_{n+1}) = y(t_n + h)$$

by approximating

$$y(t_{n+1}) \cong y(t_n) + hy'(t_n)$$

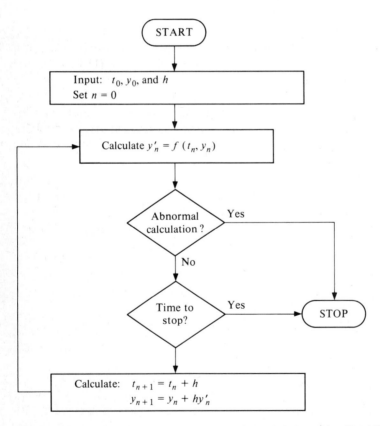

Figure 3 Flow chart for Euler's method with step length h, for $y'(t) = f[t, y(t)]$, $y(t_0) = y_0$.

Two modifications of the terminology are usual, however. The *estimated value* is called y_n, to distinguish it from the *true value* $y(t_n)$, and the estimated value of the derivative is called

$$y'_n = f(t_n, y_n) \tag{4}$$

to distinguish it from the true value $y'(t_n)$. Thus modified, the two steps of Euler's method are combined in the single formula

$$y_{n+1} = y_n + hf(t_n, y_n) \tag{5}$$

A table of the results should contain n, t_n, y_n, and y'_n.
The method is easy to flowchart, as in Figure 3.

3. Reducing the Step Size

To some extent, reducing the step size h may be expected to improve the accuracy, but at the cost of increasing the number of steps. For the example (1), reducing the step size to $h = 0.025$ increases the number of steps to 54. The results are given in Table 4, but only every eighth step is reported.

Because each step introduces an error due to rounding off, even the increased labor of decreasing the step size cannot be expected to reduce the error arbitrarily. Possible improvements come from "smoothing" and "correcting," to be taken up later in the section.

Table 4 Numerical solution for (1) with $h = 0.025$ (every eighth step reported)

n	t_n	y_n	y'_n
0	0	2	−2.8284
8	0.2	1.4696	−2.4246
16	0.4	1.0201	−2.0200
24	0.6	0.6516	−1.6144
32	0.8	0.3643	−1.2071
40	1.0	0.1587	−0.7968
48	1.2	0.0357	−0.3781
53	1.325	0.0022	−0.0937
54	1.35	−0.0001	
		$t_z = 1.349$	

4. Sample Problems with Answers

Use Euler's method with the given step size to approximate the solution to each initial value problem.

A. $y' + y = 0$, $y(1) = 2$; step size $h = 0.1$, on the interval $1 \leq t \leq 2$

Answer:

n	t_n	y_n	y_n'
0	1	2	-2
1	1.1	1.8	-1.8
2	1.2	1.62	-1.62
3	1.3	1.458	-1.458
4	1.4	1.3122	-1.3122
5	1.5	1.1810	-1.1810
6	1.6	1.0629	-1.0629
7	1.7	0.9566	-0.9566
8	1.8	0.8609	-0.8609
9	1.9	0.7748	-0.7748
10	2.0	0.6974	-0.6974

B. $ty' - y^2 = 1$, $y(2) = 0$; step size $h = 0.1$, on the interval $2 \leq t \leq 3$

Answer:

n	t_n	y_n	y_n'
0	2	0	0.5
1	2.1	0.05	0.4774
2	2.2	0.0977	0.4589
3	2.3	0.1436	0.4438
4	2.4	0.1880	0.4314
5	2.5	0.2311	0.4214
6	2.6	0.2733	0.4133
7	2.7	0.3146	0.4070
8	2.8	0.3553	0.4022
9	2.9	0.3955	0.3988
10	3.0	0.4354	0.3965

5. Problems

Use Euler's method with the given step size to approximate the solution to each initial value problem.

1. $y' - y = 1$, $y(0) = 0$; step size $h = 0.1$, on the interval $0 \leq t \leq 1$

2. $y' = -\sqrt{1 - y^2}$, $y(1) = \frac{3}{4}$; step size $h = 0.1$, on the interval $1 \leq t \leq 2$

3. $y' = t + y$, $y(0) = 2$; step size $h = 0.05$, on the interval $0 \leq t \leq 1$

6. Smoothing

It is important to realize that in the example (1), the value y_n' has been used as the slope of the segment joining the points (t_n, y_n) and (t_{n+1}, y_{n+1}) of the estimated solution. (See Figure 2.) That is Euler's method. In other methods, one may use estimates for the slope of the solution between (t_n, y_n) and (t_{n+1}, y_{n+1}), which may make (t_{n+1}, y_{n+1}) lie closer to the exact solution. (See Figure 4.)

A look at Figure 2 will remind you that in the example (1) the value for y' at t_n is necessarily an extreme value on the interval between t_n and t_{n+1}. Whenever a graph is concave up on the whole of an interval (or else concave down), then y' will be largest at one end and smallest at the other end. If the step length is short, this situation will be the usual one.

Processes for reducing local extremes are often referred to as *smoothing*. The two methods we give now are examples of this kind of process, reducing the extremes of y'. They are named for their inventors.

The *Runge-Kutta second-order* method has many variants. The one we will use is also called the *modified Euler method*. In this method we take the average of y'_n and y'_{n+1} to be more accurate than y'_n itself as an estimate of the average rate of change of y on the step interval from t_n to t_{n+1}. The formula is therefore

$$y_{n+1} = y_n + h \frac{y'_n + f(t_{n+1}, y_n + hy'_n)}{2} \tag{6}$$

The formula looks complicated, but observe that if

$$\bar{y}_{n+1} = y_n + hy'_n \tag{7}$$

is the estimate for $y(t_{n+1})$ which Euler's method would have produced, and

$$\bar{y}'_{n+1} = f(t_{n+1}, y_n + hy'_n) = f(t_{n+1}, \bar{y}_{n+1}) \tag{8}$$

is the corresponding estimate for the derivative, then Equation (6) becomes

$$y_{n+1} = y_n + h \frac{y'_n + \bar{y}'_{n+1}}{2}$$

Table 5 gives the result of applying the method to example (1).

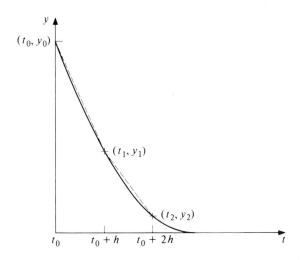

Figure 4 Using an average value for the slope.

Table 5 Runge-Kutta second-order method with $h = 0.1$, applied to example (1)

n	t_n	y_n	y'_n	$\bar{y}_{n+1}$	$\bar{y}'_{n+1}$	$\dfrac{y'_n + \bar{y}'_{n+1}}{2}$
0	0	2	−2.8284	1.7172	−2.6208	−2.7246
1	0.1	1.7275	−2.6287	1.4647	−2.4205	−2.5246
2	0.2	1.4751	−2.4291	1.2322	−2.2201	−2.3246
3	0.3	1.2426	−2.2295	1.0197	−2.0196	−2.1245
4	0.4	1.0302	−2.0299	0.8272	−1.8190	−1.9245
5	0.5	0.8377	−1.8305	0.6547	−1.6182	−1.7244
6	0.6	0.6653	−1.6313	0.5022	−1.4173	−1.5243
7	0.7	0.5129	−1.4323	0.3696	−1.2159	−1.3241
8	0.8	0.3804	−1.2336	0.2571	−1.0141	−1.1238
9	0.9	0.2681	−1.0355	0.1645	−0.8112	−0.9233
10	1.0	0.1757	−0.8384	0.0919	−0.6063	−0.7223
11	1.1	0.1035	−0.6434	0.0392	−0.3957	−0.5196
12	1.2	0.0515	−0.4540	0.0061	−0.1566	−0.3053
13	1.3	0.0210	−0.2899	−0.0080		
	1.4	−0.0080 ←	−0.2899		$t_z = 1.372$	

Only 13 steps were required, but since each point is evaluated twice, the amount of labor is comparable to 26 steps of Euler's method. Comparison with Table 3 shows the improvement obtained by smoothing.

The averaging of slopes, $(y'_n + y'_{n+1})/2$, is called the *trapezoidal rule* from the formula of that name for the numerical integration of the problem

$$y(t) = \int_{t_0}^{t} f(z)\, dz \qquad \text{or} \qquad y'(t) = f(t)$$

where it would take the form

$$y_{n+1} = y_n + h\,\frac{f(t_n) + f(t_{n+1})}{2}$$

Except for Euler's method, all the numerical methods we present use the trapezoidal rule in one form or another. (In effect, it allows us to replace the straight line segments of the tangents with chords that cut across parabolas. See Figure 4.)

The other smoothing method we consider is *Nyström's*, in which smoothing is obtained by joining together two step intervals to give an interval of length $2h$ from t_{n-1} to t_{n+1}, and evaluating the derivative at the midpoint t_n:

$$y_{n+1} = y_{n-1} + 2hy'_n \tag{9}$$

This requires no more work per point than Euler's method does, but starting the method requires a value for y_1 as well as y_0 before one can calculate y_2.

It is recommended that y_1 be found as the sum of the first three terms of the power series for y about t_0:

$$y_1 = y_0 + hy_0' + h^2 \frac{y''(t_0)}{2}$$

where

$$y'' = \frac{d}{dt} f[t, y(t)]$$

We apply it to example (1),

$$y' = -2\sqrt{y} \Big|$$
$$y(0) = 2$$

for which

$$y'' = \frac{d}{dt} y' = \frac{d}{dt} [-2y^{1/2}] = -y^{-1/2} y' = 2$$

so that if $h = 0.1$, then

$$y_1 = y_0 + hy_0' + h^2 \frac{y''(0)}{2} = 2 + (0.1)(-2.8284) + 0.01 = 1.7272$$

The calculation proceeds as in the left-hand part of Table 6. The right-hand part of the table gives a comparison with Euler's method, the Runge-Kutta second-order method, and the exact solution.

Table 6 Nyström's method with $h = 0.1$, applied to (1) and compared with other solutions

n	t_n	y_n	y_n'	$(\sqrt{2} - t)^2$	Euler	Runge-Kutta
0	0	2		2	2	2
1	0.1	1.7272	−2.6285	1.7272	1.7172	1.7275
2	0.2	1.4743	−2.4284	1.4743	1.4551	1.4751
3	0.3	1.2415	−2.2285	1.2415	1.2138	1.2426
4	0.4	1.0286	−2.0284	1.0286	0.9935	1.0302
5	0.5	0.8356	−1.8285	0.8358	0.7941	0.8377
6	0.6	0.6629	−1.6284	0.6629	0.6159	0.6653
7	0.7	0.5102	−1.4285	0.5101	0.4589	0.5129
8	0.8	0.3772	−1.2284	0.3773	0.3235	0.3804
9	0.9	0.2645	−1.0286	0.2644	0.2097	0.2681
10	1.0	0.1715	−0.8283	0.1716	0.1181	0.1757
11	1.1	0.0988	−0.6287	0.0987	0.0494	0.1035
12	1.2	0.0458	−0.4278	0.0459	0.0049	0.0515
13	1.3	0.0133	−0.2303	0.0130	−0.0091	0.0210
14	1.4	−0.0003		0.0002		−0.0080
		$t_z \cong 1.398$		1.4142	1.235	1.372

7. A Predictor-Corrector Method

Our last method, *Milne's*, illustrates the important idea of "correcting." Nyström's method is used to begin the calculation, to make a first estimate $y_{n+1}^{\{0\}}$, and then the trapezoidal rule for integration is used to base a second estimate $y_{n+1}^{\{1\}}$ upon the first.

The results of applying the method to example (1) are tabulated in Table 7. The first and second lines come from Nyström's method, as before. Thereafter, the first estimate is

$$y_{n+1}^{\{0\}} = y_{n-1}^{\{1\}} + 2hy_n^{\{1\}\prime}$$

$$y_{n+1}^{\{0\}\prime} = f(t_{n+1}, y_{n+1}^{\{0\}})$$

The improved estimate of the rate of change is the average of $y_{n+1}^{\{0\}\prime}$ and $y_n^{\{1\}\prime}$, and so the second estimate is

$$y_{n+1}^{\{1\}} = y_n^{\{1\}} + h\frac{y_{n+1}^{\{0\}\prime} + y_n^{\{1\}\prime}}{2}$$

$$y_{n+1}^{\{1\}\prime} = f(t_{n+1}, y_{n+1}^{\{1\}})$$

The column headed C_0 contains the corrections

$$C_0 = y_n^{\{1\}} - y_n^{\{0\}}$$

which are a valuable check on the precision of the method. The values there should change only slowly. The rapid change beginning at $n = 14$ shows the difficulties Nyström's method has near $\sqrt{y} = 0$. Sudden changes may also signal a calculational mistake.

The sample problems which follow illustrate the value of the corrector in

Table 7 Milne's method with $h = 0.1$, applied to (1)

n	t_n	$y_n^{\{0\}}$	$y_n^{\{0\}\prime}$	$\dfrac{y_n^{\{0\}\prime} + y_{n-1}^{\{1\}\prime}}{2}$	$y_n^{\{1\}}$	$y_n^{\{1\}\prime}$	C_0
0	0				2	−2.8284	
1	0.1				1.7272	−2.6284	
2	0.2	1.4743	−2.4284	−2.5284	1.4744	−2.4285	0.0001
3	0.3	1.2415	−2.2285	−2.3285	1.2416	−2.2285	0.0001
4	0.4	1.0287	−2.0285	−2.1285	1.0288	−2.0285	0.0001
5	0.5	0.8359	−1.8786	−1.9285	0.8359	−1.8286	0
6	0.6	0.6631	−1.6286	−1.7286	0.6630	−1.6285	−0.0001
7	0.7	0.5102	−1.4286	−1.5286	0.5101	−1.4285	−0.0001
8	0.8	0.3773	−1.2285	−1.3285	0.3773	−1.2284	0
9	0.9	0.2644	−1.0284	−1.1284	0.2645	−1.0285	0.0001
10	1.0	0.1716	−0.8285	−0.9285	0.1717	−0.8286	0.0001
11	1.1	0.0988	−0.6285	−0.7286	0.0988	−0.6285	0
12	1.2	0.0460	−0.4290	−0.5287	0.0459	−0.4286	−0.0001
13	1.3	0.0131	−0.2287	−0.3287	0.0130	−0.2283	−0.0001
14	1.4	0.0002	−0.0310	−0.1296	0	0	−0.0002
15	1.5	0.0130	−0.2280	−0.1140	−0.0114		−0.0244

actually improving the solution. It can be shown that if $5C_0$ is more than the allowable error, then the step length h should be reduced.

8. Sample Problems with Answers

C. Use the Runge-Kutta second-order method on Sample Problem A.
  Answer:

n	t_n	y_n	y'_n	$\bar{y}_{n+1}$	$\bar{y}'_{n+1}$	$\dfrac{y'_n + \bar{y}'_{n+1}}{2}$
0	1	2	−2	1.8		−1.9
1	1.1	1.81	−1.81	1.629	−1.629	−1.7195
2	1.2	1.6381	−1.6381	1.4742	−1.4742	−1.5561
3	1.3	1.4824	−1.4824	1.3342	−1.3342	−1.4083
4	1.4	1.3416	−1.3416	1.2074	−1.2074	−1.2745
5	1.5	1.2142	−1.2142	1.0927	−1.0927	−1.1534
6	1.6	1.0988	−1.0988	0.9889	−0.9889	−1.0439
7	1.7	0.9944	−0.9944	0.8950	−0.8950	−0.9447
8	1.8	0.9000	−0.9000	0.8100	−0.8100	−0.8550
9	1.9	0.8145	−0.8145	0.7330	−0.7330	−0.7737
10	2	0.7371	−0.7371	0.6634	−0.6630	−0.7002

D. For Sample Problem B, use three terms of the power series for y to evaluate $y(2.1)$ as though preparatory to using Nyström's method.

  Answer:
$$ty' - y^2 = 1$$

$$y' = \frac{1}{t}(1 + y^2)$$

$$y'' = \frac{t(2yy') - (1 + y^2)}{t^2}$$

$$y(2.1) \cong y(2) + (0.1)y'(2) + (0.1)^2 \frac{y''(2)}{2}$$

$$= 0 + (0.1)\tfrac{1}{2} + 0.01(-\tfrac{1}{4})(\tfrac{1}{2}) = 0.0488$$

E. Use Nyström's method on Sample Problem A.
  Answer: $y'' = y$, and so $y_1 = 2 + (0.1)(-2) + (0.01) = 1.81$

n	t_n	y_n	y'_n
0	1	2	−2
1	1.1	1.81	−1.81
2	1.2	1.638	−1.638
3	1.3	1.4824	−1.4824
4	1.4	1.3415	−1.3415
5	1.5	1.2141	−1.2141
6	1.6	1.0987	−1.0987
7	1.7	0.9946	−0.9946
8	1.8	0.8998	−0.8998
9	1.9	0.8146	−0.8146
10	2.0	0.7369	−0.7369

F. Use Milne's method on Sample Problem A.

Answer:

n	t_n	$y_n^{(0)}$	$y_n^{(0)'}$	$\dfrac{y_n^{(0)'} + y_{n-1}^{(1)'}}{2}$	$y_n^{(1)}$	$y_n^{(1)'}$	C_0
0	1				2	-2	
1	1.1				1.81	-1.81	
2	1.2	1.638	-1.638	-1.724	1.6376	-1.6376	-0.0004
3	1.3	1.4825	-1.4825	-1.5600	1.4816	-1.4816	-0.0009
4	1.4	1.3413	-1.3413	-1.4114	1.3405	-1.3405	-0.0008
5	1.5	1.2135	-1.2135	-1.2770	1.2128	-1.2128	-0.0007
6	1.6	1.0979	-1.0979	-1.1554	1.0973	-1.0973	-0.0006
7	1.7	0.9933	-0.9933	-1.0453	0.9928	-0.9928	-0.0005
8	1.8	0.8987	-0.8987	-0.9458	0.8982	-0.8982	-0.0005
9	1.9	0.8132	-0.8132	-0.8557	0.8126	-0.8126	-0.0006
10	2.0	0.7357	-0.7357	-0.7741	0.7352	-0.7352	-0.0005

9. Problems

Use the Runge-Kutta second-order method, Nyström's method, or Milne's method on each of the following:

4. Sample Problem B   **5.** Problem 1

6. Problem 2   **7.** Problem 3

8. $ty' + y = t$, $y(-2) = -3$; step size $h = 0.05$, on the interval $-2 \le t \le -1$

9. $y' = te^{-y}$, $y(1) = \frac{1}{2}$; step size $h = 0.05$, on the interval $1 \le t \le 2$.

MODELS OF ENGINEERING SYSTEMS

In the previous chapter we discussed the application of differential equations to fluid mixing and concentration problems and various single-element systems. In this chapter we shall describe how simple differential equations are used to model multiple-element systems, beginning with electric circuit systems and translational mechanical systems. We shall discuss how to resolve a complex system into simple components so that equations can be written for each component. When the equations have been solved simultaneously, the behavior of the system can be predicted.

The differential relations which describe the behavior of electrical or mechanical elements are called *models*. The models describe only idealized behavior for idealized elements. For example, the actual object which we call a spring exhibits not only springlike behavior but also masslike behavior and frictionlike behavior, yet the model we use for a spring describes only the springlike part of its behavior. Should we need to describe a real spring, with masslike and frictionlike behavior as well, we would model the spring as a mechanical system having ideal elements of a mass and a dashpot (for friction) in addition to the ideal spring element. We would, of course, expect that the composite model we develop for the real spring would, like the real spring, show a predominantly springlike behavior.

The models are also idealized in another sense. The mathematical formulas which we use to represent the behavior of a system of idealized elements are the simplest possible equations which could describe their behavior to a reasonable degree of accuracy. Although they are only first approximations, they lead to reasonably accurate descriptions of even quite complicated systems. They are very widely used. You will need to use them and, therefore, we shall study them in detail. However, there are other models which are more accurate but also more complicated, which we shall not present.

The engineering models we shall discuss are similar to each other in important ways. Our description will identify the variables of the system and the elements of the system, the relation between the variables for each system element, and the relations between the variables when elements are connected together. We shall see that: (1) each system has two variables, an "across" variable and a "through" variable; (2) each system has three kinds of elements, two of which store energy and one of which dissipates energy; (3) when elements are connected together, the across variable satisfies a physical-compatibility constraint, and the through variable a conservation constraint.

Each model describes an engineering system by means of a collection of equations. The chapter also includes an algorithm for reducing such a collection to a single equation containing a single unknown variable. The following two chapters contain methods for solving such equations.

A. ELECTRICAL MODEL (VOLTAGE-CURRENT)

The two most commonly used variables for electric circuits are voltage and current. Current through a conductor is the rate of flow of positive charge (positive for historical reasons), which is algebraically the same as the rate of flow of negatively charged electrons in the opposite direction. The voltage difference between one point and another is the work done per unit positive charge in passing from the first point to the second. Current is measured with an ammeter, voltage with a voltmeter.

For the simplest circuits it is possible to set up the models directly from physical descriptions as given above, but for systems of arbitrary complexity it is much easier to think of current and voltage in terms of the way they are measured, and we shall do so.

1. Definition of Variables and Direction Conventions

a. Current (the "through" variable) An ammeter is an instrument which measures current. It has two terminals, one positive, labeled $+$, and one negative, labeled $-$. The current through a wire is measured by cutting the wire and attaching one cut end to the positive terminal, and the other cut end to the negative. Current is measured in units called *amperes*. We represent an ammeter symbolically with a labeled arrow pointing from the positive terminal to the negative one. The label by the arrow indicates the amount of the current flow, and the direction of the arrow shows the direction in which the indicated amount of current flows. Thus, Figures 1 and 2a illustrate the same situation, a current of positive charges flowing from left to right at the rate of 2 amperes.

From an ammeter, one can read negative numbers as well as positive. If the ammeter is reversed in the circuit, a positive indication will change to a negative one. Thus, Figure 2b represents exactly the same current-flow situation as Figure 2a.

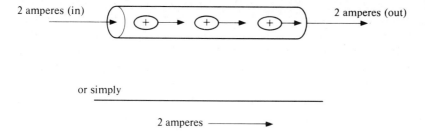

2 amperes (in)   2 amperes (out)

or simply

2 amperes ⟶

Figure 1 The diagram we use to show the flow of 2 amperes.

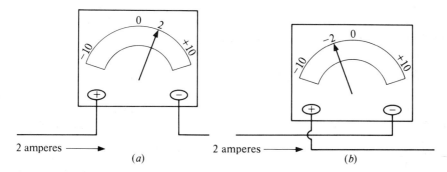

2 amperes ⟶   2 amperes ⟶

(a)   (b)

Figure 2 The same 2 amperes of current can be measured as positive or negative, depending on how the ammeter is connected.

When we wish to indicate that an unknown current $i(t)$ is flowing through a wire, we shall think of connecting an ammeter onto the wire and letting the current $i(t)$ be the value which the ammeter reads. If the meter indicates a positive number, the positive charges are actually flowing into the positive terminal and out of the negative terminal. However, if the meter indicates a negative number, the positive charges are actually flowing into the negative terminal. (See Figure 3.)

Thus, we *define a current* in the model of the circuit by drawing on the circuit diagram an arrow indicating where and in what direction an ammeter would be placed in order to measure it, and then providing a name (a letter variable) for what that ammeter will read. (See Figure 4.) The arrow itself will be referred to as a *reference arrow*, giving the *positive reference direction*.

The point is that the unknown current $i(t)$ can be positive or negative. The actual flow of positive charge through the wire or circuit element can be in the direction of the current referrence arrow or in the direction opposite to the current reference arrow.

b. Voltage (the "across" variable) A voltmeter also has two terminals, positive ($+$) and negative ($-$). The voltage difference from A to B across the circuit element (box) shown in Figures 5 and 6a is measured by connecting the positive terminal to A and the negative to B, as shown in Figure 6a. (This is a parallel

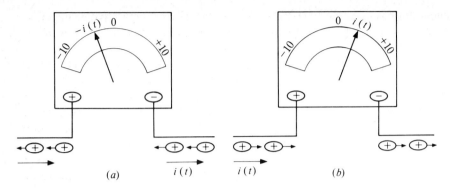

(a)

(b)

Figure 3 The current $i(t)$ shown with a specified reference direction can be measured as a positive or negative current, depending on the direction in which the charge is flowing.

Figure 4 Two ways of defining the same unknown current $i(t)$.

Figure 5 The diagram we use to show a voltage difference of 3 volts.

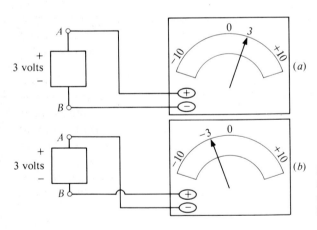

(a)

(b)

Figure 6 A 3-volt difference can be measured as a positive difference or a negative difference, depending on how the voltmeter is connected.

connection.) The unit in which voltage is measured is called the *volt*. Figures 5 and 6a both show a voltage difference of 3 volts from A to B.

From a voltmeter, one can read negative numbers as well as positive. If the voltmeter is reversed in the circuit, as in Figure 6b, then the sign of the reading is also reversed.

Hence, the voltage difference between two points can be positive or negative, depending on whether we take the voltage of point A relative to point B or the voltage of point B relative to A.

When we wish to indicate that an unknown voltage difference $v(t)$ exists across a circuit element, we shall think of connecting a voltmeter across the circuit element and letting $v(t)$ be the value which the meter reads. That is, we take the voltage $v(t)$ to be the voltage of the point of the circuit connected to the positive terminal of the voltmeter with respect to the voltage of the point of the circuit connected to the negative terminal. If the meter indicates a positive number, then the point of the circuit connected to the positive terminal is at a higher voltage than that connected to the negative terminal. However, if the meter indicates a negative number, then the point of the circuit connected to the negative terminal is at a higher voltage than that connected to the positive terminal.

Thus, we *define a voltage difference* in the model of the circuit by giving the signs $+$ and $-$ to indicate where and in what direction a voltmeter would be connected to measure it, and by providing a name (a letter variable) for what the voltmeter will read. (See Figure 7.) The $+$ and $-$ signs will be called *reference signs*, giving the positive reference direction from $+$ to $-$.

These examples illustrate that an unknown voltage difference can be positive or negative. The point of the circuit having the positive $(+)$ voltage reference associated with it is not necessarily at a higher voltage than the point of the circuit associated with the negative $(-)$ voltage reference. (See Figure 8.)

A voltage difference taken in the positive reference direction is called a *voltage drop*, regardless of the sign of the number. A voltage difference taken in the opposite direction is called a *voltage rise*. Figure 6a shows a voltage drop of 3 volts from A to B, and a voltage rise of 3 volts from B to A. Figure 6b shows a voltage drop of -3 volts from B to A, and a voltage rise of -3 volts from A to B.

c. Electric circuit connections There are two basic ways in which circuit elements can be connected: (1) series and (2) parallel. Two elements are connected in series if terminal 2 of the first element is connected to terminal 1 of the second element. This produces a composite circuit element whose first terminal is terminal 1 of the first element, and whose second terminal is terminal 2 of the second element. See Figure 9.

Two elements are connected in parallel if terminal 1 of the first element is connected to terminal 1 of the second element, and terminal 2 of the first element

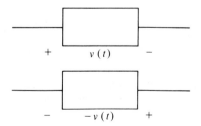

Figure 7 Two ways of defining the same unknown voltage difference.

A is at a higher voltage

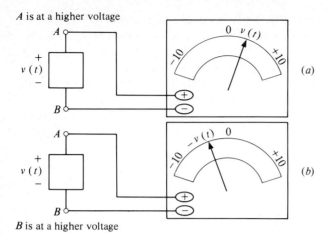

B is at a higher voltage

Figure 8 A voltage difference of $v(t)$ can be measured as a positive voltage difference or a negative voltage difference, depending on whether point *A* or point *B* is at the higher voltage.

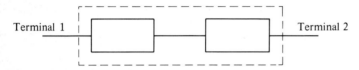

Figure 9 The series connection.

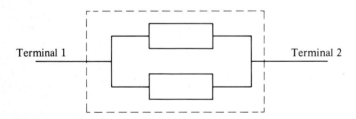

Figure 10 The parallel connection.

is connected to terminal 2 of the second element. This produces a composite element whose terminal 1 is the common terminal 1 of the two elements, and whose terminal 2 is the common terminal 2 of the two elements. See Figure 10.

Voltage is always measured as a voltage difference between two terminals of an element; current is always measured as amperes flowing from one terminal of an element, through the element, to another terminal of the element. Voltage is an "across" variable. Current is a "through" variable. (The third common variable is *power*, which is always measured as the time rate of change of the energy delivered to an element by the combination of the current flowing through it and the voltage difference across it.)

Suppose we have two circuit elements connected in series, as shown in Figure 11, which at time t have voltage drops of $v_1(t)$ and $v_2(t)$, respectively, across them from terminal $T1$ to $T2$. Then the voltage drop from terminal 1 to terminal 2 across the composite two-terminal series element is $v_1(t) + v_2(t)$.

If the "beginning" of any series circuit is connected to the "end" of the series circuit, then the circuit is said to be a *loop*. See Figure 12.

Examples of electric circuit elements which can be connected in series or parallel are diodes, capacitors, inductors, batteries, transformers, tubes, motors,

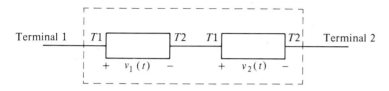

Figure 11 How the voltage drops are computed for a series connection.

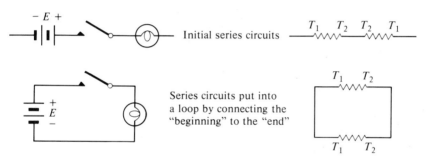

Figure 12 How loop connections are formed.

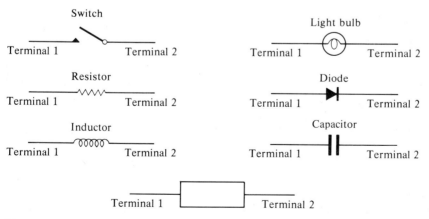

Figure 13 Some two-terminal electric circuit elements.

and resistors. Many of these elements are two-terminal devices, as shown in Figure 13. Not all electrical elements are two-terminal devices. Transistors, for example, are three-terminal devices, and transformers are four-terminal devices.

2. Electrical Model Definition

Now that we have specified how voltage and current are referenced and measured, we are ready to introduce the electrical model definitions for the resistor, capacitor, and inductor. See Figures 14 to 16.

a. Resistor The current flowing into the side of the resistor with the positive voltage-difference reference is proportional to the voltage across the resistor with constant of proportionality equal to $1/R$, where R is the resistance in ohms.

b. Capacitor The current flowing into the side of the capacitor with the positive voltage-difference reference is proportional to the time rate of change of the voltage difference across the capacitor with constant of proportionality equal to C, where C is the capacitance in farads.

c. Inductor With the positive voltage reference taken on the side of the inductor into which the current is flowing, the voltage difference across the inductor is proportional to the time rate of change of current flowing through the inductor with constant of proportionality equal to L, where L is the inductance in henries.

d. Sources of voltage and current A voltage source is one kind of element for providing a circuit with power. The source provides a voltage difference across the two points to which it is connected in parallel. The voltage difference itself is considered to be a known function of time which is independent of the current flowing through the voltage source. The voltage source is represented in Figure 17, where $v(t)$ is labeled as the known function. The current through a voltage source is determined by the voltage difference across the source and by the

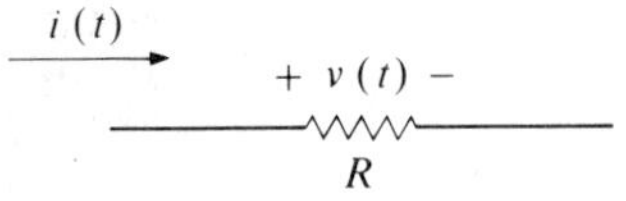

$$i(t) = \frac{v(t)}{R}$$

Figure 14 The model for the resistor.

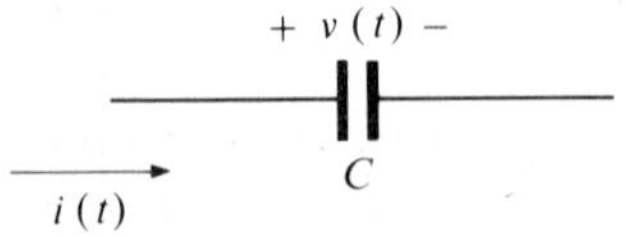

$$i(t) = C\,\frac{dv(t)}{dt}$$

Figure 15 The model for the capacitor.

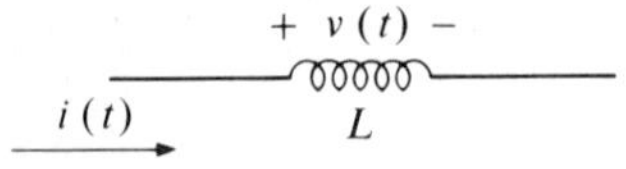

$$v(t) = L\,\frac{di(t)}{dt}$$

Figure 16 The model for the inductor.

Figure 17 Voltage source with known voltage $v(t)$.

Figure 18 Current source with known current
$i(t)$.

other elements in the circuit to which it is connected. This fact is illustrated in
Example 4c.

A current source is another kind of element for providing a circuit with power.
The source provides a current flow through the element it is connected in series
with. The current itself is considered to be a known function of time which is
independent of the voltage across the current source. It is represented in
Figure 18, where $i(t)$ is labeled as the known function. The voltage across a
current source is determined by the source and the other elements in the circuit to
which it is connected.

3. Principles for Using the Electrical Model

a. Technique

Step 1. Isolate each electric circuit element, specifying the voltage and current
associated with each electric circuit element. If the voltage or current for some
resistor, capacitor, or inductor is not defined, then you should define it by
writing onto the diagram the missing reference signs (choose either direction,
arbitrarily) and a suitable function name (not used as the name of some other
quantity in the circuit). (The process of defining missing voltages and currents
will be illustrated in Example 4c.)

Step 2. Establish the correspondence* between the voltages and currents associated with each specified electric circuit element and the model for that
element. Then use the model to write an equation relating the specified voltage and current variables. (The procedure is illustrated in Example 4a.)

Step 3. Use Kirchhoff's laws (stated below) to relate the voltages for the circuit
elements in each loop, and to relate the currents for the circuit elements in
each loop.

b. Kirchhoff's laws

1. *Kirchhoff's voltage law: across-variable law (physical-constraint or compatibility condition).* Around any closed path or loop, the algebraic sum of the

---

* A correspondence between elements of one system and a model satisfies the property that if
elements *a* and *b* have interrelationship *r*, then the elements in the model corresponding to *a* and *b*,
respectively, also have this interrelationship *r*.

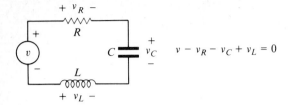

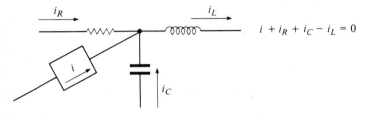

$$v - v_R - v_C + v_L = 0$$

Figure 19 The across-variable law for electric circuit elements.

$$i + i_R + i_C - i_L = 0$$

Figure 20 The through-variable law for electric circuit elements.

voltage drops across the circuit elements is zero. The sum of the voltage rises is also zero. In the loop illustrated in Figure 19, proceeding clockwise from v, the voltage rises are v, $-v_R$, $-v_C$, and v_L. Thus we obtain the equation

$$v - v_R - v_C + v_L = 0$$

2. *Kirchhoff's current law: through-variable law (conservation or continuity principle).* The algebraic sum of the currents flowing into any junction or node where two or more electric circuit elements are connected together must be zero. The same is true of the currents flowing out. The currents flowing into the node in Figure 20 are i_R, i, i_C, and $-i_L$, and so we obtain the equation

$$i_R + i + i_C - i_L = 0$$

4. Examples

a. Writing the model for a given circuit Consider the circuit shown in Figure 21a. We consider the capacitor first. The current reference in the circuit is shown pointing out of the minus voltage reference. This is equivalent to a current reference pointing into the plus voltage reference. Since this is the relationship indicated in the model for the capacitor, we have the correspondence and may use the model to write the equation

$$i_2(t) = C \frac{dv_2}{dt}(t)$$

Now consider the inductor. In our circuit and in the model, the current reference is shown flowing from the top of the inductor to the bottom of the inductor. However, in our circuit the voltage reference for the inductor is minus at

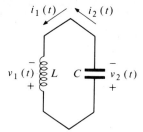

(a) The given *LC* circuit, showing voltage and current definitions

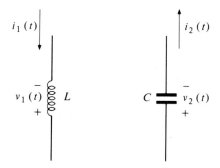

(b) The elements of the circuit, isolated

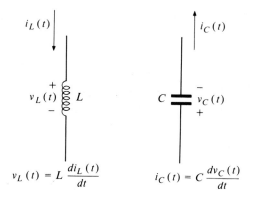

$$v_L(t) = L\,\frac{di_L(t)}{dt}$$ $$i_C(t) = C\,\frac{dv_C(t)}{dt}$$

(c) The models for capacitor and inductor

Figure 21 Rearranging the references to give an exact correspondence with the models of circuit elements.

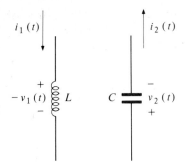

(*d*) The given circuit elements after the references have been rearranged so that an exact correspondence can be made between the circuit and the model

Figure 21 Rearranging the references to give an exact correspondence with the models of circuit elements. (*Continued.*)

the top of the inductor and plus at the bottom, while in the model the voltage reference is plus at the top and minus at the bottom. To make them correspond, we change $v_1(t)$ to $-v_1(t)$ and reverse the voltage-reference direction. Now, using the model, we can write the equation

$$-v_1(t) = L\frac{di_1}{dt}(t)$$

Note that our discussion distinguished between top and bottom. If the inductor were shown in a left-right orientation, we would rotate the diagram to make the correspondence with the model. A simple way of establishing the correspondence is to see that the relationship between the voltage and current references of the model is that the current reference is shown flowing into the plus side of the voltage reference; this is true regardless of the angular orientation of the inductor. To set up the correspondence using this algorithm, we see that, for the inductor, the current reference is shown pointing into the minus side of the voltage reference. If we change $v_1(t)$ to $-v_1(t)$ and reverse the voltage references, then the current reference will be pointing into the plus side of the voltage reference for $-v_1(t)$, and we have the correspondence.

Having taken care of the equations for the elements, we must now take care of the equations describing the interconnection relations. For the through variable, the sum of all the currents flowing into any node must be zero. Examining the node at the top of the circuit, we find a current $i_2(t)$ flowing into the node and a current $i_1(t)$ flowing out of the node. If $i_1(t)$ flows out of the node, then $-i_1(t)$ flows into the node. Thus our equation is

$$-i_1(t) + i_2(t) = 0$$

For the across variable, the sum of all the voltage rises around any closed loop must be zero. Proceeding clockwise around the loop from the bottom of the inductor, there is a voltage drop of $v_1(t)$ across the inductor. A voltage drop of $v_1(t)$ is equivalent to a voltage rise of $-v_1(t)$. Thus we have a voltage rise of $-v_1(t)$ across the inductor and a voltage rise of $v_2(t)$ across the capacitor:

$$-v_1(t) + v_2(t) = 0$$

The system of equations which describe the behavior of the circuit is

$$\left. \begin{aligned} -v_1(t) &= L\frac{di_1}{dt}(t) \\[2mm] i_2(t) &= C\frac{dv_2}{dt}(t) \\[2mm] -i_1(t) + i_2(t) &= 0 \\[2mm] -v_1(t) + v_2(t) &= 0 \end{aligned} \right\}$$

b. Finding the time dependence of voltages and currents in a circuit containing only resistor and capacitor (RC circuit) Consider the circuit shown in Figure 22. Given that the voltage across the capacitor at time t_0 is $v_C(t_0)$, determine the currents $i_C(t)$ and $i_R(t)$ and the voltages $v_C(t)$ and $v_R(t)$ for all $t \geq t_0$.
The resistor equation is

$$v_R(t) = Ri_R(t) \tag{1}$$

The capacitor equation is

$$i_C(t) = C\frac{dv_C}{dt}(t) \tag{2}$$

The sum of all currents leaving any node must be zero:

$$i_C(t) + i_R(t) = 0 \tag{3}$$

The sum of all voltage rises around any closed loop must be zero:

$$v_C(t) - v_R(t) = 0 \tag{4}$$

Observe that there are four variables, $v_R(t)$, $v_C(t)$, $i_R(t)$, and $i_C(t)$, related by the four equations given above. In order to find the time dependence of one of these, we must write and solve a differential equation containing that variable but no

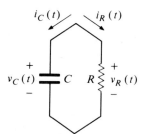

Figure 22 Simple *RC* circuit.

other unknowns. Thus, to obtain an equation for $v_C(t)$, we must eliminate the other three variables. Here is how it is done. The procedure is explained in more generality later in the chapter, in Section D.

By (1) and (2), we have

$$i_R(t) = \frac{v_R(t)}{R} \quad \text{and} \quad i_C(t) = C \frac{dv_C}{dt}(t)$$

Substituting these relations into (3) eliminates $i_R(t)$ and $i_C(t)$, and we obtain

$$C \frac{dv_C}{dt}(t) + \frac{v_R(t)}{R} = 0 \tag{5}$$

By (4), $v_R(t) = v_C(t)$. Substituting this into (5) eliminates $v_R(t)$ as well:

$$C \frac{dv_C}{dt}(t) + \frac{v_C(t)}{R} = 0 \tag{6}$$

This is the desired differential equation containing no unknowns other than $v_C(t)$. It is solved by separation of variables, as in Chapter I, beginning with transposition and division:

$$C \frac{dv_C}{dt}(t) = -\frac{v_C(t)}{R}$$

$$\frac{(dv_C/dt)(t)}{v_C(t)} = -\frac{1}{RC}$$

We make a substitution of variables

$$\int_{t_0}^{t} \frac{1}{v_C(z)} \frac{dv_C}{dz}(z) \, dz = \int_{t_0}^{t} -\frac{1}{RC} \, dz$$

$$dy = \frac{dv_C}{dz}(z) \, dz$$

$$\int_{v_C(t_0)}^{v_C(t)} \frac{dy}{y} = -\int_{t_0}^{t} \frac{dz}{RC}$$

$$\ln |y| \Big|_{v_C(t_0)}^{v_C(t)} = -\frac{1}{RC} z \Big|_{t_0}^{t}$$

$$\ln |v_C(t)| - \ln |v_C(t_0)| = -\frac{1}{RC}(t - t_0)$$

$$\ln \left| \frac{v_C(t)}{v_C(t_0)} \right| = -\frac{1}{RC}(t - t_0)$$

$$\frac{v_C(t)}{v_C(t_0)} = e^{-(1/RC)(t - t_0)}$$

$$v_C(t) = v_C(t_0) e^{-(1/RC)(t - t_0)} \qquad t \geq t_0 \tag{7}$$

The solution

$$v_C(t) = v_C(t_0)e^{-(1/RC)(t-t_0)}$$

to the differential equation

$$C\frac{dv_C(t)}{dt} + \frac{v_C(t)}{R} = 0$$

with initial value $v_C(t_0)$ at time $t = t_0$ is checked by verifying that it satisfies: (1) the differential equation and (2) the initial value. To check that it satisfies the differential equation, we shall substitute the formula given for $v_C(t)$ in Equation (7) in each place that the function $v_C(t)$ appears in the left-hand side of Equation (6). It satisfies the equation if the left-hand side reduces to zero, which is the value of the right-hand side:

$$C\frac{dv_C(t)}{dt} + \frac{v_C(t)}{R} = C\frac{-v_C(t_0)}{RC}e^{-(1/RC)(t-t_0)} + \frac{1}{R}v_C(t_0)e^{-(1/RC)(t-t_0)}$$

$$= -\frac{v_C(t_0)}{R}e^{-(1/RC)(t-t_0)} + \frac{v_C(t_0)}{R}e^{-(1/RC)(t-t_0)} = 0$$

Hence, it satisfies the differential equation. To check that it satisfies the initial condition at t_0, we shall substitute t_0 for t in Equation (7) and see if the resulting left- and right-hand sides are identical:

$$v_C(t_0) = v_C(t_0)e^{-(1/RC)(t-t_0)} = v_C(t_0)e^0 = v_C(t_0)$$

Thus, it satisfies the initial condition.

Having solved for $v_C(t)$, it is easy to determine $v_R(t)$, $i_R(t)$, and $i_C(t)$. From (4),

$$v_R(t) = v_C(t)$$

Hence,

$$v_R(t) = v_C(t_0)e^{-(1/RC)(t-t_0)} \qquad t \geq t_0$$

From (1),

$$i_R(t) = \frac{v_R(t)}{R}$$

Hence,

$$i_R(t) = \frac{v_C(t_0)}{R}e^{-(1/RC)(t-t_0)} \qquad t \geq t_0$$

Finally, from (3),

$$i_C(t) = -i_R(t)$$

Hence,

$$i_C(t) = -\frac{v_C(t_0)}{R}e^{-(1/RC)(t-t_0)} \qquad t \geq t_0$$

By working harder, we can obtain $i_C(t)$ in an alternative manner. From (2),

$$i_C(t) = C\frac{dv_C(t)}{dt}$$

But
$$v_C(t) = v_C(t_0)e^{-(1/RC)(t-t_0)}$$

and so
$$\frac{dv_C}{dt}(t) = v_C(t_0)e^{-(1/RC)(t-t_0)}\left(-\frac{1}{RC}\right)$$

$$i_C(t) = C\frac{dv_C(t)}{dt} = -\frac{v_C(t_0)}{R}e^{-(1/RC)(t-t_0)} \qquad t \geq t_0$$

A graph of the solutions shows their behavior.

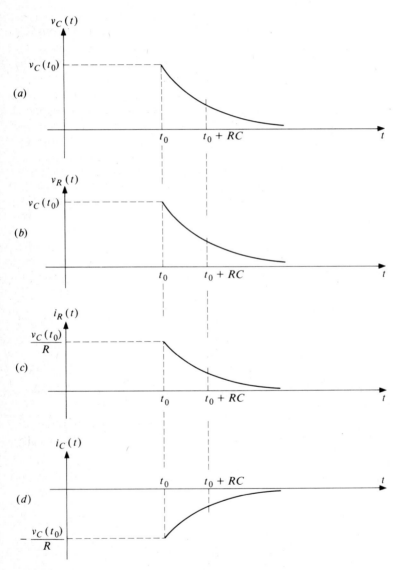

Figure 23 The graphs of the solutions to the circuit and initial value problems shown in Figure 22.

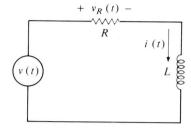

Figure 24 The *RL* circuit of Example c, as given. The voltage $v(t)$ is known. Not all currents and voltages have been defined.

c. Series circuit with an impressed voltage, with some variables undefined Consider the circuit shown in Figure 24. Suppose $v(t)$ and initial condition $i(t_0)$ are given. Determine the time dependence of $v_R(t)$ and $i(t)$.

The situation is typical in that the information given is enough to determine the behavior of the system but not enough to perform all the steps of the calculations systematically. In particular, the voltage across the inductor is undefined, and so is the current through the resistor. We shall define them by *writing onto the circuit diagram* our choices of reference directions and variables, as in Figure 25.

In Figure 25 our choices were made so that the circuit resembles the model as closely as possible. With all quantities defined, we may proceed to write down the equations which model the circuit, beginning with the resistor and the capacitor.

$$v_R(t) = i_R(t)R \tag{8}$$

$$v_L(t) = L\frac{di}{dt}(t) \tag{9}$$

The sum of the currents entering the node between resistor and inductor is zero, and so

$$i_R(t) - i(t) = 0 \tag{10}$$

The sum of the voltage rises clockwise around the loop is zero:

$$v(t) - v_R(t) - v_L(t) = 0 \tag{11}$$

(Notice that it is irrelevant whether a voltage or current is known or unknown. The model is unaffected.)

There are five variables, one of which represents a known function and four of which represent unknown functions, and four equations. We must determine $v_R(t)$

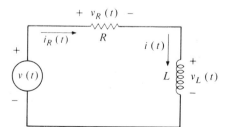

Figure 25 The *RL* circuit of Example c, with all currents and voltages defined.

and $i(t)$. Let us do the latter first. Substitute (8) and (9) into (11), eliminating $v_R(t)$ and $v_L(t)$:

$$v(t) - i_R(t)R - L\frac{di}{dt}(t) = 0 \tag{12}$$

Now substitute (10) into (12), eliminating $i_R(t)$:

$$v(t) - i(t)R - L\frac{di}{dt}(t) = 0 \tag{13}$$

Upon rearranging, we obtain

$$L\frac{di}{dt}(t) + Ri(t) = v(t)$$

or
$$\frac{di}{dt}(t) + \frac{R}{L}i(t) = \frac{v(t)}{L} \tag{14}$$

To solve this equation, we use the four-step integrating-factor method from Section F of Chapter I.

Step 1. Write down the corresponding homogeneous equation, and obtain any solution of it.

$$\frac{di_h(t)}{dt} + \frac{R}{L}i_h(t) = 0$$

One solution to this equation is

$$i_h(t) = e^{-(R/L)t}$$

Step 2. Divide both sides of (14) by $e^{-(R/L)t}$.

$$e^{Rt/L}\frac{di}{dt}(t) + i(t)e^{Rt/L}\frac{R}{L} = e^{Rt/L}\frac{v(t)}{L} \tag{15}$$

Step 3. Identify the left-hand side of (15) as the derivative of $i(t)e^{Rt/L}$.

$$\frac{d}{dt}[i(t)e^{Rt/L}] = e^{Rt/L}\frac{v(t)}{L} \tag{16}$$

Step 4. Perform a definite integration of both sides of Equation (16), and rearrange.

$$i(t)e^{Rt/L} - i(t_0)e^{Rt_0/L} = \int_{t_0}^{t} e^{Rz/L}\frac{v(z)}{L}dz$$

$$i(t)e^{Rt/L} = i(t_0)e^{Rt_0/L} + \int_{t_0}^{t} e^{Rz/L}\frac{v(z)}{L}dz$$

$$i(t) = i(t_0)e^{-(R/L)(t-t_0)} + e^{-(Rt/L)}\int_{t_0}^{t} e^{Rz/L}\frac{v(z)}{L}dz \qquad t \geq t_0 \tag{17}$$

Using (8) and (17), we can solve for $v_R(t)$.

$$v_R(t) = i(t)R \tag{18}$$

This finishes the solution of the specified problem, but we shall look at the behavior of the system for two possible inputs $v(t)$. First, suppose that $v(t)$ is a constant: $v(t) = V$. Then, by Equation (17),

$$i(t) = i(t_0)e^{-(R/L)(t-t_0)} + e^{-(Rt/L)} \int_{t_0}^{t} e^{Rz/L} \frac{V}{L} dz$$

$$= i(t_0)e^{-(R/L)(t-t_0)} + e^{-(Rt/L)} \frac{V}{L} \left(e^{Rz/L} \frac{L}{R} \right)\Big|_{t_0}^{t}$$

$$= i(t_0)e^{-(R/L)(t-t_0)} + \frac{V}{R} e^{-(Rt/L)} \left(e^{Rt/L} - e^{Rt_0/L} \right)$$

$$= i(t_0)e^{-(R/L)(t-t_0)} + \frac{V}{R} \left(1 - e^{-(R/L)(t-t_0)} \right)$$

$$= \underbrace{\frac{V}{R}}_{\substack{\text{Steady-} \\ \text{state} \\ \text{part}}} + \underbrace{e^{-(R/L)(t-t_0)} \left[i(t_0) - \frac{V}{R} \right]}_{\substack{\text{Transient} \\ \text{part}}}$$

The solution is graphed in Figure 26. The steady-state part appears as the asymptote; the transient part is the difference between the asymptote and the curve.

Next, suppose that $v(t) = \sin \omega t$. Then, by Equation (17),

$$i(t) = i(t_0)e^{-(R/L)(t-t_0)} + e^{-(Rt/L)} \int_{t_0}^{t} e^{Rz/L} \frac{\sin \omega z}{L} dz \tag{19}$$

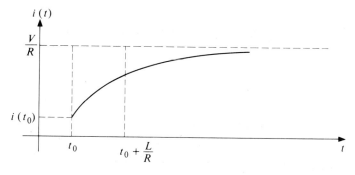

Figure 26. The graph of the solution $i(t)$ to the initial value problem of Figure 24, with $v(t) = V$.

Since R, L, and ω are constant, the integral is of the form

$$I = \int_{t_0}^{t} e^{az} \sin bz \, dz$$

By letting

$$u = e^{az} \qquad\qquad du = ae^{az} \, dz$$

$$dv = \sin bz \, dz \qquad v = \frac{-\cos bz}{b}$$

and using integration by parts, we obtain

$$I = \frac{-e^{az} \cos bz}{b}\Big|_{t_0}^{t} + \frac{a}{b} \int_{t_0}^{t} e^{az} \cos bz \, dz$$

By letting

$$u = e^{az} \qquad\qquad du = e^{az} a \, dz$$

$$dv = \cos bz \, dz \qquad v = \frac{\sin bz}{b}$$

and integrating by parts again, we obtain

$$I = \frac{-e^{az} \cos bz}{b}\Big|_{t_0}^{t} + \frac{a}{b}\left(\frac{e^{az} \sin bz}{b}\Big|_{t_0}^{t} - \int_{t_0}^{t} \frac{a}{b} e^{az} \sin bz \, dz\right)$$

$$= \left(\frac{-e^{az} \cos bz}{b} + \frac{ae^{az} \sin bz}{b^2}\right)\Big|_{t_0}^{t} - \frac{a^2}{b^2} I$$

Transposing the last term to the left-hand side gives

$$I + \frac{a^2}{b^2} I = \left(\frac{a \sin bz}{b^2} - \frac{b \cos bz}{b^2}\right)e^{az}\Big|_{t_0}^{t}$$

and solving for I yields

$$I = \frac{a \sin bz - b \cos bz}{a^2 + b^2} e^{az}\Big|_{t_0}^{t} \tag{20}$$

Evaluation and substitution into (19) give the required formula for $i(t)$. We shall, however, analyze the function into its steady-state and transient parts, beginning with the trigonometric identity

$$\cos (bz + \phi) = \cos bz \cos \phi - \sin bz \sin \phi$$

Identifying coefficients in (20),

$$\cos \phi = -\frac{b}{(a^2 + b^2)^{1/2}}$$

$$\sin \phi = -\frac{a}{(a^2 + b^2)^{1/2}}$$

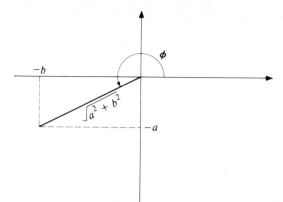

Figure 27 The relationship between a, b, and the angle ϕ, where a and b are positive.

We rewrite (20) as

$$I = \frac{e^{az}}{\sqrt{a^2 + b^2}} \cos (bz + \phi) \Big|_{z=t_0}^{z=t} \qquad \text{where } \phi = \left(\mathrm{Tan}^{-1} \frac{a}{b}\right) + \pi \qquad (21)$$

Using the expression in (21) for the integral in (19), $a = R/L$ and $b = \omega$, we obtain

$$i(t) = i(t_0)e^{-(R/L)(t-t_0)} - \frac{e^{-(Rt/L)}}{L} \left[\frac{e^{(R/L)z}}{\sqrt{(R/L)^2 + \omega^2}} \cos (\omega z + \phi)\right]_{z=t_0}^{z=t}$$

$$\phi = \left(\mathrm{Tan}^{-1} \frac{R}{\omega L}\right) + \pi$$

or

$$i(t) = i(t_0)e^{-(R/L)(t-t_0)} - \frac{e^{-(Rt/L)}}{\sqrt{R^2 + (\omega L)^2}}$$

$$\times \left[e^{Rt/L} \cos (\omega t + \phi) - e^{Rt_0/L} \cos (\omega t_0 + \phi)\right]$$

Multiplication and regrouping now show the transient and steady-state parts.

$$i(t) = i(t_0)e^{-(R/L)(t-t_0)} - \frac{1}{\sqrt{R^2 + (\omega L)^2}}$$

$$\times \left[\cos (\omega t + \phi) - e^{-(R/L)(t-t_0)} \cos (\omega t_0 + \phi)\right]$$

$$i(t) = e^{-(R/L)(t-t_0)} \underbrace{\left[i(t_0) + \frac{\cos (\omega t_0 + \phi)}{\sqrt{R^2 + (\omega L)^2}}\right]}_{\text{Transient part}} - \underbrace{\frac{\cos (\omega t + \phi)}{\sqrt{R^2 + (\omega L)^2}}}_{\text{Steady-state part}} \qquad (22)$$

The required voltage v_R is easily calculated from (22) and (18).

5. **Power and Energy in Electric Circuit Elements**

Whenever a current flows through an electric circuit element and a voltage difference exists across it, energy is being transferred to or from the element. Energy has the dimensional unit of *joules*. The rate at which energy is transferred (energy per unit time) is called *instantaneous power*, and it has the dimensional unit of *watts*. We denote power by p, and energy by w.

The instantaneous power $p(t)$ delivered to any electric circuit element at time t is the product of the current flowing through it and the voltage across it when the current reference direction is pointing into the plus side of the voltage reference direction. (See Figure 28.) Depending on the signs of $i(t)$ and $v(t)$, the circuit may actually be delivering energy to the circuit element or removing energy from the circuit element.

The amount of energy delivered to any circuit element can be computed on the basis of the relationship between energy and power: $p(t) = (dw/dt)(t)$. Integrating both sides of this relationship, we obtain

$$\int_{t_0}^{t} p(z)\, dz = \int_{t_0}^{t} \frac{d}{dz} w(z)\, dz = w(t) - w(t_0)$$

For a resistor R, as in Figure 29, we have $v(t) = i(t)R$, and the power delivered to the resistor is $p(t) = i(t)v(t) = i(t)[i(t)R] = i^2(t)R$. Since R is always positive for a real resistance, the instantaneous power delivered to a resistor is always positive. Since

$$w(t) - w(t_0) = \int_{t_0}^{t} p(z)\, dz$$

the energy transferred to the resistor from time t_0 to $t > t_0$ is always positive. This energy is not stored by the resistor (because it cannot be given up by the resistor) and is dissipated as heat.

For a capacitor C, as in Figure 30, we have $i(t) = C(dv/dt)(t)$, and the power delivered to the capacitor is

$$p(t) = i(t)v(t) = C\frac{dv}{dt}(t)v(t)$$

Depending on the function $v(t)$, $(dv/dt)(t)$ can have the same sign as $v(t)$ or the opposite sign. Therefore, power can be delivered to or transported away from a

Figure 28 The instantaneous power $p(t)$ flowing into an electric circuit element.

Figure 29 The model for the resistor.

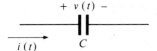

$+ \; v(t) \; -$

$i(t)$ C **Figure 30** The model for the capacitor.

capacitor. We can see this quite explicitly when we examine the energy delivered to the capacitor over a time interval (t_0, t).

$$w(t) - w(t_0) = \int_{t_0}^{t} p(u) \, du = \int_{t_0}^{t} Cv(u) \frac{dv}{du}(u) \, du$$

Letting $z = v(u)$ and $dz = (dv/du)(u) \, du$ and integrating, we obtain

$$w(t) - w(t_0) = C \int_{v(t_0)}^{v(t)} z \, dz = C \frac{z^2}{2} \Big|_{v(t_0)}^{v(t)}$$

or $$w(t) - w(t_0) = \tfrac{1}{2}C[v^2(t) - v^2(t_0)]$$

If the initial voltage across a capacitor of 1 farad is 0 volt at time t_0, and the final voltage at time t is 10 volts, then the energy delivered to the capacitor between t_0 and t is

$$w(t) - w(t_0) = \tfrac{1}{2}(1)(100 - 0) = 50 \text{ joules}$$

If at a later time t_1 the voltage drops to 2 volts, we have

$$w(t_1) - w(t) = \tfrac{1}{2}(1)(4 - 100) = -\tfrac{96}{2} = -48 \text{ joules}$$

Minus 48 joules of energy delivered to the capacitor means that 48 joules of energy was released by the capacitor. The capacitor delivered 48 joules of energy back to the circuit. Thus, the capacitor may store energy and later release it back to the circuit. The energy that a capacitor stores is stored in the capacitor's electric field.

For an inductor L, as in Figure 31, we have $v(t) = L(di/dt)(t)$, and the power delivered to the inductor is

$$p(t) = i(t)v(t) = i(t)L \frac{di}{dt}(t)$$

Depending on the function $i(t)$, $(di/dt)(t)$ can have the same sign as $i(t)$ or the opposite sign. Therefore, power can be delivered to or transported away from an inductor. We can see this quite explicitly when we examine the energy delivered to the inductor over a time interval (t_0, t).

$$w(t) - w(t_0) = \int_{t_0}^{t} p(u) \, du = \int_{t_0}^{t} Li(u) \frac{di}{du}(u) \, du$$

$i(t)$

$+ \; v(t) \; -$

L **Figure 31** The model for the inductor.

Letting $z = i(u)$ and $dz = (di/du)(u)\, du$ and integrating, we obtain

$$w(t) - w(t_0) = \int_{i(t_0)}^{i(t)} Lz\, dz = L\frac{z^2}{2}\bigg|_{i(t_0)}^{i(t)} = \tfrac{1}{2}L[i^2(t) - i^2(t_0)]$$

If the initial current at time t_0 flowing through an inductor of 2 henries is 1 ampere, and the final current at time t is 2 amperes, then the energy delivered to the inductor is $w(t) - w(t_0) = \tfrac{1}{2}(2)(4 - 1) = 3$ joules. If at a later time t_1 the current drops to 0 ampere, we have $w(t_1) - w(t) = \tfrac{1}{2}(2)(0 - 4) = -4$ joules. Minus 4 joules of energy delivered to the inductor means that 4 joules of energy was released by the inductor. The inductor delivered 4 joules of energy back to the circuit. Thus, the inductor may store energy and later release it back to the circuit. The energy that an inductor stores is stored in the inductor's magnetic field.

6. Sample Problems with Answers

A. In Figure 32, $i(t) = 5$ amperes. Determine $v(t)$.
   Answer: $v(t) = -2.5$ volts

$i(t)$

0.5 ohm

+ $v(t)$ −   **Figure 32** Simple resistor circuit.

B. In Figure 33, $v(t) = 6$ volts. Determine $i(t)$.
   Answer: $i(t) = 0$ ampere

− $v(t)$ + $i(t)$

0.01 farad   **Figure 33** Simple capacitor circuit.

C. In Figure 33, $v(t) = 0.1 \sin 1000t$. Determine $i(t)$.
   Answer: $i(t) = \cos 1000t$

D. In Figure 34, $i(t) = -2t$. Determine $v(t)$.
   Answer: $v(t) = 6$ volts

+ $v(t)$ − $i(t)$

3 henries   **Figure 34** Simple inductor circuit.

E. In Figure 34, $i(t) = 100$ amperes. Determine $v(t)$.
   Answer: $v(t) = 0$ volt

F. Consider the circuit shown in Figure 35.
   (a) Set up the equations which model the circuit.
   (b) Eliminate $v_R(t)$, $v_L(t)$, and $i_R(t)$, obtaining a differential equation in $i_L(t)$ alone.
   (c) Solve the equation for $i_L(t)$, granted that $i_L(3) = -2$.
   (d) Find $i_R(t)$, $v_L(t)$, and $v_R(t)$ from your formula for $i_L(t)$.
   Answers: (a) $v_R(t) = -i_R(t)R$; $v_L(t) = -L(di_L/dt)(t)$; $i_R(t) - i_L(t) = 0$; $v_L(t) + v_R(t) = 0$. (b) $L(di_L/dt)(t) + Ri_L(t) = 0$. (c) $i_L(t) = -2e^{-(R/L)(t-3)}$. (d) $i_R(t) = -2e^{-(R/L)(t-3)}$; $v_L(t) = -2Re^{-(R/L)(t-3)}$; $v_R(t) = 2Re^{-(R/L)(t-3)}$.

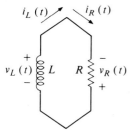

Figure 35 Simple *LR* circuit.

G. In Figure 36, at time $t = 1$, $v_C(1) = 10$ volts. Determine $v_C(t)$ when:

(a) $i(t) = 2$ amperes

(b) $i(t) = te^{-t}$ amperes

(c) $i(t) = t^2$ amperes

Answers: (a) $v_C(t) = [2(t - 1)/C] + 10$ volts. (b) $v_C(t) = [(2e^{-1} - te^{-t} - e^{-t})/C] + 10$ volts. (c) $v_C(t) = [(t^3 - 1)/3C] + 10$ volts.

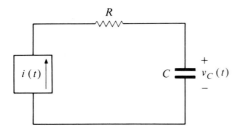

Figure 36 Simple *RC* circuit.

H. Determine the relationship between $v(t)$ and $i(t)$ for the circuit shown in Figure 37.

Answer: $i(t) = [C_1 C_2/(C_1 + C_2)](dv/dt)(t)$ amperes

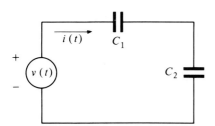

Figure 37 Series capacitor circuit.

I. Determine the power $p(t)$ dissipated in the resistor in the circuit shown in Figure 38 when $i(t) = 4$ amperes.

Answer: $p(t) = 16R$ watts

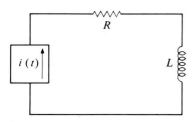

Figure 38 Series *LR* circuit.

J. Determine the resistor current $i_R(t)$ and the power $p(t)$ dissipated in the resistor in the circuit shown in Figure 39. Take the initial voltage at $v(t_0)$ to be 4 volts and $i(t) = 3$ amperes.

    Answer: $i_R(t) = 3 + e^{-(t-t_0)}$ amperes; $p(t) = (3 + e^{-(t-t_0)})^2$ watts

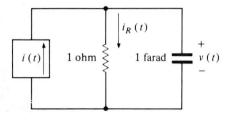

Figure 39 Simple parallel RC circuit.

7. Problems

In problems where *no value* is given for R, L, C, or t_0, give the answers in terms of these letters. If a graph is required but *no value* is given, draw the graph for the case $R = 0.1$, $L = 0.05$, $C = 5$, or $t_0 = 0$.

1. In the circuit shown in Figure 40, $i(t) = 5$ amperes. Determine $v(t)$ and the power $p(t)$ dissipated in the resistor. Graph $p(t)$, $v(t)$, and $i(t)$.

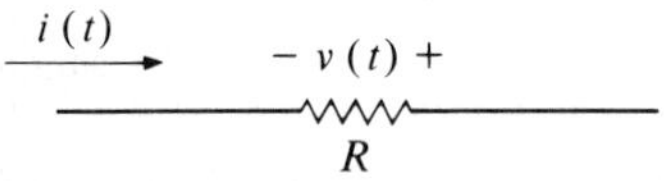

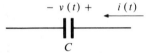

Figure 40 Simple resistor circuit.

2. In the circuit shown in Figure 41, $v(t) = t^2$ volts. Determine $i(t)$ and the power $p(t)$ being delivered to the capacitor. Graph $p(t)$, $v(t)$, and $i(t)$.

Figure 41 Simple capacitor circuit.

3. In the circuit shown in Figure 42, $i(t) = -2t$. Determine $v(t)$ and the power $p(t)$ being delivered to the conductor. Graph $p(t)$, $v(t)$, and $i(t)$.

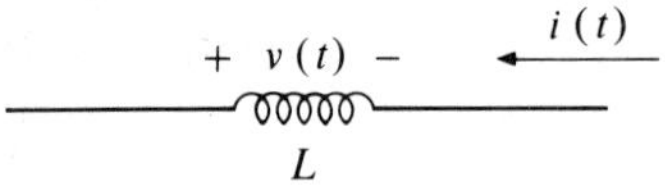

Figure 42 Simple inductor circuit.

4. Consider the capacitor circuit shown in Figure 43. Before some initial fixed time t_0, the voltage $v(t)$ is zero. That is, $v(t) = 0$ for $t < t_0$. At time t_0, the current $i(t)$ changes from 0 to 7 amperes. That is,

$$i(t) = \begin{cases} 0 & t < t_0 \\ 7 & t \geq t_0 \end{cases}$$

Determine $v(t)$ for $t \geq t_0$. Graph $i(t)$ and $v(t)$.

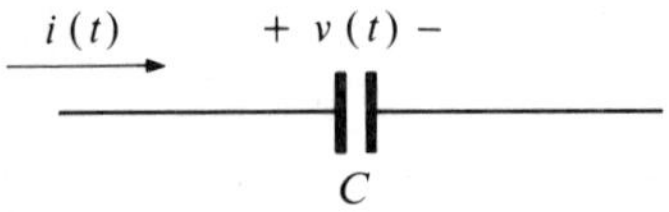

Figure 43 Simple capacitor circuit.

5. Consider the inductor circuit shown in Figure 44. Before some initial fixed time t_0, the current $i(t)$ is zero. That is, $i(t) = 0$ for $t \le t_0$. At time t_0, the voltage $v(t)$ changes from 0 to 2 volts. That is,

$$v(t) = \begin{cases} 0 & t \le t_0 \\ 2 & t > t_0 \end{cases}$$

Determine $i(t)$ for $t > t_0$. Graph $i(t)$ and $v(t)$.

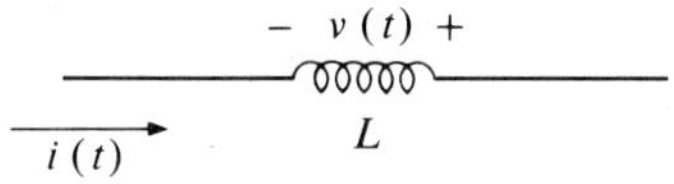

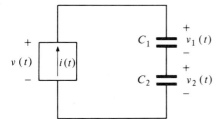

Figure 44 Simple inductor circuit.

6. At some instant the rate of change of the current through a 16-millihenry inductor is 4 milliamperes/microsecond, and the current has a value of 32 milliamperes. What is the voltage across the inductor at this instant? What is the energy stored in the inductor at this instant if no energy was stored when the current was zero?

7. A 10-kilohm resistor has no voltage across it until an instant we take as $t = 0$; thereafter, the voltage across it is 15 volts. What is the current through the resistor when $t = 10$ seconds? What is the power delivered to (and dissipated in) the resistor at $t = 10$ seconds? How much energy has been delivered to the resistor by $t = 10$ seconds?

8. Consider the series capacitor circuit shown in Figure 45.

(a) Write the differential equation describing the relationship between $v_1(t)$ and $i(t)$.

(b) Write the differential equation describing the relationship between $v_2(t)$ and $i(t)$.

(c) Using your answers to parts a and b, determine (derive) the differential equation describing the relationship between $v(t)$ and $i(t)$.

(d) Using your answers to parts a and b, determine (derive) the differential equation describing the relationship between $v_1(t)$ and $v_2(t)$.

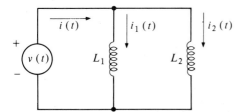

Figure 45 Simple series capacitor circuit.

9. Consider the parallel inductor circuit of Figure 46.

(a) Write the differential equation describing the relationship between (1) $v(t)$ and $i_1(t)$, and (2) $v(t)$ and $i_2(t)$.

(b) Write the equation describing the relationship between $i(t)$, $i_1(t)$, and $i_2(t)$.

(c) Using your answers to parts a and b, determine (derive) the differential equation describing the relationship between $v(t)$ and $i(t)$.

(d) Using your answers to parts a and b, determine the differential equation describing the relationship between $i_1(t)$ and $i_2(t)$.

Figure 46 Simple parallel inductor circuit.

10. Consider the parallel resistor circuit shown in Figure 47. Determine the relationship between $v(t)$ and $i(t)$.

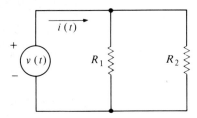

Figure 47 Simple parallel resistor circuit being driven by a voltage source.

11. Consider the series resistor circuit shown in Figure 48. Determine the relationship between $v(t)$ and $i(t)$.

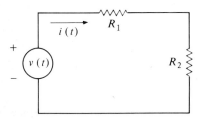

Figure 48 Series resistor circuit being driven by a voltage source.

12. Consider the series inductor circuit of Figure 49. Determine the relationship between $v(t)$ and $i(t)$.

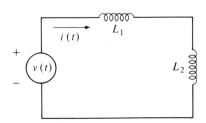

Figure 49 Series inductor circuit being driven by a voltage source.

13. Consider the circuit shown in Figure 50. At time t_0, $v_L(t_0) = 4$ volts. Determine and graph $v_L(t)$ and $p(t)$, the power being delivered to the inductor, if $v(t)$ is (a) 2 volts; (b) e^{-3t} volts; (c) sin $10t$ volts; (d) $2 - 3t$ volts.

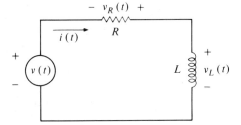

Figure 50 Series LR circuit being driven by a voltage source.

14. Consider the circuit shown in Figure 51. Suppose that $v(t) = e^{-4t}$ and $i_L(t_0) = 4$. Determine and graph (a) $i_L(t)$; (b) $i_R(t)$; (c) $i(t)$.

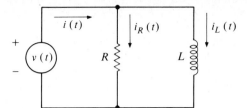

Figure 51 Parallel LR circuit being driven by a voltage source.

15. At time t_0, the circuit shown in Figure 52 has $i_L(t_0) = -1$. Determine and graph $i_L(t)$ and $v(t)$ if $i(t)$ is (a) cos $7t$ amperes; (b) 4 amperes; (c) e^{-6t} amperes.

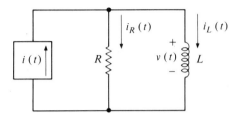

Figure 52 Parallel LR circuit being driven by a current source.

16. Consider the circuit shown in Figure 53. At time t_0, $v_R(t_0) = 3$ volts. Determine and graph $i(t)$ and $v_C(t)$ when (a) $v(t) = 4$ volts; (b) $v(t) = e^{-t} - e^{-2t}$ volts; (c) $v(t) = \sin 2\pi t$ volts.

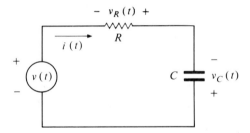

Figure 53 Series RC circuit being driven by a voltage source.

17. Consider the circuit shown in Figure 54. At time $t = 1$, $v_C(1) = 10$ volts. Determine and graph $v_C(t)$ when (a) $i(t) = 2 \cos 10t$ amperes; (b) $i(t) = te^{-3t}$ amperes; (c) $i(t) = e^{-t} \sin 5t$ amperes.

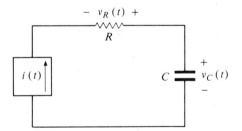

Figure 54 Series RC circuit being driven by a current source.

18. Consider the circuit shown in Figure 55. Determine $i(t)$, $i_R(t)$, and $i_C(t)$ when (a) $v(t) = 1$ volt; (b) $v(t) = t^{-2}$ volts; (c) $v(t) = t \sin 3\pi t$ volts. Take the initial point at $t = 0$.

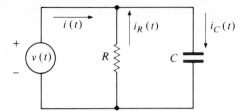

Figure 55 Parallel RC circuit being driven by a voltage source.

19. Consider the circuit shown in Figure 56. At time t_0, $v(t_0) = 4$ volts. Determine $i_R(t)$ and $i_C(t)$ when (a) $i(t) = 2$ amperes; (b) $i(t) = t$ amperes; (c) $i(t) = e^{-3t}$ amperes; (d) $i(t) = \cos 10\pi t$ amperes.

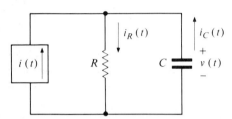

Figure 56 Parallel RC circuit being driven by a current source.

B. MECHANICAL MODEL (DISTANCE-FORCE)

In this section we describe a model for the mechanical system with translational motion in only one dimension. A rigid body moves with pure translational motion if each particle of the body undergoes the same displacement as every other particle.

Two variables which can be used to describe the mechanical model are distance and force. For each element of the model, these two variables are related by a differential equation, such as Newton's second law in the form $f = Mx''$. For any very simple system, a model can be put together from first principles. However, to work with large systems swiftly and with minimum effort, it is convenient to have easily remembered definitions, sign conventions, and techniques which allow the models to be put together routinely. These will be given and illustrated with examples in which we derive and solve the differential equations describing the motion of mechanical translation systems.

1. Definition of Variables and Direction Conventions

a. Distance (the "across" variable) Distance may be measured in the laboratory with a tape measure, the unit of distance being the *meter*. One end of the tape, the zero point, is called the *reference point*, and the direction in which the tape is extended is called the *positive reference direction*. In our diagrams the reference point will be indicated by a wall representing a fixed point, and the positive reference direction will be indicated by an arrow, as in Figure 1a. It is conventional to fix, for each problem, a single positive reference direction governing all distance measurements, and we shall do so. (See Figure 1b.)

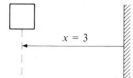

Figure 1a A reference point (wall) and positive reference direction (arrow) are required for the measurement of the distance to an object.

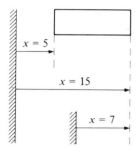

Figure 1b In any given problem, measurements may be made from several reference points, but the positive reference direction is the same for every measurement.

In the model of a mechanical system, we define the distance to a given (but possibly movable) position by writing on the diagram a reference point and positive reference direction (an arrow beginning at a wall and with its head at the desired position). We also assign a letter variable to represent that distance in subsequent calculations. In Figure 2a, the distances $x_1(t)$ and $x_2(t)$ are so defined. Calculation may ultimately show that $x_1(t)$, say, is negative at some time t. This merely means that, at that moment, the given position is on the opposite side of the reference point from the location shown in the diagram. (See Figure 2b.)

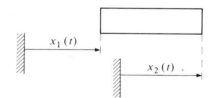

(a) Defining the distances $x_1(t)$ and $x_2(t)$

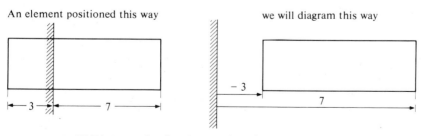

An element positioned this way   we will diagram this way

(b) Diagrams showing the meaning of negative distances

Figure 2 Interpreting variables and negative numbers as distances.

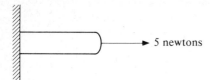

Figure 3a Externally applied force of 5 newtons pulling a rod away from a wall.

Figure 3b Free-body diagram of the rod of Figure 3a and its tension force of 5 newtons. In a free-body diagram, any forces indicated are external forces applied to the object shown.

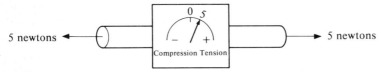

Figure 3c Free-body diagram of the rod, showing how a strain-gage load cell measures the tension force of 5 newtons through the rod.

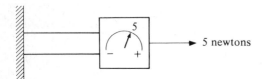

Figure 3d A load measured in this way would be diagramed as in Figure 3a.

b. Force (the "through" variable) The unit of force is the *newton*. As shown in Figure 3a, an externally applied force of 5 newtons pulling a rod away from a wall creates a tension force of 5 newtons through the rod. Figure 3b shows the free-body diagram of a rod experiencing a 5-newton tension load. Figure 3c shows a rod broken into two pieces, with a strain-gage load cell placed between the pieces. The gage is measuring the 5-newton tension force in the rod. Figure 3d shows how the force in Figure 3a would be measured, with the gage interposed between the rod and the force. Hereafter, the arrangement will be diagramed as in Figure 3a, for brevity.

In Figure 4a, an externally applied force of 5 newtons pushes a rod into a wall and thereby creates a 5-newton compression force through the rod. Figure 4b shows the free-body diagram of a rod experiencing a 5-newton compression load. Figure 4c shows a strain-gage load cell measuring the 5-newton compression force in the rod. Figure 4d shows how the force in Figure 4a would be measured. Hereafter, the simpler diagram 4a will be used.

Notice that the scale on a strain-gage load cell has both positive and negative numbers. By convention, positive readings mean tension and negative readings mean compression. The force arrow really indicates whether the force is pushing

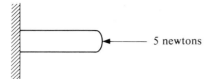

Figure 4a Externally applied force of 5 newtons pushing a rod into a wall.

Figure 4b Free-body diagram of the rod of Figure 4a and its 5-newton compression force. In a free-body diagram, any forces indicated are external forces applied to the object shown.

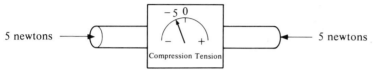

Figure 4c Free-body diagram of the rod, showing how a strain-gage load cell measures a −5-newton tension force in a rod which is experiencing a 5-newton compression.

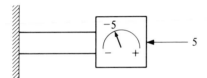

Figure 4d A load measured this way would be diagramed as in Figure 4a.

or pulling. Changing the direction of the arrow corresponds to changing the sign of any number to be read from the gage. For emphasis on this point, see Figure 5.

Several forces may act upon a body simultaneously. The sum of such forces is called the *resultant force*. To determine the resultant force, we first choose a direction. Then we add all the forces in the chosen direction, and subtract from the sum the total of all the forces in the opposite direction. Figure 6a shows three individual forces. Figure 6b shows two possible representations of the resultant.

2. Translational Mechanical Model Definition

Now that we have specified how distance and force are referenced and measured, we may define the elements of the mechanical model—the mass, spring, and dashpot. It is important to remember that only the mass element has mass. Thus, a physical spring, which has mass, must be represented in the model by two separate elements, a spring and a mass.

a. Mass A pure translational mass is a mechanical element whose particles are all rigidly connected together so that they translate with identical velocities and

$$\xrightarrow{\quad 5 \quad}$$ defines the same force as $$\xleftarrow{\quad -5 \quad}$$

(a) A force acting toward the right

$$\xleftarrow{\quad 3 \quad}$$ defines the same force as $$\xrightarrow{\quad -3 \quad}$$

(b) A force acting toward the left

$$\xrightarrow{\quad f(t) \quad}$$ defines the same force as $$\xleftarrow{\quad -f(t) \quad}$$

(c) Two ways of defining the same unknown or unspecified force

Figure 5 In defining forces, the arrow represents a gage and its polarity. The number or variable at the arrow represents the number to be read from the meter.

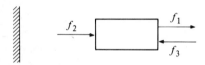

(a) Three externally applied forces

(b) Two representations for the resultant of the forces in Figure 6 a

Figure 6 Representing the resultant of forces acting upon a body.

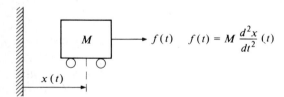

$$f(t) \quad f(t) = M \frac{d^2 x}{dt^2}(t)$$

Figure 7 The model for the mass.

accelerations at all times, and which obeys Newton's second law. If the direction of acceleration is taken in the same direction as the resultant force acting on the mass, then the resultant force is proportional to the acceleration of the mass with constant of proportionality M. M is called the *mass constant*, having units of kilograms.

b. Spring A pure translational spring is a mechanical element of zero mass which stretches or compresses in proportion to the applied tension or compression force.

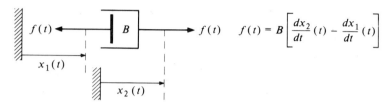

$f(t) - f_0 = K[x_2(t) - x_1(t)]$,
where f_0 is the tension force
in the spring when $x_2(t) - x_1(t) = 0$

Figure 8 The model for the spring.

$f(t) = B\left[\dfrac{dx_2}{dt}(t) - \dfrac{dx_1}{dt}(t)\right]$

Figure 9 The model for the dashpot.

The tension force through the spring is proportional to the distance across the spring minus the distance across the spring when the spring has no tension force. The constant of proportionality is called the *spring constant K* and has units of newtons per meter. Thus, stretching the spring by 1 additional meter increases the tension force in the spring by K newtons.

To keep our equations simple, unless otherwise stated we shall always choose to measure the position of each end of the spring from its own fixed point such that the tension force through the spring is zero when the ends of the spring are positioned at their respective reference points.

c. Dashpot A pure translational dashpot or damper has no mass or spring effect and represents only the effect of resistance to the rate of deformation. The tension force through the dashpot is proportional to the time rate of change of distance across it with constant of proportionality B. B is called the *damping constant*, having units of newton-seconds per meter.

d. Driving mechanical systems A mechanical system can be driven by applying forces to some of the system elements or by forcing the ends of some of the system elements to specified positions. In this way, input distance or force variables can be introduced into the system. These are indicated on a diagram in the same way as all other distance and force variables. Where a force is applied, the corresponding distance will be an unknown; where a distance is known, the force produced will be unknown. This fact will be illustrated in the examples in Parts 3 and 4.

3. Principles for Using the Mechanical Model

a. Technique

Step 1. Isolate each mechanical system element, drawing a free-body diagram. Specify the distance from each end of the element to its appropriate fixed

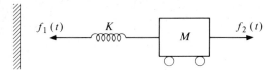

Figure 10 Mass-spring system.

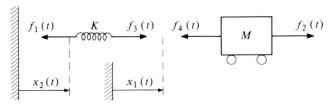

Figure 11 Free-body diagrams for the mass-spring system of Figure 10.

reference point. Specify any external forces and the forces acting on each part of the element where another element is connected to it, as in Figure 11.

Step 2. Establish the correspondence* between the free-body diagram of each mechanical system element and the model for that element. Then use the model to write an equation relating the specified force and distance variables.

Step 3. Use the interconnection law to relate the forces acting at each connection point.

b. Interconnection law In the mechanical system using force and distance as the variables, there is an interconnection law for the forces. This is a through-variable law describing a conservation or continuity principle: The algebraic sum of all forces acting at a massless point is zero. Stated in other words, it is Newton's third law, the action-reaction principle. Consider the system shown in Figure 10. Isolate the mass from the spring, draw the free-body diagram for each, and specify the interconnection forces as shown in Figure 11. The interconnection law states that $f_3(t) - f_4(t) = 0$.

We shall illustrate the correspondence using the mechanical system shown in Figure 12a. The vertical lines joining the elements are rigid, massless connectors.

---

* The correspondence between the elements of a system and the elements of a model satisfies the property that if elements a and b have interrelationship r, then the elements corresponding to a and b, respectively, in the model also have interrelationship r.

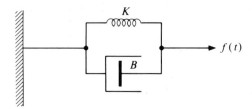

Figure 12a Spring-dashpot system.

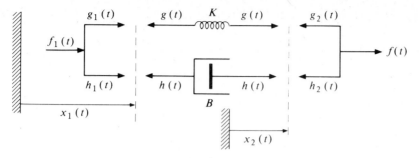

Figure 12b Free-body diagram for the elements of the system of Figure 12a.

We can isolate each element in its own free-body diagram as shown in Figure 12b. The sum of the forces on the left massless connector is zero, and so

$$f_1(t) + g_1(t) + h_1(t) = 0 \tag{1}$$

The sum of the forces on the right massless connector is also zero, and so

$$f(t) - g_2(t) - h_2(t) = 0 \tag{2}$$

Next we can define distances to the ends of the spring and dashpot. Since the spring is massless, the forces we define upon it must be equal and opposite, in order to sum to a resultant zero force. We have chosen to name the tension force in the spring g. The element equation is

$$g(t) - g_0 = K[x_2(t) - x_1(t)] \tag{3}$$

We name the tension force in the dashpot h and obtain

$$h(t) = B[x'_2(t) - x'_1(t)] \tag{4}$$

The interconnection law, applied in turn at each end point of each element, gives the equations

$$g_1(t) - g(t) = 0 \tag{5}$$
$$g(t) - g_2(t) = 0 \tag{6}$$
$$h_1(t) - h(t) = 0 \tag{7}$$
$$h(t) - h_2(t) = 0 \tag{8}$$

The set of equations (1) through (8) constitutes a model for the mechanical system. A much simpler one can be obtained from it by using Equations (5) through (8) to eliminate $g_1(t)$, $g_2(t)$, $h_1(t)$, and $h_2(t)$ from equations (1) through (4):

$$\left.\begin{array}{l} f_1(t) + g(t) + h(t) = 0 \\ f(t) - g(t) - h(t) = 0 \\ \qquad g(t) - g_0 = K[x_2(t) - x_1(t)] \\ \qquad\qquad h(t) = B[x'_2(t) - x'_1(t)] \end{array}\right\}$$

4. Examples

a. The time-dependence of force and distance in a mass-dashpot system Given that the force f in Figure 13 is 3 newtons at time t_0, determine the force $f(t)$ when the dashpot is being moved at a constant speed so that $x(t) = 2t$.

To determine the differential equation relating the force $f(t)$ and the distance $x(t)$ for the system of Figure 13, we first isolate the mass from the dashpot. The free-body diagrams are shown in Figure 14. The element equations are

$$f(t) = M \frac{d^2 x_2}{dt^2}(t) \tag{9}$$

and

$$f(t) = B \left[\frac{dx}{dt}(t) - \frac{dx_2}{dt}(t) \right] \tag{10}$$

Since $x(t)$ is known, there are two unknowns, $f(t)$ and $x_2(t)$, and two equations relating the variables.

To determine the relation between $f(t)$ and $x(t)$, we solve Equation (10) for $(dx_2/dt)(t)$

$$\frac{dx_2}{dt}(t) = \frac{dx}{dt}(t) - \frac{f(t)}{B} \tag{11}$$

and substitute the formula for $(dx_2/dt)(t)$ into Equation (9), eliminating $x_2(t)$:

$$f(t) = M \frac{d^2 x_2}{dt^2}(t) = M \frac{d}{dt} \frac{dx_2}{dt}(t) = M \frac{d}{dt} \left[\frac{dx}{dt}(t) - \frac{f(t)}{B} \right]$$

Simplification gives

$$f(t) = M \frac{d^2 x}{dt^2}(t) - \frac{M}{B} \frac{df}{dt}(t)$$

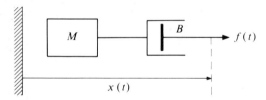

$x(t)$

Figure 13 Mass-dashpot system.

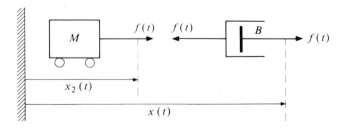

Figure 14 Free-body diagram for the mass and dashpot of Figure 13.

Thus
$$\frac{df}{dt}(t) + \frac{B}{M}f(t) = B\frac{d^2x}{dt^2}(t) \tag{12}$$

The distance function $x(t)$ is $2t$ meters. Suppose that the mass M is 4 kilograms, the damping constant B is 1 newton-second/meter, and at time t_0 the initial force $f(t_0)$ is 3 newtons. Equation (12) gives us an initial value problem,

$$\left. \begin{array}{l} \dfrac{df}{dt}(t) + \tfrac{3}{4}f(t) = 0 \\[2ex] f(t_0) = 3 \end{array} \right\} \tag{13}$$

which we solve by separation of variables, as in Chapter I.

Step 1. Get the left-hand side as $(df/dt)(t)$ divided by $f(t)$.

$$\frac{df}{dt}(t) = -\tfrac{3}{4}f(t)$$

$$\frac{1}{f(t)}\frac{df}{dt}(t) = -\tfrac{3}{4}$$

Step 2. Perform a definite integration.

$$\int_{t_0}^{t} \frac{1}{f(u)}\frac{df}{du}(u)\,du = \int_{t_0}^{t} -\tfrac{3}{4}\,du$$

To do the definite integration, we make a change of variables. Let $y = f(u)$; then $dy = (df/du)(u)\,du$.

$$\int_{t_0}^{t} \frac{1}{f(u)}\frac{df}{du}(u)\,du = \int_{f(t_0)}^{f(t)} \frac{1}{y}\,dy = \int_{t_0}^{t} -\tfrac{3}{4}\,du$$

$$\ln|y|\Big|_{f(t_0)}^{f(t)} = -\tfrac{3}{4}(t - t_0)$$

$$\ln\left|\frac{f(t)}{f(t_0)}\right| = -\tfrac{3}{4}(t - t_0)$$

$$\frac{f(t)}{f(t_0)} = e^{-(3/4)(t-t_0)}$$

$$f(t) = 3e^{-(3/4)(t-t_0)} \tag{14}$$

The solution $f(t) = 3e^{-(3/4)(t-t_0)}$ to the differential equation $(df/dt)(t) + \tfrac{3}{4}f(t) = 0$, with initial value $f(t_0) = 3$, is checked by verifying that it satisfies (1) the differential equation and (2) the initial value. To check that it satisfies the differential equation, we shall substitute the formula given for $f(t)$ in Equation (14) in each place where the function $f(t)$ appears in the left-hand side of Equation

(13). If it checks the equation, the left-hand side will reduce to zero, which is the value of the right-hand side.

$$\frac{df}{dt}(t) + \tfrac{3}{4}f(t) = \frac{d}{dt}(3e^{-(3/4)(t-t_0)}) + \tfrac{3}{4}3e^{-(3/4)(t-t_0)}$$

$$= 3(-\tfrac{3}{4})e^{-(3/4)(t-t_0)} + \tfrac{3}{4}3e^{-(3/4)(t-t_0)}$$

$$= 0$$

Hence, it satisfies the differential equation. To check that it satisfies the initial condition at t_0, we shall substitute t_0 for t in the left- and right-hand sides of Equation (14).

$$f(t_0) = 3e^{-(3/4)(t_0-t_0)} = 3$$

The equation $f(t_0) = 3$ is the initial condition, and so we know that $f(t)$ satisfies the initial condition.

Finally, it is useful to graph the solution, as in Figure 15.

b. Finding the time-dependence of forces in a spring-dashpot system (separation of variables method) Given that the initial force $f(t)$ is 4 newtons at time t_0 and that the distance $x(t) = 10t$ meters for all times t, determine the force $f(t)$ for $t \geq t_0$, for the system in Figure 16.

To determine the relation between the force $f(t)$ and the distance $x(t)$ for the system of Figure 16, we must first isolate the spring from the dashpot. The free-body diagram is shown in Figure 17. We assume that the fixed point from which $x_1(t)$ is measured is placed in such a position that when $x_1(t) = 0$, then $f(t) = 0$. By the model,

$$f(t) = Kx_1(t) \tag{15}$$

$$f(t) = B\left[\frac{dx}{dt}(t) - \frac{dx_1}{dt}(t)\right] \tag{16}$$

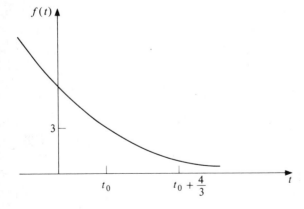

Figure 15 The solution of the initial value problem of Equation (13).

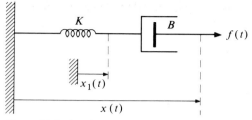

Figure 16 Spring-dashpot system.

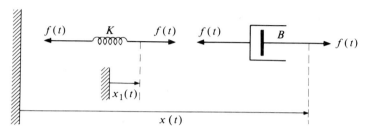

Figure 17 Free-body diagram for the system of Figure 16.

Check: Two unknowns, $f(t)$ and $x_1(t)$; two equations.

To determine the relation between $f(t)$ and $x(t)$, we solve Equation (15) for $x_1(t)$:

$$x_1(t) = \frac{f(t)}{K} \tag{17}$$

Now we substitute the expression for $x_1(t)$ into Equation (16):

$$f(t) = B\left[\frac{dx}{dt}(t) - \frac{d}{dt}x_1(t)\right] = B\left[\frac{dx}{dt}(t) - \frac{d}{dt}\frac{f(t)}{K}\right] \tag{18}$$

$$f(t) = B\frac{dx}{dt}(t) - \frac{B}{K}\frac{df}{dt}(t)$$

$$\frac{df}{dt}(t) + \frac{K}{B}f(t) = K\frac{dx}{dt}(t)$$

Since $x(t) = 10t$ and $f(t_0) = 4$, the initial value problem we must solve is

$$\left.\begin{array}{c}\dfrac{df}{dt}(t) + \dfrac{K}{B}f(t) = 10K \\[2mm] f(t_0) = 4\end{array}\right\} \tag{19}$$

To solve the initial value problem (19), we proceed as follows:

Step 1. Get the left-hand side as the derivative of $f(t) - 10B$ divided by $f(t) - 10B$:

$$\frac{df}{dt}(t) = -\frac{K}{B}[f(t) - 10B]$$

$$\frac{1}{f(t) - 10B}\frac{d}{dt}[f(t) - 10B] = -\frac{K}{B}$$

Step 2. Perform a definite integration:

$$\int_{t_0}^{t}\frac{1}{f(u) - 10B}\frac{d}{du}[f(u) - 10B]\,du = \int_{t_0}^{t}-\frac{K}{B}\,du$$

To do the definite integration, we make a change of variables. Let $y = f(u) - 10B$; then $dy = (d/du)[f(y) - 10B]\,du$ and

$$\int_{t_0}^{t}\frac{1}{f(u) - 10B}\frac{d}{du}[f(u) - 10B]\,du = \int_{f(t_0) - 10B}^{f(t) - 10B}\frac{1}{y}\,dy = \int_{t_0}^{t}-\frac{K}{B}\,du$$

$$\ln|y|\Big|_{f(t_0) - 10B}^{f(t) - 10B} = -\frac{K}{B}(t - t_0)$$

$$\ln\left|\frac{f(t) - 10B}{f(t_0) - 10B}\right| = -\frac{K}{B}(t - t_0)$$

$$e^{-(K/B)(t - t_0)} = \left|\frac{f(t) - 10B}{f(t_0) - 10B}\right|$$

$$f(t) - 10B = [f(t_0) - 10B]e^{-(K/B)(t - t_0)} \qquad (20)$$

Substituting for $f(t_0)$,

$$f(t) = 10B + (4 - 10B)e^{-(K/B)(t - t_0)}$$

Finally, it is useful to graph the solution, which is shown in Figure 18.

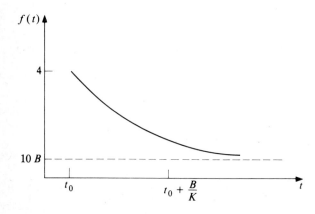

Figure 18 The solution of the initial value problem of Equation (19).

c. Solving a nonhomogeneous equation for a mass-dashpot system by the integrating-factor method Consider the system of Figure 13. Given that the force f is 5 newtons at time t_0, determine the force $f(t)$ when the dashpot is being moved at a speed increasing quadratically with time so that $x(t) = t^3$. Equation (12) is the differential equation relating $f(t)$ and $x(t)$.

$$\frac{d^2x}{dt^2}(t) = 6t$$

so that the initial value problem is

$$\left. \begin{aligned} \frac{df}{dt}(t) + \frac{B}{M}f(t) &= 6Bt \\ f(t_0) &= 5 \end{aligned} \right\} \tag{21}$$

To solve the initial value problem (21), proceed as follows:

Step 1. Write down the corresponding homogeneous equation, and obtain any solution of it.

$$\frac{df_h}{dt}(t) + \frac{B}{M}f_h(t) = 0 \tag{22}$$

One solution to (22) is $f_h(t) = e^{-(B/M)t}$.

Step 2. Divide both sides of (22) by $e^{-(B/M)t}$, the integrating factor.

$$e^{(B/M)t}\frac{df}{dt}(t) + \frac{B}{M}e^{(B/M)t}f(t) = 6Bte^{(B/M)t} \tag{23}$$

Step 3. Identify the left-hand side of (23) as the derivative of $e^{(B/M)t}f(t)$.

$$\frac{d}{dt}[e^{(B/M)t}f(t)] = 6Bte^{(B/M)t} \tag{24}$$

Step 4. Perform a definite integration of both sides of Equation (24), and rearrange.

$$\int_{t_0}^{t}\frac{d}{du}[e^{(B/M)u}f(u)]\,du = \int_{t_0}^{t}6Bue^{(B/M)u}\,du$$

$$f(t)e^{(B/M)t} - f(t_0)e^{(B/M)t_0} = \int_{t_0}^{t}6Bue^{(B/M)u}\,du$$

$$f(t) = f(t_0)e^{-(B/M)(t-t_0)} + e^{-(B/M)t}\int_{t_0}^{t}6Bue^{(B/M)u}\,du$$

$$f(t) = f(t_0)e^{-(B/M)(t-t_0)}$$

$$+ 6M\left[\left(t - \frac{M}{B}\right) - \left(t_0 - \frac{M}{B}\right)e^{-(B/M)(t-t_0)}\right]$$

d. Finding the time-dependence of forces in a spring-dashpot system The system in Figure 19 is driven by two forces, $f(t)$ and $g(t)$. They are unknown, but the resulting distances $x_1(t) = t^2$ and $x_2(t) = t$ are known, and the spring has zero tension at $t = 0$. The problem is to find $f(t)$ and $g(t)$. Even the massless link to which $f(t)$ is applied will require a free-body diagram, as shown in Figure 20.

The equations for the model are

$$h(t) - h_0 = K[x_2(t) - x_3(t)]$$
$$g(t) = B[x'_1(t) - x'_2(t)]$$
$$f(t) + g(t) - h(t) = 0$$

The left-hand side of the spring is fixed, so $x_3(t)$ is constant. Since $h(0)$ is given as zero and $x_2(0) = 0$, we have

$$-h_0 = -Kx_3$$

Hence, $x_3(0) = 0$ also. Substituting known values into the equations for the model now gives

$$h(t) = Kt$$
$$g(t) = B(2t - 1)$$
$$f(t) + g(t) - h(t) = 0$$

Elimination of $h(t)$ and $g(t)$ from the last equation gives

$$f(t) = Kt - B(2t - 1)$$
$$= (K - 2B)t + B$$

Thus we have found formulas for both $g(t)$ and $f(t)$, as required.

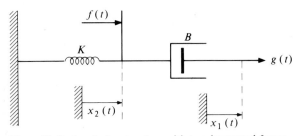

Figure 19 Spring-dashpot system with two impressed forces.

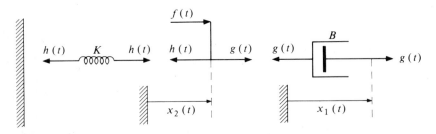

Figure 20 Free-body diagram for the system in Figure 19.

5. Power and Energy in Mechanical System Elements

Whenever a mass is being accelerated, or the ends of a spring or dashpot are moving apart or together, energy is being transferred to or from the mechanical system element. Energy has the dimensional unit of joules. The rate at which energy is transferred, energy per unit time, is called *instantaneous power*, and it has the dimensional unit of watts. We denote power by p and energy by w.

The instantaneous power $p(t)$ delivered to any mechanical system element at time t is the product of the force acting on the element and the derivative of the distance across it for the spring and dashpot, or the derivative of the displacement for the mass. Figure 21 illustrates the computation of the power being delivered to any mechanical system element. Note that, depending on whether the distances or forces are positive or negative, energy may actually be delivered to or removed from the element.

The amount of energy delivered to any mechanical system element can be obtained by integrating the power, since $p(t) = (dw/dt)(t)$, and so

$$\int_{t_0}^{t} p(z)\, dz = \int_{t_0}^{t} \frac{dw}{dz}(z)\, dz = w(t) - w(t_0)$$

For a dashpot, as shown in Figure 21, the force is

$$f(t) = B \left[\frac{dx_2}{dt}(t) - \frac{dx_1}{dt}(t) \right]$$

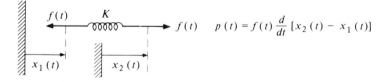

$$f(t) \qquad p(t) = f(t) \frac{d}{dt}[x_2(t) - x_1(t)]$$

$$f(t) \qquad p(t) = f(t) \frac{d}{dt}[x_2(t) - x_1(t)]$$

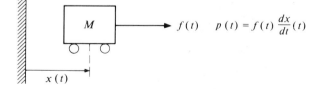

$$f(t) \qquad p(t) = f(t) \frac{dx}{dt}(t)$$

Figure 21 The computation of the instantaneous power being delivered to any mechanical system element.

and the power delivered to the dashpot is

$$p(t) = f(t) \frac{d}{dt} [x_2(t) - x_1(t)]$$

$$= B \frac{d}{dt} [x_2(t) - x_1(t)] \frac{d}{dt} [x_2(t) - x_1(t)]$$

$$= B \left| \frac{d}{dt} [x_2(t) - x_1(t)] \right|^2$$

Since B is always positive for a real dashpot, we see that the instantaneous power delivered to a dashpot is always positive. Since $w(t) - w(t_0) = \int_{t_0}^t p(z) \, dz$, the energy transferred to the dashpot from time t_0 to t is always positive. Because this energy cannot be transferred out of the dashpot back into the mechanical system, the energy cannot be stored by the dashpot. It is dissipated as heat.

For a spring, as shown in Figure 21, $f(t) = K[x_2(t) - x_1(t)]$ if there is zero residual tension. The instantaneous power delivered to the spring is

$$p(t) = f(t) \frac{d}{dt} [x_2(t) - x_1(t)]$$

$$= K[x_2(t) - x_1(t)] \frac{d}{dt} [x_2(t) - x_1(t)]$$

The derivative can have the same sign as $[x_2(t) - x_1(t)]$ or the opposite sign. Therefore, power can be delivered to or transported away from a spring. We can also see this when we examine the energy delivered to the spring over a time interval (t_0, t):

$$w(t) - w(t_0) = \int_{t_0}^t p(u) \, du = \int_{t_0}^t K[x_2(u) - x_1(u)] \frac{d}{du} [x_2(u) - x_1(u)] \, du$$

If we let $z = x_2(u) - x_1(u)$, then $dz = (d/du)[x_2(u) - x_1(u)] \, du$, and so

$$w(t) - w(t_0) = \int_{x_2(t_0) - x_1(t_0)}^{x_2(t) - x_1(t)} Kz \, dz$$

$$= \tfrac{1}{2} K \{ [x_2(t) - x_1(t)]^2 - [x_2(t_0) - x_1(t_0)]^2 \}$$

If the initial distance across a spring having a spring constant K of 1 newton/meter is 2 meters at time t_0, and the final distance across the spring at time t_1 is 10 meters, then over the time interval (t_0, t_1) the energy delivered to the spring is

$$w(t_1) - w(t_0) = \tfrac{1}{2}(1)[(10)^2 - (2)^2] = 48 \text{ joules}$$

If at a later time t_2 the distance across the spring decreases to 4 meters, then over the time interval (t_1, t_2) the energy delivered to the spring is

$$w(t_2) - w(t_1) = \tfrac{1}{2}(1)[(4)^2 - (10)^2] = -42 \text{ joules}$$

Minus 42 joules of energy delivered to the spring means that 42 joules of energy was released by the spring and delivered back to the mechanical system. Thus, the spring can store energy and release it. The energy which a spring stores is called *potential energy.*

For a mass, as shown in Figure 21, $f(t) = M(d^2x/dt^2)(t)$, and the instantaneous power delivered to the mass is

$$p(t) = f(t)\frac{dx}{dt}(t) = M\frac{d^2x}{dt^2}(t)\frac{dx}{dt}(t)$$

Depending on the function $x(t)$, $(dx/dt)(t)$ can have the same sign as $(d^2x/dt^2)(t)$ or the opposite sign. Therefore, power can be delivered to or transported away from a mass. We can also see this when we examine the energy delivered to the mass over a time interval (t_0, t_1):

$$w(t_1) - w(t_0) = \int_{t_0}^{t_1} p(u)\, du = \int_{t_0}^{t_1} Mx'(u)\frac{d}{du}x'(u)\, du$$

If we let $z = x'(u)$, then $dz = (d/du)x'(u)\, du$, and so

$$w(t_1) - w(t_0) = \int_{x'(t_0)}^{x'(t_1)} Mz\, dz = \tfrac{1}{2}M[x'(t_1)^2 - x'(t_0)^2]$$

If the initial velocity of a mass of 2 kilograms at time t_0 is 1 meter/second, and the final velocity of the mass at time t_1 is 4 meters/second, then the energy delivered to the mass is

$$w(t_1) - w(t_0) = \tfrac{1}{2}(2)[(4)^2 - (1)^2] = 15 \text{ joules}$$

If at a later time t_2 the velocity of the mass drops to 0 meter/second, we see that the energy delivered to the mass is

$$w(t_2) - w(t_1) = \tfrac{1}{2}(2)[(0)^2 - (4)^2] = -16 \text{ joules}$$

Minus 16 joules of energy delivered to the mass means that 16 joules of energy was released by the mass. The mass delivered 16 joules of energy back to the mechanical system. Thus the mass can store energy and release it. The energy which a mass stores is called *kinetic energy.*

6. Sample Problems with Answers

A. In Figure 22, $x(0) = 1$ meter. Determine $x(t)$ when: (a) $f(t) = 6$ newtons; (b) $f(t) = 10t$ newtons; (c) $f(t) = e^{-t}$ newtons.
   Answers: (a) $x(t) = 3t + 1$ meters. (b) $x(t) = 2.5t^2 + 1$ meters. (c) $x(t) = -\tfrac{1}{2}e^{-t} + \tfrac{3}{2}$ meters

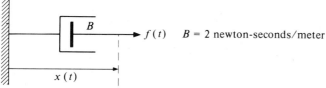

Figure 22 Mechanical system consisting of a dashpot.

B. In Figure 23, when $f(t) = 0$, $x(t) = 1$. Determine $x(t)$ when: (a) $f(t) = 10$ newtons; (b) $f(t) = 20t$ newtons; (c) $f(t) = \cos t$ newtons.

    Answers: (a) $x(t) = 6$ meters. (b) $x(t) = 10t + 1$ meters. (c) $x(t) = \frac{1}{2}\cos t + 1$ meters

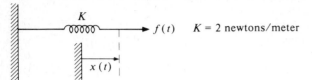

$K = 2$ newtons/meter

Figure 23 Simple mechanical system consisting of a spring.

C. In Figure 24, determine $f(t)$ when: (a) $x(t) = 1000$ meters, for all t; (b) $x(t) = 1000t$ meters; (c) $x(t) = 1000t^2$ meters.

    Answer: (a) $f(t) = 0$ newton. (b) $f(t) = 0$ newton. (c) $f(t) = 2000$ newtons

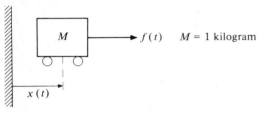

$M = 1$ kilogram

Figure 24 Simple mechanical system consisting of a mass.

D. In Figure 25, determine $f(t)$ when:

    (a) $x_1(t) = 10t$ meters and $x_2(t) = 10t - 5$ meters.
    (b) $x_1(t) = 18e^{-t}$ meters and $x_2(t) = 18(1 - e^{-t}) + 2$ meters.
    (c) Use the information in part b to determine the value M of the mass.

    Answers: (a) $f(t) = 0$ newton. (b) $f(t) = -36Be^{-t}$ newtons. (c) $M = 2B$ kilograms

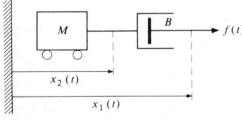

Figure 25 Mass-dashpot mechanical system.

E. In Figure 25, let $f(0) = 5$. Find $f(t)$ if $M = \frac{1}{2}$ kilogram, $B = 1$ newton-second/meter, and $x_1(t) = t^3$.

    Answer: $f(t) = 5e^{-2t} + 3t - \frac{3}{2}(1 - e^{-2t})$

F. In Figure 25, given that $x_1(0) = 10$ meters, $B = 1$ newton-second/meter, and $M = 5$ kilograms, determine $x_1(t)$ when

    (a) $f(t) = 10$ newtons for all t, and $x_2'(0) = 0$
    (b) $f(t) = e^{-t}$ newtons and $x_2'(0) = 1$
    Answers: (a) $x_1(t) = 10(t + 1) + t^2$ meters. (b) $x_1(t) = (54 + 6t - 4e^{-t})/5$ meters

G. Determine the relationship between $f(t)$ and $x(t)$ for the mechanical system of Figure 26.

    Answer: $f(t) = [B_1 B_2/(B_1 + B_2)](dx/dt)(t)$

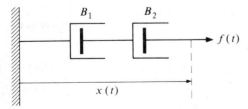

Figure 26 Mechanical system consisting of two dashpots in series.

H. Determine the power $p(t)$ dissipated in the dashpot for the system shown in Figure 25 when $f(t) = \cos 10t$ and $B = 1$ newton-second/meter.

Answer: $p(t) = \cos^2 10t$ watts

I. Determine the position $x_2(t)$ for the system shown in Figure 25 when:
  (a) $x_1(t) = 100$ meters, $x_2'(0) = 1$ meter/second, and $x_2(0) = 0$ meter
  (b) $x_1(t) = 100t$ meters, $x_2'(0) = 2$ meters/second, and $x_2(0) = 1$ meter
  Answers: (a) $x_2(t) = (M/B)(1 - e^{-(B/M)t})$ meters. (b) $x_2(t) = 100t + (M/B)(98e^{-(B/M)t} - 97)$ meters

J. For the system shown in Figure 27, the tension force in the spring is zero when $x(t) = 0$. Determine $x(t)$ given that $f(t) = 10$ newtons and $x(0) = 5$ meters.
  Answer: $x(t) = (10/K)(1 - e^{-(K/B)t}) + 5e^{-(K/B)t}$

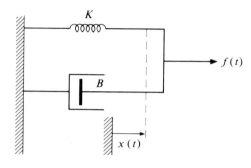

Figure 27 Mechanical system consisting of a spring and dashpot in parallel.

7. Problems

1. A force $f(t)$ is applied to the mass of Figure 24. If the initial conditions are $x(0) = 10$ meters and $x'(0) = 2$ meters/second, determine and graph the position $x(t)$ and the power $p(t)$ delivered to the mass when (a) $f(t) = 20$ newtons; (b) $f(t) = 20 \cos 30\pi t$ newtons; (c) $f(t) = 20(1 + \cos 30\pi t)$ newtons; (d) $f(t) = e^{-t}$ newtons.

2. A force $f(t)$ is applied to the spring of Figure 23. When the position $x(t)$ equals 0 meter, the tension force in the spring is 0 newton. Determine and graph the position $x(t)$ and the energy $w(t)$ stored in the spring when (a) $f(t) = 10$ newtons; (b) $f(t) = 10 \sin 20\pi t$ newtons; (c) $f(t) = 5te^{-t}$ newtons; (d) $f(t) = t^2$ newtons.

3. In Figure 22, determine and graph the tension force $f(t)$ acting on the dashpot and the power $p(t)$ delivered to the dashpot when (a) $x(t) = 1000$ meters; (b) $x(t) = 0.1 \sin 1000t$ meters; (c) $x(t) = 1 - e^{-10t}$ meters; (d) $x(t) = 5t$ meters.

4. Consider the dashpot of Figure 22. Before some initial fixed time t_0, $x(t) = 10$ newtons; that is, $x(t) = 10$ for $t < t_0$. At time t_0, the tension force $f(t)$ changes from 0 to te^{-t} newtons; that is,

$$f(t) = \begin{cases} 0 & t < t_0 \\ te^{-t} & t \geq t_0 \end{cases}$$

Determine $x(t)$ for $t \geq t_0$. Graph $x(t)$ and $f(t)$.

5. At some instant the velocity of a 10-kilogram mass is 2 meters/second, and its acceleration is 0.1 meter/second/second. What is the energy stored in the mass, and what is the force being applied to the mass at this instant?

6. Consider the double-mass system of Figure 28.
  (a) Write the differential equation describing the relationship between $x_1(t)$, $f_1(t)$, and $f_2(t)$.
  (b) Write the differential equation describing the relationship between $x_2(t)$ and the tension force $f_2(t)$ between the masses.

(c) Write the relationship between the derivatives of $x_1(t)$ and $x_2(t)$.

(d) Using your answers to parts a to c, determine (derive) the differential equation describing the relationship between $f_1(t)$ and $x_1(t)$.

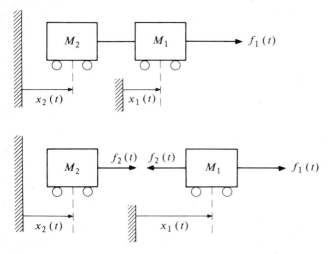

Figure 28 Two-mass system and the corresponding free-body diagram for the system.

C. TRANSLATIONAL MECHANICAL FORCE-VELOCITY MODEL

An alternative mechanical model which uses the variables force and velocity is sometimes preferred by systems engineers. For completeness, we shall give a brief discussion, but it is not necessary to master the model before continuing.

1. The Model Equations and Interconnection Laws

Figures 1, 2, and 3 illustrate the force-velocity model for the mass, spring, and dashpot. Notice that these equations differ from those of the force-distance model by one differentiation only.

The across-variable law represents a physical constraint or compatibility condition. It says that any bodies which are rigidly connected together move with the same velocity. Hence, the velocity difference across any rigid body must be zero, and the sum of the velocity rises around any closed path must be zero. In

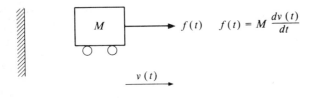

Figure 1 The model for the mass.

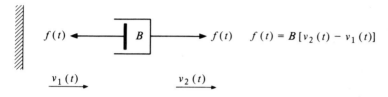

$$v_2(t) - v_1(t) = \frac{1}{K}\frac{df}{dt}(t)$$

Figure 2 The model for the spring.

$$f(t) = B[v_2(t) - v_1(t)]$$

Figure 3 The model for the linear dashpot.

Figure 4a the right end of the spring is moving at a velocity of v_1 meters/second, and the left end of the dashpot is moving at a velocity of v_2 meters/second. Applying the across-variable law to this situation, where the spring and the dashpot are rigidly connected, we must have $v_1 = v_2$. The difference $v_2 - v_1$ may be called the *velocity rise* from the left end to the right. In Figure 4b we illustrate a mechanical system having a closed path taken around two springs and a dashpot. Proceeding clockwise around the path, we sum the velocity rises; by the across-variable law, the sum must be zero.

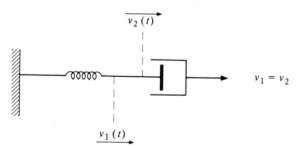

Figure 4a Application of the across-variable law

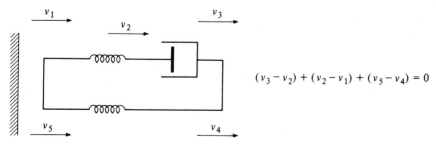

$$(v_3 - v_2) + (v_2 - v_1) + (v_5 - v_4) = 0$$

Figure 4b Application of the across-variable law.

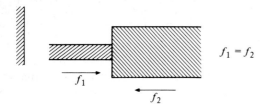

$$f_1 = f_2$$

Figure 5 Application of the through-variable law.

The through-variable law for the force variable is a conservation or continuity principle. For the mechanical system, its classical name is *Newton's third law*: Every action has an equal and opposite reaction. Hence, when an element pushes to the right with force f_1 and it touches an element pushing to the left with force f_2, then $f_1 = f_2$, as illustrated in Figure 5.

2. Examples

a. The time-dependence of force and distance in a spring-dashpot system Consider the mechanical system shown in Figure 6. Given that the tension force through the spring at time t_0 is a known number $f_K(t_0)$, determine the tension force f_B through the dashpot and the velocity $v(t)$ for all $t \geq t_0$.

The through-variable law, applied where the spring connects to the dashpot, implies that

$$f_K(t) = f_B(t) \tag{1}$$

Using the model for the spring gives

$$v(t) - 0 = \frac{1}{K} \frac{df_K}{dt}(t) \tag{2}$$

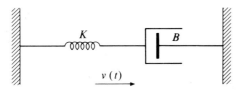

Figure 6 Simple spring-dashpot mechanical system.

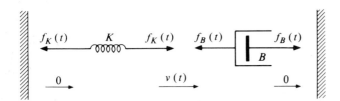

Figure 7 Free-body diagram for the elements of the system shown in Figure 6.

Using the model for the dashpot gives

$$B[0 - v(t)] = f_B(t) \qquad (3)$$

Checking our equations, we have three unknown functions, f_K, f_B, and v, and three equations. We are, therefore, ready to solve this system of equations for f_K. Equation (1), $f_K(t) = f_B(t)$, may be substituted into (3) to eliminate f_B, giving

$$-Bv(t) = f_K(t) \qquad \qquad (4)$$

$$v(t) = \frac{1}{K}\frac{df_K}{dt}(t) \qquad (5)$$

Substituting (5) into (4) to eliminate $v(t)$ gives

$$\frac{-B}{K}\frac{df_K}{dt}(t) = f_K(t)$$

or
$$\frac{df_K}{dt}(t) + \frac{K}{B}f_K(t) = 0 \qquad (6)$$

We can solve (6) by the separation of variables method given in Chapter I, beginning with a transposition, division, and integration.

Step 1. Get the left-hand side as the derivative of $f_K(t)$ divided by $f_K(t)$. We can do this, providing $f_K(t) \neq 0$ for each value of t.

$$\frac{1}{f_K(t)}\frac{df_K}{dt}(t) = -\frac{K}{B} \qquad (7)$$

Step 2. Perform a definite integration of both sides of (7).

$$\int_{t_0}^{t}\frac{1}{f_K(z)}\frac{df_K}{dz}(z)\,dz = \int_{t_0}^{t}-\frac{K}{B}\,dz$$

To do the integration, we identify $[1/f_K(z)][df_K(z)/dz]$ as the derivative of $\ln|f_K(z)|$.

$$\int_{t_0}^{t}\frac{d}{dz}\ln|f_K(z)|\,dz = \int_{t_0}^{t}-\frac{K}{B}\,dz = -\frac{K}{B}(t - t_0)$$

By the Fundamental Theorem of Calculus,

$$\int_{t_0}^{t}\frac{d}{dz}\ln|f_K(z)|\,dz = \ln|f_K(z)|\Big|_{t_0}^{t}$$

Hence,
$$\ln|f_K(z)|\Big|_{t_0}^{t} = \ln\left|\frac{f_K(t)}{f_K(t_0)}\right| = -\frac{K}{B}(t - t_0) \qquad (8)$$

Exponentiating both sides of (8) gives

$$\left|\frac{f_K(t)}{f_K(t_0)}\right| = e^{-(K/B)(t - t_0)} \qquad (9)$$

For each value of t, the right-hand side of (9), being an exponential, can never be zero. Hence, for each value of t, $f_K(t)$ can never be zero. Since f_K is differentiable (the model for the spring states that it must be), f_K must be continuous. But a continuous function which can never be zero must always take the same sign. Therefore, the sign of $f_K(t)$ and the sign of $f_K(t_0)$ are the same, which implies that

$$\left| \frac{f_K(t)}{f_K(t_0)} \right| = \frac{f_K(t)}{f_K(t_0)}$$

Using this in Equation (9), we can solve for $f_K(t)$:

$$f_K(t) = f_K(t_0)e^{-(K/B)(t-t_0)} \tag{10}$$

The solution (10) is valid even if $f_K(t_0) = 0$.

The solution $f_K(t) = f_K(t_0)e^{-(K/B)(t-t_0)}$ to the differential equation $f'_K(t) + (K/B)f_K(t) = 0$, with initial value $f_K(t_0)$, can be checked by verifying that it satisfies (1) the differential equation and (2) the initial value. To check that it satisfies the differential equation, we shall substitute the formula given for f_K in each place where the function f_K appears in the left-hand side of Equation (6). If it satisfies the equation, the left-hand side will reduce to zero, which is the value of the right-hand side:

$$f'_K(t) + \frac{K}{B}f_K(t) = -\frac{K}{B}f_K(t_0)e^{-(K/B)(t-t_0)} + \frac{K}{B}f_K(t_0)e^{-(K/B)(t-t_0)} = 0$$

Hence, it satisfies the differential equation. To check that it satisfies the initial condition at t_0, we shall substitute t_0 for t in Equation (10) and see if the resulting left- and right-hand sides are identical.

$$f_K(t_0) = f_K(t_0)e^{-(K/B)(t_0-t_0)} = f_K(t_0)$$

Thus, it satisfies the initial value. Having solved for $f_K(t)$, we can determine $v(t)$ and $f_B(t)$. By (1), $f_K(t) = f_B(t)$ so that

$$f_B(t) = f_K(t_0)e^{-(K/B)(t-t_0)}$$

By (2),

$$v(t) = \frac{1}{K}\frac{d}{dt}f_K(t)$$

so that

$$v(t) = \frac{1}{K}\left(-\frac{K}{B}\right)f_K(t_0)e^{-(K/B)(t-t_0)}$$

$$= -\frac{1}{B}f_K(t_0)e^{-(K/B)(t-t_0)}$$

Finally, it is useful to graph the solutions. The graphs are shown in Figure 8.

b. A mass-dashpot system with impressed force Consider the mechanical system shown in Figure 9. Given that the velocity of the right end of the dashpot at time

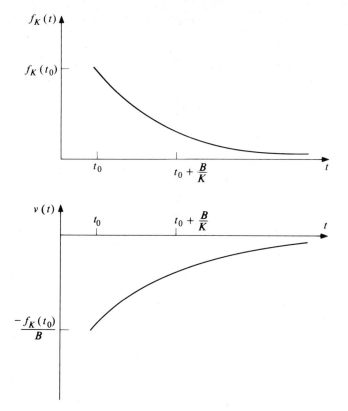

Figure 8 The graphs of $f_K(t)$ and $v(t)$, the solution to the mechanical system of Figure 6.

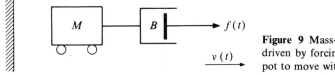

Figure 9 Mass-dashpot mechanical system driven by forcing the right end of the dashpot to move with velocity $v(t)$.

t_0 is $v(t_0)$, and it is driven by an unknown driving force $f(t)$, determine the velocity $v_M(t)$ of the mass. (See Figure 10.)

Using the model for the mass gives

$$f(t) = M \frac{dv_M}{dt}(t) \tag{11}$$

Using the model for the dashpot gives

$$f(t) = B[v(t) - v_M(t)] \tag{12}$$

Since the driving velocity $v(t)$ is known, we have the two unknown functions $f(t)$ and $v_M(t)$ and two equations. We are, therefore, ready to solve this system of

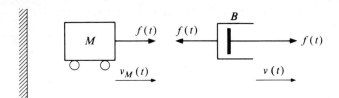

Figure 10 The free-body diagram for the elements of the system shown in Figure 9.

equations for v_M, as required. Substituting (11) into (12) to eliminate f, we obtain

$$Mv'_M(t) = B[v(t) - v_M(t)]$$

Rearranging gives

$$v'_M(t) + \frac{B}{M} v_M(t) = \frac{B}{M} v(t) \tag{13}$$

We can solve (13) using the four-step integrating-factor method.

Step 1. Write down the corresponding homogeneous equation, and obtain any nonzero solution of it.

$$v'_h(t) + \frac{B}{M} v_h(t) = 0$$

One nonzero solution to this equation is $v_h(t) = e^{-(B/M)t}$.
Step 2. Divide both sides of (13) by the solution $e^{-(B/M)t}$.

$$e^{(B/M)t} v'_M(t) + v_M(t) \frac{B}{M} e^{(B/M)t} = \frac{B}{M} e^{(B/M)t} v(t) \tag{14}$$

Step 3. Identify the left-hand side of (14) as the derivative of $v_M(t)e^{(B/M)t}$.

$$\frac{d}{dt} [v_M(t)e^{(B/M)t}] = \frac{B}{M} e^{(B/M)t} v(t) \tag{15}$$

Step 4. Perform a definite integration of both sides of Equation (15), and rearrange.

$$v_M(t)e^{(B/M)t} - v_M(t_0)e^{(B/M)t_0} = \frac{B}{M} \int_{t_0}^{t} e^{(B/M)z} v(z) \, dz$$

$$v_M(t) = v_M(t_0)e^{-(B/M)(t-t_0)} + \frac{B}{M} e^{-(B/M)t} \int_{t_0}^{t} e^{(B/M)z} v(z) \, dz \tag{16}$$

Equation (16) provides a formula for $v_M(t)$ for any driving function v. For instance, if $v(t) = V$, a constant, then

$$v_M(t) = v_M(t_0)e^{-(B/M)(t-t_0)} + \frac{B}{M}e^{-(B/M)t}V \int_{t_0}^{t} e^{(B/M)z} \, dz$$

$$= v_M(t_0)e^{-(B/M)(t-t_0)} + V\frac{B}{M}e^{-(B/M)t}\frac{e^{(B/M)t} - e^{(B/M)t_0}}{B/M}$$

$$= \underbrace{[v_M(t_0) - V]e^{-(B/M)(t-t_0)}}_{\text{Transient part}} + \underbrace{V}_{\substack{\text{Steady-state} \\ \text{part}}} \qquad (17)$$

Notice that the solution (17) has two terms. The first part is a transient part which gets smaller and smaller as t gets larger. It is also a solution to the homogeneous equation corresponding to (13). The second part is the steady-state part giving the kind of function we can expect $v_M(t)$ to get closer to as t gets larger. It is also a particular solution to Equation (13).

Since Equation (16) provides a formula for $v_M(t)$ for any driving velocity function, we can just as easily suppose $v(t) = t$ and compute $v_M(t)$.

$$v_M(t) = v_M(t_0)e^{-(B/M)(t-t_0)} + \frac{B}{M}e^{-(B/M)t} \int_{t_0}^{t} ze^{(B/M)z} \, dz$$

$$= v_M(t_0)e^{-(B/M)(t-t_0)} + e^{-(B/M)t}\left[te^{(B/M)t} - \frac{M}{B}e^{(B/M)t} - t_0 e^{(B/M)t_0} + \frac{M}{B}e^{(B/M)t_0} \right]$$

$$v_M(t) = \underbrace{\left[v_M(t_0) + \frac{M}{B} - t_0 \right]e^{-(B/M)(t-t_0)}}_{\text{Transient part}} + \underbrace{t - \frac{M}{B}}_{\substack{\text{Steady-state} \\ \text{part}}}$$

Notice that the solution has two parts. The first part is a transient part which gets smaller and smaller as t gets larger. It is also a solution to the homogeneous equation corresponding to (13). The second part is the steady-state part, giving the kind of function we can expect $v_M(t)$ to get closer and closer to as t gets larger. It is also a particular solution to Equation (13).

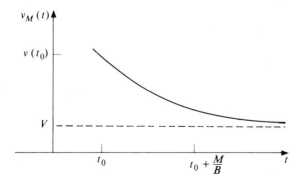

Figure 11 The graph of the solution $v_M(t)$ to the initial value problem of Figure 9, where $v(t) = V$.

3. Sample Problems with Answers

A. In Figure 12, determine $v(t)$ when (a) $f(t) = 6$ newtons; (b) $f(t) = 10t$ newtons; (c) $f(t) = e^{-t}$ newtons.

    Answers: (a) $v(t) = 3$ meters/second. (b) $v(t) = 5t$ meters/second. (c) $v(t) = \frac{1}{2}e^{-t}$ meters/second

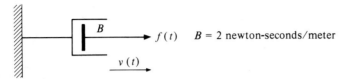

Figure 12 Dashpot connected to a fixed point.

B. In Figure 13, determine $v_2(t)$ when $v_1(t) = \sin 5t$ and (a) $f(t) = 6$ newtons; (b) $f(t) = 10t$ newtons; (c) $f(t) = e^{-t}$ newtons.

    Answers: (a) $v_2(t) = 3 - \sin 5t$ meters/second. (b) $v_2(t) = 5t - \sin 5t$ meters/second. (c) $v_2(t) = \frac{1}{2}e^{-t} - \sin 5t$ meters/second

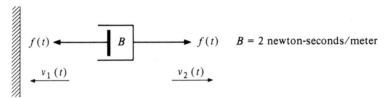

Figure 13 Dashpot whose ends are moving with different velocities.

C. In Figure 14, $v_1(t) = 3t$. Determine $v_2(t)$ when (a) $f(t) = 3 \cos 5t$ newtons; (b) $f(t) = 9e^{-7t}$ newtons; (c) $f(t) = 4/t$ newtons.

    Answers: (a) $v_2(t) = -3t + 5 \sin 5t$ meters/second. (b) $v_2(t) = -3t + 21e^{-7t}$ meters/second. (c) $v_2(t) = -3t + 12t^{-2}$ meters/second

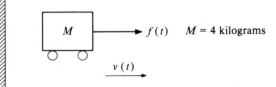

Figure 14 Spring whose ends are moving with different velocities.

D. In Figure 15, determine $f(t)$ when (a) $v(t) = e^{-t} \cos 5t$ meters/second; (b) $v(t) = 100,000$ meters/second; (c) $v(t) = 4t$ meters/second.

    Answers: (a) $f(t) = 4e^{-t}(-5 \sin 5t - \cos 5t)$ newtons. (b) $f(t) = 0$ newton. (c) $f(t) = 16$ newtons

Figure 15 A mass

E. In Figure 16, determine $v(t)$ when (a) $v(-5) = 3$ meters/second; (b) $v(10) = -2$ meters/second; (c) $v(0) = 1$ meter/second; (d) $v(100) = 0$ meter/second.

   Answers: (a) $v(t) = 3e^{-(K/B)(t+5)}$ meters/second. (b) $v(t) = -2e^{-(K/B)(t-10)}$ meters/second. (c) $v(t) = e^{-(K/B)t}$ meters/second. (d) $v(t) = 0$ meter/second

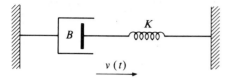

Figure 16 The given mechanical system.

F. Determine the relationship between $f(t)$ and $v(t)$ for the mechanical system of Figure 17.

   Answer: $f(t) = (B_1 + B_2)v(t)$ newtons

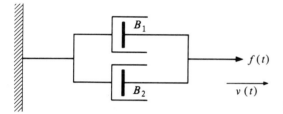

Figure 17 Two dashpots in parallel.

G. Determine the velocity $v(t)$ for the mass shown in Figure 18 when $v(0) = 0$ and $v_1(t) = 10$ meters/second.

   Answer: $v(t) = (10 - \frac{5}{8})(1 - e^{-(B/M)t})$ meters/second

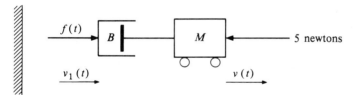

Figure 18 Mass-dashpot system.

H. Consider the mechanical system shown in Figure 19. Given that the velocity of the left end of the dashpot at time t_0 is the number $v(t_0)$, determine the velocity $v(t)$ and the tension force $f(t)$ acting through the dashpot for all $t \geq t_0$.

   Answers: $v(t) = v(t_0)e^{-(B/M)(t-t_0)}$ meters/second. $f(t) = -Bv(t_0)e^{-(B/M)(t-t_0)}$ newtons

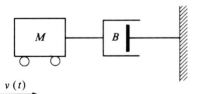

Figure 19 Simple MB mechanical system.

4. Problems

1. A force $f(t)$ is applied to the mass of Figure 15. If the initial condition is $v(0) = 10$ meters/second, determine and graph the velocity $v(t)$ when (a) $f(t) = 20$ newtons; (b) $f(t) = 20 \cos 3\pi t$ newtons; (c) $f(t) = 20(1 + \cos 30\pi t)$ newtons; (d) $f(t) = e^{-t}$ newtons.

2. In Figure 14, the velocity $v_1(t) = 5$ meters/second. If the initial tension force through the spring is 2 newtons at time $t_0 = -2$ seconds, determine the tension force $f(t)$ when (a) $v_2(t) = 10$ meters/second; (b) $v_2(t) = 10 \sin 15\pi t$ meters/second; (c) $v_2(t) = 6te^{-2t}$ meters/second; (d) $v_2(t) = t^3$ meters/second.

3. In Figure 12, determine and graph the tension force $f(t)$ acting on the dashpot when (a) $v(t) = 1000$ meters/second; (b) $v(t) = 0.1 \sin 1000t$ meters/second; (c) $v(t) = 100 \sin 10t$ meters/second; (d) $v(t) = 5t$ meters/second.

4. Determine the velocity $v(t)$ for the mass shown in Figure 18 when $v(3) = 5$ meters/second and $v_1(t) = e^{-2t}$ meters/second.

5. In Figure 20, determine the relationship between $f(t)$ and $v(t)$. The initial condition is $f(t_0) = 3$ newtons. Determine $f(t)$ when

$$v(t) = \begin{cases} 10 \text{ meters/second} & t_0 \le t < 10 \\ 20 \text{ meters/second} & 10 \le t \end{cases}$$

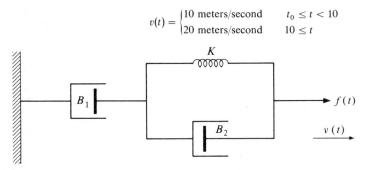

Figure 20 Spring-dashpot system.

6. Consider the double-mass system of Figure 21. Determine the tension force in the dashpot and the velocities $v_1(t)$ and $v_2(t)$ when the initial velocity of mass M_1 is $v_1(0) = 5$ meters/second, and the initial velocity of mass M_2 is $v_2(0) = 0.9$ meter/second.

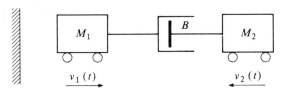

Figure 21 Dashpot slowing the movement of two masses.

D. ELIMINATING UNWANTED VARIABLES FROM SYSTEMS OF EQUATIONS

In the engineering systems discussed in the previous sections, we encountered a common situation: Using the models we wrote equations relating the through variable to the across variable for each of the system elements, and using the interconnection laws we wrote equations relating the through variables and relating the across variables. We then had to operate on the resulting system of equations to determine the differential equation relating the variables of interest. Although there is no simply described algorithm for deriving this differential equation in general, there is a simply described fundamental principle: All variables other than those desired must be eliminated.

In this section we illustrate this idea of variable elimination. Consider the electrical system of Figure 1, and suppose that we want to derive the differential equation relating $v(t)$ and $v_1(t)$. We shall first write the system of equations for the circuit, and then follow the fundamental principle: All variables other than $v(t)$ and $v_1(t)$ must be eliminated. The technique is to eliminate one variable at a time, and *write out in full* the resulting system of equations after each variable has been eliminated.

1. Example

Find the differential equation relating $v(t)$ and $v_1(t)$ in the circuit shown in Figure 1.

Writing the system of equations for the circuit, we have

$$i_R(t) = i_C(t) + i_L(t) \tag{1}$$

$$v(t) = -v_R(t) + v_1(t) \tag{2}$$

$$v_R(t) = -Ri_R(t) \tag{3}$$

$$i_C(t) = C\frac{dv_1}{dt}(t) \tag{4}$$

$$v_1(t) = L\frac{di_L}{dt}(t) \tag{5}$$

We shall use one equation to eliminate one variable from *all* the other equations in which it may appear. The variable eliminated should not be $v(t)$ or $v_1(t)$, which are required in the problem. For instance, we can eliminate $i_R(t)$ by using (1). This leaves one fewer equation and one fewer variable.

$$v(t) = -v_R(t) + v_1(t) \tag{6}$$

$$v_R(t) = -R[i_C(t) + i_L(t)] \tag{7}$$

$$i_C(t) = C\frac{dv_1}{dt}(t) \tag{8}$$

$$v_1(t) = L\frac{di_L}{dt}(t) \tag{9}$$

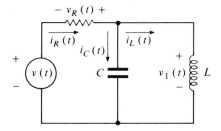

Figure 1 *LRC* circuit being driven by a voltage source.

Notice that i_R has now been eliminated *completely.* We eliminate $v_R(t)$ by using (7), substituting the right side of (7) for $v_R(t)$ in (6), (8), and (9).

$$v_1(t) - v(t) = -R[i_C(t) + i_L(t)] \tag{10}$$

$$i_C(t) = C\frac{dv_1}{dt}(t) \tag{11}$$

$$v_1(t) = L\frac{di_L}{dt}(t) \tag{12}$$

We eliminate $i_C(t)$ using (11).

$$v_1(t) - v(t) = -R\left[C\frac{dv_1}{dt}(t) + i_L(t)\right] \tag{13}$$

$$v_1(t) = L\frac{di_L}{dt}(t) \tag{14}$$

Because the problem requires an equation relating $v(t)$ and $v_1(t)$, the next variable to be eliminated *must* be $i_L(t)$. We may solve for this variable in (13), obtaining

$$i_L(t) = -C\frac{dv_1}{dt}(t) + \frac{1}{R}[v(t) - v_1(t)]$$

and then substitute into (14), differentiating as required. The resulting equation is

$$v_1(t) = -LC\frac{d^2v_1}{dt}(t) + \frac{L}{R}\frac{dv}{dt}(t) - \frac{L}{R}\frac{dv_1}{dt}(t)$$

Upon rearranging, we obtain a second-order differential equation relating $v_1(t)$ and $v(t)$:

$$RLC\frac{d^2v_1}{dt^2}(t) + L\frac{dv_1}{dt}(t) + Rv_1(t) = L\frac{dv}{dt}(t) \tag{15}$$

This is the desired equation relating v_1 and its derivatives with v. It was obtained by four successive applications of the *technique of elimination*: Select a variable to be eliminated and an equation in which it appears, and then use that equation to eliminate that variable from all the other equations in the system. Each application of the technique leaves a new system (*which should be written out in full*) containing one fewer variable and one fewer equation than the previous system of equations.

In this example we began with Equations (1) through (5), a system of five equations containing six variables, i_R, v, v_R, v_1, i_C, and i_L (and their derivatives). In the first application we eliminated i_R, using Equation (1). This left a system of four equations and five variables. The new system of equations, (6) through (9), actually contained some of the old equations unchanged. These were (2), (4), and (5), which became (6), (8), and (9). *At each stage, all equations not yet used in the elimination are repeated in the new system, and the old system is never returned to,* lest we reintroduce a variable already eliminated. At the second stage, therefore, the technique of elimination was applied to the system of equations (6) through

(9). Using (7), we eliminated the variable v_R from all the equations in which it appeared, leaving (10) through (12) as the new system. With (11), i_C was eliminated, leaving (13) and (14), two equations containing the three variables v_1, v, and i_L and their derivatives, as the new system.

The task originally set was to find a differential equation relating $v(t)$ and $v_1(t)$. Therefore, it followed (from the fundamental principle stated at the beginning of the section) that the final application of the technique must be the elimination of $i_L(t)$ and its derivatives from (13) and (14). For this we used the result of solving for $i_L(t)$ in (13), where it occurs undifferentiated only. Differentiation allowed substitution into (14), yielding (15).

A more complicated situation arises if we are asked to find an equation relating $v(t)$ with $i_L(t)$ rather than $v_1(t)$. The elimination is the same down through Equations (13) and (14). Then, the next step must be the elimination of $v_1(t)$.

We choose to eliminate using (14), for $v_1(t)$ appears *only once* in this equation. We substitute it into (13) at *each* place where $v_1(t)$ or its derivative appears, differentiating (14) where needed. The result of the substitution is

$$L\frac{di_L}{dt}(t) - v(t) = -RCL\frac{d^2i_L}{dt^2}(t) - Ri_L(t)$$

Rearranging gives a second-order equation relating $i_L(t)$ and $v(t)$ as required.

$$RLC\frac{d^2i_L}{dt^2}(t) + L\frac{di_L}{dt}(t) + Ri_L(t) = v(t)$$

Let us illustrate the procedure again, this time seeking a differential equation relating $v(t)$ and $v_R(t)$ rather than $v(t)$ and $i_L(t)$. We start again from the initial system of equations

$$i_R(t) = i_C(t) + i_L(t) \tag{1}$$

$$v(t) = -v_R(t) + v_1(t) \tag{2}$$

$$v_R(t) = -Ri_R(t) \tag{3}$$

$$i_C(t) = C\frac{dv_1}{dt}(t) \tag{4}$$

$$v_1(t) = L\frac{di_L}{dt}(t) \tag{5}$$

We must keep $v(t)$ and $v_R(t)$. We eliminate $i_R(t)$ using (1).

$$v(t) = -v_R(t) + v_1(t) \tag{16}$$

$$v_R(t) = -R[i_C(t) + i_L(t)] \tag{17}$$

$$i_C(t) = C\frac{dv_1}{dt}(t) \tag{18}$$

$$v_1(t) = L\frac{di_L}{dt}(t) \tag{19}$$

We eliminate $i_C(t)$ using (18).

$$v(t) = -v_R(t) + v_1(t) \tag{20}$$

$$v_R(t) = -R\left[C\frac{dv_1}{dt}(t) + i_L(t)\right] \tag{21}$$

$$v_1(t) = L\frac{di_L}{dt}(t) \tag{22}$$

We eliminate $v_1(t)$ using (22).

$$v(t) = -v_R(t) + L\frac{di_L}{dt}(t) \tag{23}$$

$$v_R(t) = -R\left[LC\frac{d^2i_L}{dt^2}(t) + i_L(t)\right] \tag{24}$$

We eliminate i_L and its derivatives as follows. First differentiate (23), obtaining

$$\frac{dv}{dt}(t) = -\frac{dv_R}{dt}(t) + L\frac{d^2i_L}{dt^2}(t)$$

and solve the result for the second derivative of i_L:

$$\frac{d^2i_L}{dt^2}(t) = \frac{1}{L}\left[\frac{dv}{dt}(t) + \frac{dv_R}{dt}(t)\right]$$

This may now be substituted into (24), which gives

$$v_R(t) = -R\left[C\left(\frac{dv}{dt}(t) + \frac{dv_R}{dt}(t)\right) + i_L(t)\right] \tag{25}$$

$$v(t) = -v_R(t) + L\frac{di_L}{dt}(t) \tag{23}$$

This is really a new system of equations, but while the old system (23) and (24) contained the second derivative of i_L, this new system contains only the first derivative. To eliminate it, differentiate (25) and solve (23) for the derivative of i_L, obtaining a new system:

$$\frac{dv_R}{dt}(t) = -R\left[C\left(\frac{d^2v}{dt^2}(t) + \frac{d^2v_R}{dt^2}(t)\right) + \frac{di_L}{dt}(t)\right]$$

$$\frac{di_L}{dt}(t) = \frac{v(t) + v_R(t)}{L}$$

Substitution of the latter equation into the former eliminates i_L:

$$\frac{dv_R}{dt}(t) = -R\left[C\frac{d^2v}{dt^2}(t) + C\frac{d^2v_R}{dt^2}(t) + \frac{v(t) + v_R(t)}{L}\right]$$

Now i_L has been eliminated. Rearranging gives the desired equation:

$$RC\frac{d^2v_R}{dt^2}(t) + \frac{dv_R}{dt}(t) + \frac{R}{L}v_R(t) = -RC\frac{d^2v}{dt^2}(t) - \frac{R}{L}v(t) \tag{26}$$

The procedure consists of this: Each unwanted variable is, in its turn, *completely eliminated from the system before another variable is taken up.* We have illustrated it three times, by finding the differential equations relating $v(t)$ to $v_1(t)$, $i_L(t)$, and $v_R(t)$, respectively.

2. Sample Problems with Answers

A. Begin with Equations (1) through (5), the model for the circuit in Figure 1, and find the differential equation relating $v(t)$ and $i_L(t)$ by performing the eliminations in the following order: $v_1(t)$ [use (5) to eliminate it from both (2) and (4)], $i_c(t)$, $i_R(t)$, and $v_R(t)$. The successive systems of equations you get should be written out fully. They will not be the same systems as those in the first part of the example, but the equation you derive will be the same equation derived earlier relating $v(t)$ and $i_L(t)$.

    Answer: $RLC(d^2i_L/dt^2)(t) + L(di_L/dt)(t) + Ri_L(t) = v(t)$

B. The system of equations

$$v(t) = Ri_R(t) + L\frac{di_R}{dt}(t) - L\frac{di_C}{dt}(t) \quad \left.\right\} \tag{a}$$

$$i_c(t) = CL\frac{d^2i_R}{dt^2}(t) - CL\frac{d^2i_C}{dt^2}(t) \quad \left.\right\} \tag{b}$$

results from eliminating $v_R(t)$, $v_1(t)$, and $i_L(t)$ successively from (1) through (5), the model for the circuit shown in Figure 1. Use the following steps to eliminate i_R and its derivatives from Equations (a) and (b), finding the differential equation relating $i_c(t)$ and $v(t)$.

Step 1. Use the derivative of Equation (a) to eliminate $(d^2i_R/dt^2)(t)$ from Equation (b). The resulting equation contains $(di_R/dt)(t)$ but neither $i_R(t)$ nor $(d^2i_R/dt^2)(t)$.

Step 2. Take Equation (b) and the result of step 1 for the new system of equations. Each equation now contains *only one* term with the variable to be eliminated.

Step 3. In the new system, eliminate $(d^2i_R/dt^2)(t)$ from Equation (b) again, thereby eliminating $i_R(t)$ and all its remaining derivatives from the new system of equations and obtaining the desired differential equation relating $i_c(t)$ and $v(t)$.

    Answer: $RLC(d^2i_C/dt^2)(t) + L(di_C/dt)(t) + Ri_C(t) = LC(d^2v/dt^2)(t)$

C. In Problems A and B we started from the model, Equations (1) through (5). In practice it is usually possible to avoid doing this long elimination more than once. Take as your system of equations the answer to Problem A plus Equations (4) and (5) of the original model; eliminate $v_1(t)$ and $i_L(t)$ from the system, and thus derive (more easily?) the answer to Problem B.

3. Problems

All problems relate to the mechanical system in Figure 2, which may be modeled by the following system of equations:

$$g'(t) = Kv_1(t) \quad \left.\right\}$$
$$g(t) = B[v(t) - v_1(t)]$$
$$f(t) - g(t) = Mv'(t) \quad \left.\right\}$$

Use the elimination of variables method to determine the differential equation describing the relationship between:

1. $f(t)$ and $v(t)$, by first eliminating $g(t)$ and then eliminating $v_1(t)$.
2. $f(t)$ and $v(t)$, by first eliminating $v_1(t)$ and then eliminating $g(t)$.
3. $f(t)$ and $g(t)$, by first eliminating $v_1(t)$ and then eliminating $v(t)$.
4. $f(t)$ and $g(t)$, by first eliminating $v(t)$ and then eliminating $v_1(t)$.

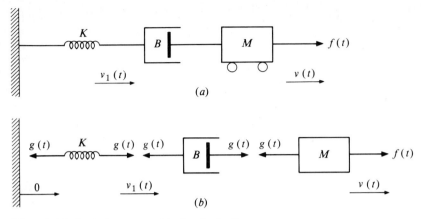

Figure 2 Mechanical system and its free-body diagram.

5. $f(t)$ and $v_1(t)$, by first eliminating $g(t)$ and then eliminating $v(t)$.

6. $f(t)$ and $v_1(t)$, by first eliminating $v(t)$ and then eliminating $g(t)$.

7. $v(t)$ and $g(t)$, by first eliminating $f(t)$ and then eliminating $v_1(t)$.

8. $v(t)$ and $g(t)$, by first eliminating $v_1(t)$ and then eliminating $f(t)$.

9. $v(t)$ and $v_1(t)$, by first eliminating $f(t)$ and then eliminating $g(t)$.

10. $v(t)$ and $v_1(t)$, by first eliminating $g(t)$ and then eliminating $f(t)$.

11. $g(t)$ and $v_1(t)$, by first eliminating $f(t)$ and then eliminating $v(t)$.

12. $g(t)$ and $v_1(t)$, by first eliminating $v(t)$ and then eliminating $f(t)$.

Check each derived relationship to make sure that the dimensional units are correct.

4. Initial Conditions

The result of eliminating unwanted variables from a system of equations is a single differential equation containing two variables. Typically, one represents a known function, or "input," and the other an unknown function, or "output." For instance, Equation (15) contains v and v_1 as defined in Figure 1. If v_1 is a known function, then (15) is a differential equation for v; if, instead, v is known, then (15) is a differential equation for v_1. Let us consider the latter case. Equation (15) cannot be solved for a unique function v_1, because many different functions satisfy (15). However, an initial value problem such as those described in Chapter I will determine v_1 completely. The appropriate form for such a problem is

$$\left.\begin{array}{r} RLCv_1''(t) + Lv_1'(t) + Rv_1(t) = Lv'(t) \\ v_1(t_0) \text{ given} \\ v_1'(t_0) \text{ given} \end{array}\right\} \tag{27}$$

where the differential equation is just (15).

A technique for solving (27) will be given in Chapter III. The problem which faces us now is specifying the initial values $v_1(t_0)$ and $v_1'(t_0)$, which is sometimes more difficult than would appear at first sight. For instance, simple voltmeters and ammeters would not measure $v_1'(t_0)$ directly. However, if we used a voltmeter to measure $v_1(t_0)$ and an ammeter to measure $i_C(t_0)$, then a quick reference to the model, in particular to Equation (4),

$$i_C(t) = C\frac{dv_1}{dt}(t) \tag{4}$$

shows us that

$$i_C(t_0) = C\frac{dv_1}{dt}(t_0)$$

or $v_1'(t_0) = i_C(t_0)/C$. Now the initial value problem is properly set up:

$$\left. \begin{array}{c} RLCv_1''(t) + Lv_1'(t) + Rv_1(t) = Lv'(t) \\[4pt] v_1(t_0) \text{ given} \\[4pt] v_1'(t_0) = \dfrac{1}{C}i_C(t_0) \end{array} \right\} \tag{28}$$

where $i_C(t_0)$ is given. The problem can be solved for a unique function v_1, as we shall see in Chapter III.

It may also happen that the circuit of Figure 1 comes equipped with ammeters only, measuring i_C and i_L. Now the situation is a little more complicated. Of course, $v_1'(t_0) = i_C(t_0)/C$ as before, but to determine $v_1(t_0)$ requires a little serious thought as well as the realization that all the equations of the model, (1) through (5), hold for all times t; in particular, they hold at t_0:

$$\left. \begin{array}{c} i_R(t_0) = i_C(t_0) + i_L(t_0) \\[4pt] v(t_0) = -v_R(t_0) + v_1(t_0) \\[4pt] v_R(t_0) = -i_R(t_0)R \\[4pt] i_C(t_0) = Cv_1'(t_0) \\[4pt] v_1(t_0) = Li_L'(t_0) \end{array} \right\} \begin{array}{l} (29) \\[28pt] (30) \end{array}$$

We have used (30) to determine $v_1'(t_0)$ in terms of $i_C(t_0)$. Our problem right now is to determine $v_1(t_0)$ in terms of $i_C(t_0)$ and $i_L(t_0)$, eliminating unwanted equations (and values) from the system as needed. For instance, $i_R(t_0)$ may be eliminated using (29), leaving the new system

$$\left. \begin{array}{c} v(t_0) = -v_R(t_0) + v_1(t_0) \\[4pt] v_R(t_0) = -R[i_C(t_0) + i_L(t_0)] \\[4pt] i_C(t_0) = Cv_1'(t_0) \\[4pt] v_1(t_0) = Li_L'(t_0) \end{array} \right\} \tag{31}$$

Now $v_R(t_0)$ may be eliminated using (31), leaving the new system

$$v(t_0) = Ri_C(t_0) + Ri_L(t_0) + v_1(t_0) \tag{32}$$
$$i_C(t_0) = Cv_1'(t_0) \tag{33}$$
$$v_1(t_0) = Li_L'(t_0) \tag{34}$$

We have used Equation (33) to evaluate $v_1'(t_0)$, and we cannot use Equation (34), for we do not know $i_L'(t_0)$; but (32) does give the desired information, since v is a known function:

$$v_1(t_0) = v(t_0) - Ri_C(t_0) - Ri_L(t_0) \tag{35}$$

Thus, from the given values $i_C(t_0)$ and $i_L(t_0)$ and the function value $v(t_0)$ of the known function v, we can compute the required initial values, $v_1(t_0)$ and $v_1'(t_0)$. For instance, if $R = 2$, $L = 5$, $C = 3$, $v(t) = -2 \cos \pi t$, and if $t_0 = 0$, $i_C(0) = -4$, and $i_L(0) = 6$, then the initial values would be

$$v_1(0) = v(0) - 2i_C(0) - 2i_L(0)$$
$$= -2 - 2(-4) - 2(6)$$
$$= -6$$
$$v_1'(0) = \tfrac{1}{3}i_C(0) = -\tfrac{4}{3}$$

and the initial value problem would be

$$\left.\begin{array}{c} 30v_1''(t) + 5v_1'(t) + 2v_1(t) = -10\pi \sin \pi t \\ v_1(0) = -6 \\ v_1'(0) = -\tfrac{4}{3} \end{array}\right\}$$

5. Example

Consider the mechanical system of Figure 3. Given the position $x_2(t_0)$ and velocity $x_2'(t_0)$ of the mass M at time t_0, the tension force in the spring at time t_0, and the fact that the tension force is zero when x_1 is zero, determine the initial value $x_2''(t_0)$ and the differential equation describing the behavior of x_2.

First we must draw the free-body diagram shown in Figure 4. The system of equations which describes the system is

$$\left.\begin{array}{c} f(t) = K[x_1(t) - 0] \\ f(t) = B[x_2'(t) - x_1'(t)] \\ f(t) = -Mx_2''(t) \end{array}\right.$$

$$\begin{array}{l} (36) \\ (37) \\ (38) \end{array}$$

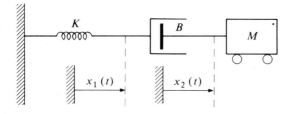

Figure 3 Spring-mass-dashpot system.

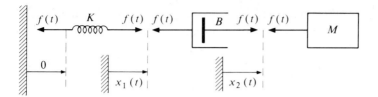

Figure 4 Free-body diagram for the system in Figure 3.

To find the required differential equation in x_2, both f and x_1 must be eliminated from the system of equations. To eliminate f, use Equation (36) and substitute into (37) and (38).

$$Kx_1(t) = B[x_2'(t) - x_1'(t)] \tag{39}$$

$$Kx_1(t) = -Mx_2''(t) \tag{40}$$

To eliminate x_1, solve Equation (40) for x_1 and substitute for it in (39) at each place where x_1 appears.

$$-Mx_2''(t) = B\left\{x_2'(t) - \frac{d}{dt}\left[-\frac{M}{K}x_2''(t)\right]\right\}$$

Upon rearranging, we obtain a third-order differential equation in x_2:

$$\frac{BM}{K}x_2'''(t) + Mx_2''(t) + Bx_2'(t) = 0 \tag{41}$$

We are given the initial values $x_2(t_0)$ and $x_2'(t_0)$, the position and velocity of the mass. We require, in addition, the acceleration $x_2''(t_0)$ of the mass at time t_0, and we are given the tension force in the spring $f(t_0)$ at time t_0. Equation (38) relates the tension force in the spring to the acceleration of the mass for any time t. In particular, Equation (38) holds for time t_0. Hence,

$$x_2''(t_0) = -\frac{f(t_0)}{M}$$

The third-order differential equation (41) and the three initial values $x_2(t_0)$, $x_2'(t_0)$, and $x_2''(t_0)$ constitute the initial value problem

$$\left.\begin{array}{r} BMx_2'''(t) + MKx_2''(t) + BKx_2'(t) = 0 \\ x_2(t_0) \text{ given} \\ x_2'(t_0) \text{ given} \\ x_2''(t_0) = -\dfrac{f(t_0)}{M} \end{array}\right\}$$

where $f(t_0)$ is the tension in the spring at t_0 (given).

6. More Sample Problems with Answers

D. For the mechanical system in Figure 3, find the differential equation for $x_1(t)$, and find the initial value $x_1(t_0)$ if $x_1'(t_0) = 3$ and $x_2'(t_0) = 5$.

    Answers: $MBx_1'' + MKx_1' + BKx_1 = 0$; $x_1(t_0) = 2B/K$

E. For the mechanical system whose free-body diagram is shown in Figure 4, find the differential equation for f, the tension in the spring, and the initial values $f(t_0)$ and $f'(t_0)$ if $x_1'(t_0) = 3$ and $x_2'(t_0) = 5$.

    Answers: $MBf'' + MKf' + KBf = 0$; $f(t_0) = 2B$, $f'(t_0) = 3K$

F. For the circuit in Figure 1, if $v(t) = 3 \cos(t - t_0)$, $v_R(t_0) = 4$, and $i_C(t_0) = 5$, find $i_L(t_0)$ and $i_L'(t_0)$.

    Answers: $i_L(t_0) = -5 - 4/R$, $i_L'(t_0) = 7/L$

7. Problems

13. For the circuit in Figure 1, the model consists of Equations (1) through (5). From these equations, find each of the following pairs of initial values in terms of $i_C(t_0)$, $i_R(t_0)$, and the value of the (known) function v and its derivatives at t_0.

    (a) $v_1(t_0)$ and $v_1'(t_0)$ (b) $i_L(t_0)$ and $i_L'(t_0)$
    (c) $i_R(t_0)$ and $i_R'(t_0)$ (d) $v_R(t_0)$ and $v_R'(t_0)$
    (e) $i_C(t_0)$ and $i_C'(t_0)$

14. In Figure 5, the springs with constants K_1 and K_2 are under tensions $f(t)$ and $g(t)$, respectively, at time t. The model for the system is the following system of equations:

$$f' = K_1 v_1$$

$$f = B(v - v_1)$$

$$g - f = Mv'$$

$$g' = K_2(-v_2 - v)$$

    (a) Find the differential equation relating f and v_2 by systematic elimination of variables.

    (b) If the values of $v_1(0)$, $v(0)$, and $v_2(0)$ are given, then what are the initial values $f(0), f'(0)$, and $f''(0)$?

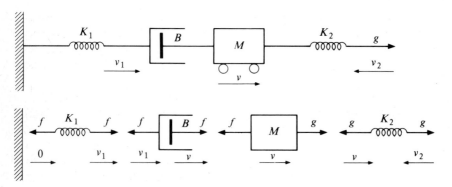

Figure 5 Mechanical system and its free-body diagram.

TRANSFORM METHODS FOR SOLVING LINEAR DIFFERENTIAL EQUATIONS WITH CONSTANT COEFFICIENTS

The transforms to be studied are the Laplace transform and the J_ω operator.

The Laplace transform is a very versatile tool. With it, one finds solutions to initial value problems and, therefore, also general solutions. It is well adapted to handling homogeneous and nonhomogeneous equations alike, including "instantaneous" impulses. With it, we study several properties of solutions of linear differential equations in Sections A through H.

Sections I and J are concerned with the J_ω operator, a much simpler transform but very specialized. Its use requires complex arithmetic but gives sine and cosine solutions to nonhomogeneous equations. Such solutions give the "steady-state" behavior of many systems. The sections on the Laplace transform are not prerequisite to those on the J_ω operator.

The study of Laplace transforms will require the use of techniques from calculus, particularly improper integration, integration by parts, L'Hospital's rule, and decomposition into partial fractions. For the J_ω operator, polar coordinates are necessary. These topics should be reviewed carefully, in advance.

The examples which appear in this chapter are all based on models which have appeared earlier, but it is not essential to have studied these models, for in each case the differential equation to be derived will be presented when needed. The models are from sections of Chapter II, as follows:

The electric circuit model (Section A)
The force-displacement translation model (Section B)
The force-velocity translation model (Section C)

In addition, the lake problem from Chapter I appears. Readers familiar with these models may derive the differential equations for themselves or verify the derivation in those places where it is derived on the spot. Other readers should simply identify the point at which the initial value problem is stated and proceed from there.

THE LAPLACE TRANSFORM

A. DEFINITION, EXAMPLES, AND METHOD

The *Laplace transform* of a function f defined on the positive real axis is given by

$$\mathcal{L}[f](s) = \int_0^\infty f(t)e^{-st}\,dt \tag{1}$$

This formula requires comment. First, $\mathcal{L}[f]$ stands for "Laplace transform of f." It is a function. Its value at the number s is $\mathcal{L}[f](s)$, just as $F(s)$ is the value at s of the function F. Thus, (1) is a formula for computing the value at the point s of the Laplace transform $\mathcal{L}[f]$.

The second point is that the value is computed by an integral, an improper integral since the region of integration is unbounded.

1. Example

Let $f(t) = e^{at}$. We shall compute the Laplace transform of f.

$$\mathcal{L}[e^{at}](s) = \int_0^\infty e^{at}e^{-st}\,dt$$

$$= \int_0^\infty e^{(a-s)t}\,dt = \lim_{T\to\infty} \int_0^T e^{(a-s)t}\,dt$$

$$= \lim_{T\to\infty} \frac{e^{(a-s)t}}{a-s}\bigg|_0^T = \lim_{T\to\infty}\left(\frac{e^{(a-s)T}}{a-s} - \frac{1}{a-s}\right)$$

$$= \frac{1}{s-a} + \lim_{T\to\infty}\frac{e^{(a-s)T}}{a-s}$$

$$= \frac{1}{s-a}$$

if $a - s < 0$; that is, if $s > a$. That is,

$$\mathcal{L}[e^{at}](s) = \frac{1}{s-a} \qquad \text{for } s > a$$

Here is the third comment: $\mathscr{L}[f](s)$ depends on s and not upon t, for t is only a dummy variable, a variable of integration.

Special cases of the example include these:

$$\mathscr{L}[e^{-t}](s) = \frac{1}{s+1} \qquad \text{for } s > -1$$

$$\mathscr{L}[e^{3t}](s) = \frac{1}{s-3} \qquad \text{for } s > 3$$

$$\mathscr{L}[1](s) = \frac{1}{s} \qquad \text{for } s > 0$$

where " 1 " denotes the function which assigns to each number t the number 1, that is, the function which is constantly 1.

The fourth comment is that $\mathscr{L}[f](s)$ may be defined for some values of s and not for others, as in the example. This property will be ignored, mainly, and dealt with only at the end of the section. We need the formula for $\mathscr{L}[f](s)$, and when we have it we shall be willing to ignore its region of definition, the domain of the function $\mathscr{L}[f]$, and consider only very large positive values of s.

A related point is that not all functions have Laplace transforms. This point we shall ignore in part, as in the definition given at the beginning of the section, but where it seems important hereafter we shall insert the proviso, " if f has a Laplace transform, then...." The matter will be discussed again at the end of the section.

2. Example

Let $f(t) = te^{at}$. We shall compute the Laplace transform of f.

$$\mathscr{L}[te^{at}](s) = \int_0^\infty te^{at}e^{-st}\, dt = \int_0^\infty te^{(a-s)t}\, dt$$

$$= \lim_{T \to \infty} \left(\frac{te^{(a-s)t}}{a-s}\bigg|_0^T - \int_0^T \frac{e^{(a-s)t}}{a-s}\, dt \right)$$

$$= \lim_{T \to \infty} \left[\frac{te^{(a-s)t}}{a-s} - \frac{e^{(a-s)t}}{(a-s)^2} \right]_0^T$$

$$= \lim_{T \to \infty} \left\{ \frac{Te^{(a-s)T}}{a-s} - \frac{0}{a-s} - \left[\frac{e^{(a-s)T}}{(a-s)^2} - \frac{1}{(a-s)^2} \right] \right\}$$

$$= \frac{1}{(a-s)^2} + \lim_{T \to \infty} \frac{Te^{(a-s)T}}{a-s} - \lim_{T \to \infty} \frac{e^{(a-s)T}}{a-s}$$

$$= \frac{1}{(a-s)^2} \qquad s > a$$

That is,
$$\mathscr{L}[te^{at}](s) = \frac{1}{(a-s)^2} \qquad s > a$$

Special cases of the example include these:

$$\mathscr{L}[te^{-t}](s) = \frac{1}{(s+1)^2} \qquad s > -1$$

$$\mathscr{L}[te^{3t}](s) = \frac{1}{(s-3)^2} \qquad s > 3$$

$$\mathscr{L}[t](s) = \frac{1}{s^2} \qquad s > 0$$

where t denotes the identity function, which assigns to each number that number itself.

The points to notice are the uses of integration by parts and L'Hospital's rule. First consider integration by parts. You will want to be able to use it to integrate *at least* the following functions:

$$te^{at} \qquad t^2e^{at} \qquad e^{at}\sin bt \qquad e^{at}\cos bt$$

In Example 2, the formula $\int u\, dv = uv - \int v\, du$ is utilized as follows:

$$u = t \qquad v = \frac{e^{(a-s)t}}{a-s} \qquad du = dt \qquad dv = e^{(a-s)t}\, dt$$

Any doubts about integration by parts should be settled at once by review and practice, using your own calculus text.

Now consider L'Hospital's rule, in the following form: If f and g are defined "near ∞" and have continuous derivatives there, and if f/g is indeterminate "at ∞," then

$$\lim_{t\to\infty}\frac{f(t)}{g(t)} = \lim_{t\to\infty}\frac{f'(t)}{g'(t)}$$

provided either limit exists. In the present example, $f(T) = T$, $g(T) = e^{(s-a)T} = 1/e^{(a-s)T}$, and

$$\lim_{T\to\infty} Te^{(a-s)T} = \lim_{T\to\infty}\frac{T}{e^{(s-a)T}} = \lim_{T\to\infty}\frac{f(T)}{g(T)} = \lim_{T\to\infty}\frac{f'(T)}{g'(T)}$$

$$= \lim_{T\to\infty}\frac{1}{(s-a)e^{(s-a)T}} = 0 \qquad \text{if } s > a$$

Before using Laplace transforms to solve an initial value problem, we must have a formula for the Laplace transform of the derivative of an unknown function. This formula is crucial, and it is dignified with the name "Theorem."

Theorem *If f has a Laplace transform, if f has a continuous derivative f', and if f'
has a Laplace transform, then*

$$\mathscr{L}[f'](s) = s\mathscr{L}[f](s) - f(0) \tag{2}$$

For now we shall assume also that $f(t)$ is continuous at $t = 0$.

Notice that if $\mathscr{L}[f]$ is known and, if $f(0)$ is known, then $\mathscr{L}[f']$ is known.
Formula (2) is surprisingly easy to compute from (1). It is a good exercise in the
use of integration by parts, and so the derivation of (2) (the proof of the theorem)
is given in Part 6 below.

Now we have enough information at hand to solve an initial value problem.

3. Example

Solve
$$\left. \begin{array}{r} y'(t) + y = e^{-t} \\ y(0) = 0 \end{array} \right\}$$

The function y is unknown, but the left-hand side of the differential equation
is the same function as that on the right. Therefore, we may take the Laplace
transforms of both sides of the equation, giving

$$\mathscr{L}[y' + y](s) = \mathscr{L}[e^{-t}](s) \tag{3}$$

which is called the *transformed equation.* We evaluate the right-hand side:

$$\mathscr{L}[e^{-t}](s) = \int_0^\infty e^{-t}e^{-st}\, dt = \frac{1}{s+1} \tag{4}$$

from the definition and Example 1. The left-hand side is evaluated as

$$\mathscr{L}[y'(t) + y(t)](s) = \int_0^\infty [y'(t) + y(t)]e^{-st}\, dt$$

$$= \int_0^\infty y'(t)e^{-st}\, dt + \int_0^\infty y(t)e^{-st}\, dt$$

$$= \mathscr{L}[y'](s) + \mathscr{L}[y](s) \tag{5}$$

Of course, y and y' are unknown, and so are $\mathscr{L}[y]$ and $\mathscr{L}[y']$ as a result; but the
theorem says that

$$\mathscr{L}[y'](s) = s\mathscr{L}[y](s) - y(0)$$

$$= s\mathscr{L}[y](s)$$

where we have now used the initial value, $y(0) = 0$. Substituting this crucial fact
back into (5) gives

$$\mathscr{L}[y' + y](s) = s\mathscr{L}[y](s) + \mathscr{L}[y](s)$$

$$= (s + 1)\mathscr{L}[y](s) \tag{6}$$

Substituting (6) and (4) into the transformed equation (3) gives

$$(s + 1)\mathscr{L}[y](s) = \frac{1}{s + 1}$$

$$\mathscr{L}[y](s) = \frac{1}{(s + 1)^2}$$

We have not yet found y, but we have found $\mathscr{L}[y]$, and a glance at Example 2 shows that y and te^{-t} have the same transform. On this basis, let us guess that $y(t) = te^{-t}$ is a solution of the initial value problem and use substitution to check that it is so:

$$y(t) = te^{-t}$$
$$y'(t) = e^{-t} - te^{-t}$$
$$y'(t) + y(t) = e^{-t} - te^{-t} + te^{-t}$$
$$= e^{-t}$$

which checks, and

$$y(0) = 0e^0 = 0$$

which also checks.

The initial value problem which we have just solved describes part of the following situation. Two tanks, each with constant volume V cubic meters capacity, are arranged as in Figure 1. At time $t = 0$, the first tank is full of fluid A, and the second is full of fluid B. Thereafter, fluid B flows into the first tank at constant flow rate $I = V$ cubic meters/minute, and the mixture in each tank is stirred continuously and thoroughly. From the first tank the mixture flows at the same rate into the second tank. What is the concentration of fluid A in each tank, as a function of time? See Figure 2.

As described in Section A of Chapter I, the concentration r of fluid A in the first tank satisfies the initial value problem

$$\left. \begin{aligned} r' + \frac{I}{V}r = 0 \\ r(0) = 1 \end{aligned} \right\}$$

so that if I is constant, then

$$r(t) = e^{-(I/V)t}$$

Fluid B First tank   Fluid A Second tank   Fluid B

Figure 1 Two tanks with the same capacity.

In this problem $I/V = 1$. The initial value problem can be solved by Laplace transforms quite as readily as by separation of variables:

$$\mathcal{L}[r' + r](s) = 0$$

$$s\mathcal{L}[r](s) - 1 + \mathcal{L}[r](s) = 0$$

$$(s + 1)\mathcal{L}[r](s) = 1$$

$$\mathcal{L}[r](s) = \frac{1}{s + 1}$$

$$\mathcal{L}[r](s) = \mathcal{L}[e^{-t}](s)$$

from Example 1, so that

$$r(t) = e^{-t}$$

The concentration of fluid A in the flow into the second tank is therefore $r(t) = e^{-t}$. As described in Section E of Chapter I, then, the concentration y of fluid A in the second tank satisfies the equation

$$y' + \frac{I}{V}y = r\frac{I}{V}$$

In this example, then, y satisfies the initial value problem

$$\left. \begin{array}{l} y'(t) + y(t) = e^{-t} \\ y(0) = 0 \end{array} \right\}$$

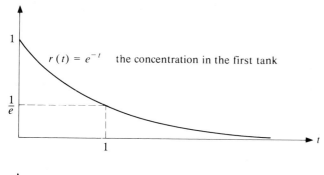

$r(t) = e^{-t}$ the concentration in the first tank

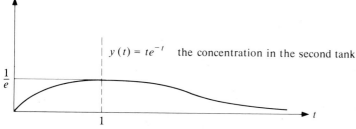

$y(t) = te^{-t}$ the concentration in the second tank

Figure 2 Concentration in the two tanks.

The solution to this problem, as shown above, is

$$y(t) = te^{-t}$$

4. Method for Solving Initial Value Problems

The example illustrates all the following steps for using Laplace transforms to solve linear differential equations with constant coefficients:

1. Taking the Laplace transform of the nonhomogeneous term (forcing function), in this case e^{-t}, and of the other side of the equation.
2. Using a theorem on Laplace transforms of derivatives to write the Laplace transform of the left-hand side of the equation in terms of a single unknown function $\mathscr{L}[y]$ and the initial values.
3. Finding a formula for $\mathscr{L}[y](s)$ in terms of s alone (just a matter of algebraic manipulation, but complicated in some problems).
4. Finding y, once $\mathscr{L}[y]$ is known (taking the inverse Laplace transform). This is often difficult, and it depends in practice on having a table of Laplace transforms.

5. The Project

Preparation for using Laplace transforms to solve initial value problems thus consists of the following parts:

1. The computation of the transforms of a few very important functions, to develop a short table of transforms
2. The development of formulas for combining these, to make the short table into a very versatile tool
3. The acquisition of skill in the decomposition of a rational function into partial fractions, which is the algebra usually needed to simplify the formula for $\mathscr{L}[y]$
4. Practice in finding inverse transforms, using the decomposition of item 3 and the table of items 1 and 2

Begin with a review of the following parts of calculus:

Improper integrals
Integration by parts
L'Hospital's rule
Decomposition into partial fractions

6. Transforms of Derivatives

The formula for the transform of a derivative, called a theorem and stated above, will be derived now, along with the corresponding formula for the Laplace transforms of higher-order derivatives.

In the statement of the theorem, both f and f' have Laplace transforms. Therefore, all the limits appearing in the following equations actually exist:

$$\mathcal{L}[f'](s) = \int_0^\infty f'(t)e^{-st}\, dt = \lim_{T \to \infty} \int_0^T f'(t)e^{-st}\, dt$$

$$= \lim_{T \to \infty} \left[f(t)e^{-st} \Big|_0^T - \int_0^T f(t)(-s)e^{-st}\, dt \right]$$

$$= \lim_{T \to \infty} f(T)e^{-sT} - f(0) + \lim_{T \to \infty} s \int_0^T f(t)e^{-st}\, dt$$

$$= -f(0) + s\mathcal{L}[f](s) \tag{7}$$

as required. The evaluation of the last integral,

$$\lim_{T \to \infty} s \int_0^T f(t)e^{-st}\, dt = s \int_0^\infty f(t)e^{-st}\, dt = s\mathcal{L}[f](s)$$

uses just the definition of Laplace transform. The other limit,

$$\lim_{T \to \infty} f(T)e^{-sT}$$

which we know exists, must equal zero, since $f(t)e^{-st}$ must become small if its integral, the transform of f, is going to be finite. That finishes the derivation.

Of course, if f' has the same properties as f, then we can apply the theorem to f'':

$$\mathcal{L}[f''](s) = s\mathcal{L}[f'](s) - f'(0)$$

$$= s\{s\mathcal{L}[f](s) - f(0)\} - f'(0)$$

$$= s^2 \mathcal{L}[f](s) - sf(0) - f'(0) \tag{8}$$

If f'' has the same properties, we can apply the theorem to f''':

$$\mathcal{L}[f'''](s) = s\mathcal{L}[f''](s) - f''(0)$$

$$= s\{s^2 \mathcal{L}[f](s) - sf(0) - f'(0)\} - f''(0)$$

$$= s^3 \mathcal{L}[f](s) - s^2 f(0) - sf'(0) - f''(0) \tag{9}$$

The same can be said of the other derivatives f may have: $f^{(4)}, f^{(5)}, \ldots, f^{(n)}, \ldots$. These facts can be written into a single formula, a generalization of the earlier formula, used in the solution of differential equations of all orders. Memorize it *now*.

Theorem (the Laplace transforms of derivatives) *If f has a Laplace transform and n continuous derivatives with Laplace transforms, and if $f, f', f'', \ldots, f^{(n-2)}$, and $f^{(n-1)}$ are all continuous at $t = 0$, then*

$$\mathcal{L}[f^{(n)}](s) = s^n \mathcal{L}[f(t)](s) - s^{n-1}f(0) - s^{n-2}f'(0) - \cdots - f^{(n-1)}(0) \tag{10}$$

Note that (7) through (9) are special cases of (10).

7. Sample Problems with Answers

For each of the following, use the definition (1) to find the Laplace transform of the function f. Then use one of the formulas for the Laplace transform of a derivative to find the transform of f' and that of f''.

A. $f(t) \equiv 1$ *Answer:* $\mathcal{L}[1](s) = 1/s$; $\mathcal{L}[0](s) \equiv 0$

B. $f(t) = e^{-3t}$

     Answer: $\mathcal{L}[e^{-3t}](s) = 1/(s+3)$; $\mathcal{L}[-3e^{-3t}](s) = -3/(s+3)$; $\mathcal{L}[9e^{-3t}](s) = 9/(s+3)$

C. $f(t) = t$ *Answer:* $\mathcal{L}[t](s) = 1/s^2$; $\mathcal{L}[1](s) = 1/s$; $\mathcal{L}[0](s) \equiv 0$

D. $f(t) = \cos bt$, where b is some number

     Answer: $\mathcal{L}[\cos bt](s) = s/(s^2 + b^2)$; $\mathcal{L}[-b \sin bt](s) = -b^2/(s^2 + b^2)$; $\mathcal{L}[-b^2 \cos bt](s) = -b^2 s/(s^2 + b^2)$

Let y be a solution to one of the following initial value problems. Find the Laplace transform of y, using the first three steps of the method given for solving initial value problems.

E. $\begin{aligned} y' + 4y &= 0 \\ y(0) &= 7 \end{aligned}$ *Answer:* $\mathcal{L}[y](s) = \dfrac{7}{s+4}$

F. $\begin{aligned} y' + 6y &= 0 \\ y(0) &= -3 \end{aligned}$ *Answer:* $\mathcal{L}[y](s) = -\dfrac{3}{s+6}$

G. $\begin{aligned} y' + 6y &= 7 \\ y(0) &= -3 \end{aligned}$ *Answer:* $\mathcal{L}[y](s) = \dfrac{7 - 3s}{s(s+6)}$

H. $\begin{aligned} 3y' - 2y &= 7 \\ y(0) &= 4 \end{aligned}$ *Answer:* $\mathcal{L}[y](s) = \dfrac{7 + 12s}{s(3s - 2)}$

I. $\begin{aligned} 4y'' + y &= 0 \\ y(0) &= 0 \\ y'(0) &= 2 \end{aligned}$ *Answer:* $\mathcal{L}[y](s) = \dfrac{8}{4s^2 + 1}$

Solve the following initial value problems.

J. $\begin{aligned} y' &= 0 \\ y(0) &= 1 \end{aligned}$ *Answer:* $y(t) = 1$

K. $\begin{aligned} y'' &= 0 \\ y(0) &= 0 \\ y'(0) &= 2 \end{aligned}$ *Answer:* $y(t) = 2t$

L. $\begin{aligned} y' + 4y &= 0 \\ y(0) &= 7 \end{aligned}$ *Answer:* $y(t) = 7e^{-4t}$

8. Problems

For each of the following, use the definition (1) to find the Laplace transform of the function f. Then use one of the formulas for the Laplace transform of a derivative to find the transform of f' and that of f''.

1. $f(t) \equiv C$, where C is some number

2. $f(t) = 1 + e^{-3t}$ **3.** $f(t) = e^{7t}$

4. $f(t) = 2 + 3t$ **5.** $f(t) = te^{-4t}$

6. $f(t) = \sin bt$, where b is some number

7. $f(t) = \sin 4t$ **8.** $f(t) = \sin(4t + 3)$

9. $f(t) = t^2$ **10.** $f(t) = t^2 - 1$

Let y be a solution to each of the following initial value problems. Find $\mathcal{L}[y]$, the Laplace transform of y, using the first three steps of the method given for solving initial value problems.

11. $\begin{aligned} y' - 3y &= 0 \\ y(0) &= 2 \end{aligned}$ **12.** $\begin{aligned} y' + 4y &= 0 \\ y(0) &= -3 \end{aligned}$

13. $\begin{aligned} 3y' + y &= 4e^{-3t} \\ y(0) &= 0 \end{aligned}$ **14.** $\begin{aligned} 3y'' + 6y' + 2y &= 0 \\ y(0) &= 4 \\ y'(0) &= 1 \end{aligned}$

Solve the following initial value problems, using only techniques discussed in this section and the factoring of polynomials.

15. $\begin{aligned} y' - 3y &= 0 \\ y(0) &= -4 \end{aligned}$ **16.** $\begin{aligned} y' + 4y &= 0 \\ y(0) &= 3 \end{aligned}$ **17.** $\begin{aligned} y' + 3y &= 6e^{-3t} \\ y(0) &= 0 \end{aligned}$

18. $\begin{aligned} y' + 3y &= -2e^{-5t} \\ y(0) &= 1 \end{aligned}$ **19.** $\begin{aligned} y'' + 8y' + 16y &= 0 \\ y(0) &= 0 \\ y'(0) &= 1 \end{aligned}$ **20.** $\begin{aligned} y'' + 4y' + 3y &= 0 \\ y(0) &= 2 \\ y'(0) &= -6 \end{aligned}$

21. $\begin{aligned} y'' + 5y' + 6y &= 0 \\ y(0) &= 3 \\ y'(0) &= -6 \end{aligned}$ **22.** $\begin{aligned} y'' + 2y' + y &= 0 \\ y(0) &= 3 \\ y'(0) &= -3 \end{aligned}$

Find a function f whose Laplace transform is given. (Compare the transforms with examples given earlier in the section.)

23. $\mathscr{L}[f](s) = \dfrac{2}{s+4}$ **24.** $\mathscr{L}[f](s) = -\dfrac{1}{s}$ **25.** $\mathscr{L}[f](s) = -\dfrac{3}{s+8}$

26. $\mathscr{L}[f](s) = \dfrac{4}{s-7}$ **27.** $\mathscr{L}[f](s) = \dfrac{1}{s} + \dfrac{1}{s+1}$ **28.** $\mathscr{L}[f](s) = -\dfrac{3}{(s+2)^2}$

29. $\mathscr{L}[f](s) = -\dfrac{4}{s^2 + 4s + 4}$ **30.** $\mathscr{L}[f](s) = \dfrac{6}{s^2 - 8s + 16}$

9. Remarks on the Validity of Laplace-Transform Formulas

There are two difficulties to be faced at once if one wants to go beyond the mechanics of using the formulas. The first is the fact that the Laplace transforms of different functions are defined on different domains. For instance,

$$\mathscr{L}[1](s) = \frac{1}{s} \qquad \text{is defined for } s > 0$$

$$\mathscr{L}[e^{-3t}](s) = \frac{1}{s+3} \qquad \text{is defined for } s > -3$$

$$\mathscr{L}[te^{6t}](s) = \frac{1}{(s-6)^2} \qquad \text{is defined for } s > 6$$

(Each of these transforms is defined on an interval which is unbounded on the right, but the intervals are different.) The second difficulty is the question of whether all the functions of interest to us have Laplace transforms.

The first step toward the resolution of such difficulties is the observation that if f is a continuous function on the right half line, $0 \le t < \infty$, and if f grows no faster than some exponential function of the form Me^{St}, that is, if

$$|f(t)| \le Me^{St} \qquad 0 \le t \tag{11}$$

then f has a Laplace transform $\mathcal{L}[f]$ defined on the interval $s > S$ by formula (1),

$$\mathcal{L}[f](s) = \int_0^\infty f(t)e^{-st}\, dt = \lim_{T \to \infty} \int_0^T f(t)e^{-st}\, dt \qquad \text{for } s > S$$

A function satisfying (11) for some value of S is said to be of *exponential order* S. The observation is that a function continuous and of exponential order S on the right half line has a Laplace transform defined on the interval $s > S$. To see that the integral is convergent (the limit exists) we compute the inequality:

$$\left| \int_0^T f(t)e^{-st}\, dt - \int_0^{T_1} f(t)e^{-st}\, dt \right| = \left| \int_T^{T_1} f(t)e^{-st}\, dt \right|$$

$$\leq \frac{M}{s - S} e^{-(s-S)T} \tag{12}$$

$$\text{if } T_1 > T \text{ and } s - S > 0$$

Specifically

$$\left| \int_T^{T_1} f(t)e^{-st}\, dt \right| \leq \int_T^{T_1} |f(t)| e^{-st}\, dt \qquad \text{for } T \leq T_1$$

$$\leq \int_T^{T_1} Me^{St} e^{-st}\, dt$$

$$= M \int_T^{T_1} e^{(S-s)t}\, dt$$

$$= \frac{M}{S-s} e^{(S-s)t} \Big|_T^{T_1}$$

$$= \frac{M}{S-s} \left(e^{(S-s)T_1} - e^{(S-s)T} \right)$$

$$= \frac{M}{s-S} \left(e^{(S-s)T} - e^{(S-s)T_1} \right)$$

$$\leq \frac{M}{s-S} e^{-(s-S)T} \qquad \text{if } s > S \text{ and } T_1 \geq T$$

Since the right-hand side of (12) goes to zero as T increases, the integrals from 0 to T and from 0 to T_1 cannot differ by much if only T and T_1 are large. Hence the integral from 0 to T actually has a limit as T increases:

$$\mathcal{L}[f](s) = \lim_{T \to \infty} \int_0^T f(t)e^{-st}\, dt \qquad \text{exists for all } s > S \tag{13}$$

That is, each function continuous and of (some) exponential order on the right half line has a Laplace transform.

The collection of all such functions therefore forms a subset of the collection of all functions having Laplace transforms. Some discontinuous functions with Laplace transforms will be presented in Section D. Functions not having exponential order may or may not have Laplace transforms, but for many purposes they may be replaced with functions which do, using ideas developed in Sections D and F. For now, therefore, we consider only continuous functions of exponential order.

When writing linear differential equations with constant coefficients, we need only two operations, differentiation and taking linear combinations. We must see how the class of continuous functions of exponential order relates to each of these operations.

A function f which is continuous need not have a derivative at any point, but if it has a derivative f' which is continuous and of exponential order S on the right half line, then f itself is necessarily also of exponential order. These are the inequalities which show that it is so:

$$f(t) = f(0) + \int_0^t f'(z)\, dz$$

$$|f(t)| \le |f(0)| + \left| \int_0^t f'(z)\, dz \right|$$

$$\le |f(0)| + \int_0^t |f'(z)|\, dz$$

$$\le |f(0)| + \int_0^t Me^{Sz}\, dz$$

If $S > 0$, then integration gives

$$|f(t)| \le |f(0)| + \left. \frac{Me^{Sz}}{S} \right|_0^t$$

$$= |f(0)| + \frac{Me^{St}}{S} - \frac{M}{S}$$

$$\le |f(0)| + \frac{M}{S} e^{St}$$

$$\le \frac{S|f(0)| + M}{S} e^{St}$$

If $S \le 0$, then $\qquad |f'(t)| \le M$

and $\qquad\qquad |f(t)| \le |f(0)| + Mt$

$$\le [|f(0)| + M]e^t$$

In either case, f is seen to be of some exponential order.

By using this fact several times, one can show that if f has an nth derivative $f^{(n)}$ which is continuous and of exponential order on the right half line, then f and its first $n - 1$ derivatives have the same property. That is, integration preserves both continuity and having exponential order, even though differentiation does not.

Taking linear combinations also preserves continuity and having exponential order. For instance,

$$4 + 2e^{-3t} \text{ is of exponential order } 0$$

In fact, 4 is of exponential order 0, and $2e^{-3t}$ is of exponential order -3, and so their sum has the maximum of 0 and -3 as exponential order. In general, if f_1 and f_2 are functions, and if a_1 and a_2 are any constants whatsoever, if, further,

$$|f_1(t)| \le M_1 e^{S_1 t} \qquad \text{and} \qquad |f_2(t)| \le M_2 e^{S_2 t} \qquad 0 \le t$$

and if $S_1 \ge S_2$, say, then

$$|a_1 f_1(t) + a_2 f_2(t)| \le |a_1| |f_1(t)| + |a_2| |f_2(t)|$$
$$\le |a_1| M_1 e^{S_1 t} + |a_2| M_2 e^{S_2 t}$$
$$\le |a_1| M_1 e^{S_1 t} + |a_2| M_2 e^{S_1 t}$$
$$= (|a_1| M_1 + |a_2| M_2) e^{S_1 t}$$

If a_1 and a_2 are constants, then $a_1 f_1 + a_2 f_2$ is a *linear combination*. Thus we have shown that a linear combination of functions continuous and of exponential order on the right half line also has a Laplace transform, defined at least on an unbounded interval, in this case the interval $s > S_1$. Linear combinations of three or more functions can be treated in the same way. However, as was noted earlier, when solving differential equations using Laplace transforms, one uses the formula for the transform and rarely needs to pay any attention to its interval of definition.

B. LAPLACE-TRANSFORM PROPERTIES AND A SHORT TABLE OF LAPLACE TRANSFORMS

From now on, the sine and cosine will be as important as the exponential.

1. Example: $\mathscr{L}[\cos bt]$ and $\mathscr{L}[\sin bt]$

$$\mathscr{L}[\cos bt](s) = \int_0^\infty (\cos bt) e^{-st} \, dt = \lim_{T \to \infty} \int_0^T (\cos bt) e^{-st} \, dt \tag{1}$$

The integration begins with two integrations by parts.

$$\int_0^T (\cos bt)e^{-st}\, dt = \frac{(\cos bt)e^{-st}}{-s}\bigg|_0^T - \int_0^T \frac{-b}{-s}(\sin bt)e^{-st}\, dt$$

$$= \frac{(\cos bt)e^{-st}}{-s}\bigg|_0^T - \left[\frac{b \sin bt}{s} \frac{e^{-st}}{-s}\bigg|_0^T - \int_0^T \frac{b^2}{-s^2}(\cos bt)e^{-st}\, dt \right]$$

Transposing the remaining integral from right to left shows that

$$\left(1 + \frac{b^2}{s^2}\right)\int_0^T (\cos bt)e^{-st}\, dt = \frac{(\cos bt)e^{-st}}{-s}\bigg|_0^T + \frac{b(\sin bt)e^{-st}}{s^2}\bigg|_0^T$$

$$= \left(\frac{\cos bT}{-s} + \frac{b \sin bT}{s^2}\right)e^{-sT} + \frac{1}{s}$$

or

$$\int_0^T (\cos bt)e^{-st}\, dt = \frac{s^2}{s^2 + b^2}\left(\frac{\cos bT}{-s} + \frac{b \sin bT}{s^2}\right)e^{-sT} + \frac{s}{s^2 + b^2}$$

For any $s > 0$, e^{-sT} goes to zero as T increases unboundedly, while the sine and cosine remain bounded. Therefore, substitution of this formula into (1) shows that

$$\mathscr{L}[\cos bt](s) = \frac{s}{s^2 + b^2}$$

Using the formula for the Laplace transform of the derivative from Section A, one finds that

$$\mathscr{L}[\sin bt](s) = \mathscr{L}\left[\frac{d}{dt}\left(-\frac{1}{b}\cos bt\right)\right](s)$$

$$= s\mathscr{L}\left[-\frac{1}{b}\cos bt\right](s) - \left(-\frac{1}{b}\cos 0\right)$$

$$= s\left(-\frac{1}{b}\right)\mathscr{L}[\cos bt](s) + \frac{1}{b} \qquad (2)$$

$$= s\left(-\frac{1}{b}\right)\frac{s}{s^2 + b^2} + \frac{1}{b}$$

$$= \frac{1}{b}\left(\frac{-s^2}{s^2 + b^2} + 1\right)$$

$$= \frac{1}{b}\frac{-s^2 + s^2 + b^2}{s^2 + b^2}$$

$$= \frac{b}{s^2 + b^2}$$

Equality (2) is shown by the following argument:

$$\mathscr{L}\left[-\frac{1}{b}\cos bt\right](s) = \int_0^\infty -\frac{1}{b}(\cos bt)e^{-st}\,dt$$

$$= -\frac{1}{b}\int_0^\infty (\cos bt)e^{-st}\,dt = -\frac{1}{b}\mathscr{L}[\cos bt](s)$$

2. The Property of Linearity

The argument just used exposes part of a very important property of Laplace transformation. If f_1 and f_2 are functions with Laplace transforms, and if a_1 and a_2 are any constants whatsoever, then

$$\mathscr{L}[a_1 f_1 + a_2 f_2](s) = \int_0^\infty [a_1 f_1(t) + a_2 f_2(t)]e^{-st}\,dt$$

$$= \int_0^\infty [a_1 f_1(t)e^{-st} + a_2 f_2(t)e^{-st}]\,dt$$

$$= \int_0^\infty a_1 f_1(t)e^{-st}\,dt + \int_0^\infty a_2 f_2(t)e^{-st}\,dt$$

$$= a_1 \int_0^\infty f_1(t)e^{-st}\,dt + a_2 \int_0^\infty f_2(t)e^{-st}\,dt$$

$$= a_1 \mathscr{L}[f_1](s) + a_2 \mathscr{L}[f_2](s)$$

That is, if a_1 and a_2 are any constants, then

$$\mathscr{L}[a_1 f_1 + a_2 f_2] = a_1 \mathscr{L}[f_1] + a_2 \mathscr{L}[f_2] \tag{3}$$

Special cases include

$$\mathscr{L}\left[-\frac{1}{b}\cos bt\right] = -\frac{1}{b}\mathscr{L}[\cos bt]$$

from Equation (2) above, and

$$\mathscr{L}[y' + y] = \mathscr{L}[y'] + \mathscr{L}[y]$$

from Equation (5) of Section A. The general case (3) is called the *linearity* of Laplace transformation. It may be remembered simply in the words, "The Laplace transform of a linear combination is the linear combination of the Laplace transforms," for the left-hand side of (3) is the Laplace transformation of $a_1 f_1 + a_2 f_2$, a linear combination of f_1 and f_2, while the right-hand side is the same linear combination but of $\mathscr{L}[f_1]$ and $\mathscr{L}[f_2]$ instead of f_1 and f_2.

3. Example: Solving a Second-Order Equation

According to Section B of Chapter II, the system illustrated in Figure 1, with $M = 1$ and $K = b^2$, with initial displacement $y(0) = 1$ and initial velocity $y'(0) = 0$, may be modeled by the initial value problem

$$\left. \begin{array}{r} y'' + b^2 y = 0 \\ y(0) = 1 \\ y'(0) = 0 \end{array} \right\} \tag{4}$$

whose solution is now well within our abilities. Take the Laplace transform of both sides of the differential equation in (4), observing that the transform of the right-hand side is zero and using linearity on the left-hand side:

$$\mathscr{L}[y''](s) + b^2 \mathscr{L}[y](s) = 0 \tag{5}$$

From the formula for the Laplace transform of a second derivative, it follows that

$$\begin{aligned} \mathscr{L}[y''](s) &= s^2 \mathscr{L}[y](s) - sy(0) - y'(0) \\ &= s^2 \mathscr{L}[y](s) - s \end{aligned} \tag{6}$$

The initial conditions from (4) have been substituted into (6), and so when (6) is in turn substituted into (5), the resulting equation can properly be called *the transformed problem*. It is this:

$$s^2 \mathscr{L}[y](s) - s + b^2 \mathscr{L}[y](s) = 0 \tag{7}$$

The transformed problem is easy to solve for $\mathscr{L}[y](s)$ as follows:

$$(s^2 + b^2)\mathscr{L}[y](s) = s$$

$$\mathscr{L}[y](s) = \frac{s}{s^2 + b^2} \tag{8}$$

Since the right-hand side of (8) is the Laplace transform of $\cos bt$, we guess that $y(t) = \cos bt$ is the solution of the initial value problem (4). The guess should be checked by substitution.

$$y''(t) + b^2 y(t) = -b^2 \cos bt + b^2 \cos bt = 0$$

$$y(0) = \cos 0 = 1$$

$$y'(t) = -b \sin bt$$

$$y'(0) = -b \sin 0 = 0$$

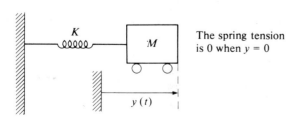

The spring tension is 0 when $y = 0$

Figure 1 Spring-mass system.

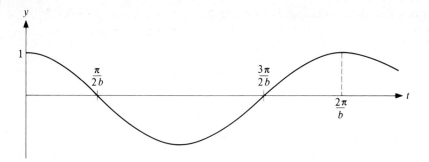

Figure 2 Graph of the function $y(t) = \cos bt$.

Thus, as might be expected, the model predicts that a mass attached to a spring will oscillate sinusoidally. (See Figure 2.)

A similar initial value problem is the model for the system illustrated in Figure 3. If $C = 1$ and $L = 1/b^2$, if the initial voltage is $v(0) = 1$, and if the initial current is $i(0) = 0$, then $v'(0) = (1/C)i(0) = 0$, and

$$
\left.
\begin{aligned}
v'' + b^2 v &= 0 \\
v(0) &= 1 \\
v'(0) &= 0
\end{aligned}
\right\}
$$

This problem is the same as (4), and so solution by Laplace transform (as before) gives $v(t) = \cos bt$. However, if the initial conditions are changed to $v(0) = 2$ and $i(0) = 1$, then $v'(0) = (1/C)i(0) = 1$, and

$$
\left.
\begin{aligned}
v'' + b^2 v &= 0 \\
v(0) &= 2 \\
v'(0) &= 1
\end{aligned}
\right\}
\tag{9}
$$

The transformed problem becomes this:

$$\mathscr{L}[v''](s) + b^2 \mathscr{L}[v](s) = 0$$
$$s^2 \mathscr{L}[v](s) - sv(0) - v'(0) + b^2 \mathscr{L}[v](s) = 0$$
$$s^2 \mathscr{L}[v](s) - 2s - 1 + b^2 \mathscr{L}[v](s) = 0$$

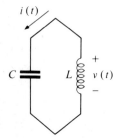

Figure 3 LC circuit.

Linearity gives the first equation, the formula for the transforms of derivatives the second, and the initial values the third. The transformed problem is solved for $\mathscr{L}[v](s)$ by this series of algebraic steps:

$$(s^2 + b^2)\mathscr{L}[v](s) = 2s + 1$$

$$\mathscr{L}[v](s) = \frac{2s + 1}{s^2 + b^2}$$

The last expression should be seen as a sum of two terms:

$$\frac{2s + 1}{s^2 + b^2} = \frac{2s}{s^2 + b^2} + \frac{1}{s^2 + b^2} \tag{10}$$

The first term is clearly 2 times the Laplace transform of the cosine; the second is, in fact, $1/b$ times the Laplace transform of the sine. Thus,

$$\mathscr{L}[v](s) = \frac{2s}{s^2 + b^2} + \frac{1}{s^2 + b^2}$$

$$= 2\frac{s}{s^2 + b^2} + \frac{1}{b}\frac{b}{s^2 + b^2} \tag{11}$$

$$= 2\mathscr{L}[\cos bt](s) + \frac{1}{b}\mathscr{L}[\sin bt](s)$$

$$= \mathscr{L}\left[2\cos bt + \frac{1}{b}\sin bt\right](s)$$

the last equality being an instance of the linearity property (3). Now we should check the possibility that

$$2\cos bt + \frac{1}{b}\sin bt$$

is a solution of the initial value problem (9). Substituting into (9) gives

$$v''(t) + b^2 v(t) = \left(-2b^2\cos bt - b^2\frac{1}{b}\sin bt\right) + b^2\left(2\cos bt + \frac{1}{b}\sin bt\right) = 0$$

$$v(0) = 2\cos 0 + \frac{1}{b}\sin 0 = 2$$

$$v'(t) = -2b\sin bt + \frac{b}{b}\cos bt$$

$$v'(0) = -2b\sin 0 + \cos 0 = 1$$

Thus, $$v(t) = 2\cos bt + \frac{1}{b}\sin bt$$

is a solution of (9). It is valid on the whole real line, $-\infty < t < \infty$.

The steps from (10) to (11) deserve attention, for they appear very often. Their purpose is to identify the Laplace transforms of the cosine and the sine. The denominator is seen to be a *sum of squares*, and the computation begins with a separation of the numerator into a multiple of s plus a multiple of b. The multiple of s is read off immediately, as in (11); the second term is identified most readily if its numerator is multiplied by b while the fraction is multiplied by $1/b$. Be sure to check, though, that (11) is actually equal to the right-hand side of the previous equation.

4. The Inverse Laplace Transform

Hitherto, each time we have used Laplace transforms to solve a differential equation, we have come at the end to an observation that the unknown function has the same Laplace transform as some known function, followed with a "guess" that the known function is a solution of the problem. Of course, any such "guess" must be properly checked to see whether it satisfies the differential equation and the initial conditions. However, there is an important result called *Lerch's theorem* which obviates the guessing and allows one to *assert* that the known function is a solution of the problem. In a simple form, the theorem is this:

Among all the functions which are continuous and of exponential order on the half line $t \geq 0$, there is only one whose Laplace transform is identically zero, and that is the function which is itself identically zero.

It follows that if v and w are continuous and of exponential order on the right half line and if they have the same Laplace transform, then the computation

$$0 = \mathcal{L}[v] - \mathcal{L}[w] = \mathcal{L}[v - w]$$

leads to an application of Lerch's theorem to the function $v - w$. Specifically, the theorem implies that

$$v(t) - w(t) = 0 \quad \text{or} \quad v(t) = w(t) \quad \text{for all } t \geq 0 \tag{12}$$

By virtue of this fact, we can define the inverse Laplace transformation $\mathcal{L}^{-1}$ as follows: If f is a function defined on the right half line and nowhere else, is continuous there, and is of exponential order, and if

$$F = \mathcal{L}[f]$$

then the *inverse Laplace transform* of F is the function

$$\mathcal{L}^{-1}[F] = f$$

(defined on the right half line).

For example,

$$\mathcal{L}^{-1}\left[\frac{1}{s^2 + b^2}\right](t) = \frac{1}{b}\sin bt \qquad t \geq 0$$

$$\mathcal{L}^{-1}\left[\frac{1}{s + a}\right](t) = e^{-at} \qquad t \geq 0$$

$$\mathcal{L}^{-1}\left[\frac{1}{s^2}\right](t) = t \qquad t \geq 0$$

Note carefully that $\mathcal{L}^{-1}$ gives only functions defined on the right half line. Here is an illustration of that fact. Let $v(t) = t$ for all t, and let

$$w(t) = |t| = \begin{cases} t & t \geq 0 \\ -t & t < 0 \end{cases}$$

so that

$$w(t) = -t = -v(t) \qquad t < 0 \qquad (13)$$

but

$$w(t) = t = v(t) \qquad t \geq 0 \qquad (14)$$

From (14) it follows that w and v have the same Laplace transform, to wit,

$$\mathcal{L}[w](s) = \mathcal{L}[v](s) = \int_0^\infty te^{-st}\, dt = \frac{1}{s^2} \qquad s > 0$$

for the integral is taken over an interval where w and v agree. On the other hand, (13) shows that w and v do not agree everywhere; in fact, they are negatives of each other on the left half line. Thus, it certainly would be unreasonable to take either as *the* inverse transform of $1/s^2$. We take the restriction to the right half line.

Here is an example to illustrate the use of the inverse Laplace transform in solving an initial value problem: Find y such that

$$\left. \begin{array}{r} y' + 4y = 7e^{3t} \\ y(0) = 1 \end{array} \right\}$$

This is the transformed problem, followed by the steps for finding $\mathcal{L}[y]$:

$$s\mathcal{L}[y](s) - 1 + 4\mathcal{L}[y](s) = \frac{7}{s - 3}$$

$$(s + 4)\mathcal{L}[y](s) = \frac{7}{s - 3} + 1 = \frac{s + 4}{s - 3}$$

$$\mathcal{L}[y](s) = \frac{s + 4}{(s + 4)(s - 3)} = \frac{1}{s - 3}$$

Now the inverse transform may be taken on both sides of the equation:

$$y(t) = \mathcal{L}^{-1}\left[\frac{1}{s - 3}\right](t) = e^{3t} \qquad t \geq 0$$

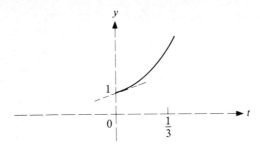

Figure 4 Graph of the function $y(t) = e^{3t}$ for $t \geq 0$.

Actually, there is more information to be obtained. In the process of checking that e^{3t} satisfies the differential equation for $t \geq 0$, one sees that it satisfies it also for $t < 0$, and that it is differentiable on the whole real line:

$$\frac{d}{dt}(e^{3t}) + 4e^{3t} = 3e^{3t} + 4e^{3t} = 7e^{3t} \qquad \text{for all } t$$

Thus e^{3t} is a solution of the initial value problem valid on the whole line. Substitution shows this, *but the inverse Laplace transform does not show it.*

5. Sample Problems with Answers

Find the inverse Laplace transform of each of the following functions.

A. $F(s) = \dfrac{1}{s}$ *Answer:* $\mathscr{L}^{-1}\left[\dfrac{1}{s}\right](t) = 1, t \geq 0$

B. $F(s) = \dfrac{1}{(s-3)^2}$ *Answer:* $\mathscr{L}^{-1}\left[\dfrac{1}{(s-3)^2}\right](t) = te^{3t}, t \geq 0$

C. $F(s) = \dfrac{2}{s^2+9}$ *Answer:* $\mathscr{L}^{-1}\left[\dfrac{2}{s^2+9}\right](t) = \frac{2}{3}\sin 3t, t \geq 0$

D. $F(s) = \dfrac{4s}{s^2+7}$ *Answer:* $\mathscr{L}^{-1}\left[\dfrac{4s}{s^2+7}\right](t) = 4\cos\sqrt{7}\,t, t \geq 0$

E. $F(s) = \dfrac{4s+2}{s^2+4}$

    Answer: $\mathscr{L}^{-1}\left[\dfrac{4s+2}{s^2+4}\right](t) = 4\cos 2t + \sin 2t, t \geq 0$

F. $F(s) = \dfrac{4s+3}{2s^2+4}$

    Answer: $\mathscr{L}^{-1}\left[\dfrac{4s+3}{2s^2+4}\right](t) = 2\cos\sqrt{2}\,t + \dfrac{3}{2\sqrt{2}}\sin\sqrt{2}\,t, t \geq 0$

G. $F(s) = \dfrac{4s+3}{s^2+7}$

    Answer: $\mathscr{L}^{-1}\left[\dfrac{4s+3}{s^2+7}\right](t) = 4\cos\sqrt{7}\,t + \dfrac{3}{\sqrt{7}}\sin\sqrt{7}\,t, t \geq 0$

Solve, using the inverse Laplace transform and finding a solution valid on the right half line.

H.
$$\left.\begin{array}{l} y'' + 4y = 0 \\ y(0) = 0 \\ y'(0) = 1 \end{array}\right\}$$
Answer: $y(t) = \frac{1}{2} \sin 2t, \ t \geq 0$

I.
$$\left.\begin{array}{l} y'' + 4y = 0 \\ y(0) = 3 \\ y'(0) = 1 \end{array}\right\}$$
Answer: $y(t) = 3 \cos 2t + \frac{1}{2} \sin 2t, \ t \geq 0$

J.
$$\left.\begin{array}{l} 2y'' + y = 0 \\ y(0) = 3 \\ y'(0) = -2 \end{array}\right\}$$
Answer: $y(t) = 3 \cos \dfrac{1}{\sqrt{2}} t - 2\sqrt{2} \sin \dfrac{1}{\sqrt{2}} t, \ t \geq 0$

K.
$$\left.\begin{array}{l} 6y' + y = 0 \\ y(0) = -4 \end{array}\right\}$$
Answer: $y(t) = -4e^{-(t/6)}, \ t \geq 0$

6. Problems

Find the inverse Laplace transform for each of the following functions.

1. $F(s) = -\dfrac{3}{s}$ **2.** $F(s) = \dfrac{1}{s^2}$ **3.** $F(s) = \dfrac{2s}{s^2 + 3}$

4. $F(s) = \dfrac{3s}{2s^2 + 6}$ **5.** $F(s) = \dfrac{-2s}{3s^2 + 4}$ **6.** $F(s) = \dfrac{4}{s^2 + 3}$

7. $F(s) = \dfrac{1}{s^2 + 8}$ **8.** $F(s) = \dfrac{s + 3}{s^2 + 9}$ **9.** $F(s) = \dfrac{-s + 1}{s^2 + 8}$

10. $F(s) = \dfrac{3s - 7}{2s^2 + 6}$ **11.** $F(s) = \dfrac{14}{3s^2 + 2}$ **12.** $F(s) = \dfrac{4s + 4}{3s^2 + 2}$

Solve the following initial value problems, using the inverse Laplace transform and finding solutions valid on the right half line.

13. (a) $\left.\begin{array}{l} y'' + 3y = 0 \\ y(0) = 1 \\ y'(0) = 0 \end{array}\right\}$ (b) $\left.\begin{array}{l} y'' + 3y = 0 \\ y(0) = 0 \\ y'(0) = 0 \end{array}\right\}$ (c) $\left.\begin{array}{l} y'' + 3y = 0 \\ y(0) = 1 \\ y'(0) = 1 \end{array}\right\}$

14. (a) $\left.\begin{array}{l} y'' + 4y = 0 \\ y(0) = 3 \\ y'(0) = 0 \end{array}\right\}$ (b) $\left.\begin{array}{l} y'' + 4y = 0 \\ y(0) = 0 \\ y'(0) = \sqrt{2} \end{array}\right\}$ (c) $\left.\begin{array}{l} y'' + 4y = 0 \\ y(0) = -2 \\ y'(0) = 4 \end{array}\right\}$

15. $\left.\begin{array}{l} y'' + 8y = 0 \\ y(0) = 1 \\ y'(0) = 3 \end{array}\right\}$ **16.** $\left.\begin{array}{l} 6y'' + 7y = 0 \\ y(0) = 4 \\ y'(0) = 1 \end{array}\right\}$ **17.** $\left.\begin{array}{l} 3y' + 5y = 0 \\ y(0) = 2 \end{array}\right\}$

18. $\left.\begin{array}{l} 3y' + 5y = 0 \\ y(0) = 0 \end{array}\right\}$ **19.** $\left.\begin{array}{l} 2y' + y = e^{-t/2} \\ y(0) = 0 \end{array}\right\}$

20. $\left.\begin{array}{l} y'' + 12y = 0 \\ y(0) = \sqrt{3} \\ y'(0) = 2 \end{array}\right\}$ **21.** $\left.\begin{array}{l} y'' + 9y = 0 \\ y(0) = y_0 \\ y'(0) = y_1 \end{array}\right\}$ where y_0 and y_1 are arbitrary

7. The First Shifting Property

The property is this: If f is a function having a Laplace transform, then $e^{at}f$ also has a Laplace transform, and

$$\mathcal{L}[e^{at}f](s) = \mathcal{L}[f](s - a) \tag{15}$$

The derivation consists of the following application of the definition of Laplace transforms:

$$\mathcal{L}[e^{at}f](s) = \int_0^\infty e^{at}f(t)e^{-st}\,dt$$

$$= \int_0^\infty f(t)e^{(-s+a)t}\,dt$$

$$= \int_0^\infty f(t)e^{-(s-a)t}\,dt$$

$$= \mathcal{L}[f(t)](s - a)$$

Notice that the property is consistant with the examples from Section A. For instance,

$$\mathcal{L}[1](s) = \frac{1}{s}$$

and so

$$\mathcal{L}[1](s + 3) = \frac{1}{s + 3}$$

$$= \mathcal{L}[e^{-3t}](s)$$

or

$$\mathcal{L}[1](s + 3) = \mathcal{L}[e^{-3t} \cdot 1](s)$$

which is the special case of (15) for which $f(t) \equiv 1$.

Notice also that the first shifting property gives transforms of some more functions:

$$\mathcal{L}[e^{at} \sin bt](s) = \mathcal{L}[\sin bt](s - a)$$

$$= \frac{b}{(s - a)^2 + b^2}$$

$$\mathcal{L}[e^{at} \cos bt](s) = \mathcal{L}[\cos bt](s - a)$$

$$= \frac{s - a}{(s - a)^2 + b^2}$$

For instance, in case $a = -1$ and $b = 3$,

$$\mathcal{L}[e^{-t} \sin 3t](s) = \frac{3}{(s + 1)^2 + 3^2}$$

$$= \frac{3}{s^2 + 2s + 10}$$

Recall that $\mathscr{L}[f]$ is the name of a function, and that $\mathscr{L}[f](s)$ is its value at s. Thus,

$$\left.\begin{aligned}
\mathscr{L}[\cos 3t](s) &= \frac{s}{s^2 + 3^2} = \frac{s}{s^2 + 9} \\[2mm]
\mathscr{L}[\cos 3t](s + 1) &= \frac{s + 1}{(s + 1)^2 + 3^2} = \frac{s + 1}{s^2 + 2s + 10} \\[2mm]
\mathscr{L}[\cos 3t](s - a) &= \frac{s - a}{(s - a)^2 + 3^2} = \frac{s - a}{s^2 - 2as + a^2 + 9}
\end{aligned}\right\} \qquad (16)$$

The right-hand sides of the equations in (16) are evaluations of one function at three points, s, $s - 1$, and $s - a$. This fact is made plain on the left-hand sides of the equations. On the other hand, the first shifting property turns these three equations into formulas for the Laplace transforms of three different functions:

$$\mathscr{L}[\cos 3t](s) = \frac{s}{s^2 + 3^2} = \frac{s}{s^2 + 9}$$

$$\mathscr{L}[e^{-t} \cos 3t](s) = \frac{s + 1}{(s + 1)^2 + 3^2} = \frac{s + 1}{s^2 + 2s + 10}$$

$$\mathscr{L}[e^{at} \cos 3t](s) = \frac{s - a}{(s - a)^2 + 3^2} = \frac{s - a}{s^2 - 2as + a^2 + 9}$$

8. Example: A Problem Whose Solution Is a Damped Sinusoid

The function $e^{-t} \cos 3t$ which appeared above is called a *damped sinusoid*. In general, the damped sinusoids are of the form

$$A_0 e^{at} \sin (bt + \phi) = A_0 e^{at}(\cos bt \sin \phi + \sin bt \cos \phi)$$
$$= e^{at}(A_1 \cos bt + A_2 \sin bt) \qquad (17)$$

where A_0 and ϕ (or A_1 and A_2) may be arbitrary constants. (Section I of Chapter I contains some graphs of such functions.) The corresponding Laplace transforms occur frequently and must be handled routinely, for engineering systems described by such functions are exceedingly common. Figure 5 illustrates such a system.

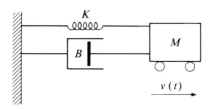

Figure 5 Damped mass on a spring.

Proceeding as in Section C of Chapter II, we may derive the differential equation which models the system, which is

$$Mv'' + Bv' + Kv = 0$$

where v is the velocity of the mass. If $M = 1$, $B = 2$, and $K = 10$, and if $v(0) = 0$ and $v'(0) = 1$, then

$$\left. \begin{array}{r} v'' + 2v' + 10v = 0 \\ v(0) = 0 \\ v'(0) = 1 \end{array} \right\}$$

The transformed problem is

$$s^2 \mathcal{L}[v](s) - sv(0) - v'(0) + 2s\mathcal{L}[v](s) - 2v(0) + 10\mathcal{L}[v](s) = 0$$

or

$$s^2 \mathcal{L}[v](s) + 2s\mathcal{L}[v](s) + 10\mathcal{L}[v](s) = 1$$

$$(s^2 + 2s + 10)\mathcal{L}[v](s) = 1$$

$$\mathcal{L}[v](s) = \frac{1}{s^2 + 2s + 10}$$

To identify the last expression as the Laplace transform of a damped sinusoid, start by completing the square in the denominator:

$$\frac{1}{s^2 + 2s + 10} = \frac{1}{s^2 + 2s + 1 + 10 - 1}$$

$$= \frac{1}{(s^2 + 2s + 1) + 9} \tag{18}$$

$$= \frac{1}{(s + 1)^2 + 3^2}$$

$$= \frac{1}{3} \frac{3}{(s + 1)^2 + 3^2}$$

$$= \tfrac{1}{3}\mathcal{L}[e^{-t} \sin 3t](s)$$

Taking the inverse transform gives

$$v(t) = \tfrac{1}{3}e^{-t} \sin 3t \qquad t \geq 0$$

9. Recognizing the Laplace Transforms of Damped Sinusoids

Since all damped sinusoids are of the form

$$e^{at}(A_1 \cos bt + A_2 \sin bt)$$

that is,

$$A_1 e^{at} \cos bt + A_2 e^{at} \sin bt$$

it follows that all damped sinusoids have Laplace transforms of the form

$$\frac{A_1(s-a)+A_2 b}{(s-a)^2+b^2}$$

in which the numerator is a polynomial of degree 1, and the denominator is of degree 2. The present task is to find out whether a given function of the form

$$\frac{Ds+E}{As^2+Bs+C} \tag{19}$$

is the Laplace transform of a damped sinusoid, and if so, which. The answers are largely determined by the properties of the polynomial As^2+Bs+C; finding the answers begins with factoring out the coefficient of s^2 and completing the square. For instance,

$$\frac{7s+4}{4s^2+4s+9} = \frac{1}{4}\frac{7s+4}{s^2+s+\frac{9}{4}} \tag{20}$$

$$= \frac{7s/4+1}{s^2+s+\frac{9}{4}}$$

$$= \frac{7s/4+1}{[s^2+s+(\frac{1}{2})^2]+[\frac{9}{4}-(\frac{1}{2})^2]}$$

$$= \frac{7s/4+1}{(s+\frac{1}{2})^2+2}$$

$$= \frac{7s/4+1}{(s+\frac{1}{2})^2+\sqrt{2}^2} \tag{21}$$

In the first equation the coefficient of s^2 is factored out of the denominator, and in the second it is divided into the numerator, so that it will not be forgotten in the sequel. In the third equation the square is completed in the denominator by adding and subtracting the square of half the coefficient of s. In the last two equations the denominator is written as a sum of squares.

The next step is identifying the transform of $e^{-t/2}\cos\sqrt{2}t$. The term $\sqrt{2}^2$ and the term $(s+\frac{1}{2})^2$ determine the coefficient of t in the exponential and the cosine, respectively. Since

$$\mathscr{L}[e^{-t/2}\cos\sqrt{2}t](s) = \frac{s+\frac{1}{2}}{(s+\frac{1}{2})^2+\sqrt{2}^2}$$

we add enough to the numerator of (21) to make a multiple of $s+\frac{1}{2}$, and subtract the same amount so as not to change the value of the numerator:

$$\frac{7s/4+1}{(s+\frac{1}{2})^2+\sqrt{2}^2} = \frac{(\frac{7}{4})(s+\frac{1}{2})-(\frac{7}{4})(\frac{1}{2})+1}{(s+\frac{1}{2})^2+\sqrt{2}^2}$$

$$= \tfrac{7}{4}\mathscr{L}[e^{-t/2}\cos\sqrt{2}t](s) + \frac{-\frac{7}{8}+1}{(s+\frac{1}{2})^2+\sqrt{2}^2}$$

$$= \mathscr{L}[\tfrac{7}{4}e^{-t/2}\cos\sqrt{2}t](s) + \frac{1}{8}\frac{1}{(s+\frac{1}{2})^2+\sqrt{2}^2} \tag{22}$$

The last step is identifying the Laplace transform of $e^{-t/2} \sin \sqrt{2}t$. Since

$$\mathcal{L}[e^{-t/2} \sin \sqrt{2}t](s) = \frac{\sqrt{2}}{(s + \frac{1}{2})^2 + \sqrt{2}^2}$$

the last term of (22) should be multiplied and divided by $\sqrt{2}$, as in Example 8. Thus, the complete computation is this:

$$\frac{7s + 4}{4s^2 + 4s + 9} = \frac{7s/4 + 1}{s^2 + s + \frac{9}{4}}$$

$$= \frac{7s/4 + 1}{(s + \frac{1}{2})^2 + 2}$$

$$= \frac{7s/4 + 1}{(s + \frac{1}{2})^2 + \sqrt{2}^2}$$

$$= \frac{(\frac{7}{4})(s + \frac{1}{2}) + 1 - \frac{7}{8}}{(s + \frac{1}{2})^2 + \sqrt{2}^2}$$

$$= \mathcal{L}[\tfrac{7}{4}e^{-t/2} \cos \sqrt{2}t](s) + \frac{1}{8} \frac{1}{(s + \frac{1}{2})^2 + \sqrt{2}^2}$$

$$= \mathcal{L}[\tfrac{7}{4}e^{-t/2} \cos \sqrt{2}t](s) + \frac{1}{8\sqrt{2}} \frac{\sqrt{2}}{(s + \frac{1}{2})^2 + \sqrt{2}^2}$$

$$= \mathcal{L}\left[\tfrac{7}{4}e^{-t/2} \cos \sqrt{2}t + \frac{1}{8\sqrt{2}}e^{-t/2} \sin \sqrt{2}t\right](s)$$

The result may also be written using the inverse Laplace transform:

$$\mathcal{L}^{-1}\left[\frac{7s + 4}{4s^2 + 4s + 9}\right](t) = e^{-t/2}\left(\tfrac{7}{4} \cos \sqrt{2}t + \frac{1}{8\sqrt{2}}\sin \sqrt{2}t\right) \qquad t \geq 0$$

The technique illustrated above should be mastered *now*. It will be needed.

The damped sinusoids are characterized by having Laplace transforms whose denominators can be written as the sums of squares, as above. The polynomials which can be so written are those quadratics $As^2 + Bs + C$ which have no real roots, those such that $B^2 - 4AC$ is negative. In the present example,

$$As^2 + Bs + C = 4s^2 + 4s + 9$$

so that
$$B^2 - 4AC = 16 - 144 < 0$$

A quadratic with real equal roots will have $B^2 - 4AC = 0$. Completion of the square is unnecessary in this case, for the polynomial is already a square. For example,

$$4s^2 + 4s + 1 = 4(s^2 + s + \tfrac{1}{4})$$

$$= 4(s + \tfrac{1}{2})^2$$

A quadratic with real unequal roots will have $B^2 - 4AC$ positive. Completion of the square yields a difference of squares, not a sum, in this case. For example,

$$4s^2 + 4s - 6 = 4(s^2 + s - \tfrac{3}{2})$$

$$= 4[s^2 + s + (\tfrac{1}{2})^2 - \tfrac{3}{2} - (\tfrac{1}{2})^2]$$

$$= 4[(s + \tfrac{1}{2})^2 - \tfrac{7}{4}]$$

$$= 4\left[(s + \tfrac{1}{2})^2 - \left(\frac{\sqrt{7}}{2}\right)^2\right]$$

If the denominator is a quadratic with real roots, equal or unequal, then the function is not the Laplace transform of a damped sinusoid. Finding the inverse transform for that case will be treated in Section C, in the discussion of the partial-fraction decomposition.

10. Laplace Transforms of Powers of t

Examples 1 and 2 in the previous section have already provided the Laplace transforms of two powers of t, namely

$$\mathscr{L}[t^0](s) = \mathscr{L}[e^{0t}](s) = \frac{1}{s}$$

$$\mathscr{L}[t](s) = \mathscr{L}[te^{0t}](s) = \frac{1}{s^2}$$

There is also a formula for the transform of t^n if n is a positive integer:

$$\mathscr{L}[t^n](s) = \frac{n!}{s^{n+1}} \qquad \text{if } n \geq 0 \tag{23}$$

The formula is derived by integrating by parts, as follows:

$$\mathscr{L}[t^n](s) = \int_0^\infty t^n e^{-st}\, dt$$

$$= \lim_{T \to \infty} \frac{t^n e^{-st}}{-s}\bigg|_0^T + \frac{n}{s}\int_0^\infty t^{n-1} e^{-st}\, dt$$

$$= \frac{n}{s}\, \mathscr{L}[t^{n-1}](s)$$

$$= \frac{n}{s}\frac{(n-1)}{s}\, \mathscr{L}[t^{n-2}](s)$$

$$= \frac{n \cdot (n-1) \cdots 2 \cdot 1}{s \cdot s \cdots s \cdot s}\, \mathscr{L}[t^0](s)$$

$$= \frac{n!}{s^{n+1}} \qquad \text{for } s > 0$$

Thus, for example,

$$\mathcal{L}[t^2](s) = \frac{2!}{s^{2+1}} = \frac{2}{s^3} \qquad s > 0$$

and

$$\mathcal{L}[t^3](s) = \frac{3!}{s^{3+1}} = \frac{6}{s^4} \qquad s > 0$$

Applying the first shifting property gives the following formula as well:

$$\mathcal{L}[t^n e^{at}](s) = \mathcal{L}[t^n](s - a)$$

$$= \frac{n!}{(s-a)^{n+1}} \qquad s > a \tag{24}$$

Here is an example of taking an inverse transform:

$$\mathcal{L}^{-1}\left[\frac{3s+2}{(s-1)^5}\right](t) = \mathcal{L}^{-1}\left[\frac{3(s-1)+3+2}{(s-1)^5}\right](t)$$

$$= \mathcal{L}^{-1}\left[\frac{3(s-1)}{(s-1)^5} + \frac{5}{(s-1)^5}\right](t)$$

$$= 3\mathcal{L}^{-1}\left[\frac{1}{(s-1)^4}\right](t) + 5\mathcal{L}^{-1}\left[\frac{1}{(s-1)^5}\right](t)$$

$$= \frac{3}{3!}t^3 e^t + \frac{5}{4!}t^4 e^t \qquad t \geq 0$$

11. Laplace Transforms of Integrals

At the end of the previous section, it was shown that if f is a continuous function of exponential order, then any antiderivative g is also of exponential order. It follows that g has a Laplace transform, and we can find it using the formula for the Laplace transform of a derivative. That is, $g' = f$ and

$$\mathcal{L}[f](s) = \mathcal{L}[g'](s) = s\mathcal{L}[g](s) - g(0)$$

Solving for $\mathcal{L}[g]$ gives

$$\mathcal{L}[g](s) = \frac{\mathcal{L}[f](s) + g(0)}{s} \qquad \text{if } g' = f$$

The particular antiderivative of f which has value zero at $t = 0$ is the definite integral from 0 to t. Applying the previous formula to it shows that

$$\mathcal{L}\left[\int_0^t f(z)\,dz\right](s) = \frac{\mathcal{L}[f](s)}{s} \tag{25}$$

For example, $(d/dt)[t^2] = 2t$, or

$$t^2 = \int_0^t 2z \, dz$$

and it follows that

$$\mathscr{L}[t^2](s) = \frac{2\mathscr{L}[t](s)}{s} = \frac{2}{s^3} \qquad s > 0$$

Similarly, $\mathscr{L}[t^n](s) = n(1/s)\mathscr{L}[t^{n-1}](s)$. In fact,

$$\mathscr{L}[t^n](s) = \mathscr{L}\left[n \int_0^t z^{n-1} \, dz\right](s)$$

$$= n \frac{\mathscr{L}[t^{n-1}](s)}{s}$$

$$= \frac{n(n-1)}{s \cdot s} \mathscr{L}[t^{n-2}](s)$$

$$= \frac{n \cdot (n-1) \cdots 2 \cdot 1}{s \cdot s \cdots s \cdot s} \mathscr{L}[t^0](s)$$

$$= \frac{n!}{s^n} \frac{1}{s} = \frac{n!}{s^{n+1}} \qquad s > 0$$

entirely in agreement with (23).

12. Differentiating a Laplace Transform

We shall now see that if f is continuous and of exponential order, then so is tf, and

$$\mathscr{L}[tf](s) = -\frac{d}{ds} \mathscr{L}[f](s) \tag{26}$$

In fact, under the same circumstance, $t^n f$ has a Laplace transform too:

$$\mathscr{L}[t^n f](s) = (-1)^n \frac{d^n}{ds^n} \mathscr{L}[f](s)$$

For example, if $f(t) \equiv 1$, then (26) shows that

$$\mathscr{L}[t](s) = \mathscr{L}[t \cdot 1](s) = -\frac{d}{ds} \mathscr{L}[1](s)$$

$$= -\frac{d}{ds}\left(\frac{1}{s}\right) = \frac{1}{s^2}$$

in agreement with (23). In fact, (23) may be derived from (26):

$$\mathcal{L}[t^n](s) = \mathcal{L}[t^n \cdot 1](s)$$

$$= (-1)^n \frac{d^n}{ds^n} \mathcal{L}[1](s)$$

$$= (-1)^n \frac{d^n}{ds^n} \left(\frac{1}{s}\right)$$

$$= \frac{n!}{s^{n+1}}$$

Formula (26) allows us to calculate

$$\mathcal{L}[t \sin bt](s) = -\frac{d}{ds} \mathcal{L}[\sin bt](s)$$

$$= -\frac{d}{ds} \left(\frac{b}{s^2 + b^2}\right)$$

$$= \frac{2bs}{(s^2 + b^2)^2} \tag{27}$$

and

$$\mathcal{L}[t \cos bt](s) = -\frac{d}{ds} \mathcal{L}[\cos bt](s)$$

$$= -\frac{d}{ds} \left(\frac{s}{s^2 + b^2}\right)$$

$$= \frac{s^2 - b^2}{(s^2 + b^2)^2} \tag{28}$$

Formula (27) shows at once that

$$\mathcal{L}^{-1} \left[\frac{s}{(s^2 + b^2)^2}\right](t) = \frac{1}{2b} t \sin bt \qquad t \geq 0 \tag{29}$$

a fact of use in applying the partial-fraction decomposition in more complicated cases. A companion fact is derived from (28) and the formula for the Laplace transform of the sine:

$$\mathcal{L}\left[\frac{1}{2b^2}\left(\frac{1}{b}\sin bt - t\cos bt\right)\right](s) = \frac{1}{2b^2}\left[\frac{1}{s^2 + b^2} - \frac{s^2 - b^2}{(s^2 + b^2)^2}\right]$$

$$= \frac{1}{2b^2}\left[\frac{s^2 + b^2}{(s^2 + b^2)^2} - \frac{s^2 - b^2}{(s^2 + b^2)^2}\right]$$

$$= \frac{1}{2b^2}\frac{2b^2}{(s^2 + b^2)^2} = \frac{1}{(s^2 + b^2)^2}$$

Thus, $\qquad \mathcal{L}^{-1}\left[\dfrac{1}{(s^2 + b^2)^2}\right](t) = \dfrac{1}{2b^2}\left(\dfrac{1}{b}\sin bt - t\cos bt\right) \qquad t \geq 0 \qquad$ (30)

Finally, we give the computation by which (26) is derived.

$$-\frac{d}{ds}\mathcal{L}[f](s) = -\frac{d}{ds}\int_0^\infty f(t)e^{-st}\,dt$$

$$= -\int_0^\infty f(t)\frac{d}{ds}e^{-st}\,dt$$

$$= -\int_0^\infty f(t)(-t)e^{-st}\,dt$$

$$= \int_0^\infty tf(t)e^{-st}\,dt$$

$$= \mathcal{L}[tf](s)$$

(The computation is natural, but the mathematically advanced reader may observe that the exponential order of f justifies differentiation under the integral sign.)

13. Sample Problems with Answers

Find the inverse Laplace transform of each.

L. $\dfrac{s + 3}{(s + 3)^2 + 4}$ $\qquad$ *Answer:* $e^{-3t}\cos 2t,\ t \geq 0$

M. $\dfrac{s + 4}{s^2 + 8s + 20}$ $\qquad$ *Answer:* $e^{-4t}\cos 2t,\ t \geq 0$

N. $\dfrac{2}{s^2 + 6s + 18}$ $\qquad$ *Answer:* $\frac{2}{3}e^{-3t}\sin 3t,\ t \geq 0$

O. $\dfrac{s + 2}{s^2 + 6s + 18}$ $\qquad$ *Answer:* $e^{-3t}(\cos 3t - \frac{1}{3}\sin 3t),\ t \geq 0$

P. $\dfrac{\sqrt{2}s + 4\sqrt{2}}{9s^2 + 6s + 9}$ $\qquad$ *Answer:* $e^{-t/3}\left(\dfrac{\sqrt{2}}{9}\cos\dfrac{2\sqrt{2}\,t}{3} + \dfrac{11}{\sqrt{8}}\sin\dfrac{2\sqrt{2}\,t}{3}\right),\ t \geq 0$

Q. $\dfrac{1}{(s + 1)^3}$ $\qquad$ *Answer:* $\dfrac{t^2}{2}e^{-t},\ t \geq 0$

R. $\dfrac{7}{s^4}$ $\qquad$ *Answer:* $\frac{7}{6}t^3,\ t \geq 0$

S. Find $\mathcal{L}[te^{3t}\cos 7t]$. $\qquad$ *Answer:* $\dfrac{(s - 3)^2 - 49}{[(s - 3)^2 + 49]^2}$

14. Problems

Find the inverse Laplace transform of each.

22. $\dfrac{s+4}{(s+4)^2+9}$ **23.** $\dfrac{s-2}{(s-2)^2+16}$ **24.** $\dfrac{s+1}{s^2+2s+10}$

25. $\dfrac{s-3}{s^2-6s+10}$ **26.** $\dfrac{3}{(s+1)^2+9}$ **27.** $\dfrac{1}{(s+2)^2+9}$

28. $\dfrac{1}{s^2+2s+10}$ **29.** $\dfrac{1}{s^2+6s+11}$ **30.** $\dfrac{s+2}{s^2+6s+13}$

31. $\dfrac{2s+3}{s^2+4s+6}$ **32.** $\dfrac{3s+4}{3s^2+3s+1}$ **33.** $\dfrac{-7}{(s-2)^3}$

34. $\dfrac{1}{(s+3)^4}$ **35.** $\dfrac{s}{(s-2)^3}$

Find each Laplace transform twice, once by integrating first and then taking the transform, and once by finding the transform of the integrand and then using the formula for the transform of an integral.

36. $\displaystyle\int_0^t z\,dz$ **37.** $\displaystyle\int_0^t \sin 2z\,dz$ **38.** $\displaystyle\int_0^t e^{4z}\,dz$

39. $\displaystyle\int_0^t (z^3 - \cos \tfrac{1}{2}z)\,dz$ **40.** $e^{-3t}\cos 2t - \displaystyle\int_0^t \cos 2z\,dz$

Solve the following initial value problems.

41. $\begin{aligned} y'' - 2y' + 5y &= 0\\ y(0) &= 0\\ y'(0) &= 2 \end{aligned}$ **42.** $\begin{aligned} y'' - 2y' + 5y &= 0\\ y(0) &= 3\\ y'(0) &= 3 \end{aligned}$ **43.** $\begin{aligned} y'' + 2y' + 5y &= 0\\ y(0) &= 0\\ y'(0) &= 3 \end{aligned}$

44. $\begin{aligned} y'' + 4y' + 13y &= 0\\ y(0) &= 1\\ y'(0) &= 1 \end{aligned}$ **45.** $\begin{aligned} 2y'' + y' + 3y &= 0\\ y(0) &= -1\\ y'(0) &= 0 \end{aligned}$ **46.** $\begin{aligned} 3y'' + 2y' + 3y &= 0\\ y(0) &= \tfrac{1}{2}\\ y'(0) &= 2 \end{aligned}$

Find $\mathcal{L}[y]$ if y is the solution of one of the following initial value problems. (Finding the inverse transform, and hence finding y, is the subject of Section C.)

47. $\begin{aligned} y'' - 2y' + 5y &= 1\\ y(0) &= 0\\ y'(0) &= 2 \end{aligned}$ **48.** $\begin{aligned} y'' - 2y' + 5y &= e^{3t}\\ y(0) &= 3\\ y'(0) &= 3 \end{aligned}$ **49.** $\begin{aligned} y'' + 2y' + 5y &= \sin t\\ y(0) &= 0\\ y'(0) &= 0 \end{aligned}$

50. $\begin{aligned} y'' + 4y' + 13y &= t\\ y(0) &= 0\\ y'(0) &= 0 \end{aligned}$ **51.** $\begin{aligned} 3y'' + 2y' + 3y &= te^{-2t}\\ y(0) &= 1\\ y'(0) &= -1 \end{aligned}$ **52.** $\begin{aligned} 2y'' + y' - 3y &= \cos 2t\\ y(0) &= 0\\ y'(0) &= 1 \end{aligned}$

53. $\begin{aligned} 4y'' + 4y' + y &= 0\\ y(0) &= 1\\ y'(0) &= -1 \end{aligned}$ **54.** $\begin{aligned} y'' + 3y' + 2y &= 0\\ y(0) &= -1\\ y'(0) &= 1 \end{aligned}$

15. Making a Short Table of Laplace Transforms

You should make your own table, from information given in the previous section and this one. Add to it as you progress through the chapter. Begin with Equation (1) of Section A. Be sure to include such properties as linearity and the first shifting property. You may want to use the following format.

$f(t)$	$\mathscr{L}[f](s)$
$f(t)$	$\displaystyle\int_0^\infty f(t)e^{-st}\,dt$
e^{at}	$\dfrac{1}{s-a}$
....	....

16. Initial Value Problems and the General Solution of a Differential Equation

Most elementary problems solved by Laplace transforms are initial value problems. It was explained in Section H of Chapter I, the synopsis on linear differential equations, that an initial value problem consists of a differential equation to be satisfied on an interval and a collection of values for the unknown function and its derivatives, specified at a single point of that interval. In the present chapter all the differential equations will have constant coefficients, and their solutions ordinarily will be valid on the whole real line, $-\infty < t < \infty$.

From the fact that the Laplace-transform formula for the nth derivative requires exactly n initial values, it is clear that n initial values are required for an initial value problem featuring an nth-order linear differential equation with constant coefficients. In fact, the values of the higher derivatives can be computed from the differential equation and the first n initial values. For instance, if

$$\left.\begin{aligned} y''(t) + b^2 y(t) &= 0 \\ y(0) &= 1 \\ y'(0) &= 0 \end{aligned}\right\} \tag{31}$$

then

$$y''(0) + b^2 y(0) = 0$$

from the differential equation, with $t = 0$. Substituting the initial values gives

$$y''(0) + b^2 = 0$$

and therefore $y''(0) = -b^2$. Differentiating both sides of the differential equation gives

$$y'''(t) + b^2 y'(t) = 0$$

and substituting the initial values gives

$$y'''(0) = 0$$

Proceeding similarly, one can show that

$$y^{(k)}(0) = \begin{cases} 0 & \text{if } k \text{ is odd} \\ (-1)^{k/2} b^k & \text{if } k \text{ is even} \end{cases}$$

It follows, then, that exactly n initial values may be chosen arbitrarily, while the remaining ones are entirely determined by them and are not arbitrary.

It happens occasionally that a prediction is required for the behavior of a system even though the initial conditions are unknown. Suppose, for instance, that the mechanical system in Figure 1 of Example 3 were set in motion under emergency conditions. One might not know in advance what the initial conditions would be when the emergency arose, but one would want to predict something about the behavior of the system thereafter and perhaps indicate the worst possible behavior. To make such a prediction, one needs to discover just how the solution of an initial value problem depends on the initial values. The idea is to find the differential equation's general solution by simply leaving the initial conditions unspecified, as new parameters.

For example, the mechanical system in Figure 1 of Example 3 is modeled by the following differential equation:

$$y'' + b^2 y = 0 \tag{32}$$

Let us write and solve the initial value problem

$$\left. \begin{array}{l} y'' + b^2 y = 0 \\ y(0) = y_0 \\ y'(0) = y_1 \end{array} \right\} \tag{33}$$

where the initial values y_0 and y_1 are not specified numbers but new parameters. The transformed problem is

$$s^2 \mathscr{L}[y](s) - sy_0 - y_1 + b^2 \mathscr{L}[y](s) = 0$$

Solving it shows that

$$(s^2 + b^2)\mathscr{L}[y](s) = sy_0 + y_1$$

$$\mathscr{L}[y](s) = y_0 \frac{s}{s^2 + b^2} + \frac{y_1}{b} \frac{b}{s^2 + b^2} \tag{34}$$

Thus, $$y(t) = y_0 \cos bt + y_1 \left(\frac{1}{b} \sin bt\right) \qquad t \geq 0 \tag{35}$$

The two functions $\cos bt$ and $(\sin bt)/b$ which appear in (35) are necessarily solutions of (32) themselves. The first is the solution of the initial value problem

$$\left. \begin{array}{l} y'' + b^2 y = 0 \\ y(0) = 1 \\ y'(0) = 0 \end{array} \right\}$$

while the second is the solution of the problem

$$\left. \begin{array}{l} y'' + b^2 y = 0 \\ y(0) = 0 \\ y'(0) = 1 \end{array} \right\}$$

Together, these two solutions constitute *the basic solution system for* (32) *with initial point* 0, which we shall study further in Section E of Chapter IV. If we call them v_0 and v_1, then we may rewrite (35) as

$$y(t) = y_0 v_0(t) + y_1 v_1(t) \qquad t \geq 0 \tag{36}$$

Equation (35) or (36) gives the general solution of (32) on the right half line, $t \geq 0$. For each choice of the pair of parameters y_0 and y_1, (35) or (36) gives a solution, and there are no other solutions because of the uniqueness of the solutions of initial value problems. (For this and a definition of "general solution," see Section H of Chapter I.) It can be said at once that the general solution of the equation is the collection of sinusoids with angular velocity b, and any information one may have about conditions in the system at time $t = 0$ may be turned at once into information about the behavior of the system, using (35) or (36).

Now let us consider a nonhomogeneous equation and find its general solution. The differential equation is first order, and so one nonspecified initial value is required:

$$\left. \begin{array}{l} y'(t) + y(t) = e^{-t} \\ y(0) = y_0 \end{array} \right\} \tag{37}$$

The transformed problem

$$s\mathscr{L}[y](s) - y_0 + \mathscr{L}[y](s) = \frac{1}{s+1}$$

can be solved as usual:

$$(s+1)\mathscr{L}[y](s) = y_0 + \frac{1}{s+1}$$

$$\mathscr{L}[y](s) = y_0 \frac{1}{s+1} + \frac{1}{(s+1)^2}$$

Thus, the general solution of (37) is

$$y(t) = y_0 e^{-t} + t e^{-t} \qquad t \geq 0 \tag{38}$$

It is traditional to point out at this stage that the term $t e^{-t}$ which appears in (38) is a solution of (37), the one for which $y(0) = 0$; but the term $y_0 e^{-t}$ is not a solution of (37). It is, instead, the general solution of the equation

$$y' + y = 0 \tag{39}$$

which differs from (37) by lacking a nonhomogeneous term. Equation (39) is called the *homogeneous equation associated with* (37). The term $y_0 e^{-t}$ is called the *general homogeneous solution of* (37), despite the fact that it is a solution of (39), not of (37). Thus, the general solution of (37) is the sum of *some* solution of (37) and the general homogeneous solution. The term which is a solution of (37) is sometimes referred to as a *particular solution* and designated y_p; the general homogeneous solution is sometimes designated y_h; and the general solution can be written as

$$y = y_p + y_h$$

Here is another example to illustrate the same point. The problem

$$\left. \begin{array}{l} y''(t) + 4y'(t) + 4y(t) = 7e^{-2t} \\ y(0) = y_0 \\ y'(0) = y_1 \end{array} \right\} \tag{40}$$

has the following transformed problem:

$$s^2 \mathscr{L}[y](s) - sy_0 - y_1 + 4\{s\mathscr{L}[y](s) - y_0\} + 4\mathscr{L}[y](s) = \frac{7}{s+2}$$

$$(s^2 + 4s + 4)\mathscr{L}[y](s) = \frac{7}{s+2} + y_0(s + 4) + y_1$$

$$\mathscr{L}[y](s) = \frac{7}{(s+2)^3} + y_0 \frac{s+4}{(s+2)^2} + y_1 \frac{1}{(s+2)^2} \tag{41}$$

$$= \frac{7}{(s+2)^3} + y_0 \frac{(s+2)+2}{(s+2)^2} + y_1 \frac{1}{(s+2)^2}$$

$$= \frac{7}{(s+2)^3} + y_0 \left[\frac{1}{s+2} + \frac{2}{(s+2)^2} \right] + y_1 \frac{1}{(s+2)^2} \tag{42}$$

Thus, the general solution of (40) is this:

$$y(t) = \tfrac{7}{2}t^2 e^{-2t} + y_0[e^{-2t} + 2te^{-2t}] + y_1 te^{-2t} \qquad t \ge 0 \tag{43}$$

which is of the form

$$y(t) = y_p(t) + y_h(t) \tag{44}$$

where $\qquad\qquad\qquad y_p = \tfrac{7}{2}t^2 e^{-2t}$

is a solution of the given equation (40), and

$$y_h(t) = y_0[e^{-2t} + 2te^{-2t}] + y_1[te^{-2t}] \tag{45}$$

$$= y_0 v_0(t) + y_1 v_1(t)$$

is the general solution of the associated homogeneous equation

$$y_h'' + 4y_h' + 4y_h = 0 \tag{46}$$

Specifically, (43) shows that y_p is the unique solution of the given equation for which

$$y(0) = y_0 = 0 \qquad \text{and} \qquad y'(0) = y_1 = 0$$

while substitution into (46) shows that y_h is that solution of (46) with the initial values y_0 and y_1, which are arbitrary.

The algebraic rearrangement between (41) and (42) is a special case of the partial-fraction decomposition, to be explained in the next section. Most problems require its use as a practical matter before the inverse Laplace transform is taken. However, the analysis of the general solution into $y_p + y_h$ does not really depend on it, as we shall see in Chapter IV.

For the record, the functions v_0 and v_1, in brackets in (45), together constitute the basic solution system for (40) with initial point zero. Each is a solution of (46), the homogeneous equation associated with (40), and they satisfy the initial conditions

$$v_0(0) = 1 \qquad v_1(0) = 0$$
$$v_0'(0) = 0 \qquad v_1'(0) = 1$$

The problems given below can be solved before studying the next section.

17. More Sample Problems with Answers

Find the general solution of each.

T. $y' - 16y = 0$ *Answer:* $y(0)e^{16t}$, $t \geq 0$

U. $y'' + 4y' + 4y = 0$ *Answer:* $y(0)(1 + 2t)e^{-2t} + y'(0)te^{-2t}$, $t \geq 0$

18. More Problems

Find the general solution of each.

55. $y'' - 2y' + 5y = 0$ **56.** $y'' + 4y' + 13y = 0$

57. $3y'' + 2y' + 3y = 0$ **58.** $3y' + 2y = 4e^{-2t/3}$

C. THE INVERSE LAPLACE TRANSFORM OF A RATIONAL FUNCTION

1. The Rational Function with Quadratic Denominator

This is the moment to list and compare the inverse Laplace transforms of all three special cases of the rational function of the form

$$\frac{Ds + E}{As^2 + Bs + C} \tag{1}$$

in which the denominator has degree two.

In the first special case, $B^2 - 4AC < 0$ and so the denominator does not have real roots. Completion of the square in the denominator (as explained in the previous section) leads to a decomposition into terms whose inverse Laplace transforms we can identify. For example,

$$\frac{s+3}{4s^2+4s+9} = \frac{s/4 + \frac{3}{4}}{s^2+s+\frac{9}{4}}$$

$$= \frac{s/4 + \frac{3}{4}}{(s+\frac{1}{2})^2 + (\frac{9}{4} - \frac{1}{4})}$$

$$= \frac{s/4 + \frac{3}{4}}{(s+\frac{1}{2})^2 + 2}$$

$$= \frac{(s+\frac{1}{2})/4 + (3 - \frac{1}{2})/4}{(s+\frac{1}{2})^2 + \sqrt{2}^2}$$

$$= \frac{1}{4}\frac{s+\frac{1}{2}}{(s+\frac{1}{2})^2 + \sqrt{2}^2} + \frac{1}{4}\frac{3-\frac{1}{2}}{\sqrt{2}}\frac{\sqrt{2}}{(s+\frac{1}{2})^2 + \sqrt{2}^2}$$

$$= \frac{1}{4}\mathscr{L}[e^{-t/2}\cos\sqrt{2}\,t](s) + \frac{1}{4}\frac{5}{2\sqrt{2}}\mathscr{L}[e^{-t/2}\sin\sqrt{2}\,t](s)$$

and therefore

$$\mathscr{L}^{-1}\left[\frac{s+3}{4s^2+4s+9}\right](t) = \frac{1}{4}e^{-t/2}\cos\sqrt{2}\,t + \frac{5}{8\sqrt{2}}e^{-t/2}\sin\sqrt{2}\,t \qquad t \geq 0 \quad (2)$$

The first special case always yields sinusoids or damped sinusoids.

In the second special case, $B^2 - 4AC = 0$ and so the denominator in (1) is already a perfect square. The numerator is treated in much the same way as in the previous case, but there is cancellation in one term, and the taking of the inverse transform is different. For instance,

$$\frac{s+3}{4s^2+4s+1} = \frac{s/4 + \frac{3}{4}}{s^2+s+\frac{1}{4}} \tag{3}$$

$$= \frac{s/4 + \frac{3}{4}}{(s+\frac{1}{2})^2} \tag{4}$$

$$= \frac{(s+\frac{1}{2})/4 + (3 - \frac{1}{2})/4}{(s+\frac{1}{2})^2} \tag{5}$$

$$= \frac{(s+\frac{1}{2})/4 + \frac{5}{8}}{(s+\frac{1}{2})^2}$$

$$= \frac{1}{4}\frac{s+\frac{1}{2}}{(s+\frac{1}{2})^2} + \frac{5}{8}\frac{1}{(s+\frac{1}{2})^2} \tag{6}$$

$$= \frac{1}{4}\frac{1}{s+\frac{1}{2}} + \frac{5}{8}\frac{1}{(s+\frac{1}{2})^2} \tag{7}$$

$$= \frac{1}{4}\mathscr{L}[e^{-t/2}](s) + \frac{5}{8}\mathscr{L}[te^{-t/2}](s)$$

and therefore

$$\mathscr{L}^{-1}\left[\frac{s+3}{4s^2+4s+1}\right](t) = \tfrac{1}{4}e^{-t/2} + \tfrac{5}{8}te^{-t/2} \qquad t \geq 0 \qquad (8)$$

The steps were these: In (3) the coefficient of s^2 was divided out of numerator and denominator, and in (4) the denominator was factored into the form $(s+\alpha)^2$. In (5) the numerator was rewritten into two terms, the first a multiple of $s+\alpha = s + \tfrac{1}{2}$, the second a constant. In (6), the decomposition of the numerator became a decomposition of the fraction. In (7), the first term was simplified by elimination of the common factor $s + \alpha = s + \tfrac{1}{2}$ from numerator and denominator, and the terms became identifiable as Laplace transforms of multiples of $e^{-\alpha t}$ and $te^{-\alpha t}$, respectively, with $\alpha = \tfrac{1}{2}$. *The second case always yields t times an exponential.*

There is a third case of (1) to consider, in which $B^2 - 4AC > 0$ and in which the denominator therefore has distinct real roots. That is,

$$\frac{Ds+E}{As^2+Bs+C} = \frac{Ds/A + E/A}{(s+\alpha)(s+\beta)} \qquad \text{where } \alpha \neq \beta \qquad (9)$$

In this situation one uses the partial-fraction decomposition to rewrite (9) as a linear combination of $1/(s+\alpha)$ and $1/(s+\beta)$. *In the third case, the inverse transform is always a linear combination of two different exponentials.*

To begin the decomposition, recall that

$$\frac{A}{X} + \frac{B}{Y} = \frac{AY}{XY} + \frac{BX}{XY} = \frac{AY+BX}{XY}$$

Our present problem is to reverse such a computation. For instance, since

$$\frac{s+3}{4s^2+4s-3} = \frac{s/4 + \tfrac{3}{4}}{s^2+s-\tfrac{3}{4}}$$

$$= \frac{s/4 + \tfrac{3}{4}}{(s+\tfrac{3}{2})(s-\tfrac{1}{2})} \qquad (10)$$

it must be asked whether we can find numbers A and B such that

$$\frac{s/4 + \tfrac{3}{4}}{s^2+s-\tfrac{3}{4}} = A\frac{1}{s+\tfrac{3}{2}} + B\frac{1}{s-\tfrac{1}{2}} \qquad (11)$$

allowing us to take the inverse transform of the left-hand side of (10) by taking that of the right-hand side of (11). The answer is that we can, and here is the computation. First consider arbitrary numbers A and B, and put the right-hand side of (11) over a common denominator:

$$A\frac{1}{s+\tfrac{3}{2}} + B\frac{1}{s-\tfrac{1}{2}} = \frac{As - A/2 + Bs + 3B/2}{s^2+s-\tfrac{3}{4}}$$

$$= \frac{(A+B)s + (-A/2 + 3B/2)}{s^2+s-\tfrac{3}{4}} \qquad (12)$$

Equation (12) holds for *all* values of A and B, but we seek values such that

$$\frac{s/4 + \frac{3}{4}}{s^2 + s - \frac{3}{4}} = \frac{(A + B)s + (-A/2 + 3B/2)}{s^2 + s - \frac{3}{4}} \tag{13}$$

that is, such that

$$\frac{s}{4} + \frac{3}{4} = (A + B)s + \left(-\frac{A}{2} + \frac{3B}{2}\right) \qquad \text{for all } s \tag{14}$$

If there are to be such A and B, then the polynomial on the right-hand side of (14) must be the same as that on the left and must have the same coefficients. Therefore, we equate the coefficients of terms for each power of s:

$$\left. \begin{array}{ll} \text{Constant term:} & \dfrac{3}{4} = -\dfrac{A}{2} + \dfrac{3B}{2} \\[2mm] s \text{ term:} & \frac{1}{4} = A + B \end{array} \right\} \tag{15}$$

In order to complete the decomposition indicated in (11), we must solve the simultaneous equations (15):

$$\begin{array}{r} -A + 3B = \frac{6}{4} \\ \underline{A + B = \frac{1}{4}} \\ 4B = \frac{7}{4} \end{array}$$

$$B = \tfrac{7}{16}$$

$$A + B = \tfrac{1}{4}$$

$$A = \tfrac{1}{4} - B = \tfrac{4}{16} - \tfrac{7}{16}$$

$$= -\tfrac{3}{16}$$

Checking the solution, we obtain

$$-A + 3B = \tfrac{3}{16} + \tfrac{21}{16} = \tfrac{24}{16} = \tfrac{6}{4}$$

$$A + B = -\tfrac{3}{16} + \tfrac{7}{16} = \tfrac{4}{16} = \tfrac{1}{4}$$

Substituting the values of A and B into (11) gives

$$\frac{s + 3}{4s^2 + 4s - 3} = \frac{s/4 + \frac{3}{4}}{s^2 + s - \frac{3}{4}}$$

$$= A \frac{1}{s + \frac{3}{2}} + B \frac{1}{s - \frac{1}{2}}$$

$$= -\frac{3}{16} \frac{1}{s + \frac{3}{2}} + \frac{7}{16} \frac{1}{s - \frac{1}{2}}$$

$$= \mathcal{L}\left[-\tfrac{3}{16} e^{-3t/2} + \tfrac{7}{16} e^{t/2}\right](s)$$

and it follows that

$$\mathscr{L}^{-1}\left[\frac{s+3}{4s^2+4s-3}\right](t) = -\tfrac{3}{16}e^{-3t/2} + \tfrac{7}{16}e^{t/2} \qquad t \geq 0 \tag{16}$$

2. The General Form of the Partial-Fraction Decomposition

A quotient of polynomials is called a rational function. Every rational function has a partial-fraction decomposition. We shall not prove that theorem, but merely state the form of the decomposition suitable to our purposes and give directions for finding it in all cases.

If $F(s)$ is a rational function whose decomposition is desired, then the degree of numerator and denominator should be checked, and polynomial division performed if the degree of the denominator is less than or equal to that of the numerator. For instance,

$$\frac{s^2-2}{s+1} = s - 1 - \frac{1}{s+1}$$

The result of this division is a quotient (which is a polynomial) plus a remainder which is a rational function whose denominator has degree strictly greater than that of the numerator. The polynomial has an inverse Laplace transform found using definitions to be given in Sections E and F. It is the remainder term which must be further decomposed. Therefore, in the rest of this section it will be presumed that *the degree of the denominator is (strictly) greater than that of the numerator.*

The form of the decomposition is entirely determined by the denominator, which must be factored as a product. When fully factored, the product may contain factors of only three different forms: a constant factor, factors of the form $(s+\alpha)^k$, and factors of the form $[(s+a)^2 + b^2]^m$. The factors of the form $(s+\alpha)^k$, if there are any, must all have different roots, that is, different values of α. Likewise, those of the form $[(s+a)^2 + b^2]^m$ must also be distinct, no two having both the same value of a and the same value of b^2. In the *first form of the decomposition*, each nonconstant factor of the denominator is taken as the denominator of a rational function whose numerator, although perhaps unknown, is of degree at least one less. For example,

$$\frac{s^2+2s+3}{(s+1)^2(s+2)[(s+1)^2+2^2]^2}$$

$$= \frac{A_1 s + A_2}{(s+1)^2} + \frac{A_3}{s+2} + \frac{A_4 s^3 + A_5 s^2 + A_6 s + A_7}{[(s+1)^2+2^2]^2} \tag{17}$$

It must be remembered that the coefficients A_1 through A_7 have not been computed yet, and that (17) is only the form of the decomposition—actually the first of two forms. The coefficients will be computed last.

In the next (and final) form of the decomposition, terms with denominators raised to powers higher than 1 are themselves decomposed. Those with denominator of the form $(s + \alpha)^k$ are decomposed as in Equations (4) through (7) above. For instance,

$$\frac{A_1 s + A_2}{(s + 1)^2}$$

can always be written

$$\frac{B_1(s + 1) + B_2}{(s + 1)^2} = B_1 \frac{1}{s + 1} + B_2 \frac{1}{(s + 1)^2}$$

In general, if $(s + \alpha)^k$ is one of the factors of the denominator, then the second form of the decomposition will contain the k terms

$$B_1 \frac{1}{s + \alpha} + B_2 \frac{1}{(s + \alpha)^2} + \cdots + B_k \frac{1}{(s + \alpha)^k} \qquad (18)$$

in which the numerators are constant, but the values of $B_1, B_2, \ldots, B_k$ have yet to be determined. The inverse Laplace transform of each of these terms was given in Section B.

The terms like the last term of (17) have also to be decomposed:

$$\frac{A_4 s^3 + A_5 s^2 + A_6 s + A_7}{[(s + 1)^2 + 2^2]^2} = C_1 \frac{s + 1}{(s + 1)^2 + 2^2} + C_2 \frac{2}{(s + 1)^2 + 2^2}$$

$$+ C_3 \frac{s + 1}{[(s + 1)^2 + 2^2]^2} + C_4 \frac{2}{[(s + 1)^2 + 2^2]^2}$$

In general, if $[(s + a)^2 + b^2]^m$ is one of the factors of the denominator, then the second form of the decomposition will contain the $2m$ terms

$$C_1 \frac{s + a}{(s + a)^2 + b^2} + C_2 \frac{b}{(s + a)^2 + b^2} + \cdots$$

$$+ C_{2m-1} \frac{s + a}{[(s + a)^2 + b^2]^m} + C_{2m} \frac{b}{[(s + a)^2 + b^2]^m} \qquad (19)$$

For the first two terms of (19) one may find an inverse Laplace transform using techniques practiced in the previous section. Inverse Laplace transforms of the other terms should be looked up in tables rather than memorized. The technique of deriving them is that given at the end of the previous section.

Here are five examples of the decomposition in its final form. In each case the constants A, B, etc., have yet to be determined by techniques to be given below.

$$\frac{s + 2}{s^2 + 5s + 4} = \frac{s + 2}{(s + 1)(s + 4)} = A \frac{1}{s + 1} + B \frac{1}{s + 4}$$

$$\frac{s^2 + 2s - 4}{2s^4 + 4s^3 + 2s^2} = \frac{(s^2 + 2s - 4)/2}{s^2(s + 1)^2} = A \frac{1}{s} + B \frac{1}{s^2} + C \frac{1}{s + 1} + D \frac{1}{(s + 1)^2}$$

$$\frac{1}{5s^4 + 10s^3 + 15s^2} = \frac{\frac{1}{5}}{s^2[(s+1)^2 + 2]}$$

$$= A\frac{1}{s} + B\frac{1}{s^2} + C\frac{s+1}{(s+1)^2 + \sqrt{2}^2} + D\frac{\sqrt{2}}{(s+1)^2 + \sqrt{2}^2}$$

$$\frac{s}{(s^2 + 4s + 3)(s^2 + 2s + 10)} = \frac{s}{(s+1)(s+3)[(s+1)^2 + 9]}$$

$$= A\frac{1}{s+1} + B\frac{1}{s+3} + C\frac{s+1}{(s+1)^2 + 3^2} + D\frac{3}{(s+1)^2 + 3^2}$$

$$\frac{3s^3 + 2s}{4(s^2 + 2s + 10)(s^2 + 6s + 13)} = \frac{3s^3/4 + 2s/4}{[(s+1)^2 + 9][(s+3)^2 + 4]}$$

$$= A\frac{s+1}{(s+1)^2 + 3^2} + B\frac{3}{(s+1)^2 + 3^2}$$

$$+ C\frac{s+3}{(s+3)^2 + 2^2} + D\frac{2}{(s+3)^2 + 2^2}$$

3. Determining the Constants in the Partial-Fraction Decomposition

The basic idea is to turn the partial-fraction decomposition, unknown constants and all, back into a rational function by putting all fractions over a common denominator. One example of this occurred between Equations (11) and (13). Here is another example:

$$\frac{s^2 + 2s - 4}{s^4 + 2s^3 + s^2} = \frac{s^2 + 2s - 4}{s^2(s+1)^2} \tag{20}$$

$$= A\frac{1}{s} + B\frac{1}{s^2} + C\frac{1}{s+1} + D\frac{1}{(s+1)^2} \tag{21}$$

$$= \frac{As(s+1)^2}{s^2(s+1)^2} + \frac{B(s+1)^2}{s^2(s+1)^2} + \frac{Cs^2(s+1)}{s^2(s+1)^2} + \frac{Ds^2}{s^2(s+1)^2} \tag{22}$$

$$= \frac{A(s^3 + 2s^2 + s) + B(s^2 + 2s + 1) + C(s^3 + s^2) + Ds^2}{s^2(s+1)^2}$$

$$= \frac{s^3(A+C) + s^2(2A + B + C + D) + s(A + 2B) + 1(B)}{s^2(s+1)^2} \tag{23}$$

Since (20) and (23) are equal and have the same denominator, they must also have the same numerator. That is,

$$s^2 + 2s - 4 = s^3(A+C) + s^2(2A + B + C + D)$$

$$+ s(A + 2B) + 1(B) \qquad \text{for all } s \tag{24}$$

Since the left- and right-hand sides of (24) are the same polynomial, it follows that the coefficients of the corresponding terms must be equal:

$$
\left.
\begin{aligned}
\text{Constant term:} \quad & -4 = B \\
s \text{ term:} \quad & 2 = A + 2B \\
s^2 \text{ term:} \quad & 1 = 2A + B + C + D \\
s^3 \text{ term:} \quad & 0 = A + C
\end{aligned}
\right\}
\tag{25}
$$

Equations (25) must be solved for A, B, C, and D. Clearly $B = -4$, and so the second equation gives the information

$$
2 = A + 2B = A - 8
$$

$$
A = 10
$$

Also, the last equation shows that

$$
0 = A + C = 10 + C
$$

$$
C = -10
$$

Now the third equation from (25) implies that

$$
1 = 2A + B + C + D = 20 - 4 - 10 + D
$$

$$
= 6 + D
$$

$$
D = -5
$$

Substitution of these numbers back into (21) shows that

$$
\frac{s^2 + 2s - 4}{s^4 + 2s^3 + s^2} = 10\frac{1}{s} - 4\frac{1}{s^2} - 10\frac{1}{s+1} - 5\frac{1}{(s+1)^2}
\tag{26}
$$

Taking the inverse Laplace transform of both sides shows that

$$
\mathscr{L}^{-1}\left[\frac{s^2 + 2s - 4}{s^4 + 2s^3 + s^2}\right](t) = 10 - 4t - 10e^{-t} - 5te^{-t} \qquad t \geq 0
$$

Thus, after the final form of the decomposition has been written [as in (21)] and has been put over a common denominator [as in (22)] with the like powers in the numerator collected [as in (23)], the next step is to equate the equal numerators [as in (24)]. Then equate the coefficients of like powers of s as in (25), and solve for the unknown constants by solving the simultaneous linear algebraic equations. Once found, the values of the constants must be substituted back into the general form of the decomposition, turning (21) into (26), and the inverse Laplace transform taken.

Here is an example leading to a damped sinusoid.

$$\frac{s^2 + 2s - 4}{2s^3 + 4s^2 + 3s} = \frac{s^2/2 + s - 2}{s(s^2 + 2s + \frac{3}{2})}$$

$$= \frac{s^2/2 + s - 2}{s[(s + 1)^2 + (1/\sqrt{2})^2]} \tag{27}$$

$$= A\frac{1}{s} + B\frac{s + 1}{(s + 1)^2 + (1/\sqrt{2})^2}$$

$$+ C\frac{1/\sqrt{2}}{(s + 1)^2 + (1/\sqrt{2})^2} \tag{28}$$

$$= \frac{A(s^2 + 2s + \frac{3}{2}) + Bs(s + 1) + Cs/\sqrt{2}}{s[(s + 1)^2 + (1/\sqrt{2})^2]}$$

$$= \frac{s^2(A + B) + s(2A + B + C/\sqrt{2}) + 3A/2}{s[(s + 1)^2 + (1/\sqrt{2})^2]} \tag{29}$$

Since the constant 2 was factored out of both denominator and numerator at the first step, this last expression does not have the same denominator as the first. It has, instead, the same denominator as (27), to which it is equal. Therefore, we may equate the terms of the numerators of (27) and (29).

$$\text{Constant term:} \qquad -2 = \frac{3A}{2}$$

$$s \text{ term:} \qquad 1 = 2A + B + \frac{C}{\sqrt{2}}$$

$$s^2 \text{ term:} \qquad \tfrac{1}{2} = A + B$$

Solving the simultaneous equations gives

$$-2 = \frac{3A}{2}$$

$$A = -\tfrac{4}{3}$$

$$\tfrac{1}{2} = A + B = -\tfrac{4}{3} + B$$

$$B = \tfrac{1}{2} + \tfrac{4}{3} = \tfrac{11}{6}$$

$$1 = 2A + B + \frac{C}{\sqrt{2}} = -\tfrac{8}{3} + \tfrac{11}{6} + \frac{C}{\sqrt{2}}$$

$$\frac{C}{\sqrt{2}} = 1 + \tfrac{8}{3} - \tfrac{11}{6} = \tfrac{11}{6}$$

$$C = \frac{11\sqrt{2}}{6}$$

Substitution back into (28) gives

$$\frac{s^2 + 2s - 4}{2s^3 + 4s^2 + 3s} = -\frac{4}{3}\frac{1}{s} + \frac{11}{6}\frac{s + 1}{(s + 1)^2 + (1/\sqrt{2})^2}$$

$$+ \frac{11\sqrt{2}}{6}\frac{1/\sqrt{2}}{(s + 1)^2 + (1/\sqrt{2})^2}$$

from which it follows that

$$\mathcal{L}^{-1}\left[\frac{s^2 + 2s - 4}{2s^3 + 4s^2 + 3s}\right](t) = -\frac{4}{3} + \frac{11}{6}e^{-t}\cos\frac{t}{\sqrt{2}} + \frac{11\sqrt{2}}{6}e^{-t}\sin\frac{t}{\sqrt{2}} \qquad t \geq 0$$

4. Reducing the Number of Algebraic Equations to Be Solved

The coefficient of the term $1/(s + \alpha)^k$ containing the *highest* power of $s + \alpha$ can be found independently, before the decomposition is put over a common denominator. The technique is an efficient labor-saving device. Here is an example worked out this way, for comparison with the computation which began with (20).

$$\frac{s^2 + 2s - 4}{s^2(s + 1)^2} = A\frac{1}{s} + B\frac{1}{s^2} + C\frac{1}{s + 1} + D\frac{1}{(s + 1)^2} \qquad (30)$$

Multiply both sides of (30) by s^2, and then evaluate at $s = 0$:

$$\frac{s^2 + 2s - 4}{(s + 1)^2} = As + B + C\frac{s^2}{s + 1} + D\frac{s^2}{(s + 1)^2}$$

$$\frac{0 + 0 - 4}{(0 + 1)^2} = 0 \quad + B + 0 \qquad + 0$$

$$-4 = B$$

Again, multiply both sides of (30) by $(s + 1)^2$, and then evaluate at $s = -1$, the point where $s + 1$ is zero:

$$\frac{s^2 + 2s - 4}{s^2} = A\frac{(s + 1)^2}{s} + B\frac{(s + 1)^2}{s^2} + C(s + 1) + D$$

$$\frac{(-1)^2 - 2 - 4}{(-1)^2} = 0 \qquad + 0 \qquad + 0 \qquad + D$$

$$-5 = D$$

After this (if there remain unknown constants, as there do in this problem), put all the fractions over a common denominator, and solve for any remaining unknown constants:

$$\frac{s^2 + 2s - 4}{s^2(s+1)^2} = A\frac{1}{s} - 4\frac{1}{s^2} + C\frac{1}{s+1} - 5\frac{1}{(s+1)^2}$$

$$= \frac{As(s^2 + 2s + 1) - 4(s^2 + 2s + 1) + Cs^2(s+1) - 5s^2}{s^2(s+1)^2}$$

$$= \frac{s^3(A + C) + s^2(2A - 4 + C - 5) + s(A - 8) - 4}{s^2(s+1)^2}$$

Constant term: $-4 = -4$

s term: $2 = A - 8$

s^2 term: $1 = 2A - 4 + C - 5$

s^3 term: $0 = A + C$

from which

$$2 = A - 8$$

$$A = 10$$

$$0 = A + C = 10 + C$$

$$C = -10$$

Use the remaining equation as a check:

$$2A - 4 + C - 5 = 20 - 4 - 10 - 5$$

$$= 20 - 19 = 1$$

which checks.

5. Finding Coefficients by Evaluating the Numerator

Some people prefer to evaluate as many coefficients as possible after putting the decomposition over a common denominator but before setting up simultaneous equations. Using the same example as previously but beginning at (22), we have

$$s^2 + 2s - 4 = As(s+1)^2 + B(s+1)^2 + Cs^2(s+1) + Ds^2 \qquad (31)$$

Evaluating both sides of (31) at $s = 0$ gives

$$-4 = 0 + B + 0 + 0$$

$$B = -4$$

Evaluating both sides at $s + 1 = 0$ or $s = -1$ gives

$$(-1)^2 - 2 - 4 = 0 + 0 + 0 + D$$

$$D = -5$$

Substitution back into (31) gives

$$s^2 + 2s - 4 = As(s + 1)^2 - 4(s + 1)^2 + Cs^2(s + 1) - 5s^2$$
$$= s^3(A + C) + s^2(2A - 4 + C - 5) + s(A - 8) - 4$$

Equating coefficients as before gives the values of A and C.

6. Sample Problems with Answers

Give the form of the decomposition into partial fractions, then find the coefficients and give the decomposition, and finally take the inverse Laplace transforms. (In each problem, the given answer is that for the second of the three requirements.)

A. $\dfrac{s}{s^2 - 5s + 6}$ $Answer: 3\dfrac{1}{s - 3} - 2\dfrac{1}{s - 2}$

B. $\dfrac{2s + 3}{(s - 1)^3}$ $Answer: 5\dfrac{1}{(s - 1)^3} + 2\dfrac{1}{(s - 1)^2} + 0\dfrac{1}{s - 1}$

C. $\dfrac{s^2}{(s + 1)^2(s - 1)^3}$ $Answer: -\dfrac{1}{8}\dfrac{1}{(s + 1)^2} + \dfrac{1}{16}\dfrac{1}{s + 1} + \dfrac{1}{4}\dfrac{1}{(s - 1)^3} + \dfrac{1}{4}\dfrac{1}{(s - 1)^2} - \dfrac{1}{16}\dfrac{1}{s - 1}$

D. $\dfrac{2s^4 - 4s + 90}{2s(s^2 + 2s + 5)(s^2 + 4s + 9)}$

$Answer: \dfrac{1}{s} - \dfrac{s + 1}{(s + 1)^2 + 2^2} - 2\dfrac{2}{(s + 1)^2 + 2^2} + \dfrac{s + 2}{(s + 2)^2 + \sqrt{5}^2} - \dfrac{1}{\sqrt{5}}\dfrac{\sqrt{5}}{(s + 2)^2 + \sqrt{5}^2}$

7. Problems

Find the decomposition into partial fractions if necessary, and find the inverse Laplace transform.

1. $\dfrac{2s - 3}{s^2 + 2s + 5}$ 2. $\dfrac{s/2 + 4}{s^2 + 2s + 1}$ 3. $\dfrac{-6}{s^2 + 2s - 8}$

4. $\dfrac{4s - 7}{3s^2 + 18s + 33}$ 5. $\dfrac{s}{3s^2 + 18s + 27}$ 6. $\dfrac{-2s + 1}{3s^2 + 18s + 15}$

7. $\dfrac{s}{(s + 1)^2(s + 2)}$ 8. $\dfrac{s^2 + 4}{(s + 1)(s + 2)(s + 3)}$ 9. $\dfrac{1}{(s - 1)^2(s + 1)^2}$

10. $\dfrac{1}{s^2(s^2 + 4)}$ 11. $\dfrac{3}{s(s - 4)(s + 3)}$ 12. $\dfrac{4s^2}{(s + 1)^2(s - 3)}$

13. $\dfrac{s}{(s + 1)(s^2 + 4)}$ 14. $\dfrac{2s + 1}{s^2(2s^2 + 1)}$ 15. $\dfrac{6}{(s^2 + 1)(s^2 + 4)}$

16. $\dfrac{-3s}{(s^2 + 4)(s^2 + 6s + 13)}$ 17. $\dfrac{4s^3}{(s^2 + 6s + 13)(4s^2 + 12s + 9)}$

18. $\dfrac{s^3 + 1}{(4s^2 + 4s + 5)(s^2 + 3s + 2)}$

Solve the initial value problems. (These problems are very suitable for practicing graphing.)

19. $\left.\begin{aligned} y'' + 3y' + 2y &= 0 \\ y(0) &= 0 \\ y'(0) &= 1 \end{aligned}\right\}$ **20.** $\left.\begin{aligned} y'' + 3y' + 2y &= 0 \\ y(0) &= 1 \\ y'(0) &= 0 \end{aligned}\right\}$

21. $\left.\begin{aligned} y'' + 3y' + 2y &= e^t \\ y(0) &= 0 \\ y'(0) &= 0 \end{aligned}\right\}$ **22.** $\left.\begin{aligned} 4y' + 6y &= e^{-2t} \\ y(0) &= 3 \end{aligned}\right\}$

23. $\left.\begin{aligned} 4y'' + 4y' + 5y &= 6 \\ y(0) &= 1 \\ y'(0) &= 2 \end{aligned}\right\}$ **24.** $\left.\begin{aligned} y'' + 2y' + y &= t \\ y(0) &= -3 \\ y'(0) &= 0 \end{aligned}\right\}$

25. $\left.\begin{aligned} y'' + 2y' + y &= 2e^{-t} \\ y(0) &= 0 \\ y'(0) &= 1 \end{aligned}\right\}$ **26.** $\left.\begin{aligned} 3y' + y &= \cos 2t \\ y(0) &= 0 \end{aligned}\right\}$

27. $\left.\begin{aligned} 3y' + y &= 1 - 2t^2 \\ y(0) &= 0 \end{aligned}\right\}$ **28.** $\left.\begin{aligned} y'' + 3y' + 3y &= 0 \\ y(0) &= 1 \\ y'(0) &= 1 \end{aligned}\right\}$

29. In Figure 1, $M = 2$, $K = 4$, $f(t) = \sin t$, and $x(0) = x'(0) = 0$. Find $x(t)$.

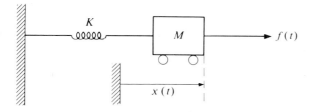

Figure 1 Distance and force defined in a mechanical system. When $x(t) = 0$, the tension in the spring is zero.

30. In Figure 1, $M = 2$, $K = 4$, and $x(t) = \cos t$. Find $f(t)$.

31. In Figure 2, $R = 1$, $C = 4$, $L = 1$, $i(t) = e^{-t/3}$, $v_R(0) = 0$, and $v'_R(0) = 0$. Find $v_R(t)$.

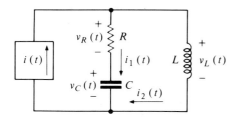

Figure 2 RLC circuit.

32. In Figure 2, $R = 2$, $C = \frac{1}{10}$, $L = 1$, $i(t) = \cos 4t$, $v_R(0) = 0$, and $i'_1(0) = 1$. Find $v_R(t)$.

33. (a) Solve the initial value problem

$$\left.\begin{aligned} y'' + By' + 2y &= 0 \\ y(0) &= 0 \\ y'(0) &= 1 \end{aligned}\right\}$$

for each of the following values of B: 0, 1, 2, $2\sqrt{2}$, and 3.

(b) Graph the five solutions on the same graph.

(c) Describe a mechanical system composed of a mass of mass 1, a spring with spring constant 2, and a dashpot with damping coefficient B modeled by the initial value problem given above, and describe the behavior of the system for each of the values of B.

The following procedure is known as *solution in the transform domain*: Write the equations for the model of a system and immediately take the transform of each equation, obtaining the *transformed model*, a system of equations. Substitute the initial values. Then solve the transformed model for the transform of the unknown, and take the inverse transform. Use solution in the transformed domain to solve the following problems:

34. Problem 29.

35. In Figure 2, let $R = 2$, $C = \frac{1}{10}$, $L = 1$, and $i(t) = \cos t$, with $v_c(0) = -1$ and $i_2(0) = 1$. Find $v_R(t)$.

8. The Characteristic Polynomial for a Homogeneous Equation

If the given equation is homogeneous, that is, of the form

$$a_n y^{(n)} + a_{n-1} y^{(n-1)} + \cdots + a_1 y' + a_0 = 0 \tag{32}$$

and all coefficients are constant, then the Laplace transform of the solution of any initial value problem is a rational function

$$\mathcal{L}[y](s) = \frac{q(s)}{p(s)}$$

where the polynomial in the numerator depends on the initial conditions $y(0)$, $y'(0)$, ..., $y^{(n-1)}(0)$, but that in the denominator,

$$p(s) = a_n s^n + a_{n-1} s^{n-1} + \cdots + a_1 s + a_0 \tag{33}$$

is determined only by the differential equation itself. The polynomial $p(s)$ is called the *characteristic polynomial* of the equation. It is easy to write down because the coefficients are the constants multiplying the various derivatives of y, and it completely determines the form of the partial-fraction decomposition of $\mathcal{L}[y](s)$. Specifically, if $(s - \alpha)^m$ appears in the factoring of $p(s)$, then

$$\frac{B_1}{s - \alpha} + \frac{B_2}{(s - \alpha)^2} + \cdots + \frac{B_m}{(s - \alpha)^m}$$

appears in the partial-fraction decomposition of $\mathcal{L}[y](s)$. Taking the inverse Laplace transform shows that

$$B_1 e^{\alpha t} + B_2 t e^{\alpha t} + \cdots + B_m t^{m-1} \frac{e^{\alpha t}}{(m-1)!}$$

appears in the general solution of the equation, whatever the numbers $B_1, \ldots, B_m$ may be. Thus, a factor $(s - \alpha)^m$ in $p(s)$ determines a portion of the general solution of the equation.

For example,

$$y''' + 5y'' + 8y' + 4y = 0 \tag{34}$$

has characteristic polynomial

$$p(s) = s^3 + 5s^2 + 8s + 4 \tag{35}$$

$$= (s + 2)^2(s + 1) \tag{36}$$

and so
$$y(t) = A_1 e^{-2t} + A_2 te^{-2t} + A_3 e^{-t} \qquad (37)$$

is the general solution of (34), where A_1, A_2, and A_3 are arbitrary.

By contrast, taking the Laplace transform of (34) leads to the following calculation:

$$s^3 \mathscr{L}[y](s) - s^2 y(0) - sy'(0) - y''(0) + 5\{s^2 \mathscr{L}[y](s) - sy(0) - y'(0)\}$$
$$+ 8\{s\mathscr{L}[y](s) - y(0)\} + 4\mathscr{L}[y](s) = 0$$

$$\mathscr{L}[y](s) = \frac{s^2 y(0) + sy'(0) + y''(0) + 5sy(0) + 5y'(0) + 8y(0)}{s^3 + 5s^2 + 8s + 4}$$

$$= \frac{B_1}{s + 2} + \frac{B_2}{(s + 2)^2} + \frac{B_3}{s + 1}$$

That is, $y(t) = A_1 e^{-2t} + A_2 te^{-2t} + A_3 e^{-t}$ for some numbers A_1, A_2, A_3

Clearly the calculation that leads from (34) to (37) requires less writing and less waste effort.

The characteristic polynomial can be used in the way illustrated to find the *general* solution of a *homogeneous* linear equation (with constant coefficients). First one writes the equation as in (32) or (34), and then the characteristic polynomial as in (33) or (35). The next step is to factor the polynomial as in (36), and then write down at once the corresponding solution functions, the inverse Laplace transforms of the rational functions for which the factors are the denominators. Each inverse transform must be given a different arbitrary coefficient. For a factor of the form $(s - \alpha)^m$, one must write the terms

$$A_1 e^{\alpha t} + A_2 te^{\alpha t} + A_3 t^2 e^{\alpha t} + \cdots + A_m t^{m-1} e^{\alpha t}$$

in the general solution. For a factor of the form

$$[(s - a)^2 + b^2]^m$$

one must write the terms

$$B_1 e^{at} \cos bt + B_2 e^{at} \sin bt + B_3 te^{at} \cos bt + B_4 te^{at} \sin bt$$

$$+ \cdots + B_{2m-1} t^{m-1} e^{at} \cos bt + B_{2m} t^{m-1} e^{at} \sin bt$$

9. Sample Problems with Answers

Write and factor the characteristic polynomial for each of the following homogeneous equations. Then write the general solution of the equation.

E. $y''' + 6y'' + 9y' = 0$
     Answers: $p(s) = s^3 + 6s^2 + 9s = s(s + 3)^2$; $y(t) = A_1 + A_2 e^{-3t} + A_2 te^{-3t}$

F. $y''' + 3y'' + 3y' + y = 0$
     Answers: $p(s) = s^3 + 3s^2 + 3s + 1 = (s + 1)^3$; $y(t) = e^{-t}(A_1 + A_2 t + A_3 t^2)$

G. $y'' + 8y' + 25y = 0$
     Answers: $p(s) = s^2 + 8s + 25 = [(s + 4)^2 + 3^2]^1$; $y(t) = A_1 e^{-4t} \cos 3t + A_2 e^{-4t} \sin 3t$

H. $y^{(4)} + 2y'' + y = 0$
     Answers: $p(s) = s^4 + 2s^2 + 1 = (s^2 + 1)^2$; $y(t) = A_1 \cos t + A_2 \sin t + A_3 t \cos t + A_4 t \sin t$

10. Problems

Write and factor the characteristic polynomial of each of the following homogeneous equations. Then write the general solution of the equation.

36. $y'' + 4y' + 4y = 0$ **37.** $y'' + 4y' + 3y = 0$

38. $y'' + 4y' + 2y = 0$ **39.** $y'' + 4y' + 5y = 0$

40. $y''' + 4y'' + 5y' = 0$ **41.** $y''' = 0$

42. $y'' + 5y = 0$ **43.** $y^{(4)} + 6y'' + 9y = 0$

Each of the following polynomials is the characteristic polynomial of some homogeneous linear differential equation with constant coefficients. Write the equation, and write the general solution.

44. $(s + 2)^4$ **45.** $(s + 3)^2(s^2 + 1)^2$ **46.** $(s^2 + 2)^3$

D. STEP FUNCTIONS AND THEIR APPLICATIONS

1. The Unit Step Function and Its Translates

The *unit step function* is defined by the formula

$$u(t) = \begin{cases} 0 & t \le 0 \\ 1 & t > 0 \end{cases} \tag{1}$$

Its graph is Figure 1. The function with unit step up at $t = t_1$ is obtained from u by horizontal translation by t_1, and it is given in these formulas:

$$u(t - t_1) = \begin{cases} 0 & t - t_1 \le 0 \\ 1 & t - t_1 > 0 \end{cases}$$

$$= \begin{cases} 0 & t \le t_1 \\ 1 & t > t_1 \end{cases}$$

(See Figure 2. In Chapter 1, see Part 1 of Section 1 and Part 6 of Section E.) From these functions can be constructed a whole collection of functions useful in applications. For instance, in the circuit of Figure 3, if the switch is open for $t \le t_1$, closed for $t_1 < t \le t_2$, and open again for $t > t_2$, then the voltage across

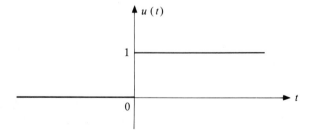

Figure 1 The unit step function.

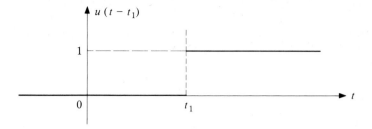

Figure 2 The function with unit step at $t = t_1$.

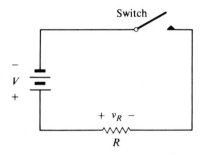

Figure 3 Circuit with battery, resistor, and switch.

the resistor is as shown in Figure 4. It is

$$v_R(t) = \begin{cases} 0 & t \leq t_1 \\ V & t_1 < t \leq t_2 \\ 0 & t_2 < t \end{cases}$$

$$= Vu(t - t_1) - Vu(t - t_2)$$

That is, at $t = t_1$ it steps up from zero to voltage V, and at $t = t_2$ it steps down by V, to voltage zero again.

If a rocket engine is fired from t_1 to t_2, producing V newtons of force during that interval and zero force at other times, then the graph of the thrust is Figure 4 also.

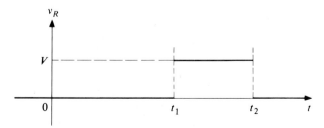

Figure 4 The graph of v_R.

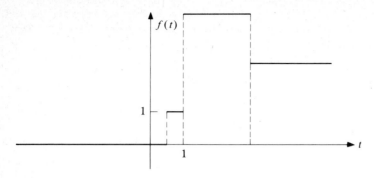

Figure 5 Function which is piecewise constant.

  The function graphed in Figure 5 is constant upon intervals (piecewise constant). It may be defined by the formula

$$f(t) = \begin{cases} 0 & t \leq \frac{1}{2} \\ 1 & \frac{1}{2} < t \leq 1 \\ 4 & 1 < t \leq 3 \\ \frac{5}{2} & 3 < t \end{cases} \tag{2}$$

giving the function values taken on the various intervals. It may also be written more concisely as a linear combination of step functions,

$$f(t) = u(t - \tfrac{1}{2}) + 3u(t - 1) - \tfrac{3}{2}u(t - 3)$$

in which the coefficient of $u(t - t_1)$ is the height of the step up at t_1. The height of each step is the difference between the values on the two adjacent intervals, that on the right minus that on the left.

  Observe that the function values are recovered by evaluating the functions in the usual way. For instance,

$$f(2) = u(2 - \tfrac{1}{2}) + 3u(2 - 1) - \tfrac{3}{2}u(2 - 3)$$
$$= 1 \quad\quad + (3)(1) \quad\quad - (\tfrac{3}{2})(0) = 4$$

which agrees with (2).

  The function u defined in (1) will appear many times, because of its usefulness. You should remember it, just as you remember t^m, cos, ln, $|t|$, and many others.

2. The Laplace Transform of a Step Function

If $t_1 < 0$, then $u(t - t_1) = 1$ for all $t > 0$. Therefore,

$$\mathscr{L}[u(t - t_1)](s) = \mathscr{L}[1](s) = \frac{1}{s}$$

This is the transform of the step function whose step occurs to the left of the origin. If $t_1 \geq 0$, then the step occurs to the right of the origin and

$$\mathscr{L}[u(t - t_1)](s) = \int_0^\infty u(t - t_1)e^{-st}\, dt$$

$$= \int_{t_1}^\infty e^{-st}\, dt$$

$$= \lim_{T\to\infty} \left. \frac{e^{-st}}{-s}\right|_{t_1}^T = \frac{e^{-st_1}}{s} - \lim_{T\to\infty} \frac{e^{-sT}}{s}$$

$$= \frac{e^{-st_1}}{s}$$

(Recall that we consider the variable s to take large positive values. See Section A.) If $t_2 > t_1 > 0$, then

$$\mathscr{L}[u(t - t_1) - u(t - t_2)](s) = \int_0^{t_1} 0\, dt + \int_{t_1}^{t_2} e^{-st}\, dt + \int_{t_2}^\infty 0\, dt$$

$$= \int_{t_1}^{t_2} e^{-st}\, dt = \frac{e^{-st_2}}{-s} - \frac{e^{-st_1}}{-s}$$

$$= \frac{1}{s}e^{-st_1} - \frac{1}{s}e^{-st_2}$$

which is $\mathscr{L}[u(t - t_1)](s) - \mathscr{L}[u(t - t_2)](s)$, as we could have seen from the linearity of $\mathscr{L}$.

3. Example: Second Shifting Property

The result of multiplying a function f by $u(t - t_1)$ is a new function whose value at t is $u(t - t_1)f(t)$. This value equals $f(t)$ unless $t \leq t_1$, in which case the value is zero. For instance, graphs of functions of the form $g(t) = A(1 - e^{-at})$ and $u(t)g(t)$ are given in Figures 7 and 8. Figure 9 shows the translate by t_1, that is, $u(t - t_1)g(t - t_1)$. Observe how the zero value of the factors $u(t)$ and $u(t - t_1)$ affects the product.

Figure 6 shows a circuit whose behavior is like that in Figure 9, as we shall see. The switch is open for $t \leq t_1$, and closed for $t_1 < t$, where $t_1 \geq 0$. Thus $i(0) = 0$ and

$$v_R + v_L = u(t - t_1)$$

The initial value problem is

$$\left. \begin{array}{l} Li'(t) + Ri(t) = u(t - t_1) \\ i(0) = 0 \end{array} \right\} \tag{3}$$

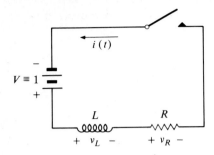

Figure 6 *RL* circuit with battery and switch.

Taking Laplace transforms gives

$$Ls\,\mathscr{L}[i](s) + R\,\mathscr{L}[i](s) = \frac{e^{-t_1 s}}{s}$$

$$\mathscr{L}[i](s) = \frac{e^{-t_1 s}}{s(sL + R)}$$

$$= \frac{e^{-t_1 s}}{L} \frac{1}{s(s + R/L)}$$

Decomposition of the rational function into partial fractions gives

$$\mathscr{L}[i](s) = e^{-t_1 s}\frac{1}{R}\left(\frac{1}{s} - \frac{1}{s + R/L}\right) \tag{4}$$

This transformation is of a very special form,

$$e^{-t_1 s}\mathscr{L}[g](s) \tag{5}$$

where $\mathscr{L}[g]$ is the Laplace transform of a function we can find; in this case,

$$g(t) = \frac{1}{R}\left(1 - e^{-(R/L)t}\right)$$

The second shifting property, which we are about to derive, will tell us that the inverse transform of (4) is

$$i(t) = u(t - t_1)g(t - t_1) = \begin{cases} 0 & t \le t_1 \\ \frac{1}{R}\left(1 - e^{-(R/L)(t - t_1)}\right) & t > t_1 \end{cases} \tag{6}$$

The inverse transform is not g itself, which is shown in Figure 7, nor g multiplied by $u(t)$, which is shown in Figure 8, but the result of translating the latter function to the right by t_1, as shown in Figure 9.

In fact, Figure 8 shows the special case for which $t_1 = 0$, the solution to the initial value problem

$$\left.\begin{array}{c} Li' + Ri = u(t) \\ i(0) = 0 \end{array}\right\} \tag{7}$$

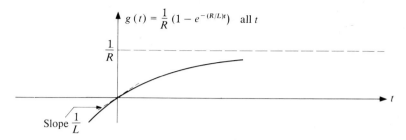

Figure 7 Graph of $g(t)$, the solution of (8).

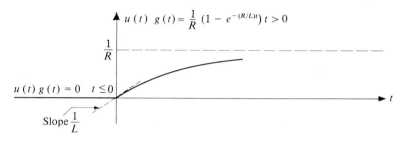

Figure 8 Graph of $u(t)g(t)$, the solution of (7).

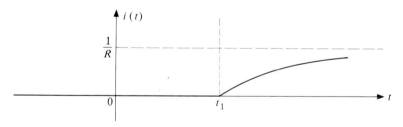

Figure 9 Graph of (6), $u(t - t_1)g(t - t_1)$, the solution of (3).

Figure 7 is the solution of the problem

$$Li' + Ri = 1 \atop i(0) = 0 \bigg\}$$

(8)

in which the input has no step at all.

Now it is time to derive the formula showing that the Laplace transform of (6) is (5).

$$\mathscr{L}[u(t - t_1)g(t - t_1)](s) = \int_0^\infty u(t - t_1)g(t - t_1)e^{-st}\,dt$$

$$= \int_{t_1}^\infty g(t - t_1)e^{-st}\,dt$$

(9)

With the change of variable

$$z = t - t_1$$
$$t = z + t_1$$
$$dt = dz$$

(9) becomes
$$\int_{t_1}^{\infty} g(t - t_1)e^{-st}\, dt = \int_0^{\infty} g(z)e^{-s(z+t_1)}\, dz$$

$$= e^{-st_1} \int_0^{\infty} g(z)e^{-sz}\, dz$$

$$= e^{-st_1} \mathscr{L}[g](s)$$

Since the derivation is also valid for other functions g, we have derived the desired formula. We will use it in the following form.

Second Shifting Theorem *If g is a function with a Laplace transform $\mathscr{L}[g]$, and if $t_1 \geq 0$, then the inverse Laplace transform of $e^{-t_1 s}\mathscr{L}[g](s)$ is $u(t - t_1)g(t - t_1)$, for $t \geq 0$.*

Notice first that if $g(t) \equiv 1$ then $\mathscr{L}[g](s) = 1/s$ and

$$\mathscr{L}[u(t - t_1)g(t - t_1)](s) = \mathscr{L}[u(t - t_1)](s) = \frac{e^{-t_1 s}}{s} = e^{-t_1 s}\mathscr{L}[g](s)$$

just as the theorem says. Notice further that if the nonhomogeneous term of a differential equation contains a step function or discontinuity, then solving the equation with Laplace transforms will always require a use of the Second Shifting Theorem.

4. Example

Find $x(t)$ for $t \geq 0$ for the system of Figure 10 if $M = K = 1, f(t) = u(t - \pi/2)$, $x(0) = 0$, and $x'(0) = 1$.

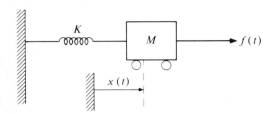

Figure 10 Spring-mass system.

The initial value problem is

$$x''(t) + x(t) = u\left(t - \frac{\pi}{2}\right)$$ (10)

$$x(0) = 0$$

$$x'(0) = 1$$

The transformed problem is this:

$$s^2 \mathcal{L}[x](s) - 1 + \mathcal{L}[x](s) = \frac{e^{-s\pi/2}}{s}$$

$$\mathcal{L}[x](s) = \frac{1}{s^2 + 1} + \frac{e^{-s\pi/2}}{s(s^2 + 1)}$$

$$= \frac{1}{s^2 + 1} + \left(\frac{1}{s} + \frac{-s}{s^2 + 1}\right) e^{-s\pi/2}$$

$$= \mathcal{L}[\sin t](s) + e^{-s\pi/2} \mathcal{L}[1 - \cos t](s)$$

Taking the inverse transform shows that

$$x(t) = \sin t + u\left(t - \frac{\pi}{2}\right)\left[1 - \cos\left(t - \frac{\pi}{2}\right)\right] \qquad t \geq 0$$

$$x(t) = \begin{cases} \sin t & 0 \leq t \leq \frac{\pi}{2} \\ \sin t + 1 - \sin t & \frac{\pi}{2} < t \end{cases}$$

or

$$x(t) = \begin{cases} \sin t & 0 \leq t \leq \frac{\pi}{2} \\ 1 & \frac{\pi}{2} < t \end{cases}$$

since

$$\cos\left(t - \frac{\pi}{2}\right) = \cos t \cos \frac{\pi}{2} + \sin t \sin \frac{\pi}{2} = \sin t$$

See Figure 11. We have not graphed x on the left half line, nor was it asked that we should.

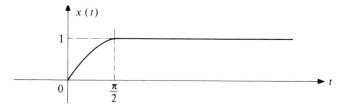

Figure 11 Graph of $x(t)$ for $t \geq 0$.

At $t = t_1 = \pi/2$, both x and x' are continuous, but x'' jumps by 1 (from $-\sin \pi/2 = -1$ to 0) as one would expect from the differential equation (10). This matter will be returned to in Part 10 of this section.

5. Method and Example: Using the Second Shifting Theorem to Find and Graph an Inverse Laplace Transform

If $\mathcal{L}[f](s) = 3e^{-2s}[1/s(s + 1)]$, then both the finding and the graphing of f follow the same steps.

Step 1. Decompose the rational function into partial fractions.

Step 2. Identify the rational function as $\mathcal{L}[g](s)$, find g, and graph g.

Step 3. Change variable, giving $g(t - t_1)$, and shift the graph.

Step 4. Multiply by $u(t - t_1)$, and alter the graph accordingly.

Notice, though, that there is no more information contained in this statement of four steps than was contained in the statement of the theorem.

Step 1 $\qquad 3\dfrac{e^{-2s}}{s(s+1)} = e^{-2s}\dfrac{3}{s(s+1)} = e^{-2s}3\left(\dfrac{1}{s} - \dfrac{1}{s+1}\right)$

Step 2 $\qquad 3\left(\dfrac{1}{s} - \dfrac{1}{s+1}\right) = \mathcal{L}[g](s) \qquad$ and $\qquad g(t) = 3(1 - e^{-t})$

Step 3 $\qquad g(t - t_1) = g(t - 2) = 3(1 - e^{-(t-2)})$

Step 4 $\qquad f(t) = u(t - t_1)g(t - t_1)$

$\qquad\qquad = u(t - 2)3(1 - e^{-(t-2)})$

$$f(t) = \begin{cases} 0 & t \le 2 \\ 3(1 - e^{-(t-2)}) & t > 2 \end{cases}$$

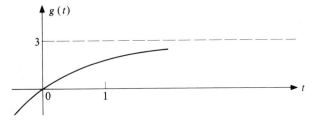

Figure 12 Graph of $g(t)$.

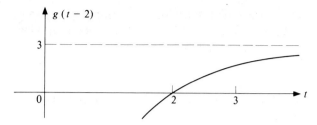

Figure 13 Graph of $g(t - 2)$.

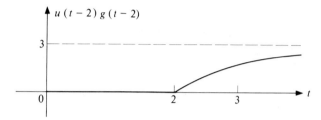

Figure 14 Graph of $u(t - 2)g(t - 2)$.

6. Example

For the circuit illustrated in Figure 15 with $i(0) = i'(0) = 0$ and

$$v(t) = \begin{cases} -16t & 0 \le t \le 3 \\ -48 & 3 < t \end{cases}$$

find and graph $i(t)$ for $t \ge 0$.

The differential equation for $i(t)$ is

$$Li''(t) + \frac{1}{C}i(t) = v'(t)$$

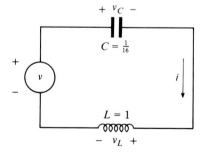

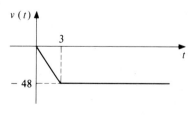

Figure 15 The LC circuit and its impressed voltage.

Since

$$v'(t) = \begin{cases} -16 & 0 < t < 3 \\ 0 & 3 < t \end{cases}$$

the initial value problem for $i(t)$ is

$$i''(t) + 16i(t) = -16 + 16u(t-3)$$
$$i(0) = 0$$
$$i'(0) = 0$$

Taking Laplace transforms gives

$$(s^2 + 16)\mathscr{L}[i](s) = 16\left(-\frac{1}{s} + \frac{e^{-3s}}{s}\right)$$

$$\mathscr{L}[i](s) = \frac{16}{s(s^2 + 16)}(-1 + e^{-s})$$

$$= \left(\frac{1}{s} - \frac{s}{s^2 + 16}\right)(-1 + e^{-3s})$$

$$= \mathscr{L}[g](s) - e^{-3s}\mathscr{L}[g](s)$$

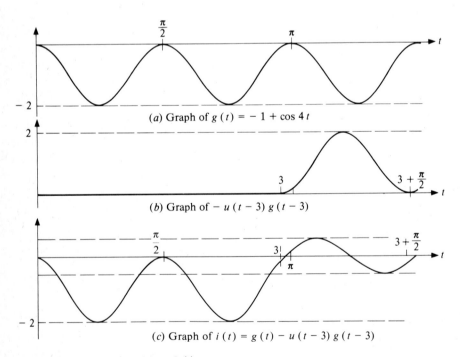

(a) Graph of $g(t) = -1 + \cos 4t$

(b) Graph of $-u(t-3)g(t-3)$

(c) Graph of $i(t) = g(t) - u(t-3)g(t-3)$

Figure 16 Steps in the graphing of $i(t)$.

where $g(t) = -1 + \cos 4t$. Thus,

$$i(t) = g(t) - u(t - 3)g(t - 3)$$
$$= -1 + \cos 4t - u(t - 3)[1 - \cos 4(t - 3)] \qquad t \geq 0$$

The graphs of g, $-u(t - 3)g(t - 3)$, and their sum i are given in Figure 16. (In Chapter I, see Parts 2 and 3 of Section I.)

7. Sample Problems with Answers

A. Write each function as a linear combination of constants and step functions, and then take its Laplace transform using the formula for the Laplace transform of a step function.

(a) $f(t) = \begin{cases} 1 & t \leq 2 \\ 7 & 2 < t \leq 3 \\ 0 & 3 < t \end{cases}$

Answers: $f(t) = 1 + 6u(t - 2) - 7u(t - 3)$; $\mathcal{L}[f](s) = \dfrac{1}{s} + \dfrac{6}{s}e^{-2s} - \dfrac{7}{s}e^{-3s}$

(b) $f(t) = \begin{cases} 1 & t \leq -1 \\ 7 & -1 < t \leq 3 \\ 0 & 3 < t \end{cases}$

Answers: $f(t) = 1 + 6u(t + 1) - 7u(t - 3)$; $\mathcal{L}[f](s) = \dfrac{7}{s} - \dfrac{7}{s}e^{-3s}$

B. Decompose each rational factor into partial fractions, and take the inverse Laplace transform.

(a) $F(s) = e^{-3s}\dfrac{1}{s^2 - 1}$

Answers: $F(s) = e^{-3s}\dfrac{1}{2}\left(\dfrac{1}{s - 1} - \dfrac{1}{s + 1}\right)$; $\mathcal{L}^{-1}[F](t) = u(t - 3)(\tfrac{1}{2}e^{t - 3} - \tfrac{1}{2}e^{-(t - 3)})$ $\qquad t \geq 0$

(b) $F(s) = (e^{-3s} + 1)\dfrac{1}{s^2 - 1}$

Answers: $F(s) = e^{-3s}\dfrac{1}{2}\left(\dfrac{1}{s - 1} - \dfrac{1}{s + 1}\right) + \dfrac{1}{2}\left(\dfrac{1}{s - 1} - \dfrac{1}{s + 1}\right)$;

$$\mathcal{L}^{-1}[F](t) = u(t - 3)(\tfrac{1}{2}e^{t - 3} - \tfrac{1}{2}e^{-(t - 3)}) + \tfrac{1}{2}e^t - \tfrac{1}{2}e^{-t} \qquad t \geq 0$$

8. Problems

For each of the following, first graph the function f and find $\mathcal{L}[f]$ by integration; then represent f as a linear combination of constants and step functions and find $\mathcal{L}[f]$ a second time using the formula for the Laplace transform of a step function.

1. $f(t) = \begin{cases} 0 & t \leq 1 \\ 1 & 1 < t \leq 2 \\ 4 & 2 < t \end{cases}$

2. $f(t) = \begin{cases} 4 & t \leq -1 \\ 1 & -1 < t \leq 1 \\ -2 & 1 < t \end{cases}$

3. $f(t) = \begin{cases} 0 & t \leq 2 \\ -5 & 2 < t \leq 3 \\ 2 & 3 < t \leq 5 \\ 0 & 5 < t \end{cases}$

4. $f(t) \begin{cases} -3 & t \leq 0 \\ 7 & 0 < t \leq 4 \\ 0 & 4 < t \end{cases}$

Find the inverse Laplace transforms.

5. (a) $\dfrac{e^{-3s}}{s}$ **(b)** $\dfrac{e^{-\pi s} + 2e^{-4s} - 4e^{-6s}}{s}$

6. (a) $\dfrac{e^{-2s}}{s+7}$ **(b)** $\dfrac{e^{-2s}}{s^2+9}$ **(c)** $e^{-2s}\dfrac{1}{s^2-9}$

7. (a) $\dfrac{e^{-\pi s}}{s^2+1}$ **(b)** $\dfrac{se^{-\pi s}}{s^2+1}$ **(c)** $\dfrac{e^{-3s}}{s+2}$ **(d)** $e^{-3s}\left(\dfrac{1}{s}-\dfrac{1}{s+2}\right)$

8. (a) $\dfrac{e^{-\pi s}}{s^2+4}$ **(b)** $\dfrac{1}{3}\dfrac{e^{-2\pi s}}{s^2+4}$ **(c)** $e^{-3s}\dfrac{s}{s^2+3s+2}$

    (d) $(1-e^{-3s})\dfrac{s}{s^2+3s+2}$ **(e)** $e^{-3s}\dfrac{2}{s^2+6s+13}$ **(f)** $e^{-3s}\dfrac{4s+3}{s^2+6s+13}$

    (g) $(1-3^{-3s})\dfrac{4s+3}{s^2+6s+13}$

Solve the following initial value problems using Laplace transforms, and graph the solutions.

9. $\begin{aligned}y'(t) + 3y(t) &= 1 - u(t-2)\\ y(0) &= -\tfrac{1}{3}\end{aligned}$ **10.** $\begin{aligned}y'(t) + 4y(t) &= u(t-1)\\ y(0) &= 1\end{aligned}$

11. $\begin{aligned}y''(t) + y(t) &= 1 - u(t-2\pi)\\ y(0) &= 0\\ y'(0) &= 0\end{aligned}$ **12.** $\begin{aligned}y''(t) + y(t) &= u(t-2\pi)\\ y(0) &= 0\\ y'(0) &= 1\end{aligned}$

Solve the following initial value problems.

13. $\begin{aligned}y''(t) + 4y'(t) + 13y(t) &= 5 - u(t-1)\\ y(0) &= 0\\ y'(0) &= 0\end{aligned}$ **14.** $\begin{aligned}y'''(t) - y(t) &= u(t-3)\\ y(0) &= 1\\ y'(0) &= 0\\ y''(0) &= 0\end{aligned}$

15. $\begin{aligned}y'(t) + 3y(t) &= (t-2)u(t-2)\\ y(0) &= 0\end{aligned}$

16. $\begin{aligned}y'(t) + 3y(t) &= f(t)\\ y(0) &= 0\end{aligned}$ where $f(t) = \begin{cases} -1 & t \le 0 \\ 1 & t > 0 \end{cases}$

9. Method and Examples: Finding (Direct) Laplace Transforms Using the Second Shifting Theorem

To emphasize the difference between the Laplace transformation and its inverse, $\mathscr{L}[f]$ is sometimes called the *direct* transform of f, while $\mathscr{L}^{-1}[f]$ is always called the *inverse* transform. So far we have used the Second Shifting Theorem only for finding inverse transforms. Now we shall use it to find direct Laplace transforms of functions defined by different formulas on different intervals, such as

$$f(t) = \begin{cases} 0 & t \le 3 \\ e^t & t > 3 \end{cases}$$

This function is of the form

$$f(t) = \begin{cases} 0 & t \le t_1 \\ h(t) & t > t_1 \end{cases}$$

where h is given and t_1 is positive. Observe that f can also be written

$$f(t) = u(t - t_1)h(t) \tag{11}$$

because when $t \leq t_1$ then the vanishing of the first factor makes the product zero, but when $t > t_1$ then the first factor is 1 and the product is $h(t)$.

Since the Second Shifting Theorem says

$$\mathscr{L}[u(t - t_1)g(t - t_1)](s) = e^{-st_1}\mathscr{L}[g](s) \tag{12}$$

it follows that (11) would be more immediately useful if it were written as

$$f(t) = u(t - t_1)g(t - t_1) \tag{13}$$

Unfortunately, g is unknown, but equating the right-hand sides of (11) and (13) gives

$$u(t - t_1)g(t - t_1) = u(t - t_1)h(t)$$

from which we conclude that

$$g(t - t_1) = h(t) \qquad \text{if } t > t_1$$
$$g(t) = h(t + t_1) \qquad \text{if } t > 0$$

Now we know g, and so these two formulas may be substituted into (12), giving the direct form of the Second Shifting Theorem, which you should memorize:

$$\mathscr{L}[u(t - t_1)h(t)](s) = e^{-st_1}\mathscr{L}[h(t + t_1)](s) \qquad \text{if } t_1 > 0 \tag{14}$$

For example, if

$$f(t) = \begin{cases} 0 & t \leq \dfrac{5\pi}{2} \\[2mm] \sin t & t > \dfrac{5\pi}{2} \end{cases}$$

$$= u\left(t - \frac{5\pi}{2}\right)\sin t$$

then $t_1 = 5\pi/2$, $h(t) = \sin t = g(t - 5\pi/2)$, and $h(t + t_1)$ is

$$h\left(t + \frac{5\pi}{2}\right) = \sin\left(t + \frac{5\pi}{2}\right)$$

$$= \sin t \cos \frac{5\pi}{2} + \cos t \sin \frac{5\pi}{2}$$

$$= \cos t$$

Therefore, $\qquad \mathscr{L}\left[u\left(t - \dfrac{5\pi}{2}\right)\sin t\right](s) = e^{(-5\pi s/2)}\mathscr{L}\left[\sin\left(t + \dfrac{5\pi}{2}\right)\right](s)$

$$= e^{(-5\pi s/2)}\mathscr{L}[\cos t](s)$$

$$= \dfrac{se^{(-5\pi s/2)}}{s^2 + 1}$$

One may make a similar analysis of

$$f(t) = \begin{cases} t^2 & t \le 3 \\ 0 & t > 3 \end{cases}$$

$$= [1 - u(t - 3)]t^2$$

$$= t^2 - u(t - 3)t^2 \qquad (15)$$

Notice first that the function $1 - u(t - 3)$ in brackets has value 1 for $t \le 3$, and 0 for $t > 3$. In fact, (15) shows that $f(t)$ is $t^2 - 0 = t^2$ for $t \le 3$ but $f(t) = t^2 - t^2 = 0$ for $t > 3$. That is, the second term of (15) subtracts the value of the first term for $t > 3$.

The direct form of the Second Shifting Theorem (14) is applied to (15) as follows:

$$\mathscr{L}[f](s) = \mathscr{L}[t^2](s) - \mathscr{L}[u(t - 3)t^2](s)$$

$$= \mathscr{L}[t^2](s) - e^{-3s}\mathscr{L}[(t + 3)^2](s)$$

$$= \mathscr{L}[t^2](s) - e^{-3s}\mathscr{L}[t^2 + 6t + 9](s)$$

$$= \dfrac{2}{s^3} - e^{-3s}\left(\dfrac{2}{s^3} + \dfrac{6}{s^2} + \dfrac{9}{s}\right)$$

In general, if a function f has values given by $h(t)$ on the interval $t_1 < t \le t_2$, then writing f as a sum of multiples of step functions will be expected to produce a corresponding expression

$$[u(t - t_1) - u(t - t_2)]h(t) \qquad (16)$$

If the interval is unbounded on the right, then $u(t - t_2)$ is replaced by zero, as in the first example. If the interval is unbounded on the left, then $u(t - t_1)$ is replaced by 1, as in the second example.

Here is a more general example.

$$f(t) = \begin{cases} t^2 & t \le 3 \\ e^{2t} & 3 < t \le \dfrac{5\pi}{2} \\ \sin t & \dfrac{5\pi}{2} < t \end{cases}$$

$$= [1 - u(t - 3)]t^2 + \left[u(t - 3) - u\left(t - \dfrac{5\pi}{2}\right)\right]e^{2t} + u\left(t - \dfrac{5\pi}{2}\right)\sin t$$

$$= t^2 + u(t - 3)(-t^2 + e^{2t}) + u\left(t - \dfrac{5\pi}{2}\right)(-e^{2t} + \sin t)$$

In this, $h_1(t) = -t^2 + e^{2t}$ and $h_2(t) = -e^{2t} + \sin t$

so that $h_1(t + 3) = -(t + 3)^2 + e^{2(t+3)} = -t^2 - 6t - 9 + e^6 e^{2t}$

and $h_2\left(t + \dfrac{5\pi}{2}\right) = -e^{2(t+5\pi/2)} + \sin\left(t + \dfrac{5\pi}{2}\right) = -e^{5\pi}e^{2t} + \cos t$

Applying the Second Shifting Theorem gives

$$\mathcal{L}[f](s) = \mathcal{L}[t^2](s) + e^{-3s}\mathcal{L}[h_1(t + 3)](s) + e^{(-5\pi s/2)}\mathcal{L}\left[h_2\left(t + \dfrac{5\pi}{2}\right)\right](s)$$

$$= \dfrac{2}{s^3} + e^{-3s}\left(\dfrac{-2}{s^3} - \dfrac{6}{s^2} - \dfrac{9}{s} + e^6\dfrac{1}{s-2}\right)$$

$$+ e^{(-5\pi s/2)}\left(-e^{5\pi}\dfrac{1}{s-2} + \dfrac{s}{s^2+1}\right)$$

The following example illustrates what happens in the presence of a step on the *left* half line.

$$f(t) = \begin{cases} 2 & t < -3 \\ t & -3 < t \le 4 \\ 7 & 4 < t \end{cases}$$

$$= 2[1 - u(t + 3)] + [u(t + 3) - u(t - 4)]t + 7u(t - 4)$$

$$= 2 + u(t + 3)(-2 + t) + u(t - 4)(-t + 7)$$

The Laplace transform is taken this way:

$$\mathcal{L}[f](s) = \mathcal{L}[2](s) + \mathcal{L}[u(t + 3)(-2 + t)](s) + \mathcal{L}[u(t - 4)(-t + 7)](s)$$

$$= \mathcal{L}[2](s) + \mathcal{L}[-2 + t](s) + e^{-4s}\mathcal{L}[-(t + 4) + 7](s)$$

$$= \dfrac{2}{s} + \dfrac{-2}{s} + \dfrac{1}{s^2} + e^{-4s}\left(\dfrac{-1}{s^2} + \dfrac{3}{s}\right)$$

$$= \dfrac{1}{s^2} + e^{-4s}\left(\dfrac{-1}{s^2} + \dfrac{3}{s}\right)$$

since $u(t + 3)(-2 + t)$ is just $-2 + t$ on the right half line.

10. Checking the Solutions of Differential Equations

Recall Example 3 earlier in the section:

$$Li'(t) + Ri(t) = u(t - t_1) \qquad (17)$$

$$i(0) = 0 \qquad (18)$$

with $t_1 > 0$. We found that the solution is

$$i(t) = u(t - t_1)\dfrac{1}{R}(1 - e^{-(R/L)(t - t_1)})$$

Let us look again at the manner in which this function satisfies the equation on the two intervals $t < t_1$ and $t > t_1$. On the left interval, the function i is identically zero, and so its derivative is also identically zero.

$$i(t) = 0 \qquad t < t_1 \tag{19}$$

$$i'(t) = 0 \qquad t < t_1 \tag{20}$$

On this interval, Equation (17) is

$$Li'(t) + Ri(t) = 0 \qquad t < t_1$$

Therefore the function i certainly satisfies the equation, and the initial condition (18) as well.

Now consider the interval to the right of t_1, on which i reduces to

$$i(t) = \frac{1}{R}\left(1 - e^{(-R/L)(t - t_1)}\right) \qquad t > t_1 \tag{21}$$

Its derivative is

$$i'(t) = \frac{1}{L} e^{(-R/L)(t - t_1)} \qquad t > t_1 \tag{22}$$

Here, too, the function i may be substituted into Equation (17) to satisfy it, for the equation here is

$$Li'(t) + Ri(t) = 1 \qquad t > t_1$$

Combining formulas (20) for $t < t_1$ and (22) for $t > t_1$ gives a single formula for i':

$$i'(t) = u(t - t_1) \frac{1}{L} e^{(-R/L)(t - t_1)} \qquad t \neq t_1$$

Substitution into the left-hand side of (17) gives

$$Li'(t) + Ri(t) = \frac{L}{L} u(t - t_1) e^{-(R/L)(t - t_1)} + \frac{R}{R} u(t - t_1)(1 - e^{-(R/L)(t - t_1)})$$

$$= u(t - t_1) \qquad t \neq t_1$$

which is the same as (17), the given equation, except at $t = t_1$. There the function i is not differentiable, and so the substitution cannot be carried out. Notice, though, that i is continuous there. See Figure 17.

Now let us consider a second-order equation, solved in Example 4.

$$\left. \begin{array}{l} x''(t) + x(t) = u\left(t - \dfrac{\pi}{2}\right) \\[2mm] x(0) = 0 \\[2mm] x'(0) = 1 \end{array} \right\} \tag{23}$$

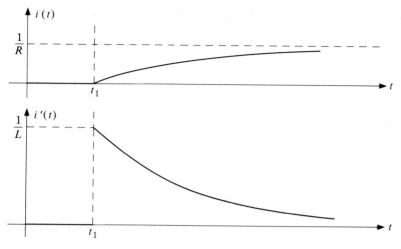

Figure 17 Graph of $i(t)$ and its derivative.

Its solution is

$$x(t) = \begin{cases} \sin t & 0 \le t \le \dfrac{\pi}{2} \\[2mm] 1 & \dfrac{\pi}{2} < t \end{cases}$$

The derivatives of x are

$$x'(t) = \begin{cases} \cos t & 0 \le t \le \dfrac{\pi}{2} \\[2mm] 0 & \dfrac{\pi}{2} < t \end{cases}$$

$$x''(t) = \begin{cases} -\sin t & 0 \le t < \dfrac{\pi}{2} \\[2mm] 0 & \dfrac{\pi}{2} < t \end{cases}$$

See Figure 18. Substitution into the left-hand side of (23) gives

$$x''(t) + x(t) = \begin{cases} -\sin t + \sin t & 0 \le t < \dfrac{\pi}{2} \\[2mm] 0 + 1 & \dfrac{\pi}{2} < t \end{cases}$$

$$= u\left(t - \frac{\pi}{2}\right) \qquad t \ne \frac{\pi}{2}$$

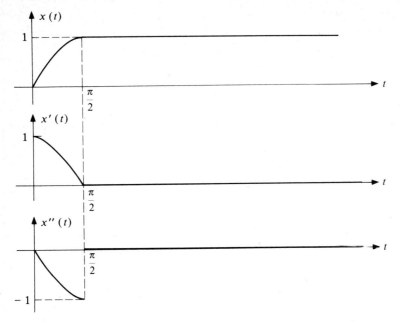

Figure 18 The graph of $x(t)$ and its derivatives.

as required. What happens at $t = \pi/2$? Since x' is not differentiable there, the substitution cannot be performed, but

x is continuous at $t = \pi/2$.
x' is continuous at $t = \pi/2$.
x'' exhibits the step at $\pi/2$ required by the differential equation.

The function x may be called a *generalized solution* of (23), and the function i is a generalized solution of (17). A generalized solution y of a similar equation of higher order, for instance order n, would satisfy the differential equation except at the stepping point t_1 and have the properties

y is continuous at t_1.
y' is continuous at t_1.
. . . .
$y^{(n-1)}$ is continuous at t_1.
$y^{(n)}$ exhibits the step at t_1 required by the differential equation.

11. Example: Excitation over a Very Short Interval

Consider the initial value problem

$$Mv'(t) + Bv(t) = [u(t) - u(t - t_2)]f_1$$
$$v(0) = 0$$

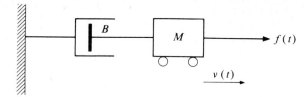

Figure 19 Mass-dashpot system.

This corresponds to a mass-dashpot system which is excited over the time interval $0 < t \le t_2$ with a constant force f_1. Taking Laplace transforms gives

$$(Ms + B)\mathscr{L}[v](s) - Mv(0) = (1 - e^{-st_2})\frac{f_1}{s}$$

$$\mathscr{L}[v](s) = (1 - e^{-st_2})\frac{f_1}{B}\left(\frac{1}{s} - \frac{1}{s + B/M}\right)$$

Using the Second Shifting Theorem with

$$g(t) = \frac{f_1}{B}\left(1 - e^{-(B/M)t}\right)$$

gives

$$v(t) = \frac{f_1}{B}\left[1 - e^{-(B/M)t} - u(t - t_2)\left(1 - e^{-(B/M)(t - t_2)}\right)\right]$$

$$v(t) = \begin{cases} \dfrac{f_1}{B}\left(1 - e^{-(B/M)t}\right) & 0 \le t \le t_2 \\[2ex] \dfrac{f_1}{B}\left(1 - e^{-(B/M)t_2}\right)e^{-(B/M)(t - t_2)} & t_2 < t \end{cases} \qquad (24)$$

The function v rises to a peak at $t = t_2$ and then goes exponentially to zero. See Figure 20.

If t_2 is small, then f_1 may be hard to measure in a practical situation. It is often easier to measure the product $P = f_1 t_2$, which in the present case has the units of

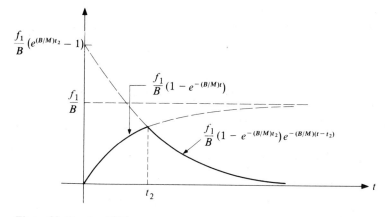

Figure 20 Graph of (24).

momentum. Let us therefore analyze the situation in the following way. M and B are fixed. Let t_2 be a parameter at our disposal, let $f_1 = P/t_2$, where P is the measurable quantity, and let us observe how the solutions of

$$Mv' + Bv = \frac{P}{t_2}[u(t) - u(t - t_2)]$$

$$v(0) = 0 \qquad\qquad (25)$$

depend upon the parameter t_2.

$$(Ms + B)\mathcal{L}[v](s) - Mv(0) = \frac{P}{t_2}\frac{1 - e^{-st_2}}{s} \qquad (26)$$

$$v(t) = \begin{cases} \dfrac{P}{B}\left(\dfrac{1 - e^{-(B/M)t}}{t_2}\right) & 0 \le t \le t_2 \quad (27) \\[3mm] \dfrac{P}{B}\left(\dfrac{1 - e^{-(B/M)t_2}}{t_2}\right)e^{-(B/M)(t - t_2)} & t_2 < t \quad (28) \end{cases}$$

The factor in parenthesis in (27) represents an increasing function, rising at $t = t_2$ to the number

$$\frac{1 - e^{-(B/M)t_2}}{t_2} \qquad (29)$$

which appears again in parenthesis in (28). (See Figure 21.) L'Hospital's rule may be applied to (29). It shows that if t_2 is small then (29) is near the value B/M, and consequently (28) is near the function

$$\frac{P}{M}e^{-(B/M)(t - t_2)} \qquad t_2 < t$$

which is also near

$$\frac{P}{M}e^{-(B/M)t} \qquad t_2 < t$$

Since P has the units of momentum, P/M has the units of a velocity; call it v_1. Our analysis leads to the conclusion that if t_2 is small, then

$$v(t) \text{ is near } v_1 e^{-(B/M)t} \qquad t_2 < t$$

Now the function $v_1 e^{-(B/M)t}$ is the solution of the problem

$$Mv' + Bv = 0$$

$$v(0) = v_1 \qquad\qquad (30)$$

Thus, if the length of the time interval is very short, but if the integral of the force over that interval is $P = v_1 M$, then the velocity is very nearly that of an unexcited system with given initial velocity v_1, that is, given momentum $P = v_1 M$ at time $t = 0$. (See Figure 21.)

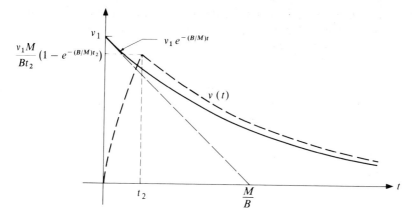

Figure 21 Comparison of $v(t)$ and $v_1 e^{-Bt/M}$, the solutions of (25) and (30), respectively.

This observation should be checked by reference to the transformed equation (26). The limit of the equation's right-hand side as t_2 goes to zero is $v_1 M$ by L'Hospital's rule, and so the limiting equation is

$$(Ms + B)\mathscr{L}[v](s) = Mv_1$$

which is the Laplace transform of (30).

12. Sample Problems with Answers

Graph each function f, write it in the form

$$\cdots + u(t - t_1)h(t) + \cdots$$

and find its Laplace transform.

C. $f(t) = \begin{cases} 0 & t \leq 2 \\ t & t > 2 \end{cases}$ *Answers:* $f(t) = tu(t - 2)$; $\mathscr{L}[f](s) = e^{-2s}\left(\dfrac{1}{s^2} + \dfrac{2}{s}\right)$

D. $f(t) = \begin{cases} e^t & t \leq \pi \\ 0 & t > \pi \end{cases}$ *Answers:* $f(t) = [1 - u(t - \pi)]e^t$; $\mathscr{L}[f](s) = \dfrac{1}{s - 1} - e^{-\pi s}\dfrac{e^{\pi}}{s - 1}$

E. $f(t) = \begin{cases} t^2 & t \leq 1 \\ t^3 & 1 < t \end{cases}$

Answers: $f(t) = [1 - u(t - 1)]t^2 + u(t - 1)t^3$;

$$\mathscr{L}[f](s) = \frac{2}{s^3} - e^{-s}\left(\frac{2}{s^3} + \frac{2}{s^2} + \frac{1}{s}\right) + e^{-s}\left(\frac{6}{s^4} + \frac{6}{s^3} + \frac{3}{s^2} + \frac{1}{s}\right)$$

F. $f(t) = \begin{cases} t^2 & t \leq 1 \\ t & 1 < t \leq 2 \\ 0 & 2 < t \end{cases}$

Answers: $f(t) = [1 - u(t - 1)]t^2 + [u(t - 1) - u(t - 2)]t$;

$$\mathscr{L}[f](s) = \frac{2}{s^3} - e^{-s}\left(\frac{2}{s^3} + \frac{2}{s^2} + \frac{1}{s}\right) + e^{-s}\left(\frac{1}{s^2} + \frac{1}{s}\right) - e^{-2s}\left(\frac{1}{s^2} + \frac{2}{s}\right)$$

13. More Problems

Graph each of the following, and find its Laplace transform.

17. $f(t) = \begin{cases} 0 & t \le 2 \\ (t-2)^2 & t > 2 \end{cases}$ **18.** $f(t) = \begin{cases} 1 - e^{-t} & t \le 1 \\ 0 & t > 1 \end{cases}$

19. $f(t) = \begin{cases} 0 & t \le \pi \\ \sin t & \pi < t \le 2\pi \\ 0 & 2\pi < t \end{cases}$ **20.** $f(t) = \begin{cases} 0 & t \le 0 \\ t & 0 < t \le 2 \\ 4 - t & 2 < t \le 4 \\ 0 & 4 < t \end{cases}$

Find the Laplace transform twice, once by integration and once by the Second Shifting Theorem.

21. $f(t) = u(t-2)e^{t-2}$ **22.** $f(t) = \begin{cases} 0 & 0 \le t \le 4 \\ 3e^{t-4} & 4 < t \end{cases}$

23. $f(t) = \begin{cases} 0 & t \le 3 \\ 2e^t & 3 < t \end{cases}$ **24.** $f(t) = \begin{cases} t & t \le 2 \\ 2 & 2 < t \end{cases}$

25. $f(t) = \begin{cases} t & t \le 2 \\ 4 - t & 2 < t \end{cases}$ **26.** $f(t) = \begin{cases} t & t \le 2 \\ 4 - t & 2 < t \le 4 \\ 0 & 4 < t \end{cases}$

27. $f(t) = \begin{cases} 0 & t \le \pi \\ -\sin t & \pi < t \le 2\pi \\ 0 & 2\pi < t \end{cases}$ **28.** $f(t) = \begin{cases} 0 & t \le \pi \\ \cos\left(t + \dfrac{\pi}{2}\right) & \pi < t \le 2\pi \\ 0 & 2\pi < t \end{cases}$

Solve *and check* these initial value problems. Graph the solutions and their derivatives (up to the order of the equation).

29. $\left. \begin{aligned} y'(t) + y(t) &= u(t-1) \\ y(0) &= 1 \end{aligned} \right|$ **30.** $\left. \begin{aligned} y'(t) + y(t) &= u(t) - u(t-1) \\ y(0) &= 0 \end{aligned} \right|$

31. $\left. \begin{aligned} y'(t) + y(t) &= 1 - tu(t-1) \\ y(0) &= 0 \end{aligned} \right|$ **32.** $\left. \begin{aligned} y''(t) + y(t) &= -\tfrac{1}{2}u\left(t - \dfrac{\pi}{2}\right) \\ y(0) &= 1 \\ y'(0) &= 0 \end{aligned} \right|$

14. Comments on Piecewise-Continuous Functions

The step function is an example of a "piecewise-continuous" function. Those functions f of interest to us are *piecewise continuous on the right half line $t \ge 0$,* satisfying the following conditions:

1. f has a limit from the right at $t_0 = 0$.
2. At each point $t_0 > 0$, f has a limit from the left, $f(t_0 -) = \lim_{t \to t_0 -} f(t)$, and also a limit from the right, $f(t_0 +) = \lim_{t \to t_0 +} f(t)$.
3. f has at most finitely many discontinuities on every (bounded) interval of the form $0 \le t \le T$, where T is finite.

These properties ensure that piecewise-continuous functions are continuous on intervals separated by *jump discontinuities.* The *jump* at a point t_0 is the right-hand limit less the left-hand limit, $f(t_0 +) - f(t_0 -)$.

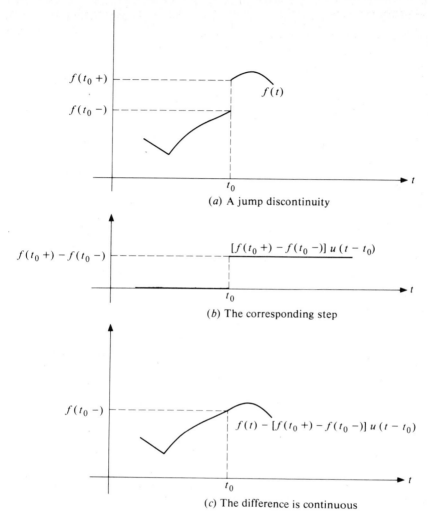

(a) A jump discontinuity

(b) The corresponding step

(c) The difference is continuous

Figure 22 Jump discontinuity decomposed into the sum of a step function and a continuous function.

Near any such discontinuity, a piecewise-continuous function f may be written as the sum of a step function

$$[f(t_0+) - f(t_0-)]u(t - t_0)$$

plus a continuous function

$$f(t) - [f(t_0+) - f(t_0-)]u(t - t_0)$$

(See Figure 22.) Thus, if f is piecewise continuous on the right half line, then the integral

$$\int_0^T f(t)e^{-st}\, dt$$

exists for all finite T. If, in addition, f is of some exponential order S, then for each $s > S$ the integrals will have a limit as T increases, so that f will have a Laplace transform

$$\mathscr{L}[f](s) = \lim_{T \to \infty} \int_0^T f(t)e^{-st}\, dt$$

$$= \int_0^\infty f(t)e^{-st}\, dt$$

defined on the interval $S < s < \infty$.

E. THE DIRAC δ AND ITS USES

Consider first an unrestrained mass. If it is subjected to a net force $f(t)$ during the time interval from t_1 to t_2, then its momentum will change. This change is called the *impulse* given to the mass. Specifically, the impulse is

$$\int_{t_1}^{t_2} f(t)\, dt = \int_{t_1}^{t_2} \frac{d(Mv)}{dt}(t)\, dt = Mv(t_2) - Mv(t_1)$$

This section concerns the differential equations describing the behavior of systems which are excited over very short time intervals, in which, for instance, it is easier to specify the change of momentum than the force which produced it. The use of the Dirac δ is motivated in an extended example (Part 1), from which the essentials are then summarized (Part 2).

1. Example

The system in Figure 1 satisfies the initial value problem

$$Mv' + Bv = 0 \tag{1}$$

$$v(0) = v_0 \tag{2}$$

and has velocity $v(t) = v_0 e^{-(B/M)t}$, as may be shown by separation of variables or the use of Laplace transforms.

Suppose that the system in Figure 1 were hit a sharp blow at time $t = t_1$ and thereby given an impulse I. If $t < t_1$, the velocity must be

$$v(t) = v_0 e^{-(B/M)t} \tag{3}$$

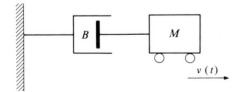

Figure 1 Unexcited mass-dashpot system.

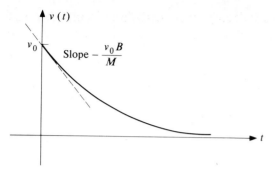

Figure 2 The velocity of the unexcited system.

for then the system has not yet been struck; but at $t = t_1$ the momentum changes by I and the velocity must change by I/M. If the mass is struck in the positive v direction, then the velocity increases by I/M at time $t = t_1$ and reaches velocity

$$v_0 e^{-(B/M)t_1} + \frac{I}{M}$$

After time t_1, though, the system is no longer excited and therefore satisfies the differential equation (1) again. Thus, if $t_1 > t$, then

$$v(t) = v_1 e^{-(B/M)(t - t_1)} \qquad t_1 < t$$

where v_1 is I/M more than $v_0 e^{-(B/M)t_1}$. That is,

$$v(t) = \left(v_0 e^{-(B/M)t_1} + \frac{I}{M}\right) e^{-(B/M)(t - t_1)}$$

$$= v_0 e^{-(B/M)t} + \frac{I}{M} e^{-(B/M)(t - t_1)} \qquad t_1 < t \qquad (4)$$

In (4) the term $v_0 e^{-(B/M)t}$ is the same as the right-hand side of (3). Therefore, (3) and (4) are conveniently combined as follows:

$$v(t) = v_0 e^{-(B/M)t} + u(t - t_1)\frac{I}{M} e^{-(B/M)(t - t_1)} \qquad 0 \le t \qquad (5)$$

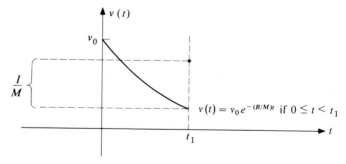

Figure 3 Jump in velocity at time t_1.

The graph of this function (Figure 4) shows the exponential decay both before and after the jump at $t = t_1$.

The analysis has involved solving two initial value problems, one to the left of t_1 with initial value $v(0) = v_0$, and the other to the right of t_1, where the initial value

$$v_1 = v_0 e^{-(B/M)t_1} + \frac{I}{M}$$

is taken. This value is such that the jump of I/M in v would yield a jump of I in the momentum, as required. The task now is to simplify this analysis to the point where we can use Laplace transforms, once only, to solve an initial value problem written this way:

$$\left. \begin{array}{l} Mv'(t) + Bv(t) = I\,\delta(t - t_1) \\ v(0) = v_0 \end{array} \right\} \tag{6}$$

(See Figure 5.) More specifically, it is required to determine a meaning of, and a Laplace transform for, the symbol $I\,\delta(t - t_1)$ such that

1. In (6), $I\,\delta(t - t_1)$ represents the sudden transfer of momentum I to the system at exactly the instant $t = t_1$.
2. The function (5) may be determined as the solution of (6) by the ordinary techniques of Laplace transformation.

Since it is easier to determine what $\mathscr{L}[I\,\delta(t - t_1)](s)$ must be than to interpret $I\,\delta(t - t_1)$ in terms of previously known mathematics, we shall find the transform first.

First note that $v(t)$ has already been found in (5), and its Laplace transform is therefore known:

$$\mathscr{L}[v(t)](s) = v_0 \frac{1}{s + B/M} + \frac{I}{M} e^{-st_1} \frac{1}{s + B/M}$$

$$= \frac{1}{s + B/M}\left(v_0 + \frac{I}{M} e^{-st_1}\right)$$

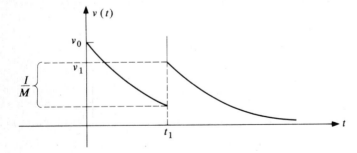

Figure 4 Graph of (5), the velocity of the excited system.

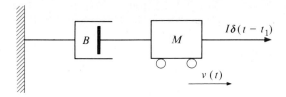

Figure 5 Mass-dashpot system, showing excitation.

Now $\mathscr{L}[\delta(t - t_1)](s)$ can be determined from (6) as follows.

$$\delta(t - t_1) = \frac{M}{I} v'(t) + \frac{B}{I} v(t)$$

$$\mathscr{L}[\delta(t - t_1)](s) = \frac{1}{I}\{(Ms + B)\mathscr{L}[v(t)](s) - v_0 M\}$$

$$= \frac{1}{I}\left[\frac{Ms + B}{s + B/M}\left(v_0 + \frac{I}{M} e^{-st_1}\right) - v_0 M\right]$$

$$= \frac{1}{I}\left[M\left(v_0 + \frac{I}{M} e^{-st_1}\right) - v_0 M\right]$$

$$= e^{-st_1}$$

That is, $\mathscr{L}[\delta(t - t_1)](s) = e^{-st_1}$ (7)

You should add (7) to your table of Laplace-transform pairs.

There remains the problem of interpreting $\delta(t - t_1)$ in terms of the mathematics we already know. If $t_1 = 0$, then we have simply $\delta(t)$, which is called the *Dirac delta*. Be warned at the outset, though, that we are dealing with a new kind of object; in particular, it is not a formula for the values of some function.

Let us start by approximating the sudden impulse by a slightly less sudden impulse of nearly the same magnitude, imagining the system as in Figure 6. Consider an excitation force

$$f(t) = \begin{cases} 0 & t \leq t_1 \\ \dfrac{I}{t_2 - t_1} & t_1 < t \leq t_2 \\ 0 & t_2 < t \end{cases} \qquad (8)$$

$$= \frac{I}{t_2 - t_1}[u(t - t_1) - u(t - t_2)]$$

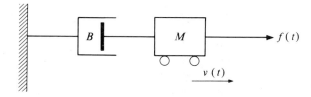

Figure 6 Mass-dashpot system, showing excitation.

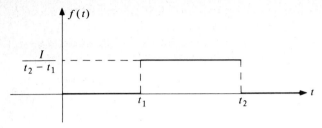

Figure 7 Graph of f, the excitation force.

Then solve the initial value problem

$$Mv' + Bv = f \Big|$$
$$v(0) = v_0 \Big|$$
(9)

and then compute the impulse delivered by f. The solution to (9) is found by Laplace transforms, using the Second Shifting Theorem (Section D).

$$v(t) = v_0 e^{-(B/M)t} + \frac{I}{B(t_2 - t_1)}$$
$$\times [u(t - t_1)(1 - e^{-(B/M)(t - t_1)}) - u(t - t_2)(1 - e^{-(B/M)(t - t_2)})] \quad (10)$$

The expression may look formidable, but $v(t_1)$ and $v(t_2)$ are in fact very easy to compute from it. Since $t_1 < t_2$, the second step function is zero at $t = t_1$. The rest of the function in brackets is also zero there, since $1 - e^{-(B/M)(t_1 - t_1)} = 0$. In fact, it follows from the latter observation that v is continuous at t_1 and

$$v(t_1) = v_0 e^{-(B/M)t_1}$$

At $t = t_2$, we have

$$u(t - t_1) = u(t_2 - t_1) = 1$$

since $t_2 > t_1$, but $1 - e^{-(B/M)(t_2 - t_2)} = 0$. Hence, v is continuous at t_2 and

$$v(t_2) = v_0 e^{-(B/M)t_2} + \frac{I}{B(t_2 - t_1)}(1 - e^{-(B/M)(t_2 - t_1)})$$

Thus the impulse, the momentum change, is

$$Mv(t_2) - Mv(t_1) = Mv_0(e^{-(B/M)t_2} - e^{-(B/M)t_1})$$

$$+ \frac{MI}{B(t_2 - t_1)}(1 - e^{-(B/M)(t_2 - t_1)}) \quad (11)$$

The first term of the right-hand side is the momentum loss due to the force exerted by the dashpot, for it is the same whatever value I may have, even if $I = 0$. Thus, the momentum change due to the excitation is

$$\frac{MI}{B} \frac{1 - e^{-(B/M)(t_2 - t_1)}}{t_2 - t_1} \quad (12)$$

This is not just I, for the reason that $f(t)$ was not the net force on the mass. [In fact, the net force was $f(t) - Bv(t)$, which is not the same as $f(t) - Bv_0 e^{-(B/M)t}$ because v is actually changed by the application of the force f.] However, if $t_2 - t_1$ is small, then (12) is very nearly I. Specifically, L'Hospital's rule shows that

$$\lim_{t_2 \to t_1} \frac{MI}{B} \frac{1 - e^{-(B/M)(t_2 - t_1)}}{t_2 - t_1} = \frac{MI}{B} \lim_{t_2 \to t_1} \frac{(-B/M)(-e^{-(B/M)(t_2-t_1)})}{1} \tag{13}$$

$$= I$$

That fact deserves a closer look. For any $t_2 > t_1$, if f is given by (8), then the impulse given by f is (12); but if t_2 is very close to t_1, then the impulse is very close to I, and it is imparted over the interval $t_1 \le t \le t_2$. That is what it means to say that I is the limit of (12) as t_2 goes to t_1.

2. Our Interpretation of $\delta(t - t_1)$

If the excitation of the system can be reasonably approximated by $1/(t_2 - t_1)$ on the interval $t_1 \le t \le t_2$ and by zero everywhere else, and if the approximation is a good one if t_2 is arbitrarily close to t_1, then

1. In any differential equation describing the system, we shall describe that excitation as $\delta(t - t_1)$.
2. In the Laplace transform of the equation, we shall take

$$\mathscr{L}[\delta(t - t_1)](s) = e^{-t_1 s}$$

3. Example: Solving an Equation Containing $\delta(t - t_1)$

Solve the initial value problem

$$\left. \begin{array}{c} y''(t) + 3y'(t) + 2y(t) = 4\delta(t - 6) \\ y(0) = 1 \\ y'(0) = 0 \end{array} \right\} \tag{14}$$

Solution

$$s^2 \mathscr{L}[y](s) - s + 3\{s\mathscr{L}[y](s) - 1\} + 2\mathscr{L}[y](s) = 4e^{-6s}$$

$$(s^2 + 3s + 2)\mathscr{L}[y](s) = s + 3 + 4e^{-6s}$$

$$\mathscr{L}[y](s) = \frac{s + 3}{s^2 + 3s + 2} + e^{-6s} \frac{4}{s^2 + 3s + 2}$$

$$= \frac{2}{s + 1} - \frac{1}{s + 2} + e^{-6s}\left(\frac{4}{s + 1} - \frac{4}{s + 2} \right)$$

$$y(t) = 2e^{-t} - e^{-2t} + u(t - 6)(4e^{-(t-6)} - 4e^{-2(t-6)}) \qquad t \ge 0$$

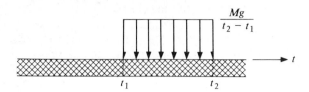

Figure 8 Load intensity due to a box of weight Mg with length $t_2 - t_1$.

The process of checking the solution of an equation such as (14) is complicated by the fact that $y'(t)$ has a jump discontinuity at $t = 6$. Indeed, $y(t)$ is not differentiable at $t = 6$, even though it is continuous there. However, $y(t)$ satisfies the equation

$$y''(t) + 3y'(t) + 2y(t) = 0 \qquad t \neq 6 \tag{15}$$

at all points t except $t = 6$, and this fact should be checked. The initial values should also be checked. The remaining portions of the check will be described at the end of the section and in the next section, for they depend on a description of the discontinuities of y' and y''.

4. Examples: Quantities and Processes Described by the Dirac δ

A box of mass M sits on a beam, extending along the beam from t_1 to t_2. Its weight Mg is distributed uniformly over the interval from t_1 to t_2. Hence, the force per unit length on the beam is $-Mg/(t_2 - t_1)$. If the box is replaced by one of the same mass M but of shorter length, then the force per unit length on the beam is increased. In each case the weight of the box Mg is the total force on the beam, and the pressure is the quotient of that divided by the length of the box. The limiting case is *point loading*, which may be described as a pressure

$$w(t) = -Mg\,\delta(t - t_1)$$

In this case the force $-Mg$ is applied entirely at the point $t = t_1$.

A capacitor holds charge q, which is proportional to the voltage v across the capacitor:

$$q = Cv$$

If the capacitor discharges at a constant rate from time t_1 to time t_2, then during that interval the current is $-q/(t_2 - t_1)$. If, instead, the interval of discharge is shortened, then the quotient $-q/(t_2 - t_1)$ becomes more negative, but the amount of charge lost during the interval is still q. The limiting case is an *instantaneous discharge* of amount q, which may be described as a current

$$i(t) = -q\,\delta(t - t_1)$$

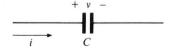

Figure 9 Current flow through a capacitor.

5. Example

In the circuit of Figure 10, the current source yields zero current except at time $t = t_1$. At that moment the source yields enough current to charge the capacitor to voltage v_1 instantaneously. We shall show that the circuit behaves in this way if $v(0) = 0$ and if $i(t)$ is taken to be $Cv_1\, \delta(t - t_1)$.

The equations for the elements and nodes are unaffected by the nature of the impressed current. Thus,

$$v = Ri_R \qquad (16)$$

$$Cv' = i_C$$

$$i = i_R + i_C \qquad (17)$$

and the result of elimination is the equation

$$Cv' + \frac{1}{R}v = i$$

or

$$v'(t) + \frac{1}{RC}v(t) = \frac{1}{C}i(t)$$

Substituting the formula for $i(t)$ yields

$$v'(t) + \frac{1}{RC}v(t) = v_1\, \delta(t - t_1)$$

Taking Laplace transforms gives

$$\left(s + \frac{1}{RC}\right)\mathscr{L}[v](s) = v_1 e^{-st_1}$$

so that

$$\mathscr{L}[v](s) = v_1 \frac{1}{s + 1/RC} e^{-st_1}$$

and

$$v(t) = u(t - t_1)v_1 e^{-(1/RC)(t - t_1)} \qquad t \geq 0$$

Notice that the voltage on the capacitor is as required: zero at $t = 0$, and jumping to v_1 at $t = t_1$. Except at t_1, the voltage satisfies the equation

$$v'(t) + \frac{1}{RC}v(t) = 0 \qquad t \neq t_1$$

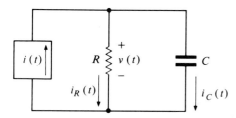

Figure 10 Parallel RC circuit.

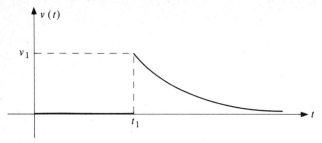

Figure 11 Graph of $v(t)$, showing the jump at t_1.

Specifically, substitution into the left-hand side of the equation gives this:

$$v'(t) + \frac{1}{RC} v(t) = \begin{cases} 0 + \dfrac{1}{RC} 0 & 0 \leq t < t_1 \\[2ex] -\dfrac{1}{RC} v_1 e^{-(1/RC)(t-t_1)} + \dfrac{1}{RC} v_1 e^{-(1/RC)(t-t_1)} & t_1 < t \end{cases}$$

$$= 0 \qquad 0 \leq t, \text{ if } t \neq t_1$$

The current generator, $i(t) = Cv_1 \, \delta(t - t_1)$, can be interpreted as delivering charge in the amount Cv_1 at the instant $t = t_1$, and with it charging the capacitor to voltage v_1. Thereafter, the charge is bled off through the resistor, as we now verify. Since $i(t) = 0$ for $t \neq t_1$, it follows that the only current is around the loop through the resistor and capacitor. By (16), the current in this circuit is

$$i_R(t) = \frac{1}{R} v(t) = \frac{1}{R} v_1 e^{-(1/RC)(t-t_1)} u(t - t_1)$$

The current i_C through the capacitor is the negative of i_R except in the following interesting respect. From (17),

$$i_C(t) = i(t) - i_R(t)$$
$$= Cv_1 \, \delta(t - t_1) - i_R(t)$$

That is, $i_C(t)$ is the negative of $i_R(t)$ as expected, except at the instant $t = t_1$, when it carries the impressed current $i(t) = Cv_1 \, \delta(t - t_1)$, delivering the charge Cv_1 with which the capacitor is charged.

6. Sample Problems with Answers

Solve the initial value problems and graph the solutions.

A. $\begin{array}{l} 2y'(t) + 3y(t) = -\delta(t - 1) \\ y(0) = 0 \end{array}$ *Answer:* $y(t) = -\frac{1}{2}u(t - 1)e^{-(3/2)(t-1)}$

B.
$$y''(t) + 5y(t) = \tfrac{1}{2}\delta(t-2)\big|$$
$$y(0) = 1$$
$$y'(0) = 0$$

Answer: $y(t) = \cos\sqrt{5}\,t + \dfrac{1}{2\sqrt{5}}u(t-2)\sin\sqrt{5}\,(t-2)$

C. Find the inverse Laplace transform of $(s-1)/(s+1)$.
Answer:

$$\frac{s-1}{s+1} = \frac{s+1-2}{s+1} = 1 - \frac{2}{s+1}$$

and so
$$\mathscr{L}^{-1}\left[\frac{s-1}{s+1}\right](t) = \mathscr{L}^{-1}[1](t) - \mathscr{L}^{-1}\left[\frac{2}{s+1}\right](t) = \delta(t) - 2e^{-t} \qquad t \geq 0$$

D. Find the inverse Laplace transform of $(s^2+1)/(4s^2-1)$.
Answer:

$$\frac{s^2+1}{4s^2-1} = \frac{1}{4}\frac{s^2+1}{s^2-\frac{1}{4}} = \frac{1}{4}\frac{s^2-\frac{1}{4}+\frac{5}{4}}{s^2-\frac{1}{4}}$$

$$= \frac{1}{4}\left(1 + \frac{\frac{5}{4}}{s^2-\frac{1}{4}}\right) = \frac{1}{4} + \frac{5}{16}\frac{1}{(s-\frac{1}{2})(s+\frac{1}{2})}$$

and so
$$\mathscr{L}^{-1}\left[\frac{s^2+1}{4s^2-1}\right](t) = \mathscr{L}^{-1}[1](t) + \tfrac{5}{16}\mathscr{L}^{-1}\left[\frac{1}{(s-\frac{1}{2})(s+\frac{1}{2})}\right](t)$$

$$= \delta(t) + \tfrac{5}{16}(e^{-t/2} - e^{t/2}) \qquad t \geq 0$$

7. Problems

Solve the initial value problems and graph the solutions.

1. *(a)*
$$y'(t) + 4y(t) = 2\delta(t-1)\big|$$
$$y(0) = 0$$

(b)
$$2y'(t) + 4y(t) = 2\delta(t-1)\big|$$
$$y(0) = 0$$

(c)
$$y'(t) + 4y(t) = 2\delta(t)\big|$$
$$y(0) = 0$$

(d)
$$y'(t) + 4y(t) = 0\big|$$
$$y(0) = 2$$

2.
$$y''(t) + 4y(t) = -4\delta(t-\pi)\big|$$
$$y(0) = 0$$
$$y'(0) = 4$$

3.
$$y'(t) + 2y(t) = 3u(t-1) - \delta(t-2)\big|$$
$$y(0) = 1$$

4.
$$y''(t) + 3y'(t) + 2y(t) = 8\delta(t-3)\big|$$
$$y(0) = 0$$
$$y'(0) = 0$$

5.
$$y''(t) + 4y'(t) + 5y(t) = \delta(t-1)\big|$$
$$y(0) = 0$$
$$y'(0) = 0$$

Find the inverse Laplace transforms.

6. $\dfrac{2}{s+2}$

7. $\dfrac{2s+1}{s+2}$

8. $\dfrac{s^2+2s}{s^2+9}$

9. $\dfrac{3s^3+s}{s^3+2s^2+9s+18}$

10. $\dfrac{2s^2-3s}{3s^2-10}$

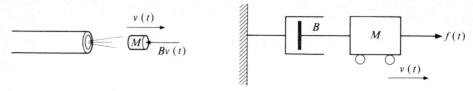

Figure 12 Gun firing a pellet: pictorial and free-body diagram.

8. Example: The Jump Caused by the Dirac δ

Figure 12 represents a gun pointed horizontally, firing a pellet of mass M at time $t = t_1 > 0$ into a gas where the viscous friction of the pellet is $-Bv(t)$. The muzzle velocity of the pellet is v_1. We shall neglect the effect of gravity, for over fairly short ranges gravity does not affect the horizontal velocity of the pellet. The force provided by firing the gun is

$$f(t) = Mv_1\,\delta(t - t_1)$$

It provides impulse Mv_1 at time $t = t_1$, in accordance with our interpretation in Part 2 of this section: The transfer of momentum at t_1 is reasonably approximated by a force of duration $t_2 - t_1$ and magnitude $Mv_1/(t_2 - t_1)$ for t_2 arbitrarily close to t_1. The initial value problem is

$$\left.\begin{aligned} Mv'(t) + Bv(t) &= Mv_1\,\delta(t - t_1) \\ v(0) &= 0 \end{aligned}\right\} \tag{18}$$

Solution

$$(Ms + B)\mathscr{L}[v](s) - M\underbrace{v(0)}_{0} = Mv_1 e^{-st_1} \tag{19}$$

$$\mathscr{L}[v](s) = e^{-st_1}\frac{Mv_1}{Ms + B} = e^{-st_1}\frac{v_1}{s + B/M}$$

$$v(t) = v_1\,u(t - t_1)e^{-(B/M)(t - t_1)} \qquad 0 \le t \tag{20}$$

The example is often done with distance $x(t)$ as the unknown, rather than velocity $v(t)$. In this formulation, the initial value problem is

$$\left.\begin{aligned} Mx''(t) + Bx'(t) &= Mv_1\,\delta(t - t_1) \\ x(0) &= 0 \\ x'(0) &= 0 \end{aligned}\right\} \tag{21}$$

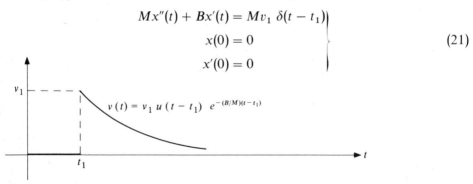

Figure 13 Velocity of a pellet fired at time $t = t_1$.

The transformed problem is

$$Ms^2\mathcal{L}[x](s) + Bs\mathcal{L}[x](s) = Mv_1 e^{-st_1}$$

$$\mathcal{L}[x](s) = e^{-st_1}\frac{Mv_1}{Ms^2 + Bs}$$

$$= e^{-st_1}\frac{Mv_1}{B}\left(\frac{1}{s} - \frac{1}{s + B/M}\right)$$

Thus,

$$x(t) = \frac{Mv_1}{B}u(t - t_1)(1 - e^{-(B/M)(t - t_1)}) \qquad 0 \le t \qquad (22)$$

Note that the function $x(t)$ is not differentiable at $t = t_1$. Its graph does not have a well-defined tangent there, but has instead a tangent from the right of slope v_1 and a tangent from the left of slope 0. (See Figure 14.) We say that at t_1 the function x' has *left-hand limit* zero:

$$x'(t_1-) = \lim_{t \to t_1-} x'(t) = 0$$

and *right-hand limit* v_1:

$$x'(t_1+) = \lim_{t \to t_1+} x'(t) = v_1$$

(These are sometimes called *limit from the left* and *limit from the right*, respectively.) However, x' fails to have any function value at all at $t = t_1$, for x is not differentiable there.

In the application this is very natural, for, in that "instant," x' changes from 0 to v_1. That is, it jumps up by v_1 (see Figure 14). The *height* of the jump is the right-hand limit less the left-hand limit:

$$x'(t_1+) - x'(t_1-) = v_1 - 0 = v_1$$

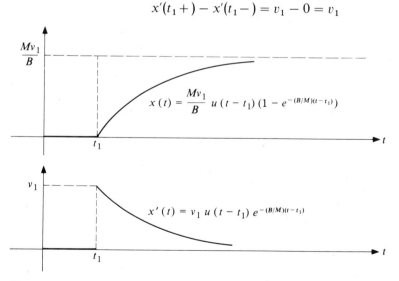

Figure 14 The gun fired at time $t = t_1$: displacement and velocity.

As long as $t_1 > 0$, all this is apparent and plainly visible on the graph. In the interval between $t = 0$ and $t = t_1$, there are no jumps or other discontinuities; in particular, all the functions are defined and continuous at $t = 0$. That is the case $t_1 > 0$. As far as the mechanics of solution are concerned, the exceptional case $t_1 = 0$ is solved in the same way. The interpretation of the initial values is given below.

Notice that $\delta(t - t_1)$ causes a jump at t_1 in only some derivatives of the unknown function. In a first-order equation, it causes a jump in the 0th derivative (the unknown function itself), as illustrated in (18), whose solution is (20). In a second-order equation, it causes a jump in the first derivative, while the 0th derivative is continuous. [See (21), whose solution is (22).] In general, in an nth-order equation, $\delta(t - t_1)$ will cause a jump in the $(n - 1)$st derivative of the unknown function, but all lower derivatives will be continuous. (What happens to the nth-order derivative will be discussed in Section F.)

9. Satisfying the Initial Conditions

Now consider the system of Example 8, but with the gun fired at $t = 0$. The Dirac δ is not required for setting up this problem, for we may imagine the bullet to have the muzzle velocity at $t = 0$ and to have only the damping force on it during the interval $t \geq 0$. Set up this way, the initial value problem is

$$\left. \begin{array}{r} Mx'' + Bx' = 0 \\ x(0) = 0 \\ x'(0) = v_1 \end{array} \right\} \tag{23}$$

The Laplace transform of (23) is

$$Ms^2 \mathscr{L}[x](s) - Mv_1 + Bs\mathscr{L}[x](s) = 0$$

$$\mathscr{L}[x](s) = \frac{Mv_1}{Ms^2 + Bs} = \frac{Mv_1}{B}\left(\frac{1}{s} - \frac{1}{s + B/M}\right) \tag{24}$$

$$x(t) = \frac{Mv_1}{B}\left(1 - e^{-(B/M)t}\right) \qquad 0 \leq t$$

$$x'(t) = v_1 e^{-(B/M)t} \qquad 0 \leq t$$

This solution is also the function which results from setting $t_1 = 0$ in (22), the solution of the problem when formulated with the Dirac δ. Unlike the case in which $t_1 > 0$, however, it is both continuous and differentiable at $t_1 = 0$. Notice that v_1 is not only the right-hand limit of x' at $t = 0$, it is also *the* limit at $t = 0$ and the value of $x'(0)$ as well, for neither $x(t)$ nor $x'(t)$ is defined for $t < 0$. That is, the discontinuity of $x'(t)$ has been removed by simply ignoring the behavior of the system for $t < 0$:

$$v_1 = \lim_{t \to 0+} x'(t) = x'(0+) \tag{25}$$

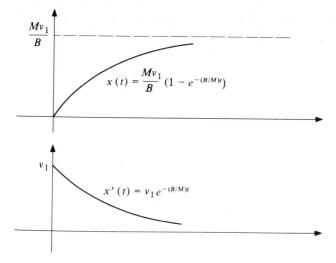

Figure 15 The gun fired at time $t = 0$: displacement and velocity.

It is possible, and often it is conventional, to use the Dirac δ when doing this kind of problem, in which case it is set up this way:

$$Mx''(t) + Bx'(t) = Mv_1\,\delta(t) \qquad 0 \le t$$
$$x(0) = 0 \qquad\qquad (26)$$
$$x'(0) = 0$$

$$Ms^2\mathscr{L}[x](s) + Bs\mathscr{L}[x](s) = Mv_1$$

$$\mathscr{L}[x](s) = \frac{Mv_1}{Ms^2 + Bs} \qquad (27)$$

Since (27), the Laplace transform of $x(t)$, is the same as (24), we have evidently set up the problem correctly. The only puzzle is the initial value $x'(0) = 0$: If it is correct, then why is it correct, and what does it mean?

Its meaning may be seen from Figures 14 and 15. The equation $x'(0) = 0$ is the element in this problem which corresponds to the equation $x'(t_1 -) = 0$ in the problem of Example 8. In the present case, we write the problem only for $t \ge 0$, as in (26), but there is a part of the problem which does not appear in Equation (26)—the fact that the pellet is at rest and the gun is unfired for certain values of $t < 0$. Thus, the problem should have been written not as in (26) but in the following form:

$$Mx''(t) + Bx'(t) = Mv_1\,\delta(t) \qquad 0 \le t$$
$$x(0) = 0 \qquad\qquad (28)$$
$$x'(0-) = 0$$

By convention, however, the difference between (26) and (28) is ignored, the problem is written as in (26), and the meaning of the limit $x'(0-) = 0$ is assumed without being written.

If, at $t = 0$, there is no need to consider the behavior for $t < 0$ of either the equation or the system it represents, in particular, if the excitation of the system is continuous or appears in the differential equation as nothing worse than a step function, then it is possible and quite natural to write the problem with the initial value, which is the right-hand limit as in (23). For an nth-order equation, it might be written this way:

$$a_n y^{(n)}(t) + a_{n-1}y^{(n-1)}(t) + \cdots + a_2 y''(t) + a_1 y'(t) + a_0 y(t) = f(t)$$
$$y(0+) = c_0$$
$$y'(0+) = c_1$$
$$y''(0+) = c_2$$
$$\cdots\cdots\cdots\cdots\cdots$$
$$y^{(n-2)}(0+) = c_{n-2}$$
$$y^{(n-1)}(0+) = c_{n-1}$$

On the other hand, if some account is to be taken of the prior behavior of the system, then the excitation may include the Dirac δ, and the initial values should be the left-hand limits:

$$y(0-) = c_0$$
$$y'(0-) = c_1$$
$$y''(0-) = c_2$$
$$\cdots\cdots\cdots\cdots\cdots$$
$$y^{(n-2)}(0-) = c_{n-2}$$
$$y^{(n-1)}(0-) = c_{n-1}$$

Such observations can be very useful in situations where the "initial instant" can be chosen at will, as in Example 5. There, since the circuit is inactive (relaxed) before $t = t_1$, we could have rephrased the problem so as to take the capacitor charged with initial voltage v_1 at time $t = 0$, then found $v(t) = v_1 e^{-(1/RC)t}$, and then translated back in time to t_1:

$$v(t) = u(t - t_1)v_1 e^{-(1/RC)(t - t_1)} \qquad t \geq 0$$

(Notice the similarity to the operations suggested by the Second Shifting Theorem.)

In Figure 8, on the other hand, it is natural to take the "initial point" at the end of the beam, or at the point of support (where one can calculate all the initial conditions *before* solving the equations). Since the point load is not applied at either end point or support point, the use of the Dirac δ to represent the point load seems very natural.

10. Integrating the Dirac δ

Finding the indefinite integral of a function f is tantamount to finding the general solution y of the equation $y' = f$. Using the same idea, we can define an indefinite integral for the Dirac δ, using Laplace transforms to solve the differential equation $y'(t) = \delta(t - t_1)$. Thus,

$$y(t) = \int \delta(t - t_1)\, dt = u(t - t_1) + C \tag{29}$$

In fact,
$$y(t) = u(t - t_1) + y(0) \qquad \text{if } t_1 \geq 0$$

We may also go on to define a definite integral,

$$\int_\alpha^\beta \delta(t - t_1)\, dt = u(\beta - t_1) - u(\alpha - t_1) \tag{30}$$

consistent with the Fundamental Theorem of Calculus.

If the mass M with which we began the section has velocity zero at time $t = 0$ and is given impulse Mv_1 at time t_1, then (as a function of time) the momentum is

$$\int_0^t Mv_1\, \delta(z - t_1)\, dz = Mv_1 u(t - t_1)$$

The impulse provided during an interval $\alpha \leq t < \beta$ may be calculated as

$$Mv(\beta) - Mv(\alpha) = \int_\alpha^\beta Mv_1\, \delta(t - t_1)\, dt$$

$$= Mv_1[u(\beta - t_1) - u(\alpha - t_1)]$$

which is Mv_1 if t_1 falls in the interval, and zero otherwise.

The right-hand side of (30) may be evaluated, giving the formula for integrating over an interval $\alpha \leq t \leq \beta$:

$$\int_\alpha^\beta \delta(t - t_1)\, dt = \begin{cases} 1 & \text{if } \alpha \leq t_1 < \beta \\ 0 & \text{otherwise} \end{cases} \tag{31}$$

This is entirely consistent with our "interpretation" of $\delta(t - t_1)$ as given in Part 2, as may be shown as follows. The function

$$g(t) = \frac{1}{t_2 - t_1}[u(t - t_1) - u(t - t_2)] \tag{32}$$

is a "reasonable approximation" for $\delta(t - t_1)$, provided $t_2 > t_1$ and $t_2 - t_1$ is small, because of the interpretation in Part 2. From (32), the definition of g, it follows that

$$\int_\alpha^\beta g(t)\, dt = \int_\alpha^{t_1} 0\, dt + \int_{t_1}^{t_2} \frac{1}{t_2 - t_1}\, dt + \int_{t_2}^\beta 0\, dt \tag{33}$$

$$= 0 + 1 + 0 = 1 \qquad \text{if } \alpha \leq t_1 < t_2 \leq \beta$$

while
$$\int_{\alpha}^{\beta} g(t)\, dt = \int_{\alpha}^{\beta} 0\, dt = 0 \qquad \text{if } t_1 \geq \beta$$

and the same is true if $t_1 < \alpha$ and $t_2 - t_1 < \alpha - t_1$. This result is consistent with (31): the integrals of $\delta(t - t_1)$ and g should be the same.

The argument just given is easily extended to evaluate the integral of a product of the Dirac δ with a continuous function f:

$$\int_{\alpha}^{\beta} f(t)\, \delta(t - t_1)\, dt = \int_{\alpha}^{t_1} f(t)\, \delta(t - t_1)\, dt + \int_{t_1}^{t_2} f(t)\, \delta(t - t_1)\, dt$$

$$+ \int_{t_2}^{\beta} f(t)\, \delta(t - t_1)\, dt \qquad (34)$$

leading to formula (36), which will be important in systems analysis (Section H). Specifically, the argument shows that, on the right-hand side of (34), the first and third integrals are zero, since the approximating function g is zero, and the product $f(t)g(t)$ is zero also. In an application this reflects the fact that "nothing happens" on the intervals $\alpha < t < t_1$ and $t_2 < t < \beta$. What happens occurs in the interval $t_1 \leq t \leq t_2$, and so

$$\int_{\alpha}^{\beta} f(t)\, \delta(t - t_1)\, dt = \int_{t_1}^{t_2} f(t)\, \delta(t - t_1)\, dt \qquad (35)$$

Since the equality holds no matter how close t_2 is to t_1, the continuity of f shows that f may be replaced by the constant function with value $f(t_1)$. (See Figure 16.) Thus, the right-hand side of (35) is

$$\int_{t_1}^{t_2} f(t_1)\, \delta(t - t_1)\, dt = f(t_1) \int_{t_1}^{t_2} \delta(t - t_1)\, dt = f(t_1)$$

since $f(t_1)$ is just a number, and (31) holds.

The arguments given in the previous paragraph also show that the integral on the left in (34) is zero unless $\alpha \leq t_1 < \beta$. Thus, the result may be summarized as

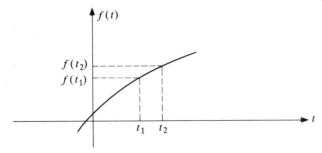

Figure 16 If f is continuous then $f(t)$ is near $f(t_1)$ on the interval $t_1 \leq t \leq t_2$, provided t_2 is close to t_1.

follows: If f is a function on the interval $\alpha \le t < \beta$, and if it is continuous at t_1, then

$$\int_\alpha^\beta f(t)\,\delta(t - t_1)\,dt = \begin{cases} f(t_1) & \text{if } \alpha \le t_1 < \beta \\ 0 & \text{otherwise} \end{cases} \tag{36}$$

Since $f(t_1)$ is just a particular value of f, the formula is remarkably easy to use. For instance,

$$\int_0^{2\pi} (\sin t)\,\delta\left(t - \frac{\pi}{3}\right) dt = \sin \frac{\pi}{3} = \frac{\sqrt{3}}{2}$$

and

$$\int_0^T \delta(t - t_1)e^{-st}\,dt = e^{-st_1} \qquad \text{if } 0 \le t_1 < T$$

Taking the limit of the latter as T increases confirms the Laplace-transform formula:

$$\mathscr{L}[\delta(t - t_1)](s) = \int_0^\infty \delta(t - t_1)\,e^{-st}\,dt$$

$$= \lim_{T \to \infty} \int_0^T \delta(t - t_1)\,e^{-st}\,dt$$

$$= e^{-st_1}$$

Finally, taking $f(z) = \sin(t - z)$ gives

$$\int_0^{2\pi} \sin(t - z)\,\delta\left(z - \frac{\pi}{3}\right) dz = \sin\left(t - \frac{\pi}{3}\right) \tag{37}$$

11. More Problems

11. Consider the system shown in Figure 17, with $M = 2$ and $K = 8$. Find the behavior of the system for $t \ge 0$ for each of the following sets of prior conditions and excitations. In each case, graph x and x'.

(a) $x(t) = 0$ for $t \le 0$, and $f(t) = 1$ for $t \ge 0$.
(b) $x(t) = 0$ for $t \le 0$, and $f(t) = \delta(t)$.
(c) $x(t) = \sin 2t$ for $t \le 0$, and $f(t) = 1$ for $t \ge 0$.
(d) $x(t) = \sin 2t$ for $t \le 0$, and $f(t) = -2\delta(t)$.
(e) $x(0-) = 0$, $x'(0-) = -2$, and, at $t = 0$, f provides momentum 4 in the direction of increasing x (in the positive x direction).

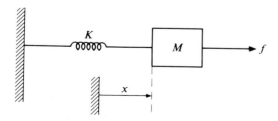

Figure 17 Spring-mass system, in which the spring tension is zero whenever $x(t) = 0$.

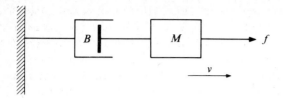

Figure 18 Mass-dashpot system.

12. Consider the system shown in Figure 18, with $M = 2$ and $B = 4$. Find the behavior of the system for $t \geq 0$ for each of the following sets of prior conditions. In each case, graph v, compute x under the restriction that $x' = v$, and graph x as well.

(a) $x(t) = 0$ for $t \leq 0$, and $f(t) = 1$ for $t \geq 0$.
(b) $x(t) = 0$ for $t \leq 0$, and $f(t) = \delta(t)$.
(c) $x(0) = 0$, $v(t) = 1$ for $t \leq 0$, and $f(t) = 1$ for $t \geq 0$.
(d) $x(0) = 0$, $v(t) = e^{-2t}$ for $t \leq 0$, and $f(t) = -\delta(t)$.
(e) $x(0-) = 0$, $v(0-) = -1$, and, at $t = 0$, f provides momentum 3 in the direction of increasing
v (in the positive v direction).

13. Consider the system shown in Figure 19, with $C_1 = 1$, $C_2 = 2$, and $R = \frac{2}{3}$. Find the behavior of the system for $t \geq 0$, giving v and its graph, for each of the following sets of prior conditions.

(a) $i_3(t) = 0$ for $t \leq 0$, and $i(t) = 0$ for $t \geq 0$.
(b) $i_3(t) = 0$ for $t \leq 0$, and $i(t) = \delta(t)$.
(c) $i_3(t) = e^{-2t}$ for $t \leq 0$, and $i(t) = 1$ for $t \geq 0$.
(d) $i_3(t) = e^{-2t}$ for $t \leq 0$, and $i(t) = 4\delta(t)$.
(e) $v(0-) = 1$, and, at $t = 0$, i carries charge 2 (in the direction shown by the arrow).

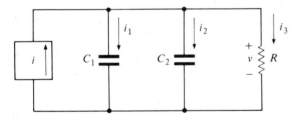

Figure 19 Parallel RC circuit.

Perform the indicated integrations.

14. (a) $\int_0^1 \delta(t) \, dt$ (b) $\int_0^1 \delta(t - \frac{1}{2}) \, dt$ (c) $\int_0^1 \delta(t - 1) \, dt$ (d) $\int_0^1 \delta(t - 2) \, dt$

15. $\int_0^2 e^{\sqrt{2}t} \, \delta(t - \sqrt{3}) \, dt$ **16.** $\int_a^b t \, \delta\left(t - \frac{a+b}{2}\right) dt$ **17.** $\int_0^\pi \sin\left(t - \frac{\pi}{3}\right) \delta\left(t - \frac{\pi}{4}\right) dt$

18. $\int_0^{10} (\sin z) \, \delta(z - t) \, dz$, for the cases (a) $0 \leq t < 10$; and (b) $10 \leq t$

19. $\int_0^t \sin(t - z) \, \delta(z - \pi) \, dz$, for the cases (a) $t > \pi$; and (b) $t \leq \pi$

12. Comments on Continuous and Discontinuous Functions

When setting up equations containing step functions or the Dirac δ as models for idealized or even real systems, it is clear that we are only approximating the excitation conditions. The voltage change as a switch is closed, the pellet-velocity change as a gun is fired, even the pressure change at the end of a box, all look discontinuous when measured with the crudest kinds of laboratory instruments. With finer measuring devices, though, we discover that these quantities all begin to look quite continuous. The engineer concludes that " what is really going on is a continuous process," and our steps and deltas are only approximations. The engineer wonders whether it is legitimate physically to make such an approximation, and whether it is legitimate mathematically.

We can attach some mathematical model to the engineer's system, no matter how the engineer may look at the system, and we must then analyze the mathematical system which results. So far, our solutions have been exact, but in general they will be approximations—approximations to an exact solution too troublesome to find. As long as the approximation suits our purposes, we need not worry about the philosophical implications of approximating a discontinuous function with a continuous one, or conversely. In fact, mathematicians very often will choose to make just such an approximation because it is convenient, but then they must satisfy themselves that the approximation is good enough to suit their purposes.

While the mathematics of differential equations, once discovered, remains fixed, its applications vary according to the ability of the engineer to make measurements on real systems. This is the way in which the engineer faces the same kinds of approximation choices which the mathematician may face. Thus, if we *knew* the forces on a pellet inside a gun, we could write a differential equation using or describing those forces. However, it may be far easier (and for many purposes quite as satisfactory) to measure the muzzle velocity, and describe the effect of firing the gun as a discontinuity in the velocity of the pellet.

But engineers, too, face a deeper question. They may conclude that "what is really going on is a continuous process," but in fact it is only continuous at certain levels of measurement. Give engineers a device that measures continuously, and they will measure continuous things; give them a device that measures discontinuously, and they will discover discontinuous things. At the molecular level even matter is at least somewhat discontinuous, and at the quantum level velocity (such as it is) is discontinuous too. On the other hand, we think of quantum processes as probabilistically distributed, and the probability distribution is continuous. Thus far, each discontinuous measurement could be superseded by a finer continuous one, and conversely; and it would take a rash person to claim that we had reached the very last level of discontinuity or continuity.

Therefore, whether in mathematics or in engineering, from time to time we shall choose to approximate one kind of function with another, when it is convenient or important for us to do so. When we choose, as choose we must, we should be conscious of the consequences of the choice we make, whether or not it is or will long remain sufficient for our purposes and efficient for our use.

For now, the step function and the Dirac δ are convenient enough to be efficient tools, and the models in which they are used can be made accurate enough to justify studying them.

F. GENERALIZED DERIVATIVES

This section concerns the way in which generalized solutions "satisfy" differential equations.

1. Example

Consider the initial value problem

$$\left.\begin{aligned} y'(t) &= u(t - t_1) \\ y(0) &= 0 \end{aligned}\right\} \qquad \text{where } t_1 > 0 \qquad (1)$$

Taking Laplace transforms gives

$$s\mathscr{L}[y](s) = \frac{e^{-st_1}}{s} \qquad (2)$$

and an application of the Second Shifting Theorem gives

$$y(t) = (t - t_1)u(t - t_1) \qquad t \geq 0$$

The function y' is the derivative of y wherever y is differentiable, that is, at each point except $t = t_1$. If $t < t_1$ then $y'(t) = 0$, and if $t > t_1$ then $y'(t) = 1$.

Notice, however, that (1) provides a function value for y' at $t = t_1$,

$$y'(t_1) = u(t_1 - t_1) = 0$$

even though y does not have a derivative at $t = t_1$. What is it, then, that (1) says, if y' cannot be said to be the derivative of y? Just as y is not the "solution" of the differential equation, but rather the generalized solution, so we should call y' not the "derivative" of y, but the *generalized derivative* of y, when we write

$$y'(t) = \begin{cases} 0 & 0 \leq t \leq t_1 \\ 1 & t_1 < t \end{cases}$$

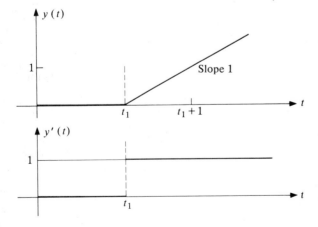

Figure 1 The solution of (1) and its derivative.

Here we have done again what we did implicitly in the previous two sections when we noted that we wanted the Laplace transform of the step function and the Dirac δ to behave as the transforms of other things do, and wrote, for the unknown function y,

$$\mathscr{L}[y'](s) = s\mathscr{L}[y](s) - y(0)$$

Notice, too, that y may be obtained from $u(t - t_1)$ by direct integration:

$$\int_0^t u(x - t_1)\,dx = \begin{cases} \int_0^t 0\,dx = 0 & \text{if } t \le t_1 \\[2ex] \int_{t_1}^t 1\,dx = x\,\Big|_{t_1}^t = t - t_1 & \text{if } t_1 < t \end{cases}$$

$$= (t - t_1)u(t - t_1) = y(t)$$

Thus, the generalized derivative of our function y is consistent both with the Laplace-transform formulas for integrals and derivatives and with the process of integration.

It must be recognized, though, that y is not differentiable at $t = t_1$, even though y has a generalized derivative. For y to be differentiable at a point t, the difference quotient

$$\frac{\Delta y}{\Delta t} = \frac{y(t + \Delta t) - y(t)}{t + \Delta t - t}$$

must have a limit as Δt goes to zero (for that is the definition of "differentiable," as is learned at the beginning of calculus). The function y of this example has this property at all points $t \ne t_1$. At $t = t_1$, however, it does not, for there

$$\frac{\Delta y}{\Delta t} = \frac{y(t_1 + \Delta t) - y(t_1)}{t_1 + \Delta t - t_1} = \frac{y(t_1 + \Delta t)}{\Delta t}$$

$$\frac{\Delta y}{\Delta t} = \begin{cases} 0 & \Delta t < 0 \\ 1 & \Delta t > 0 \end{cases}$$

and no limit exists as Δt goes to zero. This, then, is what y has:

1. A derivative at each $t \ne t_1$ as computed according to all the rules of calculus
2. No derivative at $t = t_1$
3. A generalized derivative as described above, defined on the right half line, $t \ge 0$

Note that at any point where y has a derivative, the derivative and the generalized derivative agree.

Now, however, we must reinterpret Equation (1):

$$y'(t) = u(t - t_1)$$

Equation (1) means, "The generalized derivative of y is $u(t - t_1)$." All our differential equations involving step functions will be reinterpreted in this way without changing either the generalized solutions we found for them or the Laplace-transform solution technique we have given.

2. Example: Finding the Generalized Derivative of $y(t) = u(t - 1) \sin (t - 1)$

We shall do this in two ways, the first being direct differentiation. On the half line $t > 1$, since

$$y(t) = u(t - 1) \sin (t - 1)$$
$$= \sin (t - 1) \qquad t > 1$$

differentiation shows that

$$y'(t) = \cos (t - 1) \qquad t > 1 \tag{3}$$

On the half line $t < 1$,

$$y(t) = 0 \qquad \text{and so} \qquad y'(t) = 0 \qquad t < 1 \tag{4}$$

We may combine (3) and (4) into one equation,

$$y'(t) = \begin{cases} 0 & t < 1 \\ \cos (t - 1) & t > 1 \end{cases}$$
$$= u(t - 1) \cos (t - 1) \qquad t \neq 1 \tag{5}$$

Equation (5) is the derivative of y, the calculus derivative, for we have avoided the one point at which y is not differentiable. For the generalized derivative, we shall provide, in addition, an arbitrary function value at the point $t = 1$. The choice is arbitrary, but the most convenient choice is that which simply extends (5) to the excluded point:

$$y'(t) = u(t - 1) \cos (t - 1)$$

(See Figure 2.)

Second, let us use the Laplace transform. The transform of y' should be related to that of y by the usual formula,

$$\mathscr{L}[y'](s) = s\mathscr{L}[y](s) - y(0)$$

and so $\qquad y'(t) = \mathscr{L}^{-1}[s\mathscr{L}[y](s) - y(0)](t)$

$$= \mathscr{L}^{-1}\left[se^{-s} \frac{1}{s^2 + 1} \right](t) = \mathscr{L}^{-1}\left[e^{-s} \frac{s}{s^2 + 1} \right](t)$$

$$= u(t - 1) \cos (t - 1) \qquad t \geq 0$$

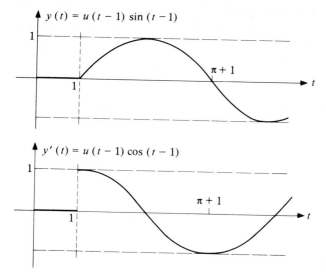

Figure 2 The graph of y and y'.

3. Definition of the Generalized Derivative

If f has a Laplace transform $\mathscr{L}[f]$, and if $s\mathscr{L}[f](s)$ has an inverse transform, then f has a *generalized derivative* f', which is

$$f' = \mathscr{L}^{-1}[s\mathscr{L}[f](s) - f(0-)]$$

If f is a function continuous at 0, then $f(0-)$ is just $f(0)$; otherwise, $f(0-)$ is, of course, the left-hand limit of f at $t = 0$. For instance, if $f(t) = u(t)$ or $f(t) = \delta(t)$, then $f(0-) = 0$. The left-hand limit appears for the same reasons as in Parts 8 and 9 of the previous section.

4. Examples

If $f(t) = \cos bt$, then

$$f'(t) = \mathscr{L}^{-1}[s\mathscr{L}[\cos bt](s) - \cos 0](t)$$

$$= \mathscr{L}^{-1}\left[\frac{s^2}{s^2 + b^2} - 1\right](t)$$

$$= \mathscr{L}^{-1}\left[\frac{s^2 - (s^2 + b^2)}{s^2 + b^2}\right](t)$$

$$= -b\mathscr{L}^{-1}\left[\frac{b}{s^2 + b^2}\right](t)$$

$$= -b \sin bt \qquad t \geq 0$$

If $f(t) = u(t - t_1)$ and $t_1 \geq 0$, then

$$f'(t) = \mathcal{L}^{-1}[s\mathcal{L}[u(t - t_1)](s) - u(0 - t_1) -](t)$$

$$= \mathcal{L}^{-1}\left[s\frac{e^{st_1}}{s} - 0\right](t)$$

$$= \delta(t - t_1)$$

Notice that if $t_1 > 0$, then $u(t - t_1)$ is continuous at $t = 0$ and takes value zero there. On the other hand, if $t_1 = 0$, then $u[(0 - 0) -] = u(0-)$ is the left-hand limit of u at $t = 0$, which is also zero.

It is worth recalling that even though $u(t - t_1)$ is a function, $u'(t - t_1) = \delta(t - t_1)$ is not a function, for its peculiarity at $t = t_1$ prevents it from being a function. However, it has a Laplace transform, it is the generalized derivative of $u(t - t_1)$, and it may be said to have function values except at $t = t_1$.

If $f(t) = u(t - t_1)\cos b(t - t_1)$, then

$$f'(t) = \mathcal{L}^{-1}\left[e^{-st_1}\frac{s^2}{s^2 + b^2}\right](t) \tag{6}$$

$$= \mathcal{L}^{-1}\left[e^{-st_1}\left(1 - \frac{b^2}{s^2 + b^2}\right)\right](t) \tag{7}$$

$$= \mathcal{L}^{-1}[e^{-st_1}](t) + \mathcal{L}^{-1}\left[e^{-st_1}\frac{b(-b)}{s^2 + b^2}\right](t)$$

$$= \delta(t - t_1) + u(t - t_1)(-b)\sin b(t - t_1) \qquad t \geq 0 \tag{8}$$

The rational function in Equations (6) and (7) deserves some comment. It can happen that the Laplace transform of a generalized derivative contains a rational function in which the degree of s in the denominator is not greater than the degree in the numerator. For instance, $\delta(t) = u'(t)$ and $\mathcal{L}[\delta(t)](s) = 1$; also, $\mathcal{L}[\delta'(t)](s) = s$. This situation occurs in Equation (6). In (7) we see the partial-fraction decomposition. As mentioned in Part 2 of Section C, the decomposition consists of two parts, a polynomial and a rational fraction in which the degree of the denominator is strictly greater than that of the numerator. In this example, the polynomial is identically 1, and the rational fraction is $-b^2/(s^2 + b^2)$.

In most practical cases the polynomial may be obtained by inspection, by adding to the numerator enough to make a multiple of the denominator and simultaneously subtracting the same. For example,

$$\frac{s^2}{s^2 + b^2} = \frac{s^2 + b^2 - b^2}{s^2 + b^2} = \frac{s^2 + b^2}{s^2 + b^2} - \frac{b^2}{s^2 + b^2} = 1 - \frac{b^2}{s^2 + b^2}$$

For another example,

$$\frac{s^3 + 2s^2}{s^2 + s + 1} = \frac{s^3 + (s^2 + s - s^2 - s) + 2s^2}{s^2 + s + 1}$$

$$= \frac{s(s^2 + s + 1) + s^2 - s}{s^2 + s + 1}$$

$$= s + \frac{s^2 - s}{s^2 + s + 1}$$

$$= s + \frac{s^2 + (s + 1 - s - 1) - s}{s^2 + s + 1}$$

$$= s + 1 + \frac{-2s - 1}{s^2 + s + 1}$$

Sample problems and problems are found in Parts 6 and 7 of Section E. In the general case, the polynomial is found by polynomial long division.

The terms of a polynomial in s are the Laplace transforms of the generalized derivatives of δ, and

$$\mathcal{L}[\delta^{(n)}(t - t_1)](s) = s^n \tag{9}$$

All these generalized derivatives can appear in differential equations, as we shall see in Section G.

There is also an application of δ' in problems concerning loads on beams, where $\delta'(t - t_1)$ is the loading corresponding to a "point moment" at t_1, just as $\delta(t - t_1)$ corresponds to a "point loading" at t_1.

The function f whose generalized derivative is given in (8) is of the form

$$f(t) = u(t - t_1)g(t - t_1) = u(t - t_1)h(t) \tag{10}$$

where h is differentiable. Such functions arise frequently in applications. For use in checking the solutions of differential equations, we now derive a general formula for the first and second derivatives of a function of the form (10).

Recall that

$$\mathcal{L}[g'(t)](s) = s\mathcal{L}[g(t)](s) - g(0)$$

from which it follows that

$$\mathcal{L}[h'(t + t_1)](s) = s\mathcal{L}[h(t + t_1)](s) - h(t_1)$$

or

$$s\mathcal{L}[h(t + t_1)](s) = \mathcal{L}[h'(t + t_1)](s) + h(t_1)$$

Thus, the Second Shifting Theorem implies that

$$\mathcal{L}[f'](s) = s\mathcal{L}[u(t - t_1)h(t)](s) - 0$$
$$= e^{-st_1}s\mathcal{L}[h(t + t_1)](s)$$
$$= e^{-st_1}\{\mathcal{L}[h'(t + t_1)](s) + h(t_1)\}$$
$$= \mathcal{L}[u(t - t_1)h'(t) + h(t_1)\,\delta(t - t_1)](s)$$

and therefore
$$f'(t) = u(t - t_1)h'(t) + h(t_1)\,\delta(t - t_1) \tag{11}$$

Notice that (8) is a special case, and that $h(t_1)$ does not depend upon t. The formula for f'' is found by applying (11) to $u(t - t_1)h'(t)$ in (11) and taking the generalized derivative of $h(t_1)\,\delta(t - t_1)$:

$$f''(t) = u(t - t_1)h''(t) + h'(t_1)\,\delta(t - t_1) + h(t_1)\,\delta'(t - t_1) \tag{12}$$

Similarly,

$$f^{(n)}(t) = u(t - t_1)h^{(n)}(t) + h^{(n-1)}(t_1)\,\delta(t - t_1)$$
$$+ h^{(n-2)}(t_1)\,\delta'(t - t_1) + \cdots + h(t_1)\,\delta^{(n-1)}(t - t_1) \tag{13}$$

5. Properties

If f_1 and f_2 have generalized derivatives f'_1 and f'_2, and if c_1 and c_2 are constants, then the linear combination $c_1 f_1 + c_2 f_2$ has a generalized derivative, which is $c_1 f'_1 + c_2 f'_2$, according to the following computation:

$$\mathcal{L}^{-1}[s\mathcal{L}[c_1 f_1 + c_2 f_2](s) - [c_1 f_1(0-) + c_2 f_2(0-)]]$$
$$= \mathcal{L}^{-1}[sc_1 \mathcal{L}[f_1](s) - c_1 f_1(0-) + sc_2\mathcal{L}[f_2](s) - c_2 f_2(0-)]$$
$$= c_1 \mathcal{L}^{-1}[s\mathcal{L}[f_1](s) - f_1(0-)] + c_2 \mathcal{L}^{-1}[s\mathcal{L}[f_2](s) - f_2(0-)]$$
$$= c_1 f'_1 + c_2 f'_2,$$

That is, $(c_1 f_1 + c_2 f_2)' = c_1 f'_1 + c_2 f'_2$

That is,
$$(c_1 f_1 + c_2 f_2)' = c_1 f'_1 + c_2 f'_2$$

This linearity property is the extension of the usual linearity of differentiation as learned in calculus:

$$\frac{d}{dt}(c_1 f_1 + c_2 f_2) = c_1 \frac{d}{dt}f_1 + c_2 \frac{d}{dt}f_2$$

if c_1 and c_2 are constants and f_1 and f_2 are differentiable functions.

Warning: The other classical "rules" of differential calculus, the product rule, power rule, quotient rule, and chain rule, are for derivatives and *do not generalize* to generalized derivatives. Where a function has both derivative and generalized derivative, these rules may be applied confidently; but at points where a function is not differentiable, these "rules" have no validity whatsoever. In fact, (12) shows the failure of the rule applicable to products of differentiable functions because applying that rule twice to a product $f = gh$ gives

$$\frac{d^2}{dt^2}[gh] = gh'' + 2g'h' + g''h$$

by contrast with (12), the generalized second derivative of $u(t - t_1)h(t)$.

6. Example: Checking the Generalized Solution of an Initial Value Problem

Now consider the gun and pellet from Example 8 of the previous section, where (21) gives the initial value problem

$$Mx''(t) + Bx'(t) = Mv_1\,\delta(t - t_1)$$
$$x(0) = 0 \tag{14}$$
$$x'(0) = 0$$

The generalized solution was given in (22):

$$x(t) = \frac{Mv_1}{B}u(t - t_1)(1 - e^{-(B/M)(t - t_1)}) \qquad 0 \le t$$

The generalized derivative can be computed from formula (11),

$$x'(t) = v_1 u(t - t_1)e^{-(B/M)(t - t_1)} \qquad 0 \le t$$

The generalized derivative of x' may be found using (13):

$$x''(t) = v_1\,\delta(t - t_1) - \frac{Bv_1}{M}u(t - t_1)e^{-(B/M)(t - t_1)} \qquad 0 \le t$$

Substituting these formulas into (14) verifies that x is a generalized solution:

$$Mx''(t) + Bx'(t) = Mv_1\,\delta(t - t_1) - Bv_1 u(t - t_1)e^{-(B/M)(t - t_1)}$$
$$+ Bv_1 u(t - t_1)e^{-(B/M)(t - t_1)}$$
$$= Mv_1\,\delta(t - t_1) \qquad 0 \le t$$

It also satisfies the initial conditions. These things it does whether $t_1 > 0$ or $t_1 = 0$.

7. Example: Setting Up an Initial Value Problem

Sometimes the excitation of a system is usefully described with a generalized derivative. Consider the recoil mechanism (shock-absorbing mechanism) for the gun firing the pellet in the previous section. Typically, the recoil mechanism consists of a slide to allow the gun to move with the momentum it receives from firing, a friction device (such as a dashpot) to absorb and dissipate the energy of recoil, and a return device (such as a spring) to restore the gun to its original position on the slide and hold it there. Let M be the mass of the gun barrel

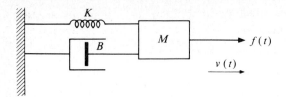

Figure 3 The recoil mechanism.

together with everything else which moves in recoil, and let K and B be the spring and damping constants for the recoil mechanism.

The differential equation for the velocity v of the recoil mechanism is

$$M \frac{d^2v}{dt^2}(t) + B \frac{dv}{dt}(t) + Kv(t) = \frac{df}{dt}(t)$$

At time $t = t_1$, the bullet is given velocity v_1 and momentum $M_1 v_1$, where M_1 is the mass of the bullet. The force on the bullet is therefore $M_1 v_1 \delta(t - t_1)$, and the reaction force on the gun is $-M_1 v_1 \delta(t - t_1)$. Therefore,

$$M \frac{d^2v}{dt^2}(t) + B \frac{dv}{dt}(t) + Kv(t) = \frac{d}{dt}[-M_1 v_1 \delta(t - t_1)]$$

$$= -M_1 v_1 \delta'(t - t_1)$$

where $\delta'(t - t_1)$ is the generalized derivative of $\delta(t - t_1)$. The Laplace transform of $\delta'(t - t_1)$ is $s\mathscr{L}[\delta(t - t_1)](s) - 0 = se^{-st_1}$. Using this information, we can solve the initial value problem

$$\left. \begin{aligned} Mv'' + Bv' + Kv &= -M_1 v_1 \delta'(t - t_1) \\ v(0) &= 0 \\ v'(0) &= 0 \end{aligned} \right\}$$

$$Ms^2 \mathscr{L}[v(t)](s) + Bs\mathscr{L}[v(t)](s) + K\mathscr{L}[v(t)](s) = -M_1 v_1 se^{-st_1}$$

$$\mathscr{L}[v(t)](s) = \frac{-M_1 v_1 se^{-st_1}}{Ms^2 + Bs + K}$$

$$v(t) = \mathscr{L}^{-1} \left[e^{-st_1} \frac{-M_1 v_1}{Ms^2 + Bs + K} \right](t) \qquad t \geq 0$$

Since the rational function on the right-hand side of the last equation has a denominator whose degree is actually greater than that of the numerator, it can be decomposed into partial fractions in the usual way and discovered to be the Laplace transform of some combination of exponentials, sines, and cosines. An application of the Second Shifting Theorem then completes the taking of the inverse transform.

It should be recognized specifically that *all* the derivatives appearing in the problem are generalized derivatives. While this fact need not be used explicitly in working out the present problem as stated, its computational importance becomes more visible in the case $t_1 = 0$, where the problem must be written this way:

$$Mv'' + Bv' + Kv = -M_1 v_1\, \delta'(t)$$
$$v(0-) = 0$$
$$v'(0-) = 0$$

That is, the initial conditions concern v and its *generalized* derivatives.

Let us go back to the general differential equation describing the system,

$$Mv'' + Bv' + Kv = \frac{df}{dt}$$

and observe how the generalized derivatives appear in the writing of the nonhomogeneous term for different choices of f and t_1.

If the time interval of interest were that during which the bullet is in the barrel of the gun, then we might try to model the system using

$$f(t) = Cu(t - t_1)$$

for some constant C:

$$\frac{df}{dt}(t) = Cu'(t - t_1) = C\,\delta(t - t_1)$$

In the special case $t_1 = 0$, we would have

$$\frac{df}{dt}(t) = C\,\delta(t)$$

Compare this description of f with what might have happened if we had said, "f is of constant value C from $t = 0$ on, so $f'(t) \equiv 0$." The initial value problem which would have resulted from such an analysis

$$Mv'' + Bv' + Kv = 0$$
$$v(0-) = 0$$
$$v'(0-) = 0$$

is clearly wrong, because it contains no excitation force at all. In fact, its solution is

$$v(t) = 0 \qquad t \geq 0$$

The fault lies in the way this analysis ignores the crucial condition that $f(t) = Cu(t)$ has limit zero from the left at $t = 0$, and therefore a jump of C at $t = 0$. Thus, f is not the function identically equal to C, and f' is not the function identically equal to zero, but

$$f'(t) = C\,\delta(t)$$

The important point is not the value of f at $t = 0$; it is the limit from the left.

Now suppose instead that

$$f(t) = Cu(t - t_1)e^{\lambda(t - t_1)}$$

where C and λ are constants. By (11),

$$f'(t) = C \, \delta(t - t_1) + C\lambda u(t - t_1)e^{\lambda(t - t_1)} \qquad t \geq 0$$

Of course, the special case $t_1 = 0$ is

$$f'(t) = C \, \delta(t) + C\lambda u(t)e^{\lambda t} \qquad t \geq 0$$

Notice that the jump in f at $t = t_1 = 0$, the difference between the right- and left-hand limits, is accounted for in the generalized derivative with the term $C \, \delta(t)$.

8. Problems

1. The systems depicted in Figure 4 are equivalent. The differential equations for the displacement x and velocity v are

$$Mx'' + Kx = f$$

and

$$Mv'' + Kv = f' \qquad (15)$$

respectively, where $M = \frac{1}{2}$ and $K = 3$.

(a) If $x(t) = 0$ for $t < t_1$, and $f(t) = -3u(t - t_1)$, then set up and solve the initial value problem determining the displacement x, for both the case $t_1 > 0$ and the case $t_1 = 0$.

(b) If $x(t) = 0$ for $t < t_1$, and $f(t) = -3u(t - t_1)$, then use (15) to set up and solve the initial value problem determining the velocity v, for both the case $t_1 > 0$ and the case $t_1 = 0$.

(c) Find x' from part a, and compare it with v as determined in part b. Graph f, x, and v on the same graph for the case $t_1 = 0$.

(d) If $x(t) = 0$ for $t < 0$, and $f(t) = e^{-t}$ for $t > 0$, find $v(t)$.

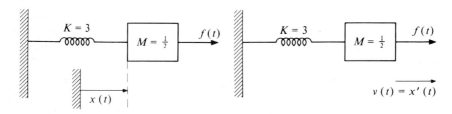

Figure 4 Two drawings of a spring-mass system.

2. In the circuit illustrated in Figure 5, $i(t) = 0$ for all t prior to some t_1.

(a) Set up and solve the differential equation relating $i(t)$ to $v_i(t)$ if $v_i(t) = 2u(t - t_1)$, for the case $t_1 > 0$ and for the case $t_1 = 0$.

(b) Set up and solve the differential equation relating $v_o(t)$ to $v_i(t)$ if $v_i(t) = 2u(t)$. (Notice that a generalized derivative will be required for the nonhomogeneous term of the differential equation.)

(c) Compare the result of part b with that of part a for the case $t_1 = 0$, using the relation between $v_o(t)$ and $i(t)$. Graph v_i, i, and v_o on the same graph.

(d) If $v_i(t) = \cos \omega t$ for $t > 0$, find $v_o(t)$.

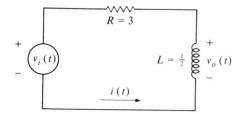

Figure 5 The RL circuit of Problem 2.

G. ANALYSIS OF SYSTEMS: THE TRANSFER FUNCTION AND THE CONVOLUTION INTEGRAL

So far, our analysis of engineering systems has been based on the following scheme:

1. Derive a differential equation to describe the system (Chapter II):
   a. Set up a system of differential equations for the elements of the system and for the connections between the elements.
   b. Eliminate from the system of differential equations all variables but two (one representing a known function, the other an unknown function) appearing in a single differential equation.
2. Solve the differential equation, using Laplace transforms (Chapter III):
   a. Take the Laplace transform of the differential equation, making use of the initial values, thus obtaining the transformed problem.
   b. Solve the transformed problem for the Laplace transform of the unknown function.
   c. Take the inverse Laplace transform to find the unknown function itself.

The scheme is very computational, and therefore routine in a manner of speaking, but it is not short. Our present task is to discover and reveal a more intimate relation between the known function (the "input" into the system) and the unknown function (the "output" of the system), a relation which summarizes the action of the system upon the input. The relation turns out to be given by the ratio of the Laplace transform of the output to that of the input. Since it is a ratio of functions of s, this ratio is itself a function of s, and it is called the *transfer function*. The important observation is that the transfer function is determined by the system, not by any formula for input. Thus (at least in the most important cases), the Laplace transform of the output can be obtained from that of the input by multiplication by the transfer function. This multiplication therefore summarizes all the computations down to 2b in the scheme given above. For a given system, the transfer function provides in this way a convenient means of analyzing the dependence of the output (the unknown function) upon the input (the known function). It is also used to analyze a succession of systems of which the output for one is the input for the next, and to compare related systems. Therefore, it is very often used in engineering systems analysis. The cases in which it is used and the terminology appropriate to its use are discussed in Part 1 below.

The transfer function is computed in the process of setting up or solving the transformed problem. In order to use it effectively, therefore, we shall look again at the transformed problem and introduce the "transform domain" in which such problems can be solved. That is the subject of Part 2 below. It is illustrated by examples in Part 3, and by sample problems with answers in Part 4.

The transfer function is itself an object in the transform domain. Its inverse Laplace transform may be interpreted as the response of a system to a unit impulse. Part 5 is concerned with this "impulse response" of the system, and it identifies the collection of all possible inputs and responses as the "time domain." In the time domain the general problem of systems analysis is solved, once and for all, with a new tool called the "convolution" integral. That is the subject of Part 6.

The convolution is introduced independently, without reference to systems concepts and notation, in Part 7. There and in the succeeding parts, it is exhibited in examples and used for solving initial value problems. These parts, which may be studied usefully even if Parts 1 to 6 are omitted entirely, may be summarized in a single formula, which should be added to your short table of Laplace transforms. The formula is (50), and it is derived in Part 8.

1. The "Systems Perspective," Input, Output, and the Quiescent State

Consider the circuit shown in Figure 1. If the voltage (whether constant or time-varying) produced by the voltage generator is called v_i, and the voltage recorded by a voltmeter placed across the capacitor is v_o, then v_i and v_o are related by the differential equation

$$RCv_o' + v_o = v_i \tag{1}$$

Of course, v_i stands for a function. If v_i changes, then v_o changes too, for v_o and v_i are related by the equation.

The first step in adopting the "systems" point of view is to think of v_i and v_o as input and output functions. Specifically, v_i (whether constant or varying) is the voltage put into the system, or the *input* voltage; v_o (whether constant or varying) is the voltage put out by the system, or the *output* voltage. To emphasize the roles of the input and output, the circuit of Figure 1 is usually redrawn as in Figure 2.

In Figure 2, v_i is represented as the voltage across a pair of terminals on the left, where the input voltage is applied, and v_o is represented as the voltage across a pair of terminals at the ends of the capacitor, to which is attached the voltmeter

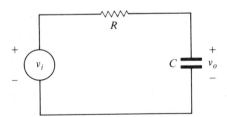

Figure 1 *RC* series circuit.

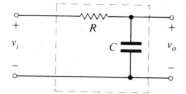

Figure 2 System representation for an RC series circuit.

that reads v_o. Thus, Figures 1 and 2 represent exactly the same circuit. In Figure 2 the dashed lines surround "the system," from which the input and output terminals emerge. When constructed as a device, this particular circuit might be arranged as in Figure 3, with two input terminals and two output terminals, each labeled $+$ or $-$.

Of course, the input into a different electrical system might be a current rather than a voltage, so the usual terminology is a little less specific. The input, whatever its character, is referred to as the input signal, input reference, or *input variable*; the output, as the output signal, controlled variable, or *output variable*. Notice that the word *variable* is used, to emphasize that much of our analysis will be done *before* the function actually to be represented by v_i is given. The function represented by v_o depends upon the system and v_i.

Figure 4 could represent a suspension and shock-absorbing system like that over one wheel of an automobile. We may displace the input rod and ask what is the resulting displacement of the mass M. The displacement x_i is the input variable, the displacement of the mass M is x_o, the output variable, and the two are related by the differential equation

$$Mx_o'' + Bx_o' + Kx_o = Bx_i' + Kx_i + f_0 - Mg$$

where f_0 is the compression in the spring when $x_o - x_i = 0$.

If we are given an input displacement function x_i and the initial values for the output x_o and its derivatives, then x_o will be the solution of an initial value problem, the kind of problem we have solved before. If, instead, our object is to study and describe the behavior of the system, then we may wish to change the zero points on the scale we use to define x_i and x_o, making the zero points correspond to the equilibrium state of the system, and simplify our description by having x_i and x_o represent displacement from equilibrium.

Engineers often do a system analysis for a system which is initially in a quiescent or equilibrium state, using variables whose zero points correspond to the equilibrium state. For the system of Figure 4, there is an equilibrium or "at-rest" position in which the system can remain motionless forever. When the

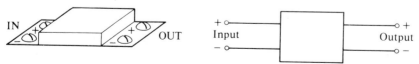

Figure 3 System device: pictorial and schematic.

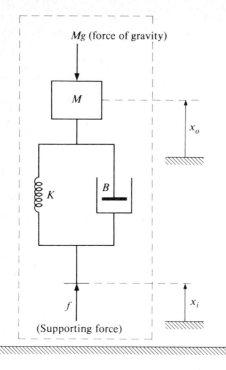

Figure 4 Simple suspension and shock-absorbing system.

system is in this position, the compression force in the spring is due only to the gravitational force affecting the mass Mg. By measuring the displacements x_i and x_o from this equilibrium state, we obtain some simplification when solving for x_o, since all the initial values will be zero and $f_0 - Mg$ will be zero.

Often the equilibrium state in an electric circuit is the "relaxed" state. A circuit is *relaxed until time* $t = 0$ if there is a time interval (of nonzero length) preceding time $t = 0$ during which all capacitors have had zero voltage across them and all inductors have had zero current running through them. A relaxed circuit is one in which the amount of energy stored is zero.

In a mechanical system, the relaxed state occurs less frequently. A real spring is usually compressed or stretched, even in the absence of excitation, and therefore stores energy in its equilibrium state. More often, the mechanical system is "at rest." A mechanical system is *at rest until time* $t = 0$ if there is a time interval (of nonzero length) preceding time $t = 0$ during which each element has been motionless.

In our system analysis in this section, we shall assume either that the system is relaxed until $t = 0$ or that the system is at rest until $t = 0$. If we assume that the system is at rest until $t = 0$, then we shall write the differential equations for new variables which are the differences between the values of the actual system variables at time t and their equilibrium or quiescent values. As a consequence, all resulting differential equations will relate input and output functions whose values at time $t < 0$ will be zero also.

As an example, consider the initial value problem for the system in Figure 4. The system is at rest until $t = 0$, or rather it is at rest on some interval $(\alpha, 0)$, and we set the scale for x_i so that $x_i(t) = 0$ for t in the interval $(\alpha, 0)$. Similarly, set the scale for x_o so that $x_o(t) = 0$ for t in the interval $(\alpha, 0)$. With x_i and x_o defined, we may proceed to write the differential equation for the system:

$$Mx_o'' + Bx_o' + Kx_o = Bx_i' + Kx_i \tag{2}$$

Next we must write the initial values. Since x_o is zero throughout the interval $(\alpha, 0)$, all the derivatives of x_o are also zero on that interval:

$$x_o^{(k)}(t) = 0 \qquad \text{for all } k, \ \alpha < t < 0 \tag{3}$$

The same is true of the input displacement x_i and all its derivatives:

$$x_i^{(k)}(t) = 0 \qquad \text{for all } k, \ \alpha < t < 0 \tag{4}$$

If x_i is continuous near $t = 0$, along with those derivatives (namely x_i') appearing in the differential equation, then x_o is also continuous at $t = 0$, along with those derivatives (x_o' and x_o'') which appear in the equation. (This fact was stated in Theorem 2 of Section H of Chapter I, the synopsis on linear equations.) In such a situation, the set of initial conditions will look like this:

$$\left. \begin{aligned} x_o(0) &= 0 \\ x_o'(0) &= 0 \\ \cdots\cdots\cdots \\ x_o^{(n-1)}(0) &= 0 \end{aligned} \right\} \tag{5}$$

In the present example, if x_i and x_i' are continuous near $t = 0$, then the initial value problem for the system in Figure 4 (at rest until $t = 0$) is

$$\left. \begin{aligned} Mx_o'' + Bx_o' + Kx_o &= Bx_i' + Kx_i \\ x_o(0) &= 0 \\ x_o'(0) &= 0 \end{aligned} \right\} \tag{6}$$

In fact, the problem could be written in the same way as long as the only discontinuity of x_i and its derivatives is a step in the *highest* derivative appearing in the equation.

In general, however, one cannot guarantee that all the possible input functions will be continuous with continuous derivatives. In the general case, the initial values of x_o and its derivatives are written as left-hand limits (as in Section E), and these left-hand limits are all zero since $x_o(t) = 0$ on the interval $(\alpha, 0)$:

$$\left. \begin{aligned} x_o(0-) &= 0 \\ x_o'(0-) &= 0 \\ \cdots\cdots\cdots \\ x_o^{(n-1)}(0-) &= 0 \end{aligned} \right\} \tag{7}$$

In fact, the simplest way to look at the initial values is the most general way: *All* derivatives are to be considered as generalized derivatives (as in Section F), and the initial values are therefore written in terms of left-hand limits as in (7), of which (5) is just the special case in which all derivatives happen to be continuous. The general initial value problem for the system in Figure 4 (at rest until $t = 0$) is therefore not (6) but

$$Mx_o'' + Bx_o' + Kx_o = Bx_i' + Kx_i$$
$$x_o(0-) = 0 \qquad\qquad (8)$$
$$x_o'(0-) = 0$$

The initial conditions are left-hand limits, summarizing the relevant part of the behavior of the system prior to time $t = 0$.

The transformed problem for (8) is the same as that for (6):

$$(Ms^2 + Bs + K)\mathscr{L}[x_o](s) = (Bs + K)\mathscr{L}[x_i](s) \qquad (9)$$

Notice particularly that the initial condition on x_i was also used, in the following way:

$$\mathscr{L}[x_i'](s) = s\mathscr{L}[x_i](s) - x_i(0-)$$
$$= s\mathscr{L}[x_i](s)$$

following Equation (4), just as the initial conditions on x_o followed from Equation (3). Because a relaxed system and a system at rest have all initial conditions zero for input and output functions, the result of dividing by the coefficient of $\mathscr{L}[x_o]$ in the equation in the transformed problem is a product of which one factor is $\mathscr{L}[x_i]$. For example,

$$\mathscr{L}[x_o](s) = \frac{Bs + K}{Ms^2 + Bs + K}\,\mathscr{L}[x_i](s) \qquad (10)$$

This fact shows that the transform of the output of such a system is a multiple of the transform of the input. The multiplier is called the "transfer function" of the system.

2. The Transfer Function and the Transform Domain

Schematically, a system may be depicted as in Figure 5. Its model consists of an initial state description

$$x_i(t) = 0$$
$$x_o(t) = 0 \qquad \text{for } \alpha < t < 0$$

Figure 5 Basic system configuration.

and a differential equation relating x_o to x_i. The problem is to find more direct descriptions of how x_o depends on x_i, and the technique is the use of Laplace transformation. To simplify the notation, we shall use capital letters as the names of the Laplace transforms of functions given by lowercase letters. Thus,

$$X_i = \mathscr{L}[x_i] \qquad X_o = \mathscr{L}[x_o] \qquad F = \mathscr{L}[f]$$

and so forth.

Now let us look at the Laplace transforms of the examples in the previous part. The electrical example (1) has the following transformed equation if the system is relaxed until $t = 0$:

$$RC(s\mathscr{L}[v_o](s) - v_o(0-)) + \mathscr{L}[v_o](s) = \mathscr{L}[v_i](s)$$
$$(RCs + 1)\mathscr{L}[v_o](s) = \mathscr{L}[v_i](s)$$
$$(RCs + 1)V_o(s) = V_i(s) \tag{11}$$

from which it follows that

$$V_o(s) = \frac{1}{RCs + 1} V_i(s) \tag{12}$$

Notice that Equation (12) relates the Laplace transform of *every* output function to that of the input function producing it. The action of the system upon the input function is entirely described by the multiplication of Laplace transforms on the right-hand side of (12).

Equation (12) is therefore quite remarkable. It shows that no matter what the input function v_i may be, if v_o is the corresponding output function, then the ratio of the Laplace transforms of v_o and v_i is the rational function $1/(RCs + 1)$. This ratio is the same for all v_i; it is independent of the input function and its transform, but determined entirely by the RC circuit.

Let us repeat that, for emphasis. The RC circuit (relaxed until $t = 0$) relates the output to the input. This relationship is described by the differential equation (1) and the transformed equation (11). The relationship between the Laplace transforms of v_o and v_i is that their ratio is $1/(RCs + 1)$, and this ratio is due to the RC circuit. Because the rational function $1/(RCs + 1)$ characterizes the relationship between the Laplace transforms of the output and input functions, it also characterizes the action of the RC circuit (relaxed until $t = 0$). Because of its importance, we give the ratio a special name: $1/(RCs + 1)$ is the "transfer function" of the circuit, relating the output and input of the system.

The same thing can be done for the mechanical system of Figure 4. In the present notation,

$$X_o(s) = \frac{Bs + K}{Ms^2 + Bs + K} X_i(s) \tag{13}$$

The action of the system upon the input displacement is entirely described by the multiplication of Laplace transforms on the right side of (13), and the rational function $(Bs + K)/(Ms^2 + Bs + K)$ is called the "transfer function" of the system.

In general, the *transfer function* of a system relaxed or at rest until $t = 0$ is defined to be the ratio of the Laplace transform of the output variable to that of the input variable and is designated H. It is a function of s:

$$H(s) = \frac{X_o(s)}{X_i(s)} \tag{14}$$

Consider, for example, the transfer function of the RC circuit system discussed above, $H(s) = 1/(RCs + 1)$, and now let us see it in operation. If the input function is the unit step function $v_i(t) = u(t)$, then $V_i(s) = 1/s$ and

$$V_o(s) = H(s)\frac{1}{s}$$

$$= \frac{1}{RCs + 1}\frac{1}{s}$$

$$= \frac{1}{RC}\frac{1}{s + 1/RC}\frac{1}{s}$$

$$= \frac{1}{s} - \frac{1}{s + 1/RC}$$

$$v_o(t) = u(t)(1 - e^{-t/RC})$$

Notice that, since the system is known to be relaxed until $t = 0$, we can write $v_o(t) = 0$ for $t < 0$.

If, instead, $v_i(t) = u(t) \sin t$, then $V_i(s) = 1/(s^2 + 1)$ and

$$V_o(s) = H(s)\frac{1}{s^2 + 1}$$

$$= \frac{1}{RCs + 1}\frac{1}{s^2 + 1}$$

$$= \frac{1/RC}{(s + 1/RC)(s^2 + 1)}$$

$$= \frac{1}{(RC)^2 + 1}\left(\frac{RC}{s + 1/RC} - RC\frac{s}{s^2 + 1} + \frac{1}{s^2 + 1}\right)$$

$$v_o(t) = u(t)\frac{1}{(RC)^2 + 1}(RCe^{-t/RC} - RC \cos t + \sin t)$$

In fact, the electric circuit of Figures 1 and 2, which we are using for an example, may be depicted as in Figure 6, because a system (relaxed until $t = 0$) acts upon an input voltage in a manner completely specified by the transfer function.

For the mechanical example of Figure 4, the transfer function is $H(s) = (Bs + K)/(Ms^2 + Bs + K)$, and the corresponding diagram is Figure 7.

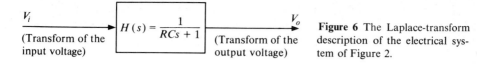

V_i

(Transform of the
input voltage)

$H(s) = \dfrac{1}{RCs + 1}$

V_o

(Transform of the
output voltage)

Figure 6 The Laplace-transform
description of the electrical system of Figure 2.

Notice that the defining property for the transfer function,

$$H(s) = \frac{X_o(s)}{X_i(s)} \quad \text{or} \quad X_o(s) = H(s)X_i(s) \tag{15}$$

is an equality relating Laplace transforms. Thus, the quotient and product indicated in (15) may be called operations in the *transform domain of the system*, which is the collection of the Laplace transforms of all possible inputs and outputs. In short, the transform domain is the collection of Laplace transforms of those functions, δ's, and so forth which are identically zero on the left half line. In the transform domain, the action of a system (relaxed or at rest until $t = 0$) is a simple multiplication.

The transform domain has many other common names. Some call it the s domain, since all functions there depend upon the variable s (although a few authors choose a different letter for the variable). Others prefer to call it the *frequency domain*, assigning to s the same dimensional units which frequency has, second^{-1}. The reason for this assignment of units is that the product st must be dimensionless if the exponential e^{-st} is to make sense dimensionally in the definition

$$X(s) = \int_0^\infty x(t)e^{-st}\,dt$$

3. Examples

(*a*) Determine the transfer function for the *RC* circuit of Figure 8, which is relaxed until $t = 0$.

From the models (Section A of Chapter II),

$$i(t) = C\frac{dv_C}{dt}(t) \tag{16}$$

$$v_o(t) = Ri(t) \tag{17}$$

$$v_i(t) = v_C(t) + v_o(t) \tag{18}$$

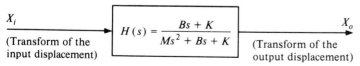

X_i

(Transform of the
input displacement)

$H(s) = \dfrac{Bs + K}{Ms^2 + Bs + K}$

X_o

(Transform of the
output displacement)

Figure 7 The Laplace-transform description of the mechanical system of Figure 4.

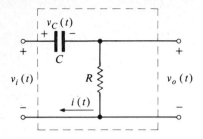

Figure 8 Relaxed *RC* circuit.

Substitution of (16) into (17) eliminates $i(t)$:

$$v_o(t) = RC \frac{dv_C}{dt}(t) \left. \right\} \tag{19}$$

$$v_i(t) = v_C(t) + v_o(t) \right| \tag{20}$$

Substitution of (20) into (19) eliminates $v_C(t)$:

$$v_o(t) = RC \frac{d}{dt}[v_i(t) - v_o(t)]$$

or $\qquad\qquad RC v_o'(t) + v_o(t) = RC v_i'(t) \tag{21}$

Taking the Laplace transforms of both sides of (21) gives the transformed problem:

$$RC[sV_o(s) - v_o(0)] + V_o(s) = RC[V_i(s) - v_i(0)]$$

$$(RCs + 1)V_o(s) = RCsV_i(s) + RC[v_o(0) - v_i(0)]$$

By Equation (18), $v_o(0) - v_i(0) = -v_C(0)$. Since the system is relaxed, $v_C(0) = 0$. Hence,

$$\frac{V_o(s)}{V_i(s)} = \frac{RCs}{RCs + 1} \tag{22}$$

By definition, the transfer function of a system (relaxed until $t = 0$) is the Laplace transform of the output divided by the Laplace transform of the input. Hence the result requested is

$$H(s) = \frac{V_o(s)}{V_i(s)} = \frac{RCs}{RCs + 1}.$$

[As noted, the Laplace transform of the output equals the transfer function times the Laplace transform of the input:

$$V_o(s) = H(s)V_i(s)$$

Thus, if the input function and transfer function are specified for a system relaxed until $t = 0$, then the output of the system can be computed using Laplace-transform methods.]

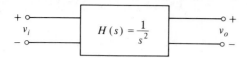

Figure 9 System represented by its transfer function.

(b) Determine the output function v_o for the system shown in Figure 9, given that the transfer function is $H(s) = 1/s^2$, that

$$v_i(t) = \begin{cases} 0 & t \le 0 \\ \sin 10t & t > 0 \end{cases}$$

and that all initial values are zero at $t = 0$.

$$\mathscr{L}[v_o](s) = V_o(s) = H(s)\mathscr{L}[v_i](s)$$

$$= \frac{1}{s^2} \frac{10}{s^2 + 10^2}$$

$$= \frac{A}{s^2} + \frac{B}{s} + \frac{Cs}{s^2 + 100} + \frac{10D}{s^2 + 100}$$

$$\frac{10}{s^2(s^2 + 100)} = \frac{A(s^2 + 100) + Bs(s^2 + 100) + Cs^3 + 10s^2D}{s^2(s^2 + 100)} \qquad (23)$$

Equating coefficients of s in the numerator of Equation (23) gives

$$\left. \begin{array}{ll} s^3: & B + C = 0 \\ s^2: & A + 10D = 0 \\ s^1: & 100B = 0 \\ s^0: & 100A = 10 \end{array} \right\} \quad \begin{array}{l} B = 0 \\ C = 0 \\ A = \frac{1}{10} \\ D = -\frac{1}{100} \end{array}$$

Thus
$$V_o(s) = \frac{1}{10} \frac{1}{s^2} - \frac{1}{100} \frac{10}{s^2 + 100} \qquad (24)$$

and
$$v_o(t) = \frac{1}{10}t - \frac{1}{100} \sin 10t \qquad t \ge 0 \qquad (25)$$

(c) Find the transfer function relating the output $v(t)$ to the input $f(t)$ for the mechanical system of Figure 10, assuming that the system is relaxed until $t = 0$.

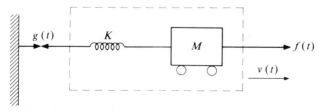

Figure 10 KM mechanical system.

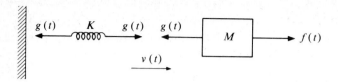

Figure 11 Free-body diagram for the system of Figure 10.

The equations for the elements of the system are

$$v(t) = \frac{1}{K} \frac{dg}{dt}(t) \tag{26}$$

$$f(t) - g(t) = M \frac{dv}{dt}(t) \tag{27}$$

Substituting (27) into (26) to eliminate $g(t)$ gives

$$v(t) = \frac{1}{K} \frac{d}{dt} \left[f(t) - M \frac{dv}{dt}(t) \right] \tag{28}$$

Rearrangement gives the differential equation for the system:

$$\frac{d^2v}{dt^2}(t) + \frac{K}{M} v(t) = \frac{1}{M} \frac{df}{dt}(t) \tag{29}$$

Taking the Laplace transform of both sides of Equation (29) gives the transformed equation:

$$s^2 V(s) - sv(0) - v'(0) + \frac{K}{M} V(s) = \frac{1}{M} [sF(s) - F(0)]$$

or

$$s^2 V(s) + \frac{K}{M} V(s) = \frac{1}{M} sF(s) \tag{30}$$

Solving (30) for $V(s)$ shows that

$$V(s) = \frac{1}{s^2 + K/M} \frac{s}{M} F(s)$$

$$= \frac{s}{Ms^2 + K} F(s) \tag{31}$$

Therefore the transfer function

$$H(s) = \frac{V(s)}{F(s)}$$

is

$$H(s) = \frac{s}{Ms^2 + K} \tag{32}$$

(d) For the relaxed mechanical system of Figures 10 and 11, find the transfer function relating the output $g(t)$, the tension force in the spring, to the input $f(t)$. Substituting (26) into (27) to eliminate $v(t)$ gives

$$f(t) - g(t) = M \frac{d}{dt} \left[\frac{1}{K} \frac{dg}{dt}(t) \right] \tag{33}$$

Upon rearranging (33), we see that

$$\frac{d^2g}{dt^2}(t) + \frac{K}{M} g(t) = \frac{K}{M} f(t) \tag{34}$$

Taking Laplace transforms on both sides of Equation (34) shows that

$$s^2 G(s) + \frac{K}{M} G(s) = \frac{K}{M} F(s)$$

Solving (34) for $G(s)$, we obtain

$$G(s) = \frac{K/M}{s^2 + K/M} F(s)$$

Hence the transfer function is

$$H(s) = \frac{G(s)}{F(s)} = \frac{K/M}{s^2 + K/M}$$

4. Sample Transfer-Function Problems with Answers

A. Find the transfer function $H(s)$ relating the output $v_1(t)$ to the input $v(t)$ for the relaxed mechanical system of Figure 12.

Answer: $H(s) = (K/M)/(s^2 + K/M)$

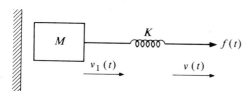

Figure 12 *KM* mechanical system.

B. Find the transfer function $H(s)$ relating the output $v_o(t)$ to the input $v_i(t)$ for the relaxed electrical system of Figure 13.

Answer: $H(s) = RCs/(LCs^2 + RCs + 1)$

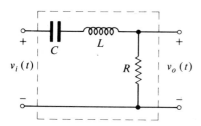

Figure 13 *LRC* electric circuit.

C. Find the transfer function $H(s)$ relating the output x_o to the input x_i for the system of Figure 14, at rest until $t = 0$.

    Answer: $H(s) = K/(Ms^2 + K)$

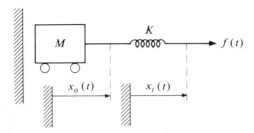

    Figure 14 KM mechanical system.

5. Impulse Response and the Time Domain

The transfer function of a given system is a quotient of Laplace transforms. It is a rational function, and it is in the transform domain. What is its inverse Laplace transform, and what significance or interpretation does that inverse transform have, if any? Is the inverse transform of H the output corresponding to some input?

    From the definition of transfer function,

$$V_o(s) = H(s)V_i(s)$$

If $\mathscr{L}^{-1}[H]$ is going to be the output v_o, then $V_o = H$ and

$$H(s) = H(s)V_i(s)$$

The input required to produce $\mathscr{L}^{-1}[H]$ as output must therefore have

$$V_i(s) = 1$$

as its Laplace transform, so the input must be $\delta(t)$, the Dirac δ. If the system is relaxed or at rest until $t = 0$, and if the input is $v_i(t) = \delta(t)$, then

$$V_o(s) = H(s)V_i(s) = H(s) \cdot 1 = H(s)$$

and the output is

$$v_o(t) = \mathscr{L}^{-1}[H](t) = h(t)$$

in the usual notation. Since it is the output corresponding to the input δ, the unit impulse, $h(t) = \mathscr{L}^{-1}[H](t)$ is called the *impulse response* of the system, or the *response to unit impulse*. Since the system is to be relaxed until $t = 0$, $h(t)$ is always assumed to be zero for $t < 0$, and of the form $h(t) = u(t)f(t)$.

$v_i(t) = \delta(t)$ $H(s) = \dfrac{s}{(s + 1)(s + 2)}$ $v_o(t) = h(t)$

Figure 15 System represented by its transfer function with the Dirac δ as input.

For example, let us find the impulse response of a system whose transfer function is $H(s) = s/(s + 1)(s + 2)$. As indicated by the figure and the definition,

$$\mathscr{L}[h](s) = V_o(s) = H(s)V_i(s)$$
$$= H(s)\mathscr{L}[\delta](s) = H(s)$$
$$= \frac{s}{(s + 1)(s + 2)}$$

$H(s)$ may be rewritten using the decomposition into partial fractions.

$$H(s) = \frac{s}{(s + 1)(s + 2)} = \frac{A}{s + 1} + \frac{B}{s + 2}$$
$$= \frac{A(s + 2) + B(s + 1)}{(s + 1)(s + 2)}$$

This time, let us find A and B by evaluating the numerators (Part 5 of Section C of Chapter III):

At $s = -1$: $-1 = A(+1) + B(0)$ or $A = -1$

At $s = -2$: $-2 = A(0) + B(-1)$ or $B = 2$

Therefore, $$H(s) = \frac{-1}{s + 1} + \frac{2}{s + 2}$$

and therefore the impulse response of the system is $h(t) = \mathscr{L}^{-1}[H](t) = u(t) \times (-e^{-t} + 2e^{-2t})$.

Notice that the impulse response h is a function of t. It is a possible response of the system, the response to one of the many possible inputs. The collection of all inputs and responses having Laplace transforms is called the *time domain or t domain*, and it is there that one finds the functions v_i, v_o, and h. It is to be contrasted with the transform domain or s domain, where one finds the functions V_i, V_o, and H.

To illustrate a system's response to a succession of impulsive inputs, let us look again at the circuit given in Figure 1, whose differential equation is (1). The transformed problem is (11). To simplify matters, let us imagine that in this circuit we have $RC = 1$, so that

$$V_o(s) = \frac{1}{s + 1} V_i(s)$$

Time domain	Transform domain
v_i	V_i
v_o	V_o
h	H

Figure 16 Correspondence between time variables and transform variables.

and the impulse response is $h(t) = u(t)e^{-t}$. Of course, the response to an input $v_i(t) = a_0\,\delta(t - t_0)$ is the inverse transform of

$$V_o(s) = \frac{1}{s + 1}\,a_0\,e^{-st_0}$$

namely

$$v_o(t) = a_0\,u(t - t_0)e^{-(t - t_0)}$$

by the Second Shifting Theorem. If, instead, v_i is a sum of two impulses,

$$v_i(t) = a_1\,\delta(t - t_1) + a_2\,\delta(t - t_2)$$

then

$$V_o(s) = H(s)(a_1\,e^{-st_1} + a_2\,e^{-st_2})$$

$$= a_1\,H(s)e^{-st_1} + a_2\,H(s)e^{-st_2}$$

and the output is

$$v_o(t) = a_1\,h(t - t_1) + a_2\,h(t - t_2)$$

$$= a_1\,u(t - t_1)e^{-(t - t_1)} + a_2\,u(t - t_2)e^{-(t - t_2)}$$

by the Second Shifting Theorem.

The significance of this is twofold. First, many systems are controlled or activated only by short impulsive bursts, the way the attitude of an artificial satellite in orbit is controlled by short "burns." The output of such a system is of the sort we saw above, a linear combination of translates of the impulse response h.

Second, any continuous input can be approximated with a quick succession of "short bursts," a linear combination of δ's, if only care is taken to do it correctly. For instance, if the function f is zero from some point T onward, for $t > T$, then it may be approximated by a combination of N δ's if N is a large number:

$$f(t) \cong \frac{T}{N}\left[f\left(\frac{T}{N}\right)\delta\left(t - \frac{T}{N}\right) + f\left(2\frac{T}{N}\right)\delta\left(t - 2\frac{T}{N}\right) + \cdots + f(T)\,\delta(t - T)\right]$$

$$= \frac{T}{N}\sum_{k=1}^{N} f\left(k\frac{T}{N}\right)\delta\left(t - k\frac{T}{N}\right)$$

Here we have used N impulses, at the points T/N, $2(T/N)$, $3(T/N)$, ..., $N(T/N) = T$, of the form $k(T/N)$. At each point we scaled up the impulse by the magnitude of f at that point, and then multiplied by the constant T/N so that the integral of our approximating function would be the same as that of f itself. Now let us see the response of our system to this input.

$$V_o(s) = \frac{T}{N}\sum_{k=1}^{N} f\left(k\frac{T}{N}\right)e^{-sk(T/N)}H(s)$$

$$v_o(t) = \frac{T}{N}\sum_{k=1}^{N} f\left(k\frac{T}{N}\right)e^{-[t - k(T/N)]}u\left(t - k\frac{T}{N}\right)$$

$$= \frac{T}{N}\sum_{k=1}^{N} f\left(k\frac{T}{N}\right)h\left(t - k\frac{T}{N}\right)$$

where the first formula for $v_o(t)$ gives the output in the example, $h(t) = u(t)e^{-t}$, and the second gives it in the general case.

If we label the origin z_0, and the points where the inputs are applied $z_1, \ldots, z_N$, and if the distance between is called Δz, then the output may be written as

$$v_o(t) = \frac{T}{N} \sum_{k=1}^{N} f(z_k)e^{-(t-z_k)}u(t-z_k)$$

for our example, and, more generally,

$$v_o(t) = \sum_{k=1}^{N} f(z_k)h(t-z_k) \, \Delta z$$

The sum on the right is actually an approximating sum for the integral

$$\int_0^T f(z)h(t-z) \, dz$$

which is equal to

$$\int_0^t f(z)h(t-z) \, dz$$

for $t > T$, since f is zero for $t > T$. This last integral is the "convolution of f and h," the subject of this section; our example shows how it is related to the response of a system to successive impulsive inputs.

6. Convolution in Systems Analysis

In the transform domain, the action of a relaxed or at-rest system on an input to produce an output is always of the same type, no matter what the system, no matter what the input: The transform of the output is the transform of the input times the transfer function of the system. This characterization of the relationship between the input and output is a transform-domain characterization. We might expect that there is a similar characterization in the time domain, and there is.

It is given by the convolution integral. Specifically, if an input v_i is put into a relaxed or at-rest system whose impulse response is h, then the output v_o can be written as the convolution of v_i with h. The symbol for convolution is $*$, so that

$$v_o = v_i * h \tag{35}$$

and the formula for computing it is

$$v_o(t) = (v_i * h)(t) = \begin{cases} 0 & t \le 0 \\ \int_0^t v_i(z)h(t-z) \, dz & t > 0 \end{cases} \tag{36}$$

$$= u(t) \int_0^t v_i(z)h(t-z) \, dz \tag{37}$$

The remainder of this section is devoted to examples and explanations of the use of convolution, and to a justification of the equality asserted in (35).

When considering the circuit of Figures 1 and 2, we determined that

$$RCv_o'(t) + v_o(t) = v_i(t) \tag{38}$$

If the circuit is relaxed until $t = 0$, then, taking Laplace transforms, we obtain

$$V_o(s) = \frac{1}{RC} \frac{1}{s + 1/RC} V_i(s) \tag{39}$$

The expression $(1/RC)[1/(s + 1/RC)]$ is the formula for the transfer function H of the system. The inverse Laplace transform of H gives the impulse response of the circuit, which therefore must be

$$h(t) = \frac{u(t)}{RC} e^{-t/RC} \tag{40}$$

Using (35), the time-domain relation between the input and the output, we can write the output $v_o(t)$ as the convolution of the input v_i with the impulse response h:

$$v_o(t) = u(t) \int_0^t v_i(z) \frac{u(t - z)}{RC} e^{-(t-z)/RC} \, dz \tag{41}$$

Notice carefully that the expression

$$\frac{u(t - z)}{RC} e^{-(t-z)/RC}$$

which appears in the integrand in (41) is exactly $h(t - z)$, the value of h not at t but at $t - z$, as required in the definition of convolution, Equation (36).

We have just asserted, without proof, that (41) is a solution to Equation (38) satisfying the initial condition $v_o(0) = 0$. We shall now justify this assertion for the special case in which $v_i(t)$ is $u(t)$, the unit step at zero, by first carrying out the integration indicated in (41), finding a nicer formula for v_o, and then comparing the result with what we find when we solve (38) using decomposition into partial fractions.

When $v_i(t) = u(t)$ the integral in (41) is

$$\int_0^t v_i(z) \frac{u(t - z)}{RC} e^{-(t-z)/RC} \, dz = \frac{1}{RC} \int_0^t u(z) u(t - z) e^{-(t-z)/RC} \, dz \tag{42}$$

Notice that for any given $t > 0$ the integration runs from 0 to t, and for z in this interval $u(z) = 1$, except for $z = 0$. Also, since $z \leq t$ there, it follows that

$$0 \leq t - z \text{ there}$$

and
$$u(t - z) = 1$$

except for $z = t$. Thus the product $u(z)u(t - z)$ has the value 1 except at the end points of the interval of integration, and it follows that

$$\int_0^t u(z)u(t - z)e^{-(t-z)/RC} \, dz = \int_0^t e^{-(t-z)/RC} \, dz = \int_0^t e^{-t/RC}e^{z/RC} \, dz$$

$$= e^{-t/RC} \int_0^t e^{z/RC} \, dz \qquad (43)$$

$$= e^{-t/RC}(RCe^{t/RC} - RC)$$

$$= RC(1 - e^{-t/RC}) \qquad (44)$$

Notice that (43) follows from the previous equation because the dummy variable is z, so that t is constant during the integration.

Substituting (44) into the right-hand side of (42) and the latter into (41) yields

$$v_o(t) = u(t)(1 - e^{-t/RC}) \qquad (45)$$

This, we claimed, is the solution of (38) satisfying the initial condition $v_o(0) = 0$. [You may check for yourself that the solution is correct, by substituting into (38) in the usual way, using generalized derivatives.]

Now, for the comparison, we shall solve (38) using the partial-fraction decomposition as we have previously. Since (in the present special case) $v_i(t) = u(t)$ and therefore $V_i(s) = 1/s$, the transformed Equation (39) becomes

$$V_o(s) = \frac{1}{RC}\left(\frac{1}{s + 1/RC}\frac{1}{s}\right)$$

$$= \frac{1}{RC}\left(\frac{RC}{s} - \frac{RC}{s + 1/RC}\right)$$

$$= \frac{1}{s} - \frac{1}{s + 1/RC}$$

Therefore, $$v_o(t) = u(t)(1 - e^{-t/RC})$$

which is the same as (45).

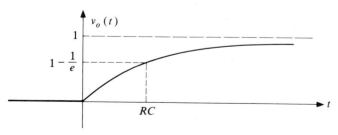

Figure 17 Graph of the solution $v_o(t) = u(t)(1 - e^{-t/RC})$.

Let us verify, for a different input v_i, that

$$v_o(t) = (v_i * h)(t)$$

This time, suppose that

$$v_i(t) = u(t)e^{-at} \qquad \text{for } a \neq \frac{1}{RC} \tag{46}$$

and perform the decomposition into partial fractions first:

$$V_o(s) = H(s)V_i(s)$$

$$= \frac{1}{RC} \frac{1}{s + 1/RC} \frac{1}{s + a}$$

$$= \frac{1/RC}{a - 1/RC}\left(\frac{1}{s + 1/RC} - \frac{1}{s + a}\right)$$

so that

$$v_o(t) = u(t)\frac{1/RC}{a - 1/RC}\left(e^{-t/RC} - e^{-at}\right) \tag{47}$$

[You may want to check that (47) does satisfy (38) in case $v_i(t) = u(t)e^{-at}$, and that $v_o(0) = 0$.] For comparison, here is the convolution of v_i and h, with the integration performed.

$$v_o(t) = u(t) \int_0^t u(z)e^{-az} \frac{u(t - z)}{RC} e^{-(t-z)/RC} \, dz$$

$$= \frac{u(t)}{RC} e^{-t/RC} \int_0^t e^{(1/RC - a)z} \, dz$$

$$= \frac{u(t)}{RC} e^{-t/RC} \frac{e^{(1/RC - a)t} - e^{(1/RC - a)0}}{1/RC - a}$$

$$= \frac{1}{RC} \frac{u(t)}{1/RC - a} \left(e^{-at} - e^{-t/RC}\right)$$

$$= \frac{1}{RC} \frac{u(t)}{a - 1/RC} \left(e^{-t/RC} - e^{-at}\right) \tag{48}$$

Notice that the solutions (47) and (48) are identical, as they must be.

7. Convolution as a Computational Technique in Laplace Transformation

The *convolution* of two functions f and h is a function $f * h$ defined by the formula

$$(f * h)(t) = \begin{cases} 0 & t \leq 0 \\ \int_0^t f(z)h(t - z) \, dz & t > 0 \end{cases}$$

$$= u(t) \int_0^t f(z)h(t - z) \, dz \tag{49}$$

Formula (49) certainly looks a little forbidding, but it will be very useful because the Laplace transform of $f * h$ is remarkably simply expressed in terms of the Laplace transforms of f and h, as their product:

$$\mathcal{L}[f * h](s) = \mathcal{L}[f](s)\mathcal{L}[h](s) \qquad (50)$$

In the notation of systems analysis as given in Part 2,

$$\mathcal{L}[f * h](s) = F(s)H(s)$$

or

$$\mathcal{L}[f * h] = FH$$

In this part of the section we shall illustrate the taking of the convolutions of pairs of functions, an operation called *convolving* the functions, and then illustrate the use of convolution in solving initial value problems. The next part of the section will be devoted to deriving Equation (50).

It is important to realize that on the right half line each value of $f * h$ is the result of an integration. Let us illustrate this by convolving $f(t) \equiv 1$ and $h(t) = e^{-at}$, calculating

$$(f * h)(t) = u(t) \int_0^t e^{-a(t-z)} \, dz$$

$$= \begin{cases} 0 & t \le 0 \\ \int_0^t e^{-a(t-z)} \, dz & t > 0 \end{cases}$$

Notice first that if $h(t) = e^{-at}$ for all t, then $h(t - z) = e^{-a(t-z)}$, and it is the latter expression which is required in the integrand, according to (49). Notice next that for each $t > 0$ we must perform the indicated integration. For our example,

$$\int_0^t f(z)h(t-z) \, dz = \int_0^t 1e^{-a(t-z)} \, dz$$

$$= \int_0^t e^{-at}e^{az} \, dz$$

Since t is fixed and the dummy variable is z, e^{-at} is constant, and the integral is therefore equal to

$$e^{-at} \int_0^t e^{az} \, dz = e^{-at} \left. \frac{e^{az}}{a} \right|_0^t$$

$$= e^{-at} \frac{1}{a} (e^{at} - 1)$$

$$= \frac{1 - e^{-at}}{a}$$

Thus,

$$1 * e^{-at} = u(t) \frac{1 - e^{-at}}{a}$$

is the required function, the convolution of 1 and e^{-at}.

Let us verify (50) in this case.

$$\mathcal{L}[1 * e^{-at}](s) = \mathcal{L}\left[u(t)\frac{1 - e^{-at}}{a}\right](s)$$

$$= \mathcal{L}\left[\frac{1}{a} - \frac{e^{-at}}{a}\right](s)$$

$$= \frac{1}{a}\left(\frac{1}{s} - \frac{1}{s+a}\right)$$

$$= \frac{1}{a}\frac{s+a-s}{s(s+a)}$$

$$= \frac{1}{s(s+a)}$$

$$= \mathcal{L}[1](s)\mathcal{L}[e^{-at}](s)$$

in accordance with (50).

For a second example, let us convolve $f(t) = u(t)e^{-bt}$ with $h(t) = e^{-at}$:

$$(f * h)(t) = u(t)\int_0^t u(z)e^{-bz}e^{-a(t-z)}\,dz$$

$$= \begin{cases} 0 & t \le 0 \\ \int_0^t u(z)e^{-bz}e^{-a(t-z)}\,dz & t > 0 \end{cases}$$

For each $t > 0$, we must perform an integration over the interval $0 \le z \le t$ to find the value of $f * h$ at t. Immediately, a simplification of the integrand should be noticed:

$$u(z) = \begin{cases} 0 & z = 0 \\ 1 & 0 < z \le t \end{cases}$$

so that the factor $u(z)$ in the integrand has value 1 except at a single point. Hence,

$$\int_0^t u(z)e^{-bz}e^{-a(t-z)}\,dz = \int_0^t e^{-bz}e^{-a(t-z)}\,dz$$

$$= \int_0^t e^{-bz}e^{-at}e^{az}\,dz$$

$$= e^{-at}\int_0^t e^{(a-b)z}\,dz$$

where we have used the fact that t is fixed during the integration in which z is the dummy variable.

Summarizing and completing the integration, we have

$$u(t)e^{-bt} * e^{-at} = (f * h)(t)$$

$$= u(t)e^{-at} \int_0^t e^{(a-b)z} \, dz \tag{51}$$

$$= u(t)e^{-at} \frac{e^{(a-b)z}}{a-b} \Big|_0^t$$

$$= u(t) \frac{e^{-bt} - e^{-at}}{a - b}$$

in the case $a \ne b$, while if $a = b$ then (51) simplifies and

$$u(t)e^{-bt} * e^{-at} = u(t)e^{-at} * e^{-at}$$

$$= u(t)e^{-at} \int_0^t dz$$

$$= u(t)te^{-at}$$

Let us verify that the assertion (50) about Laplace transforms of convolutions is also correct for the present example: If $b \ne a$, then

$$\mathscr{L}[u(t)e^{-bt} * e^{-at}](s) = \mathscr{L}\left[u(t)\frac{e^{-bt} - e^{-at}}{a-b}\right](s)$$

$$= \frac{1}{a-b} \mathscr{L}[e^{-bt} - e^{-at}](s)$$

$$= \frac{1}{a-b}\left(\frac{1}{s+b} - \frac{1}{s+a}\right)$$

$$= \frac{1}{(s+b)(s+a)}$$

$$= \mathscr{L}[e^{-bt}](s)\mathscr{L}[e^{-at}](s)$$

$$= \mathscr{L}[u(t)e^{-bt}](s)\mathscr{L}[e^{-at}](s)$$

In the case $b = a$,

$$\mathscr{L}[u(t)e^{-at} * e^{-at}](s) = \mathscr{L}[u(t)te^{-at}](s)$$

$$= \mathscr{L}[te^{-at}](s)$$

$$= \frac{1}{(s+a)^2} = \frac{1}{s+a}\frac{1}{s+a}$$

$$= \mathscr{L}[e^{-at}](s)\mathscr{L}[e^{-at}](s)$$

$$= \mathscr{L}[u(t)e^{-at}](s)\mathscr{L}[e^{-at}](s)$$

In each of these cases, (50) is therefore correct.

Now consider an initial value problem,

$$
\left.\begin{aligned}
y'' + 4y &= f \\
y(0) &= y_0 \\
y'(0) &= 0
\end{aligned}\right\}
\tag{52}
$$

The transformed problem is

$$s^2 \mathscr{L}[y](s) + 4\mathscr{L}[y](s) = \mathscr{L}[f](s) + sy_0$$

$$\mathscr{L}[y](s) = \frac{1}{s^2 + 4}\mathscr{L}[f](s) + y_0\frac{s}{s^2 + 4}$$

$$= \mathscr{L}[\tfrac{1}{2}\sin 2t](s)\mathscr{L}[f](s) + y_0\mathscr{L}[\cos 2t](s) \tag{53}$$

From Equation (53) we can write down at once a formula for y itself, even though we may not yet have specified the function f or the initial value y_0:

$$y(t) = (f * \tfrac{1}{2}\sin 2t) + y_0\cos 2t$$

$$= u(t)\int_0^t f(z)\tfrac{1}{2}\sin 2(t - z)\, dz + y_0\cos 2t$$

$$= \tfrac{1}{2}u(t)\int_0^t f(z)(\sin 2t\cos 2z - \cos 2t\sin 2z)\, dz + y_0\cos 2t$$

$$= \frac{1}{2}\left(\sin 2t\int_0^t f(z)\cos 2z\, dz - \cos 2t\int_0^t f(z)\sin 2z\, dz\right) + y_0\cos 2t \qquad t \geq 0$$

$$\tag{54}$$

Notice that $\sin 2(t - z)$ was reduced with a trigonometric identity.

Now we can observe the dependence of the solution y of (52) upon the nonhomogeneous term f and upon the initial value y_0. For instance, if $f(t) = 1$, then

$$y(t) = \frac{1}{2}\left[\sin 2t\frac{\sin 2z}{2}\Big|_0^t - \cos 2t\left(-\frac{\cos 2z}{2}\right)\Big|_0^t\right] + y_0\cos 2t \tag{55}$$

$$= \frac{1}{2}\left(\frac{\sin^2 2t}{2} + \frac{\cos^2 2t}{2} - \frac{\cos 2t}{2}\right) + y_0\cos 2t$$

$$= \tfrac{1}{2}(\tfrac{1}{2} - \tfrac{1}{2}\cos 2t) + y_0\cos 2t$$

$$= \tfrac{1}{4}(1 - \cos 2t) + y_0\cos 2t \qquad t \geq 0 \tag{56}$$

$$= \tfrac{1}{4} + (y_0 - \tfrac{1}{4})\cos 2t \qquad t \geq 0 \tag{57}$$

Equation (56) shows the dependence on y_0. It also shows, in the first term, the formula for the convolution of f with $\tfrac{1}{2}\sin 2t$, and the fact that the solution of (52) is the sum of solutions of

$$y'' + 4y = 1 \atop \begin{aligned} y(0) &= 0 \\ y'(0) &= 0 \end{aligned}} \quad \text{and} \quad y'' + 4y = 0 \atop \begin{aligned} y(0) &= y_0 \\ y'(0) &= 0 \end{aligned}}$$

because the convolution of f with $\frac{1}{2}\sin 2t$ has value zero at $t = 0$ and derivative zero there. In Equation (57), y is written as a linear combination convenient for graphing.

For another example, consider the initial value problem

$$\begin{aligned} y'(t) + 3y(t) &= \delta(t - 2) \\ y(0) &= 0 \end{aligned}} \tag{58}$$

The transformed problem is

$$s\mathscr{L}[y](s) + 3\mathscr{L}[y](s) = e^{-2s}$$

and $\mathscr{L}[y](s)$ may be written in several equivalent ways:

$$\mathscr{L}[y](s) = \frac{e^{-2s}}{s + 3}$$

$$= e^{-2s}\mathscr{L}[e^{-3t}](s) \tag{59}$$

$$= \mathscr{L}[\delta(t - 2)](s)\mathscr{L}[e^{-3t}](s) \tag{60}$$

Let us take the inverse transform in two different ways and see that we get the same result. The first way is to apply the Second Shifting Theorem (from Section D) to (59), writing at once

$$y(t) = u(t - 2)e^{-3(t-2)} \qquad t \geq 0 \tag{61}$$

The second way is to use the convolution integral, applying formula (50) to (60):

$$y(t) = u(t) \int_0^t \delta(z - 2)e^{-3(t-z)}\, dz \qquad t \geq 0$$

In order to evaluate an integral in which the Dirac δ appears in the integrand, we must use the fact brought out in Equation (36) of Section E:

$$\int_0^t \delta(z - t_0)f(z)\, dz = \begin{cases} 0 & \text{if } t \leq t_0 \\ f(t_0) & \text{if } t > t_0 \end{cases}$$

$$= f(t_0)u(t - t_0) \tag{62}$$

When we apply this fact to the integral in question, we see that $t_0 = 2$ and

$$y(t) = u(t) \int_0^t \delta(z - 2)e^{-3(t-z)} \, dz \qquad t \geq 0$$

$$= \begin{cases} 0 & 0 \leq t \leq 2 \\ u(t)e^{-3(t-2)} & t > 2 \end{cases} \qquad \qquad \qquad \circ$$

$$= u(t-2)e^{-3(t-2)} \qquad t \geq 0 \tag{63}$$

In other words, (63) and (61) define the same generalized solution of (58).

There is a fact which is not at all evident from the definition (49) of convolution but which follows from (50), the formula for its Laplace transform. It is that

$$f * h = h * f \tag{64}$$

since each of these functions has the product $\mathscr{L}[f](s)\mathscr{L}[h](s)$ as its Laplace transform, because the order of factors in a product of numbers is irrelevant. It is convenient to use (64) when evaluating a convolution integral such as

$$\int_0^t e^{-at} \cos{(t - z)} \, dz = e^{-at} * \cos t$$

$$== \cos t * e^{-at}$$

$$= \int_0^t e^{-a(t-z)} \cos z \, dz$$

$$= e^{-at} \int_0^t e^{az} \cos z \, dz \tag{65}$$

The integrand on the left requires the use of the trigonometric identity

$$\cos{(t - z)} = \cos t \cos z + \sin t \sin z$$

and the integral becomes

$$\cos t \int_0^t e^{-az} \cos z \, dz + \sin t \int_0^t e^{-az} \sin z \, dz$$

requiring two integrations, whereas the integral on the right in (65) is no more difficult but requires only one integration.

A striking example of this simplification occurs in the solution of the example in (52). You may recall from (54) that the solution may be written

$$y(t) = f * \tfrac{1}{2} \sin 2t + y_0 \cos 2t \qquad t \geq 0$$

$$= u(t) \int_0^t f(t)\tfrac{1}{2} \sin 2(t - z) \, dz + y_0 \cos 2t \qquad t \geq 0$$

and in the special case $f(t) = 1$ the integration was performed in (54) and (55), using a trigonometric identity and two integrations. Now we can do the calculation more efficiently, using the fact that $f * h = h * f$. It goes this way:

$$y(t) = 1 * \tfrac{1}{2} \sin 2t + y_0 \cos 2t$$

$$= \tfrac{1}{2} \sin 2t * 1 + y_0 \cos 2t$$

$$= u(t) \int_0^t \tfrac{1}{2} \sin 2z \cdot 1 \, dz + y_0 \cos 2t$$

$$= u(t)(-\tfrac{1}{4} \cos 2z) \Big|_0^t + y_0 \cos 2t$$

$$= u(t)\tfrac{1}{4}(1 - \cos 2t) + y_0 \cos 2t$$

$$= \tfrac{1}{4}(1 - \cos 2t) + y_0 \cos 2t \qquad t \geq 0$$

as in (56). This calculation is much quicker and simpler.

8. The Laplace Transform of the Convolution Integral

The convolution of two functions f and h is a function $f * h$. If f and h have Laplace transforms, then so does $f * h$:

$$\mathscr{L}[f * h](s) = \mathscr{L}[f](s)\mathscr{L}[h](s) \tag{66}$$

We have already made use of this fact, and we have verified it for a few choices of f and h. Now it is time to verify the formula in general. We begin with the formula defining the convolution and that defining the Laplace transform. Since

$$(f * h)(t) = u(t) \int_0^t f(z)h(t - z) \, dz$$

it follows that

$$\mathscr{L}[f * h](s) = \int_0^\infty \left[u(t) \int_0^t f(z)h(t - z) \, dz \right] e^{-st} \, dt \tag{67}$$

$$= \int_0^\infty \left[\int_0^t f(z)h(t - z) \, dz \right] e^{-st} \, dt \tag{68}$$

$$= \int_0^\infty \left[\int_0^t f(z)h(t - z)e^{-st} \, dz \right] dt \tag{69}$$

In this sequence of equalities, the first is simply the formula defining the Laplace transform. It looks complicated because the integrand is itself a multiple of an integral, namely

$$e^{-st} u(t) \int_0^t f(z)h(t - z) \, dz \tag{70}$$

Formula (70) is merely the integrand in (67). Since $u(t) = 1$ for $0 < t < \infty$ (that is, for t in the region of integration), it follows that the factor $u(t)$ need not appear in the integrand at all. Therefore, (67) equals (68), for which the integrand is

$$e^{-st} \int_0^t f(z)h(t - z) \, dz \tag{71}$$

Notice that (71) is just the value of $e^{-st}(f * h)(t)$, the value at the point t. That is, for each *fixed* t, the integration is performed with respect to a dummy variable z. In other words, e^{-st} is constant during the integration on z and may be "taken inside" the integral. The integral formula being used is, of course, this:

$$a \int g(z)\, dz = \int ag(z)\, dz \qquad \text{if } a \text{ is constant} \tag{72}$$

Therefore,
$$e^{-st} \int_0^t f(z)h(t-z)\, dz = \int_0^t e^{-st}f(z)h(t-z)\, dz$$

from which it follows that (68) equals (69):

$$\mathscr{L}[f * h](s) = \int_{t-0}^{\infty} \left[\int_{z-0}^{z=t} f(z)h(t-z)e^{-st}\, dz \right] dt \tag{73}$$

If you are acquainted with double integration, then you will know how to show that (73) is equal to

$$\int_{z=0}^{\infty} \left[\int_{t=z}^{\infty} f(z)h(t-z)e^{-st}\, dt \right] dz$$

but if you are not, then the next few paragraphs will outline the process for you.

Start with (73). For each nonnegative t, the inner integration is performed over the interval $0 \le z \le t$. This fact can be sketched as in Figure 18. For each t, the shaded region intersects the vertical line through t in exactly the interval for the corresponding integration with respect to z. The shaded region may also be

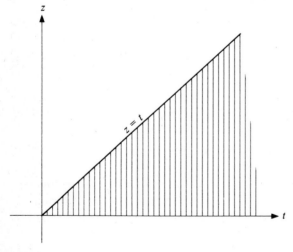

Figure 18 The region of integration for (73). For each nonnegative t, the z integration is from $z = 0$ to $z = t$.

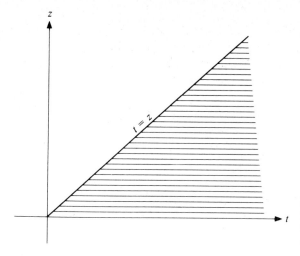

Figure 19 The region of integration for (73) and (74). For each nonnegative z, the region of integration on t is $z \le t$.

described this way: For each nonnegative z, the points $z \le t$ lie in the region. (See Figure 19.) The corresponding integral would be carried out exactly as the region is described. For each nonnegative z, integrate $f(z)h(t-z)e^{-st}$ over all $t \ge z$

$$\int_{t=z}^{\infty} f(z)h(t-z)e^{-st}\, dt$$

giving a function of z (and s) but not t, and then integrate that function of z over $z \ge 0$, giving a function of s alone:

$$\int_{z=0}^{\infty} \left[\int_{t=z}^{\infty} f(z)h(t-z)e^{-st}\, dt \right] dz \qquad (74)$$

The integrals (73) and (74) are "iterated integrals," performed exactly as indicated and described above. It is an old theorem that they are in fact equal if f and h are piecewise continuous and of some exponential order (as described in Section A of this chapter). Therefore,

$$\mathscr{L}[f * h](s) = \int_{z=0}^{\infty} \left[\int_{t=z}^{\infty} f(z)h(t-z)e^{-st}\, dt \right] dz \qquad (75)$$

This integral may be simplified in two ways. First, change the variable t, leaving z as a variable but taking

$$t - z = w$$

or
$$t = w + z \qquad \text{and} \qquad dt = dw$$

For each fixed z, then, $dt = dw$, and performing the corresponding substitutions in (75) gives this:

$$\mathscr{L}[f * h](s) = \int_{z=0}^{\infty} \left[\int_{w+z=z}^{\infty} f(z)h(w)e^{-s(w+z)} \, dw \right] dz$$

$$= \int_{z=0}^{\infty} \left[\int_{w=0}^{\infty} f(z)h(w)e^{-sw}e^{-sz} \, dw \right] dz$$

$$= \int_{z=0}^{\infty} \left[f(z)e^{-sz} \int_{w=0}^{\infty} h(w)e^{-sw} \, dw \right] dz$$

$$= \int_{z=0}^{\infty} f(z)e^{-sz} \mathscr{L}[h](s) \, dz$$

$$= \int_{z=0}^{\infty} f(z)e^{-sz} \, dz \, \mathscr{L}[h](s)$$

$$= \mathscr{L}[f](s)\mathscr{L}[h](s) \tag{76}$$

The first equality above is the result of substitution into (75). The second records the fact that $w + z = z$ (at the left end of the interval of integration with respect to w) if and only if $w = 0$. The third equality comes from the fact that the inner integral, the w integral, is performed for each *fixed* z. With z fixed, $f(z)e^{-sz}$ is not dependent upon w at all, and it may be factored out of the integral as in (72). The next equality points out that what remains of the inner integral is one of the two desired Laplace transforms. Since it depends only upon s and not upon the dummy variable z of the outer integral, it may be factored out of the integral as indicated. Finally, the remaining integral is identified as the other desired Laplace transform. That is, (76) shows exactly what was stated in (66), which should be added to your short table of Laplace transforms.

9. Examples

(a) Solve the initial value problem

$$\left. \begin{array}{c} y''(t) + 4y(t) = \cos 2t \\ y(0) = 1 \\ y'(0) = -6 \end{array} \right\}$$

Solution

$$s^2 \mathscr{L}[y](s) - s + 6 + 4\mathscr{L}[y](s) = \frac{s}{s^2 + 4} \tag{77}$$

$$\mathscr{L}[y](s) = \frac{1}{s^2 + 4} \frac{s}{s^2 + 4} + \frac{s}{s^2 + 4} - \frac{6}{s^2 + 4} \tag{78}$$

$$= \mathscr{L}[(\tfrac{1}{2} \sin 2t * \cos 2t) + \cos 2t - 3 \sin 2t](s)$$

$$y(t) = \int_0^t \tfrac{1}{2} \sin 2z \cos 2(t - z) \, dz + \cos 2t - 3 \sin 2t \qquad t \geq 0 \qquad (79)$$

$$= \tfrac{1}{2} \int_0^t \sin 2z \, (\cos 2t \cos 2z + \sin 2t \sin 2z) \, dz$$

$$+ \cos 2t - 3 \sin 2t$$

$$= \tfrac{1}{2} \cos 2t \int_0^t \sin 2z \cos 2z \, dz + \tfrac{1}{2} \sin 2t \int_0^t \sin^2 2z \, dz$$

$$+ \cos 2t - 3 \sin 2t$$

$$= \tfrac{1}{2} \cos 2t \left(\frac{\sin^2 2z}{4} \right)\Big|_0^t + \tfrac{1}{2} \sin 2t \left(\frac{z}{2} - \frac{\sin 4z}{8} \right)\Big|_0^t$$

$$+ \cos 2t - 3 \sin 2t$$

$$= \tfrac{1}{8} \cos 2t \sin^2 2t + \tfrac{1}{4} t \sin 2t - \tfrac{1}{16} \sin 2t \sin 4t$$

$$+ \cos 2t - 3 \sin 2t$$

$$= \tfrac{1}{4} t \sin 2t + \cos 2t - 3 \sin 2t \qquad t \geq 0$$

Actually, it would not have been necessary to write out the formula for the Laplace transform of the nonhomogeneous term. That is, the solution could have been started this way:

$$s^2 \mathscr{L}[y](s) - s + 6 + 4 \mathscr{L}[y](s) = \mathscr{L}[\cos 2t](s)$$

$$\mathscr{L}[y](s) = \frac{1}{s^2 + 4} \mathscr{L}[\cos 2t](s) + \frac{s}{s^2 + 4} - \frac{6}{s^2 + 4}$$

$$y(t) = \cos 2t * \tfrac{1}{2} \sin 2t + \cos 2t - 3 \sin 2t$$

$$= \int_0^t \tfrac{1}{2} \sin 2z \cos (t - z) \, dz + \cos 2t - 3 \sin 2t \qquad t \geq 0$$

The integration may, of course, proceed as before.

(b) Solve the initial value problem

$$\left. \begin{aligned} y''(t) + 4y(t) &= 3 \, \delta(t - 8) \\ y(0) &= 1 \\ y'(0) &= -6 \end{aligned} \right\}$$

Solution

$$s^2 \mathscr{L}[y](s) + 4 \mathscr{L}[y](s) = \mathscr{L}[3 \, \delta(t - 8)](s) + s - 6$$

$$\mathscr{L}[y](s) = \frac{1}{s^2 + 4} \mathscr{L}[3 \, \delta(t - 8)](s) + \frac{s}{s^2 + 4} - \frac{6}{s^2 + 4}$$

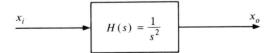

$v_i(t)$ → $H(s) = \dfrac{1}{s+3}$ → $v_o(t)$

Figure 20 System specified by its transfer function H.

$$y(t) = \int_0^t 3\,\delta(z - 8)\tfrac{1}{2}\sin 2(t - z)\,dz + \cos 2t - 3\sin 2t \qquad t \geq 0$$

$$= \tfrac{3}{2}\sin 2(t - 8)u(t - 8) + \cos 2t - 3\sin 2t \qquad t \geq 0$$

where the convolution of δ with a function is performed as explained previously in Part 7. It is also possible to write the convolution the other way, as

$$3\,\delta(t - 8) * \tfrac{1}{2}\sin 2t = \tfrac{1}{2}\sin 2t * 3\,\delta(t - 8)$$

$$= \int_0^t \tfrac{1}{2}(\sin 2z)3\,\delta[(t - 8) - z]\,dz$$

$$= \tfrac{3}{2}\sin 2(t - 8)u(t - 8)$$

the same result as above.

(c) Assuming that the system shown in Figure 20 is relaxed until $t = 0$, write as a convolution the response of the system to the input $v_i(t) = u(t)\cos 2t$.

Solution The impulse response of the system is

$$h(t) = \mathscr{L}^{-1}[H](t) = \mathscr{L}^{-1}\left[\frac{1}{s+3}\right](t) = u(t)e^{-3t}$$

The response of the system to v_i is therefore

$$v_o(t) = (v_i * h)(t)$$

$$= u(t)\cos 2t * u(t)e^{-3t}$$

$$= u(t)\int_0^t u(t)(\cos 2z)u(t - z)e^{-3(t - z)}\,dz$$

$$= u(t)\int_0^t (\cos 2z)e^{-3(t - z)}\,dz$$

(d) Assuming that the system shown in Figure 21 is at rest until $t = 0$, write as a convolution the response of the system to the input $x_i(t) = u(t)\sin 10t$.

x_i → $H(s) = \dfrac{1}{s^2}$ → x_o

Figure 21 System specified by its transfer function H.

Solution The impulse response of the system is

$$h(t) = \mathcal{L}^{-1}[H](t) = \mathcal{L}^{-1}\left[\frac{1}{s^2}\right](t) = tu(t)$$

The response of the system to x_i is therefore

$$x_o(t) = u(t) \sin 10t * tu(t)$$

$$= u(t) \int_0^t (\sin 10z)(t - z) \, dz$$

(e) The solution of the initial value problem

$$v_o'(t) + 4v_o(t) = e^{-9t}[u(t - \alpha) - u(t - \beta)] \mid$$
$$v_o(0) = 0$$

where $0 < \alpha < \beta$, is also the output of the system shown in Figure 22, which is relaxed until $t = 0$, when the input is $v_i(t) = e^{-9t}[u(t - \alpha) - u(t - \beta)]$. Write v_o as a convolution, and evaluate the integral.

Solution

$$(s + 4)\mathcal{L}[v_o](s) = \mathcal{L}[e^{-9t}\{u(t - \alpha) - u(t - \beta)\}](s)$$

$$\mathcal{L}[v_o](s) = \mathcal{L}[e^{-9t}\{u(t - \alpha) - u(t - \beta)\}](s)\frac{1}{s + 4}$$

$$v_o(t) = u(t) \int_0^t e^{-9z}[u(z - \alpha) - u(z - \beta)]e^{-4(t-z)} \, dz$$

$$= u(t)e^{-4t} \int_0^t e^{-5z}[u(z - \alpha) - u(z - \beta)] \, dz$$

$$= u(t)e^{-4t}\left[\int_0^t e^{-5z}u(z - \alpha) \, dz - \int_0^t e^{-5z}u(z - \beta) \, dz\right] \qquad (79)$$

The two integrals in (79) have the same form. We shall evaluate the first and use the resulting formula to evaluate the second.
Notice that

$$u(z - \alpha) = \begin{cases} 0 & z \leq \alpha \\ 1 & z > \alpha \end{cases}$$

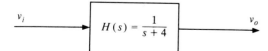

v_i — $H(s) = \dfrac{1}{s + 4}$ — v_o

Figure 22 General system specified by its transfer function.

It follows that if $t \le \alpha$, then $u(z - \alpha) = 0$ for all z in the region of integration, the interval $0 \le z \le t$, but if $t > \alpha$, then we may split the interval of integration into two parts with the integrand 0 on one of them, $0 \le z \le \alpha$. Specifically,

$$\int_0^t e^{-5z} u(z - \alpha) \, dz = \begin{cases} 0 & t \le \alpha \\ \int_0^\alpha e^{-5z} u(z - \alpha) \, dz + \int_\alpha^t e^{-5z} u(z - \alpha) \, dz & t > \alpha \end{cases}$$

$$= \begin{cases} 0 & t \le \alpha \\ 0 + \int_\alpha^t e^{-5z} \, dz & t > \alpha \end{cases}$$

$$= u(t - \alpha) \int_\alpha^t e^{-5z} \, dz$$

$$= u(t - \alpha) \frac{e^{-5z}}{-5} \Big|_\alpha^t$$

$$= u(t - \alpha) \frac{e^{-5t} - e^{-5\alpha}}{-5}$$

Similarly, $\qquad \int_0^t e^{-5z} u(z - \beta) \, dz = u(t - \beta) \dfrac{e^{-5t} - e^{-5\beta}}{-5}$

Substituting these evaluations into (79) gives this:

$$v_o(t) = u(t) e^{-4t} \left[u(t - \alpha) \frac{e^{-5t} - e^{-5\alpha}}{-5} - u(t - \beta) \frac{e^{-5t} - e^{-5\beta}}{-5} \right]$$

$$= e^{-4t} \left[u(t - \alpha) \frac{e^{-5\alpha} - e^{-5t}}{5} - u(t - \beta) \frac{e^{-5\beta} - e^{-5t}}{5} \right]$$

since the inequalities $0 < \alpha < \beta$ imply that

$$u(t)u(t - \alpha) = u(t - \alpha) \qquad \text{and} \qquad u(t)u(t - \beta) = u(t - \beta)$$

The solution may also be written this way:

$$v_o(t) = \begin{cases} 0 & t \le \alpha \\ e^{-4t} \dfrac{e^{-5\alpha} - e^{-5t}}{5} & \alpha < t \le \beta \\ e^{-4t} \dfrac{e^{-5\alpha} - e^{-5\beta}}{5} & \beta < t \end{cases}$$

10. Sample Problems with Answers

D. The system in Figure 23 is motionless until time $t = 0$.

(a) If $v(t) = tu(t)$, then find $v_1(t)$ twice: The first time, use the convolution when taking the inverse

Laplace transform; the second time, use the partial-fraction decomposition when taking the inverse Laplace transform.

$$Answer: \ v_1(t) = u(t) \int_0^t zu(z) \frac{K}{B} e^{-(K/B)(t-z)} \, dz = u(t) \left(t - \frac{B}{K} + \frac{B}{K} e^{-Kt/B} \right)$$

(b) If the system is at rest until $t = 0$, and if v is the input and v_1 the output, then find the transfer function and impulse response of the system.

$$Answer: \ H(s) = \frac{K}{Bs + K}; \ h(t) = u(t) \frac{K}{B} e^{-Kt/B}$$

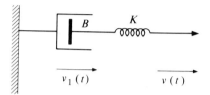

Figure 23 Spring-dashpot mechanical system.

E. The system in Figure 24 is motionless until $t = 0$, and $x(t) = 0$ for $t < 0$.
   (a) If $f(t) = u(t - 1) - u(t - 2)$, then find $x(t)$ twice, once using the convolution and once using the partial-fraction decomposition to find the inverse Laplace transform.

$$Answer: \quad x(t) = u(t) \int_0^t [u(z - 1) - u(z - 2)] \frac{1}{B} (1 - e^{-B(t-z)/M}) \, dz$$

$$= \frac{1}{B} [(t - 1)u(t - 1) - (t - 2)u(t - 2)]$$

$$+ \frac{1}{B^2} (e^{-B(t-1)/M} - 1)u(t - 1) - \frac{1}{B^2} (e^{-B(t-2)/M} - 1)u(t - 2)$$

(b) If the system is at rest until $t = 0$ and if $f(t)$ is the input then find the impulse response and the response to $f(t) = \delta(t - 1) - \delta(t - 2)$.

$$Answer: \ h(t) = \frac{1}{B} (1 - e^{-Bt/M})u(t); \ x(t) = \frac{1}{B} (1 - e^{-B(t-1)/M})u(t - 1) - \frac{1}{B} (1 - e^{-B(t-2)/M})u(t - 2)$$

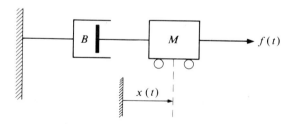

Figure 24 Mass-dashpot mechanical system.

F. For the system in Figure 25, which is at rest until $t = 0$ and for which v is the input and f the output:
   (a) Determine the transfer function.
   $Answer: \ H(s) = Bs/(s + B/M)$

(b) Determine the impulse response.

Answer: $h(t) = u(t)B \left| \delta(t) - \dfrac{B}{M} e^{-Bt/M} \right|$

(c) Determine the output *twice*, once using convolution and once using partial-fraction decomposition to find the inverse Laplace transform, if the input is $v(t) = u(t)e^{-\alpha t}$.

Answer: $\quad f(t) = u(t) \displaystyle\int_0^t B \left[\delta(t-z) - \dfrac{B}{M} e^{-B(t-z)/M} \right] e^{-\alpha z}\, dz$

$$= u(t) \left[-\dfrac{B^2/M}{\alpha - B/M} e^{-Bt/M} + \dfrac{\alpha B}{\alpha - B/M} e^{-\alpha t} \right]$$

unless $\alpha = B/M$. If $\alpha = B/M$, then what is the result?

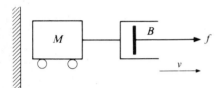

Figure 25 Mass-dashpot mechanical system.

11. Problems

Compute the convolution $f * h$ of each of the following pairs of functions. Check by multiplying the Laplace transforms of f and h and then finding the inverse transform by partial-fraction decomposition and the methods of Sections B through E.

1. (a) $f(t) = t - 2$ and $h(t) = e^{-3t}$.
   (b) $f(t) = u(t - 1)$ and $h(t) = t - 2$.
2. (a) $f(t) = e^{-3t}$ and $h(t) = u(t - 1)$.
   (b) $f(t) = u(t - 1)$ and $h(t) = e^{-3t}$.
3. (a) $f(t) = \cos 7t$ and $h(t) = e^{-3t}$.
   (b) $f(t) = e^{-3t}$ and $h(t) = \cos 7t$.
   (c) $f(t) = u(t) \cos 7t$ and $h(t) = u(t)e^{-3t}$.
4. $f(t) = (t - 1)u(t - 1)$ and $h(t) = u(t)e^{-3t}$.
5. $f(t) = u(t - 1)$ and $h(t) = u(t - 2)e^{-3t}$.
6. (a) $f(t) = \delta(t)$ and $h(t) = \cos \pi t$.
   (b) $f(t) = \delta(t - 1)$ and $h(t) = \cos \pi t$.
7. $f(t) = 3 + tu(t)$ and $h(t) = \cos \pi t$.

Use convolution to solve the following initial value problems. Solve Problems 8 to 10 a second time, taking the inverse Laplace transform by partial-fraction decomposition instead of convolution.

8. $\left. \begin{aligned} y''(t) + 4y(t) &= u(t - 1) \\ y(0) &= 0 \\ y'(0) &= 0 \end{aligned} \right\}$ 9. $\left. \begin{aligned} y''(t) + 4y'(t) + 13y(t) &= u(t)e^{2t} \\ y(0) &= 0 \\ y'(0) &= 0 \end{aligned} \right\}$

10. $\left. \begin{aligned} y''(t) + 4y'(t) + 13y(t) &= \delta(t - 1) \\ y(0) &= 0 \\ y'(0) &= 0 \end{aligned} \right\}$

11. (a)
$$y''(t) + 4y'(t) + 13y(t) = \delta(t)$$
$$y(0-) = 0$$
$$y'(0-) = 0$$

(b)
$$y''(t) + 4y'(t) + 13y(t) = u(t)$$
$$y(0-) = 0$$
$$y'(0-) = 0$$

(c)
$$y''(t) + 4y'(t) + 13y(t) = 0$$
$$y(0) = 2$$
$$y'(0) = -1$$

(d)
$$y''(t) + 4y'(t) + 13y(t) = 3\,\delta(t) + u(t)$$
$$y(0-) = 2$$
$$y'(0-) = -1$$

12. Find v_R for the circuit shown in Figure 26 if $v_R(0) = 0$ and $v(t) = u(t) - u(t - 3)$, (a) using convolution, and (b) using the partial-fraction decomposition.

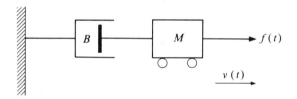

Figure 26 Series LR circuit.

13. Find v for the system shown in Figure 27 if $f(t) = u(t) \sin 20t$ and $v(0) = 0$, (a) using convolution, and (b) using the partial-fraction decomposition.

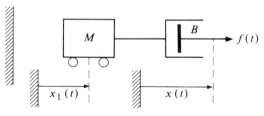

Figure 27 Mass-dashpot mechanical system.

14. Find x for the system shown in Figure 28 if $f(t) = t^2 u(t)$ and $x(0) = x'(0) = 0$, (a) using convolution, and (b) using the partial-fraction decomposition.

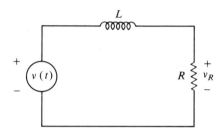

Figure 28 Mass-dashpot mechanical system.

The following problems are stated from the viewpoint of systems analysis. Problems 12 to 14 are repeated.

15. The system in Figure 29 is relaxed until $t = 0$. Its input is v_i, and its output is v_o.
(a) Find the transfer function of the system.
(b) Find the impulse response.
(c) Find v_o if $v_i(t) = u(t) \sin 20t$, once using convolution and once using the partial-fraction decomposition.

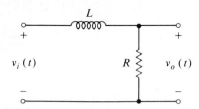

Figure 29 Series LR system.

(d) Find v_o if $v_i(t) = u(t) - u(t - 3)$, once using convolution and once using the partial-fraction decomposition.

16. Figure 27 represents a system, at rest until $t = 0$, whose input is f and whose output is v.
(a) Find the transfer function of the system.
(b) Find the impulse response.
(c) Find the output if the input is $u(t) \sin 20t$, once by convolution and once using the partial-fraction decomposition.
(d) Find the output if the input is $u(t) - u(t - 3)$, once by convolution and once using the partial-fraction decomposition.

17. Figure 28 represents a system, at rest until $t = 0$, whose input is f and whose output is x.
(a) Find the transfer function of the system.
(b) Find the impulse response.
(c) Find the output if the input is $t^2 u(t)$, once by convolution and once using the partial-fraction decomposition.
(d) Find the output if the input is $u(t)(1 - \cos 10t)$, once by convolution and once using the partial-fraction decomposition.

18. Figure 28 represents a system, at rest until $t = 0$, whose input is x and whose output is x_1.
(a) Repeat Problem 17 for this system.
(b) Find the output if the input is $u(t)$.

19. Figure 30 represents a system relaxed until $t = 0$.
(a) Find the transfer function of the system.
(b) Find the impulse response.
(c) Find the output v_o when the input is $v_i(t) = u(t)e^{-\alpha t} \sin 2\pi\beta t$, once by convolution and once using the partial-fraction decomposition.
(d) Find the output v_o when the input is $v_i(t) = u(t)e^{-\alpha t}$, once by convolution and once using the partial-fraction decomposition.

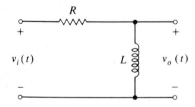

Figure 30 Series LR system.

20. Figure 31 represents a system, at rest until $t = 0$, whose input is v and whose output is v_1.
(a) Find the transfer function of the system.
(b) What must be the relation among M, B_1, B_2, and K if the impulse response is to be sinusoidal? Exponential?
(c) Let $M = B_2 = K = 1$, so that the only unspecified parameter in the system is B_1, and give the impulse response of the system for each value of B_1.

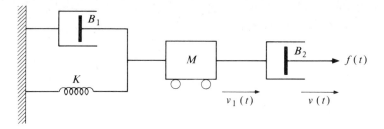

Figure 31 Mechanical system.

(d) As in part c, let $M = B_2 = K = 1$. If the input is $e^{-at}[u(t) - u(t-2)]$, find the output for each value of B_1. Find it by convolution, and find it using the partial-fraction decomposition.

12. Linearity Properties of the Convolution Integral

Consider first an initial value problem such as

$$My'' + By' + Ky = f \atop \left. \begin{aligned} y(0) &= y_0 \\ y'(0) &= y_1 \end{aligned} \right\} \tag{81}$$

whose transformed problem is

$$(Ms^2 + Bs + K)\mathscr{L}[y](s) = \mathscr{L}[f](s) + (Ms + B)y_0 + My_1 \tag{82}$$

Division by $Ms^2 + Bs + K$ gives an expression for $\mathscr{L}[y]$:

$$\mathscr{L}[y](s) = \frac{1}{Ms^2 + Bs + K}\mathscr{L}[f](s) + \frac{(Ms + B)y_0 + My_1}{Ms^2 + Bs + K} \tag{83}$$

This expression is a sum of two terms. It follows that y itself will be a sum of two terms, the first being the convolution of f with the inverse Laplace transform of the rational function $1/(Ms^2 + Bs + K)$. Thus, $y(t)$ may be written

$$y(t) = f * \mathscr{L}^{-1}\left[\frac{1}{Ms^2 + Bs + K}\right](t) + \mathscr{L}^{-1}\left[\frac{(Ms + B)y_0 + My_1}{Ms^2 + Bs + K}\right](t) \tag{84}$$

The second term of (84) contains all reference to the initial values y_0 and y_1.

From these facts, several conclusions may be drawn. The first is that in $y(t)$ the convolution term has zero initial values. It is the whole expression for $y(t)$ if $y_0 = y_1 = 0$, and so it is itself the solution of the initial value problem

$$My'' + By' + Ky = f \atop \left. \begin{aligned} y(0) &= 0 \\ y'(0) &= 0 \end{aligned} \right\} \tag{85}$$

[The limits from the left are zero at $t = 0$ in any case, because of the factor $u(t)$ in the definition of the convolution. However, if f, the right-hand side of the equation, is continuous at $t = 0$ or has a step there, then the convolution will be continuous at $t = 0$, and in fact all the initial values will be values of continuous functions.]

The other term in (84), the expression for $y(t)$, may also be interpreted as the solution of an initial value problem, the problem in which $f(t) = 0$:

$$
\left.\begin{array}{r}
My''(t) + By'(t) + Ky(t) = 0 \\
y(0) = y_0 \\
y'(0) = y_1
\end{array}\right\}
\tag{86}
$$

since this is the problem for which (84) reduces to its second term alone. The differential equation in problem (86) is a homogeneous one. Thus the solutions of (86) will be homogeneous solutions, and the terms of $y(t)$ arising from the second term of (84) will be homogeneous solutions of the differential equation in (81):

$$
y_0 \mathscr{L}^{-1}\left[\frac{Ms + B}{Ms^2 + Bs + K}\right](t) + y_1 \mathscr{L}^{-1}\left[\frac{M}{Ms^2 + Bs + K}\right](t)
\tag{87}
$$

Do you see as well that the functions

$$
v_0(t) = \mathscr{L}^{-1}\left[\frac{Ms + B}{Ms^2 + Bs + K}\right](t)
$$

and

$$
v_1(t) = \mathscr{L}^{-1}\left[\frac{M}{Ms^2 + Bs + K}\right](t)
\tag{88}
$$

must be homogeneous solutions, with the properties

$$
\left.\begin{array}{r}
v_0(0) = 1 \\
v_0'(0) = 0
\end{array}\right\} \quad \text{and} \quad \left.\begin{array}{r}
v_1(0) = 0 \\
v_1'(0) = 1
\end{array}\right\}
\tag{89}
$$

respectively? (The pair of homogeneous solutions v_0 and v_1 with these initial values constitute a "basic solution system" for the homogeneous equation. See Part 16 of Section B of Chapter III and Section E of Chapter IV.)

If the inverse Laplace transform of $1/(Ms^2 + Bs + K)$ is called h, then (84) can now be written

$$
y(t) = (f * h)(t) + y_0 v_0(t) + y_1 v_1(t)
\tag{90}
$$

where f is the right-hand side of the differential equation in problem (81), y_0 and y_1 are the initial values given in (81), and v_0 and v_1 are the functions determined by (88) [namely homogeneous solutions satisfying (89)]. That is, y is the convolution of f with h, plus a linear combination of homogeneous solutions. We are seeing here a fundamental property of *linear* differential equations, to be expanded upon in Chapter IV. In the present chapter we shall see examples.

The convolution itself has a linearity property: If f_1, f_2, and h are functions, and if α_1 and α_2 are numbers, then

$$
(\alpha_1 f_1 + \alpha_2 f_2) * h = \alpha_1(f_1 * h) + \alpha_2(f_2 * h)
\tag{91}
$$

The reason is basic:

$$[(\alpha_1 f_1 + \alpha_2 f_2) * h](t) = u(t) \int_0^t [\alpha_1 f_1(z) + \alpha_2 f_2(z)]h(t - z) \, dz$$

$$= u(t) \int_0^t [\alpha_1 f_1(z)h(t - z) + \alpha_2 f_2(z)h(t - z)] \, dz$$

$$= u(t) \left[\int_0^t \alpha_1 f_1(z)h(t - z) \, dz + \int_0^t \alpha_2 f_2(z)h(t - z) \, dz \right]$$

$$= \alpha_1 u(t) \int_0^t f_1(z)h(t - z) \, dz + \alpha_2 u(t) \int_0^t f_2(z)h(t - z) \, dz$$

$$= \alpha_1 (f_1 * h)(t) + \alpha_2 (f_2 * h)(t)$$

The usual terminology is this: Convolution with a function h is a "linear integral operator" (see Chapter IV) in which $h(t - z)$ represents the "kernel" of the operator. In the study of differential equations,

$$h(t - z) = u(t - z)\mathscr{L}^{-1} \left[\frac{1}{Ms^2 + Bs + K} \right](t - z)$$

a function of the two variables t and z, is sometimes called a *Green's function* for the problem

$$\left. \begin{aligned} My'' + By' + Ky &= f \\ y(0) &= 0 \\ y'(0) &= 0 \end{aligned} \right\}$$

Except for the reappearance of the term *linear integral operator*, we shall not make further use of the terminology introduced in the previous paragraph, which indicates the points at which our study so far fits into a larger framework, the theory of linear ordinary differential equations. In this connection, it should also be mentioned that the convolution we have used is a special case of a more general one, for which the formula is

$$(f * h)(t) = \int_{-\infty}^{\infty} f(z)h(t - z) \, dz \tag{92}$$

In our study, however, f and h may be considered as inverse Laplace transforms, and therefore defined only on the right half line:

$$f(t) = u(t)f(t) \qquad h(t) = u(t)h(t)$$

In this case,

$$f(z)h(t - z) = u(z)u(t - z)f(z)h(t - z)$$

$$= \begin{cases} 0 & z \leq 0 \\ f(z)h(t - z) & 0 < z < t \\ 0 & t \leq z \end{cases}$$

so that (92) is exactly equal to

$$u(t) \int_0^t f(z)h(t-z) \, dz$$

which was our definition.

13. Comments on the Analysis of Systems

Where we see a problem such as

$$My'' + Ky = f \left.\right\rbrace$$
$$y(0) = 0 \tag{93}$$
$$y'(0) = 0 \left.\right\rbrace$$

even if we know that it describes a system, we are not justified in asserting that $1/(Ms^2 + K)$ is the transfer function of the system unless we know that f is the input and y is the output. For instance, Figure 32 illustrates a system giving rise to just such an equation:

$$Mv'' + Kv = g' \left.\right\rbrace$$
$$v(0) = 0 \tag{94}$$
$$v'(0) = 0 \left.\right\rbrace$$

However, if g is the input and v the output, then the transfer function is

$$H(s) = \frac{s}{Ms^2 + K} \tag{95}$$

rather than $1/(Ms^2 + K)$. This is so in spite of the fact that (93) and (94) are the same initial value problem, with v and g' as the names of unknown and known functions in the latter problem, y and f as the same in the former problem.

Again, if v were the input and g the output, then the transformed problem would be

$$(Ms^2 + K)V(s) = sG(s)$$

as before, but the transfer function would be $(Ms^2 + K)/s$, the reciprocal of that in (95).

Clearly, there are natural questions to be asked about doing systems analysis only upon systems that are relaxed or at rest until $t = 0$. For instance, "What do

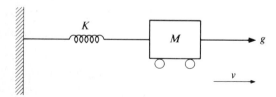

Figure 32 MK mechanical system.

we do with a system at rest only until $t = -1$?" The answer to that question is, "Reset the clock, so that it starts at the end of the at-rest period." In fact, the natural way to analyze any real system is to start the clock at the first interesting instant and continue the description until the activity of the system has ceased, at least to within the accuracy of measurement. Then shut off the clock and reset it to zero, ready for the next nonzero input.

A much more serious question concerns the fact that many interesting systems may never return to rest before a response to another signal must be analyzed. There are many ways to handle such systems, but all such techniques rest on one essential feature, linearity, known in application as the *principle of superposition.* If all the differential equations describing the system are *linear,* then the response of the system to a new input x_i is the sum of the response $x_i * h$ which it would have made if it had been at rest until $t = 0$, plus the residual behavior which it would have had in the absence of the new input. The residual behavior is determined by the initial conditions at $t = 0$.

The ways to handle the residual behavior (that is, the initial conditions) vary tremendously, but a student in doubt can always set up the initial value problem and solve it, as in Problem 11 earlier in this section.

14. Example: Formula for the Solution of the First-Order Linear Equation with Constant Coefficients

This formula is an easy consequence of our use of the convolution. The solution of the initial value problem

$$y'(t) + \alpha y(t) = f(t) \ \Big| \atop y(0) \text{ given } \Big|$$

(96)

may be obtained as follows:

$$(s + \alpha)\mathcal{L}[y](s) = y(0) + \mathcal{L}[f](s)$$

$$\mathcal{L}[y](s) = \frac{y(0)}{s + \alpha} + \frac{\mathcal{L}[f](s)}{s + \alpha}$$

$$y(t) = y(0)e^{-\alpha t} + \int_0^t f(z)e^{-\alpha(t-z)} \, dz$$

(97)

$$= y(0)e^{-\alpha t} + e^{-\alpha t} \int_0^t f(z)e^{\alpha z} \, dz$$

(98)

$$= \left[y(0) + \int_0^t f(z)e^{\alpha z} \, dz \right] e^{-\alpha t} \qquad t \geq 0$$

Formula (97) gives the general solution of the differential equation (96) in terms of a convolution. Formula (98) is the form in which it is usually used. Differentiation shows that the formula actually gives the solution on the whole real line, for

$$\frac{d}{dt}\left[e^{-\alpha t}\int_0^t f(z)e^{\alpha z}\,dz\right] = -\alpha e^{-\alpha t}\int_0^t f(z)e^{\alpha z}\,dz + e^{-\alpha t}\frac{d}{dt}\left[\int_0^t f(z)e^{\alpha z}\,dz\right]$$

$$= -\alpha e^{-\alpha t}\int_0^t f(z)e^{\alpha z}\,dz + e^{-\alpha t}f(t)e^{\alpha t}$$

$$= -\alpha e^{-\alpha t}\int_0^t f(z)e^{\alpha z}\,dz + f(t)$$

H. FURTHER PROPERTIES OF THE LAPLACE TRANSFORM

1. The Initial Value Theorem and Final Value Theorem

These two properties concern certain limiting values of a function. They can be calculated from the Laplace transform, in the transform domain, without taking the inverse Laplace transform and returning to the time domain. As in Section G, we write $F(s)$ for $\mathscr{L}[f](s)$.

The first, the *Initial Value Theorem*, concerns the jump discontinuity which may be introduced at the origin into the solution of a nonhomogeneous differential equation. Since the jump is

$$y(0+) - y(0-)$$

and since $y(0-)$ is an initial value and therefore presumably known, the point is to calculate $y(0+)$. This is the formula:

$$y(0+) = \lim_{s\to\infty} sY(s) \tag{1}$$

Later on we shall show that the formula is valid. First we calculate values for familiar functions.

If $y(t) = 3\sin 2t$, then y is continuous at $t = 0$ so that $y(0+)$ is just $y(0) = 0$. Now we use the right-hand side of (1):

$$\lim_{s\to\infty} sY(s) = \lim_{s\to\infty} s\,\frac{3}{s^2 + 4}$$

$$= \lim_{s\to\infty} \frac{3s}{s^2 + 4} = 0$$

by L'Hospital's rule or any other method of evaluation. Thus, in this example,

$$y(0+) = 0 = \lim_{s\to\infty} sY(s)$$

illustrating Equation (1).

If $y(t) = 3 \cos 2t$, then $y(0+) = y(0) = 3$ by the continuity of y. On the other hand,

$$\lim_{s \to \infty} sY(s) = \lim_{s \to \infty} s \frac{3s}{s^2 + 4}$$

$$= \lim_{s \to \infty} \frac{3s^2}{s^2 + 4} = 3$$

which is $y(0+)$, as we showed in the previous sentence. That is, in this case too,

$$y(0+) = \lim_{s \to \infty} sY(s)$$

If $y(t) = -4u(t)$, where u is the function with unit step at $t = 0$, then $y(0+) = -4$. On the other hand,

$$\lim_{s \to \infty} sY(s) = \lim_{s \to \infty} s \frac{-4}{s} = -4$$

illustrating Equation (1) again.

Here is a situation in which formula (1) might be used to evaluate $y(0+)$. Consider the system illustrated in Figure 1, which is described by the differential equation (2), and suppose that the initial conditions (3) and (4) are also satisfied. The problem is to find the behavior of y and its derivatives at $t = 0$.

$$My''(t) + By'(t) + Ky(t) = 3u(t) - 7\delta(t) \tag{2}$$

$$y(0-) = 4 \tag{3}$$

$$y'(0-) = 2 \tag{4}$$

The transformed problem and Laplace transform of y are, respectively,

$$(Ms^2 + Bs + K)Y(s) = \frac{3}{s} - 7 + 4Ms + 2M + 4B$$

$$Y(s) = \left(\frac{3}{s} + 2M + 4B - 7 + 4Ms \right) \frac{1}{Ms^2 + Bs + K}$$

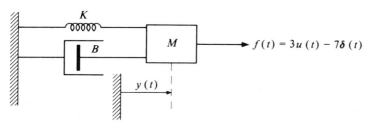

$f(t) = 3u(t) - 7\delta(t)$

Figure 1 Spring-mass-dashpot system.

From formula (1),

$$y(0+) = \lim_{s \to \infty} sY(s)$$

$$= \lim_{s \to \infty} \frac{3 + (2M + 4B - 7)s + 4Ms^2}{Ms^2 + Bs + K}$$

$$= 4$$

Since the value of $y(0-)$ is also 4 by the initial condition (3), it follows that y has no jump at $t = 0$. However, y' may behave quite differently.

$$\mathscr{L}[y'](s) = sY(s) - y(0-)$$

$$= \frac{3 + (2M + 4B - 7)s + 4Ms^2}{Ms^2 + Bs + K} - 4$$

$$= \frac{(3 - 4K) + (2M - 7)s}{Ms^2 + Bs + K}$$

Using formula (1), we see that

$$y'(0+) = \lim_{s \to \infty} s\mathscr{L}[y'](s)$$

$$= \lim_{s \to \infty} \frac{(3 - 4K)s + (2M - 7)s^2}{Ms^2 + Bs + K}$$

$$= \frac{2M - 7}{M}$$

$$= 2 - \frac{7}{M}$$

Since $y'(0-) = 2$ by (4), the jump in y' at $t = 0$ is

$$y'(0+) - y'(0-) = -\frac{7}{M}$$

and it clearly is introduced by the $-7\delta(t)$ term, not by the $3u(t)$ term.
One can also look at the calculation of the jump slightly differently:

$$y'(0+) - y'(0-) = \lim_{s \to \infty} s\mathscr{L}[y'](s) - y'(0-) = \lim_{s \to \infty} \mathscr{L}[y''](s) \qquad (5)$$

That is, the jump in y' is equal to a limiting value of the Laplace transform of the generalized derivative y''.

Now let us try to find the jump in y'' at $t = 0$.

$$\mathcal{L}[y''](s) = s\mathcal{L}[y'](s) - y'(0)$$

$$= \frac{(3 - 4K)s + (2M - 7)s^2}{Ms^2 + Bs + K} - 2$$

$$= \frac{-2K + (3 - 4K - 2B)s + (2M - 7 - 2M)s^2}{Ms^2 + Bs + K}$$

$$= \frac{-2K + (3 - 4K - 2B)s - 7s^2}{Ms^2 + Bs + K}$$

$$= \frac{-7}{M} + \frac{-2K + 7K/M + (3 - 4K - 2B + 7B/M)s}{Ms^2 + Bs + K} \tag{6}$$

The limit of s times the second term can be calculated, but s times the first term increases unboundedly, and so $s\mathcal{L}[y''](s)$ does not have a limit as s increases. In other words, formula (1) cannot be applied. In fact, (6) shows that $y''(t)$ is $-7\delta(t)/M$ plus a damped sinusoid or a sum of exponentials. Thus, y'' does not have a jump but a more complicated discontinuity at $t = 0$, and the Initial Value Theorem does not apply to it. Because of such difficulties, the formula is now stated with hypotheses which are sufficiently strong to ensure the applicability of the formula:

Initial Value Theorem *If y is a function which is piecewise-continuous (see Section D) and of some exponential order (see Section A) on the right half line $(t \geq 0)$, and if in particular $y(0+)$ exists, then*

$$y(0+) = \lim_{s \to \infty} sY(s) \tag{7}$$

With a little preparation, the formula may also be used on the sum of a function such as that described in the theorem plus a δ whose singularity is not at $t = 0$, by computing the jumps separately. For example, $\delta(t - 4)$ has no discontinuity at the origin. Thus, if

$$y(t) = u(t) \cos 2t - 3\delta(t - 4)$$

then $y(0+)$ is just the limit from the right (at $t = 0$) of $u(t) \cos 2t$, for $\delta(t - 4)$ has no effect before $t = 4$.

The formula is derived (proved) as follows. If $b > 0$ is any positive number at all, then

$$\lim_{s \to \infty} sY(s) = \lim_{s \to \infty} \left[s \int_0^b y(t)e^{-st}\, dt + s \int_b^\infty y(t)e^{-st}\, dt \right]$$

$$= \lim_{s \to \infty} \int_0^b y(t)se^{-st}\, dt + \lim_{s \to \infty} \int_b^\infty y(t)se^{-st}\, dt$$

The integrals exist because of the hypotheses of piecewise continuity and exponential order. Curious as it may seem, the second term in the last equation is zero. In fact, if y is of exponential order S, then, for some number M, $y(t) \leq Me^{St}$, which implies that

$$\left| \int_b^\infty y(t)se^{-st}\, dt \right| \leq \int_b^\infty Me^{St}se^{-st}\, dt = \frac{Mse^{-sb}}{s - S}$$

which has limit zero as s increases, by L'Hospital's rule. Thus, for each $b > 0$,

$$\lim_{s \to \infty} sY(s) = \lim_{s \to \infty} \int_0^b y(t)se^{-st}\, dt$$

If b is very small, then the interval $0 < t < b$ is very short, and all the values of $y(t)$ there are near $y(0+)$. It is apparent, then, that

$$\lim_{s \to \infty} sY(s) = \lim_{s \to \infty} \int_0^b y(0+)se^{-st}\, dt$$

$$= y(0+) \lim_{s \to \infty} \int_0^b se^{-st}\, dt$$

$$= y(0+) \lim_{s \to \infty} (1 - e^{-sb})$$

$$= y(0+)$$

(What is "apparent" may be proven with inequalities.) That gives the required formula.

As a last observation, notice that the jump at $t = 0$ is

$$y(0+) - y(0-) = \lim_{s \to \infty} sY(s) - y(0-)$$

$$= \lim_{s \to \infty} \mathcal{L}[y'](s)$$

as in (5), provided that y has a generalized derivative in addition to the properties required in the theorem. There are, however, many functions y for which the theorem ensures that formula (7) holds but which we have not shown to possess a generalized derivative.

The *Final Value Theorem* concerns the possibility that a function y may have a limiting value

$$\lim_{t \to \infty} y(t)$$

and the possibility of calculating this value directly from the Laplace transform of y. For an electrical or mechanical system, a limiting value of y describes an equilibrium state toward which the system progresses. (Other possible but more complicated "steady states" are discussed in Section J, together with a technique for calculating them.)

The formula for calculating the limiting value directly from Y, without knowing y itself, is

$$\lim_{t \to \infty} y(t) = \lim_{s \to 0} sY(s) \tag{8}$$

Before showing that the formula is valid, let us calculate values for some familiar functions. If $y(t) = 3u(t - t_0)$, then $y(t) = 3$ for all large t, and

$$\lim_{t \to \infty} 3u(t - t_0) = 3$$

Now let us calculate the right-hand side of (8):

$$\lim_{s \to 0} sY(s) = \lim_{s \to 0} s \frac{3e^{-st_0}}{s}$$

$$= \lim_{s \to 0} 3e^{-st_0} = 3$$

Thus, for this y, $\qquad \lim_{t \to \infty} y(t) = 3 = \lim_{s \to 0} sY(s)$

illustrating Equation (8).

If, instead, $y(t) = -4e^{-2t} = -4/e^{2t}$, then

$$\lim_{t \to \infty} y(t) = \lim_{t \to \infty} \frac{-4}{e^{2t}} = 0$$

while $\qquad \lim_{s \to 0} sY(s) = \lim_{s \to 0} s \frac{-4}{s + 2}$

$$= \frac{\lim_{s \to 0} -4s}{\lim_{s \to 0} (s + 2)} = 0$$

In this case also, then,

$$\lim_{t \to \infty} y(t) = \lim_{s \to 0} sY(s)$$

Here is a situation in which the formula might be applied in practice: given the initial value problem (2), (3), (4), find the limiting value of y, that is, the rest position of the system in Figure 1. As calculated before,

$$Y(s) = \left(\frac{3}{s} + 2M + 4B - 7 + 4Ms \right) \frac{1}{Ms^2 + Bs + K}$$

$$\lim_{s \to 0} sY(s) = \lim_{s \to 0} \frac{3 + (2M + 4B - 7)s + 4Ms^2}{Ms^2 + Bs + K}$$

$$= \frac{3}{K}$$

Thus, the limiting value of y is $3/K$.

The limiting velocity is going to be zero. Let us verify that.

$$\mathcal{L}[y'](s) = \frac{3 - 4K + (2M - 7)s}{Ms^2 + Bs + K}$$

$$\lim_{s \to 0} s\mathcal{L}[y'](s) = \lim_{s \to 0} \frac{(3 - 4K)s + (2M - 7)s^2}{Ms^2 + Bs + K}$$

$$= \frac{0}{K} = 0$$

Most functions do not have limiting values as t increases. For instance, e^t increases unboundedly, as do t and $e^{\alpha t}$ if α is positive. The sine and cosine, although bounded, do not have limiting values because of their periodic behavior. Nevertheless,

$$\lim_{s \to 0} s\mathcal{L}[e^t](s) = \lim_{s \to 0} \frac{s}{s - 1} = 0$$

$$\lim_{s \to 0} s\mathcal{L}[\sin t](s) = \lim_{s \to 0} \frac{s}{s^2 + 1} = 0 \qquad (9)$$

and $\qquad \lim_{s \to 0} s\mathcal{L}[\cos t](s) = \lim_{s \to 0} s\frac{s}{s^2 + 1} = 0$

Therefore the formula must be used with care, for the existence of a limit for $sY(s)$ at $s = 0$ does not assure the existence of a limit for $y(t)$ as t increases. Here is the theorem, with its traditional name:

Final Value Theorem *If y is a function which is piecewise-continuous (see Section D) on the right half line, and if it has a limit as t increases, then*

$$\lim_{t \to \infty} y(t) = \lim_{s \to 0} sY(s)$$

The formula is derived (the theorem is proved) this way:

$$\lim_{s \to 0} sY(s) = \lim_{s \to 0} \int_0^\infty y(t)se^{-st}\, dt$$

$$= \lim_{s \to 0} \int_0^\infty y(t)se^{-st}\, dt - \lim_{s \to 0} \int_0^b y(t)se^{-st}\, dt \qquad (10)$$

for each real number $b > 0$, since the last term is just zero. We see that fact as follows. If y has a limit, then it is bounded. Thus, for some M,

$$|y(t)| \leq M \qquad \text{for } t \geq 0$$

and
$$\lim_{s\to 0}\left|\int_0^b y(t)se^{-st}\,dt\right| \le \lim_{s\to 0}\int_0^b Mse^{-st}\,dt$$

$$= M\lim_{s\to 0}\int_0^b se^{-st}\,dt$$

$$= M\lim_{s\to 0}(1 - e^{-sb}) = 0$$

Therefore, continuing from (10),

$$\lim_{s\to 0} sY(s) = \lim_{s\to 0}\left[\int_0^\infty y(t)se^{-st}\,dt - \int_0^b y(t)se^{-st}\,dt\right]$$

$$= \lim_{s\to 0}\int_b^\infty y(t)se^{-st}\,dt$$

for each $b > 0$. Now $y(t)$ has a limit (call it y_1) as t increases, and so it follows that if only b is large enough, then $y(t)$ is very near y_1 for all $t \ge b$. It is apparent that

$$\lim_{s\to 0} sY(s) = \lim_{s\to 0}\int_b^\infty y_1 se^{-st}\,dt$$

$$= y_1\lim_{s\to 0}\int_b^\infty se^{-st}\,dt$$

$$= y_1\lim_{s\to 0}\left(\lim_{T\to\infty} -e^{-st}\Big|_b^T\right)$$

$$= y_1\lim_{s\to 0} e^{-sb}$$

$$= y_1 = \lim_{t\to\infty} y(t)$$

(What is "apparent" may be proven with inequalities.)

That proves the theorem. However, as remarked before, the formula must be used with caution, since the existence of

$$\lim_{s\to 0} sY(s)$$

does not imply that $y(t)$ has a limiting value, as the examples (9) show.

2. Laplace Transforms of Periodic Functions and Series

A function f is *periodic* with *period* T if

$$f(t + T) = f(t) \qquad \text{for all } t \tag{11}$$

The fact that $\cos t$ and $\sin t$ are periodic with period 2π is easy to remember:

$$\cos(t + 2\pi) = \cos t \qquad \text{for all } t$$

$$\sin(t + 2\pi) = \sin t \qquad \text{for all } t$$

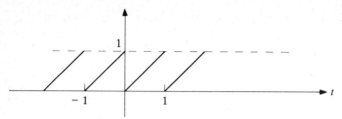

Figure 2 Graph of the fractional part of *t*.

There are many other periodic functions. For instance, Figure 2 is a graph of a periodic function of *t* known as the *fractional part of t*, for $f(t)$ is just *t* minus the largest integer less than or equal to *t*. Thus,

$$f(t) = \begin{cases} t & 0 \le t < 1 \\ t - n & n \le t < n + 1, \text{ for each integer } n \end{cases} \qquad (12)$$

The period of this function is 1.

If *f* is a function of period *T*, then any interval of length *T* is called a *period interval*.

Figure 3 is the graph of a function which has period 2. On the right half line it is given by the formula

$$f(t) = \begin{cases} 1 & 0 < t \le 1 \\ 0 & 1 < t \le 2 \\ 1 & n < t \le n + 1, \text{ for every even positive integer } n \\ 0 & n + 1 < t \le n + 2, \text{ for every even positive integer } n \end{cases} \qquad (13)$$

We shall use this function to illustrate a formula for the Laplace transform of a periodic (piecewise-continuous) function.

$$F(s) = \int_0^\infty f(t)e^{-st}\, dt$$

$$= \int_0^2 f(t)e^{-st}\, dt + \int_2^\infty f(t)e^{-st}\, dt$$

$$= \int_0^1 f(t)e^{-st}\, dt + \int_2^\infty f(t)e^{-st}\, dt$$

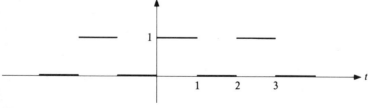

Figure 3 Graph of a periodic step function.

since f happens to be zero on the interval $1 < t \le 2$. Continuing, we have

$$F(s) = \frac{e^{-st}}{-s}\Big|_0^1 + \int_{z=0}^{\infty} f(z+2)e^{-s(z+2)}\, dz$$

where the first integral has been evaluated and the second integral has suffered the change of variable $t = z + 2$. Since our function f is periodic with period 2, we continue in this way:

$$F(s) = \frac{1-e^{-s}}{s} + \int_0^{\infty} f(z)e^{-sz}e^{-2s}\, dz$$

$$= \frac{1-e^{-s}}{s} + e^{-2s}\int_0^{\infty} f(z)e^{-sz}\, dz$$

$$= \frac{1-e^{-s}}{s} + e^{-2s}F(s) \tag{14}$$

Solving (14) for $F(s)$ gives the desired formula:

$$F(s)(1-e^{-2s}) = \frac{1-e^{-s}}{s}$$

$$F(s) = \frac{1-e^{-s}}{s}\frac{1}{1-e^{-2s}} \tag{15}$$

Observe that the first factor of the right-hand side is equal to

$$\int_0^2 f(t)e^{-st}\, dt$$

by comparing the first terms on the right in the derivation above.

If f is any function of period T and piecewise-continuous, so that it may be integrated along a period interval, then the following formula holds:

$$F(s) = \frac{\int_0^T f(t)e^{-st}\, dt}{1-e^{-sT}} \tag{16}$$

The derivation of the formula is exactly that illustrated in the example.

Now let us do the example a different way, illustrating an extension of the linearity of Laplace transformation. For each integer $k \ge 0$, define

$$f_k(t) = \begin{cases} 1 & 2k < t \le 2k+1 \\ 0 & \text{all other values of } t \end{cases}$$

a function whose graph is Figure 4. On the right half line the function f can be thought of as the sum of these functions f_k:

$$f(t) = \sum_{k=0}^{\infty} f_k(t) \qquad t \ge 0 \tag{17}$$

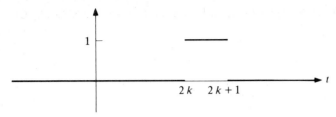

Figure 4 "Step up and down" function f_k.

You have studied series, and you know that the question must always be asked: "Does the series converge?" The series (17) does converge at each point t, for at that point all terms of the series take value zero except for one term at most.

Now let us take the Laplace transform of each term of the series:

$$F_k(s) = \mathcal{L}[f_k](s) = \int_0^\infty f_k(t)e^{-st}\, dt$$

$$= \int_{2k}^{2k+1} e^{-st}\, dt$$

$$= \frac{e^{-2ks} - e^{-(2k+1)s}}{s}$$

$$= e^{-2ks}\frac{1 - e^{-s}}{s} \qquad s > 0$$

If the Laplace transform of the sum is the sum of the transforms, does it follow that the transform of a series of functions is the sum of the series of their transform? It does if each term of the series has the same exponential order, as in this example:

$$\sum_{k=0}^\infty \mathcal{L}[f_k](s) = \sum_{k=0}^\infty e^{-2ks}\frac{1 - e^{-s}}{s}$$

$$= \frac{1 - e^{-s}}{s} \sum_{k=0}^\infty e^{-2ks}$$

$$= \frac{1 - e^{-s}}{s} \sum_{k=0}^\infty x^k \qquad \text{when } x = e^{-2s} \qquad (18)$$

$$= \frac{1 - e^{-s}}{s}\frac{1}{1 - x} \qquad \text{if } -1 < x < 1$$

$$= \frac{1 - e^{-s}}{s}\frac{1}{1 - e^{-2s}} \qquad \text{if } s > 0 \qquad (19)$$

This is exactly what we found in (15). In the calculation, the series appearing in (18) is the geometric series

$$\sum_{k=0}^{\infty} x^k = \frac{1}{1-x} \qquad \text{if } -1 < x < 1$$

If $x = e^{-2s}$, then the condition on x becomes

$$e^{-2s} = x < 1$$
$$e^{2s} > 1$$
$$2s > 0$$
$$s > 0$$

justifying the condition specified in (19).

If, instead, we take f to have the Dirac δ at each integer, i.e., for each integer k,

$$f_k(t) = \delta(t - k)$$

and

$$f(t) = \sum_{k=0}^{\infty} \delta(t - k) \qquad \text{if } t \geq 0$$

then

$$\sum_{k=0}^{\infty} \mathscr{L}[f_k](s) = \sum_{k=0}^{\infty} e^{-sk}$$

$$= \sum_{k=0}^{\infty} (e^{-s})^k$$

$$= \frac{1}{1 - e^{-s}} \qquad s > 0$$

This formula for $F(s)$ can also be deduced directly from (16).

Finally, let us consider a function of exponential order which is given by a power series but is not periodic.

$$e^{at} = \sum_{k=0}^{\infty} \frac{(at)^k}{k!}$$

The Laplace transform of the kth term of the series is

$$\mathscr{L}\left[\frac{(at)^k}{k!}\right](s) = a^k \mathscr{L}\left[\frac{t^k}{k!}\right](s)$$

$$= \frac{a^k}{s^{k+1}} \qquad \text{if } s > 0$$

The series of transforms is therefore this:

$$\sum_{k=0}^{\infty} \mathcal{L}\left[\frac{(at)^k}{k!}\right](s) = \sum_{k=0}^{\infty} \frac{a^k}{s^{k+1}}$$

$$= \frac{1}{s} \sum_{k=0}^{\infty} \left(\frac{a}{s}\right)^k$$

$$= \frac{1}{s} \frac{1}{1 - a/s} \qquad \text{if } -1 < \frac{a}{s} < 1$$

$$= \frac{1}{s - a} \qquad \text{if } s > |a|$$

This, then, is another derivation of the familiar formula for the Laplace transform of the exponential. A similar derivation can be given for the Laplace transforms of the sine and cosine.

3. Sample Problems with Answers

For the solution y of each initial value problem, find the jump at $t = 0$ and the limiting value (if any) as t increases.

A. $\begin{aligned} 7y'(t) + 5y(t) &= 4 + e^{-3t} + 2\delta(t) \\ y(0-) &= -1 \end{aligned}$ *Answers:* $y(0+) - y(0-) = \frac{2}{7}$; $\lim\limits_{t \to \infty} y(t) = \frac{4}{5}$

B. $\begin{aligned} y''(t) + y(t) &= u(t) + 3\delta'(t) \\ y(0-) &= y'(0-) = 0 \end{aligned}$ *Answers:* $y(0+) - y(0-) = 3$; no limiting value

C. If

$$Y(s) = \frac{s/(3s^2 + 5)}{1 - e^{-s}}$$

and $y(0-) = 0$, does y have (a) a jump at $t = 0$, and if so by what amount; (b) a limiting value as t increases, and if so what value?

Answer: $y(0+) - y(0-) = \frac{1}{3}$, but y has no limiting value because it is periodic but not constant.

4. Problems

Use the Initial Value and Final Value Theorems to find the jump at $t = 0$ and the limiting value as $t \to \infty$ for the solution of each initial value problem.

1. $\begin{aligned} x''(t) + 3x'(t) + 2x(t) &= -1 + 5\delta(t) - 2\delta(t - 4) \\ x(0-) &= x'(0-) = 0 \end{aligned}$

2. $\begin{aligned} 3y'(t) - 7y(t) &= 4\delta(t) + \cos \pi t \\ y(0) &= 10 \end{aligned}$

For each of the following three problems, replace the given initial value problem with a different one having the same solution but having no δ in the differential equation.

3. For the system in Figure 26 in Section G, the function $v(t) = 4u(t) + 3\delta(t) + e^{-t}$, and the value $v_R(0-) = 0$, specify the initial value problem.

4. For the system in Figure 27 in Section G, the function $f(t) = 3u(t) - 5\delta(t)$, and the value $v(0-) = -3$, specify the initial value problem.

5. For the system in Figure 28 in Section G, the function $x(t) = \sin t$, and the initial values $x_1(0-) = x_1'(0-) = 0$, specify the initial value problem in which x_1 is the unknown.

6. Graph and find the Laplace transform of the *full rectified sine wave* $f(t) = |\sin t|$.

THE J_ω OPERATOR

I. COMPLEX ARITHMETIC AND THE COMPLEX EXPONENTIAL

1. Complex Addition

The operations of complex arithmetic are addition and multiplication, and they are much like the operations of (real) arithmetic which you learned in elementary school. The difference is the presence of a new number j, called the *imaginary unit*. In mathematics books it is invariably written i, for historical reasons. Since the last century, however, it has begun to appear in engineering books, where it is called j, to avoid confusion with symbols representing electric current.

The result of adding j to j is $2j$, and so forth:

$$j + j = 2j$$

$$j + j + j = 3j$$

$$\underbrace{j + \cdots + j}_{n \text{ summands}} = nj$$

However, the result of adding $3j$ to the real number 5 is written $5 + 3j$ just as one might write "5 apples + 3 oranges." The number $5 + 3j$ is called a *complex number*. Each complex number is the sum of a real number, such as 5, called the *real part*, and an imaginary number, such as $3j$, of which the real coefficient of j is called the *imaginary part*. The real and imaginary parts of $x + yj$ are, respectively, the real numbers x and y, represented as

$$x = \text{Re } (x + yj) \qquad y = \text{Im } (x + yj)$$

For instance,

$$5 = \text{Re } (5 + 3j) \qquad 3 = \text{Im } (5 + 3j)$$

Here is the first important point: the real and imaginary parts of a complex number exactly determine that number. Thus, the complex number whose real part is -4 and whose imaginary part is 6 is the complex number $-4 + 6j$. The complex number whose real part is $\sqrt{2}$ and whose imaginary part is $-\pi$ is the complex number $\sqrt{2} - \pi j$. The sum of these two complex numbers is

$$(-4 + 6j) + (\sqrt{2} - \pi j) = (-4 + \sqrt{2}) + (6 - \pi)j$$

We can write this in column arithmetic, too:

$$
\begin{array}{r}
-4 + 6j \\
+\sqrt{2} - \pi j \\
\hline
(-4 + \sqrt{2}) + (6 - \pi)j
\end{array}
$$

In general, the *sum* of $x + yj$ and $a + bj$ has $x + a$ for its real part, and $y + b$ for its imaginary part (just as adding x apples and y oranges to a apples and b oranges gives $x + a$ apples and $y + b$ oranges). The *difference* of the two is

$$(x + yj) - (a + bj) = (x + yj) + (-a - bj)$$
$$= (x - a) + (y - b)j$$

Two complex numbers $a + bj$ and $x + yj$ are the same (that is, equal) if and only if they have the same real parts (that is, $a = x$) and the same imaginary parts (that is, $b = y$). Thus, $5 + 3j \neq 5 + 4j$, for $3 \neq 4$; and $\sqrt{2} - \pi j \neq 1.4 - \pi j$ since $\sqrt{2} \neq 1.4$.

If the imaginary part of a complex number is zero, then the number is said to be *purely real* or, more succinctly, just *real*. In this case the imaginary term is not written:

$$3 + 0j = 3$$

In this way, the real numbers are identified in the collection of complex numbers.

Similarly, if the real part of a complex number is zero, then the number is said to be *purely imaginary* or just *imaginary*. In this case the real term is not written:

$$0 + j = j$$

2. The Complex Plane (Gaussian Plane or Argand Diagram)

Real numbers are often represented as points on a line, the real line. The complex numbers are represented as points in a plane, the *complex plane*. The number $x + yj$ is represented by the point (x, y) whose first coordinate is x and second coordinate is y.

It is easy to compute the distance of a complex number $x + yj$ from the origin $0 + 0j = 0$, using the Pythagorean theorem. The distance is $(x^2 + y^2)^{1/2}$; it is called the *modulus* of $x + yj$, and it is written $|x + yj|$. Thus,

$$|x + yj| = (x^2 + y^2)^{1/2}$$

The modulus of $2 + 3j$ is $|2 + 3j| = (2^2 + 3^2)^{1/2} = (4 + 9)^{1/2} = \sqrt{13}$. Note that $x + yj$ and its negative, $-x - yj$, are symmetric about the origin and have the same modulus: $[(-x)^2 + (-y)^2]^{1/2} = (x^2 + y^2)^{1/2}$.

The *distance* between two points is the modulus of their difference:

$$|(a + bj) - (c + dj)| = |(a - c) + (b - d)j|$$
$$= ((a - c)^2 + (b - d)^2)^{1/2}$$

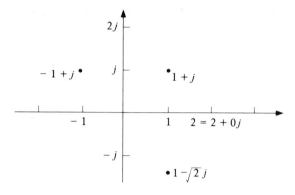

Figure 1 Points in the complex plane.

Two complex numbers are close together if the modulus of their difference is small.

When computing a modulus, the beginner sometimes wonders, "What happens to the j?" The j tells us where to look for the real and the imaginary parts, but the j does not enter into the computation of the modulus in any other way. Look at Figure 3. The lengths of the legs of the triangle are $|a - c|$ and $|b - d|$, and the length of the hypotenuse is $((a - c)^2 + (b - d)^2)^{1/2}$.

3. Complex Multiplication

The product of j with a real number y is yj, but the product of j with itself is defined to be -1. Otherwise, multiplication follows the old rule of column multiplication (known technically as the *distributive law*):

$$
\begin{array}{r}
a + bj \\
x + yj \\
\hline
ayj + byjj \\
ax + bxj \\
\hline
ax + bxj + ayj + byjj
\end{array}
$$

$$ax + bxj + ayj + byjj = ax + (bx + ay)j - by$$

$$= (ax - by) + (bx + ay)j$$

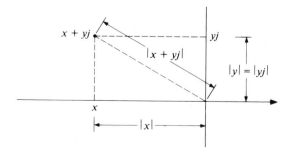

Figure 2 The modulus, distance from the origin.

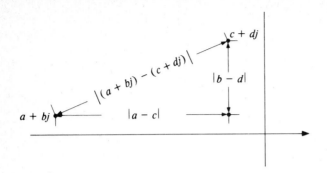

Figure 3 Distance between any two points.

Thus, for instance,

$$(1+j)(2-3j) = 2 - 3j + 2j - 3jj$$
$$= 2 - j + 3$$
$$= 5 - j$$

and

$$(1+j)(1-j) = 1 - j + j + j(-j)$$
$$= 1 - j^2$$
$$= 1 + 1$$
$$= 2$$

4. The Complex Conjugate

If $x + yj$ is a complex number with real part x and imaginary part y, then its *complex conjugate* $\overline{x + yj}$ is the number $x - yj$. This will be easy to remember if you realize that a number and its conjugate are symmetric about the real axis in the complex plane. (See Figure 4.) Thus,

$$\overline{(x + yj)} = x + yj \tag{1}$$

One of the important identities is that the product of a number and its conjugate is the square of the modulus of the number:

$$(x + yj)\overline{(x + yj)} = (x + yj)(x - yj)$$
$$= x^2 - xyj + xyj - y^2j^2$$
$$= x^2 - y^2(-1)$$
$$= x^2 + y^2$$
$$= |x + yj|^2$$

Since this is so for every complex number z, we shall write it as

$$z\bar{z} = |z|^2 \tag{2}$$

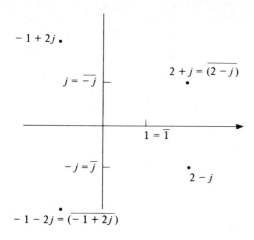

$-1 + 2j \bullet$

$j = -\bar{j}$

$2 + j = \overline{(2 - j)}$

$1 = \bar{1}$

$-j = \bar{j}$

$2 - j$

$-1 - 2j = \overline{(-1 + 2j)}$

Figure 4 Complex conjugates.

You should memorize formula (2) *right now*: "z times z-bar is the modulus of z, squared."

5. Division

There is only one complex number whose modulus is zero, and that is $0 + 0j = 0$, the complex number zero. If $z = x + yj$ is any complex number, then

$$z + 0 = x + yj + 0 + 0j = x + yj = z$$

and
$$z \cdot 0 = (x + yj)(0 + 0j) = 0 + 0j + 0j + 0j^2 = 0$$

On the other hand, if z is any complex number except zero, then the real number $1/|z|^2$ may be multiplied by both sides of (2), giving

$$z \frac{\bar{z}}{|z|^2} = 1$$

Therefore, if $z \neq 0$ we may define the *reciprocal* of z as

$$\frac{1}{z} = \frac{\bar{z}}{|z|^2} \tag{3}$$

for it will follow that

$$z \frac{1}{z} = z \frac{\bar{z}}{|z|^2} = \frac{|z|^2}{|z|^2} = 1$$

This allows us to define *division* by any nonzero complex number $z = x + yj$:

$$\frac{a + bj}{z} = (a + bj) \frac{\bar{z}}{|z|^2} = (a + bj) \frac{x - yj}{x^2 + y^2}$$

For example, the reciprocal of $2 + 3j$ is

$$\frac{1}{2 + 3j} = \frac{\overline{2 + 3j}}{|2 + 3j|^2}$$

$$= \frac{2 - 3j}{2^2 + 3^2}$$

$$= \tfrac{2}{13} - \tfrac{3}{13}j$$

Thus,
$$\frac{6 + j}{2 + 3j} = (6 + j)\frac{1}{2 + 3j}$$

$$= (6 + j)(\tfrac{2}{13} - \tfrac{3}{13}j)$$

$$= \tfrac{12}{13} + \tfrac{3}{13} + (\tfrac{2}{13} - \tfrac{18}{13})j$$

$$= \tfrac{15}{13} - \tfrac{16}{13}j$$

Many people prefer to do the calculation by multiplying numerator and denominator by the conjugate of the denominator and then applying identity (2) to the denominator, as follows:

$$\frac{6 + j}{2 + 3j} = \frac{(6 + j)(2 - 3j)}{(2 + 3j)(2 - 3j)}$$

$$= \frac{12 + 3 + (2 - 18)j}{2^2 + 3^2}$$

$$= \tfrac{15}{13} - \tfrac{16}{13}j$$

In particular, j is factored out of a complex number this way:

$$a + bj = j\frac{a + bj}{j}$$

$$= j\frac{(a + bj)(-j)}{|j|^2}$$

$$= j[(a + bj)(-j)]$$

$$= j(-aj + b)$$

$$= j(b - aj) \tag{4}$$

since $|j|^2 = 1$.

6. The Complex Exponential

For each complex number $z = x + yj$ (with x and y real), define

$$e^z = e^{x + yj} = e^x(\cos y + j \sin y)$$

$$= e^x \cos y + je^x \sin y$$

This function of z is the *complex exponential.* For each z, e^z is a complex number whose real part and imaginary part are

$$\text{Re}\,(e^z) = e^x \cos y = e^{\text{Re}\,z} \cos\,(\text{Im } z)$$

$$\text{Im}\,(e^z) = e^x \sin y = e^{\text{Re}\,z} \sin\,(\text{Im } z)$$

For example, $e^{2+3j} = e^2(\cos 3 + j \sin 3)$

and $e^{2+0j} = e^2(\cos 0 + j \sin 0)$

$$= e^2(1 + 0j) = e^2$$

and $e^{0+3j} = e^0(\cos 3 + j \sin 3)$

$$= \cos 3 + j \sin 3$$

7. Comparison of the Complex Exponential and the Real Exponential

While the definition of the complex exponential is unambiguous, one may object that it is somewhat unintuitive. There are several properties of the complex exponential which show that the definition is actually quite natural.

The first is that the complex exponential treats those complex numbers which are real in the same way as the real exponential treats real numbers; the second is that it satisfies the "rule of exponents." These two properties are very important. We now show that they hold.

If $z = x + 0j$, that is, if z is real, then

$$e^z = e^x(\cos 0 + j \sin 0)$$

$$= e^x(1 + 0j)$$

$$= e^x$$

Thus, for real numbers, the exponential function takes the usual value, and e^z has the first property.

To check the second property, note that if $z = x + yj$ and $c = a + bj$, then

$$e^{z+c} = e^{x+yj+a+bj} = e^{(x+a)+(y+b)j}$$

$$= e^{x+a}[\cos\,(y + b) + j \sin\,(y + b)]$$

$$= e^{x+a}(\cos y \cos b - \sin y \sin b + j \sin y \cos b + j \cos y \sin b)$$

$$= e^{x+a}(\cos y + j \sin y)(\cos b + j \sin b)$$

$$= e^x(\cos y + j \sin y)e^a(\cos b + j \sin b)$$

$$= e^{x+yj}e^{a+bj}$$

$$= e^z e^c$$

That is, the *rule of exponents* holds:

$$e^{z+c} = e^z e^c \qquad \text{for all complex numbers } z \text{ and } c \tag{5}$$

just as is the case for the real exponential.

The part of the definition which may have remained unclear to the intuition is

$$e^{yj} = \cos y + j \sin y \qquad \text{for } y \text{ real}$$

which is called *Euler's formula*. Let us take the power series for e^x, replace x with yj, and see what happens if we simplify the result using complex arithmetic.

$$e^{yj} = 1 + \frac{yj}{1!} + \frac{(yj)^2}{2!} + \frac{(yj)^3}{3!} + \cdots + \frac{(yj)^n}{n!} + \cdots$$

$$= 1 + j\frac{y}{1!} + j^2\frac{y^2}{2!} + j^3\frac{y^3}{3!} + \cdots + j^n\frac{y^n}{n!} + \cdots$$

Next we put all the even terms of the series together, writing the exponents in the form $n = 2k$; we also put together the odd terms, writing the odd exponents in the form $n = 2k + 1$.

$$e^{yj} = \underbrace{1 + j^2\frac{y^2}{2!} + j^4\frac{y^4}{4!} + \cdots + j^{2k}\frac{y^{2k}}{(2k)!} + \cdots}_{(6)}$$

$$\underbrace{+ j\frac{y}{1!} + j^3\frac{y^3}{3!} + j^5\frac{y^5}{5!} + \cdots + j^{2k+1}\frac{y^{2k+1}}{(2k+1)!} + \cdots}_{(7)}$$

Now let us analyze (6). The terms

$$y^0 \qquad y^4 \qquad y^8 \qquad y^{12} \qquad \cdots$$

are multiplied by powers of $j^4 = (j^2)^2 = (-1)^2 = 1$:

$$j^0 = 1 \qquad j^4 = 1 \qquad j^8 = 1 \qquad j^{12} = 1 \qquad \cdots$$

The other terms,

$$y^2 \qquad y^6 \qquad y^{10} \qquad y^{14} \qquad \cdots$$

are multiplied by j^2 times a power of j^4:

$$j^2 = -1 \qquad j^6 = j^2 j^4 = -1$$

$$j^{10} = j^2(j^4)^2 = -1 \qquad j^{14} = j^2(j^4)^3 = -1 \qquad \cdots$$

Thus, (6) may be rewritten as

$$1 - \frac{y^2}{2!} + \frac{y^4}{4!} - \frac{y^6}{6!} + \frac{y^8}{8!} - \frac{y^{10}}{10!} + \frac{y^{12}}{12!} - \frac{y^{14}}{14!} + \cdots = \cos y \qquad (8)$$

for the series is exactly the Taylor series for $\cos y$, as given in calculus books.

Now let us analyze (7) the same way, starting by factoring j out of every term:

$$j\frac{y}{1!} + j^3\frac{y^3}{3!} + j^5\frac{y^5}{5!} + \cdots + j^{2k+1}\frac{y^{2k+1}}{(2k+1)!} + \cdots$$

$$= j\left(\frac{y}{1!} + j^2\frac{y^3}{3!} + j^4\frac{y^5}{5!} + j^6\frac{y^7}{7!} + j^8\frac{y^9}{9!} + j^{10}\frac{y^{11}}{11!} + j^{12}\frac{y^{13}}{13!} + \cdots\right)$$

$$= j\left(y - \frac{y^3}{3!} + \frac{y^5}{5!} - \frac{y^7}{7!} + \frac{y^9}{9!} - \frac{y^{11}}{11!} + \frac{y^{13}}{13!}\cdots\right)$$

$$= j \sin y \tag{9}$$

for this series is exactly the Taylor series for sin y. Replacing (6) and (7) with (8) and (9) gives the formula

$$e^{yj} = \cos y + j \sin y$$

which was the part of the definition which seemed unintuitive. It may still seem unintuitive, but at least it is consistent with the Taylor-series expansion of e^x. In fact, although we shall not show it, the Taylor series for the complex exponential e^z is the same as that for the real exponential:

$$e^z = \sum_{k=0}^{\infty} \frac{z^k}{k!}$$

That fact is the basis for the similarity of the properties of the real and complex exponentials, but those properties which we need will be shown directly from the definition.

The next similarity concerns the way in which e^{ct}, a function of t, satisfies differential equations, whether c is real or complex. We begin by computing the derivative of e^{ct}, with $c = a + bj$.

$$\frac{d}{dt}e^{ct} = \frac{d}{dt}e^{(a+bj)t}$$

$$= \frac{d}{dt}(e^{at}\cos bt + je^{at}\sin bt)$$

$$= \frac{d}{dt}(e^{at}\cos bt) + j\frac{d}{dt}(e^{at}\sin bt)$$

$$= (ae^{at}\cos bt - e^{at}b\sin bt) + j(ae^{at}\sin bt + e^{at}b\cos bt)$$

$$= e^{at}[(a\cos bt - b\sin bt) + j(a\sin bt + b\cos bt)]$$

$$= e^{at}[(a+bj)\cos bt + (aj-b)\sin bt]$$

The coefficient of sin bt may have j factored out, as in (4):

$$aj - b = j\frac{aj-b}{j} = j(a+bj)$$

It follows, therefore, that the previous equation may be rewritten as

$$\frac{d}{dt} e^{ct} = e^{at}[(a + bj) \cos bt + j(a + bj) \sin bt]$$

$$= (a + bj)e^{at}(\cos bt + j \sin bt)$$

$$= ce^{(a + bj)t} = ce^{ct}$$

That is, for any complex number c,

$$\frac{d}{dt} e^{ct} = ce^{ct} \tag{10}$$

Thus, whether c is real or complex, e^{ct} is the solution of the initial value problem

$$\begin{aligned} y' - cy &= 0| \\ y(0) &= 1| \end{aligned}$$

8. Polar Coordinates for Complex Numbers

A complex number $z = x + yj$ has, as its polar coordinates, its modulus $r = |z|$ and its polar angle θ, measured counterclockwise from the real axis. When describing a complex number, one measures that angle *in radians, always,* and refers to the measure of the angle as the *argument of z.*

Thus, the argument of z is a real number $\theta = \arg z$, with the properties

$$\text{Re } z = |z| \cos \arg z = r \cos \theta \qquad \text{or} \qquad \cos \theta = \frac{\text{Re } z}{|z|}$$

$$\text{Im } z = |z| \sin \arg z = r \sin \theta \qquad \text{or} \qquad \sin \theta = \frac{\text{Im } z}{|z|} \tag{11}$$

The argument of a complex number is therefore determined only up to integer multiples of 2π. For instance, the arguments of j are the numbers $\pi/2, \pi/2 + 2\pi,$ $\pi/2 - 2\pi, \pi/2 + 4\pi, \pi/2 - 4\pi$, and, in the general case, $\pi/2 + 2n\pi$, for n an arbi-

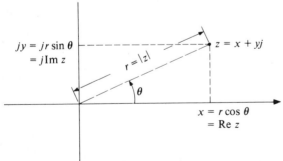

$jy = jr \sin \theta$
$= j\text{Im } z$

$r = |z|$

$z = x + yj$

θ

$x = r \cos \theta$
$= \text{Re } z$

Figure 5 Polar coordinates on the complex plane.

trary integer. Up to multiples of 2π, however, θ is determined by Re z and Im z, by (11).

9. Polar-Coordinate Representation Using the Complex Exponential

The complex exponential allows a useful simplification of the polar-coordinate representation, for if z is the complex number having modulus r and argument θ, then

$$\text{Re } z = r \cos \theta$$

$$\text{Im } z = r \sin \theta$$

Therefore,
$$z = \text{Re } z + j \text{ Im } z$$

$$= r \cos \theta + jr \sin \theta$$

$$= r(\cos \theta + j \sin \theta)$$

$$= re^{\theta j}$$

Alternatively, by replacing r with its equal $|z|$, and replacing θ with arg z, the polar-coordinate representation can be written yet another way:

$$z = re^{\theta j}$$

$$= |z| e^{j \text{ arg } z}$$

For example, the number j has polar coordinates $r = 1$ and $\theta = \pi/2$. Therefore,

$$1e^{(\pi/2)j} = 1 \left(\cos \frac{\pi}{2} + j \sin \frac{\pi}{2} \right)$$

$$= 1(0 + j \cdot 1) = j$$

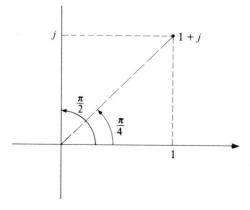

Figure 6 Finding the argument of $1 + j$.

just as expected. The number with argument $\pi/4$ and modulus $\sqrt{2}$ is

$$\sqrt{2}\,e^{(\pi/4)j} = \sqrt{2}\left(\cos\frac{\pi}{4} + j\sin\frac{\pi}{4}\right)$$

$$= \sqrt{2}\left(\frac{1}{\sqrt{2}} + j\frac{1}{\sqrt{2}}\right)$$

$$= 1 + j$$

Notice that if θ is any *real* number, then

$$e^{j\theta} = \cos\theta + j\sin\theta$$

from the definition. The point in the complex plane corresponding to $e^{j\theta}$ has coordinates $\cos\theta$ and $\sin\theta$, and it lies on the circle with unit radius about the origin. To show this we calculate the modulus:

$$|e^{j\theta}| = |\cos\theta + j\sin\theta|$$

$$= (\cos^2\theta + \sin^2\theta)^{1/2} = 1$$

Hence, $e^{j\theta}$ is the point with polar angle θ and r value 1.

The number $re^{j\theta}$ is r times $e^{j\theta}$. We can compute its real and imaginary parts:

$$re^{j\theta} = r(\cos\theta + j\sin\theta)$$

$$= r\cos\theta + jr\sin\theta$$

Thus, the polar-coordinate representation of a complex number has a geometric interpretation in the complex plane (see Figure 7): $re^{j\theta}$ lies on the ray from the origin through $e^{j\theta}$, at distance r from the origin, while $e^{j\theta}$ lies at distance 1.

In applications it is very often necessary to find modulus and argument for a

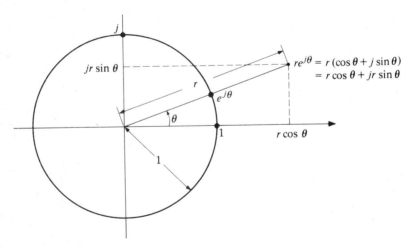

Figure 7 Polar coordinates and the complex exponential.

complex number whose real and imaginary parts are known. For example, if z is a complex number whose real part is -1 and whose imaginary part is 1, then

$$|z| = (1^2 + 1^2)^{1/2} = \sqrt{2}$$

and, for some θ,

$$z = \sqrt{2}(\cos \theta + j \sin \theta)$$

or
$$-1 + j = \sqrt{2} \cos \theta + j\sqrt{2} \sin \theta \qquad (12)$$

The left- and right-hand sides of (12) are the same complex number, and they must have the same real part:

$$-1 = \sqrt{2} \cos \theta \qquad \text{or} \qquad \cos \theta = -\frac{1}{\sqrt{2}}$$

They must have also the same imaginary part:

$$1 = \sqrt{2} \sin \theta \qquad \text{or} \qquad \sin \theta = \frac{1}{\sqrt{2}}$$

The values of θ with this property are $\theta = 3\pi/4$ and $\theta = 3\pi/4 + 2n\pi$, for each integer n. Thus, for instance, $\theta = 3\pi/4$ is an argument for $z = -1 + j$. Therefore $-1 + j = \sqrt{2}\, e^{j3\pi/4}$. (See Figure 8.)

For another example, let us find the polar coordinates of $z = -1 - \sqrt{3}j$.

$$|z| = (1^2 + \sqrt{3}\,^2)^{1/2} = 2$$

$$-1 - \sqrt{3}j = |z|(\cos \theta + j \sin \theta)$$

$$= 2 \cos \theta + 2j \sin \theta$$

Thus,
$$-1 = 2 \cos \theta \qquad \text{or} \qquad \cos \theta = -\frac{1}{2}$$

$$-\sqrt{3} = 2 \sin \theta \qquad \text{or} \qquad \sin \theta = -\frac{\sqrt{3}}{2}$$

and, therefore, $\theta = 4\pi/3$. (See Figure 9.)

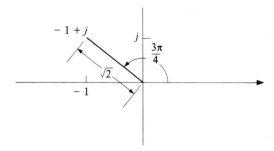

Figure **8** Finding the argument of $-1 + j$.

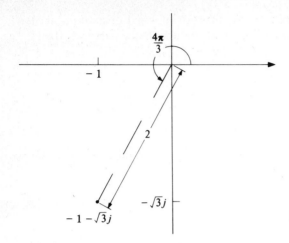

Figure 9 Finding the argument of $-1 - \sqrt{3}\,j$.

Because of the law of exponents, multiplication is very easy using the complex exponential and the polar coordinates. For instance, multiplying proceeds as in

$$6e^{2j}(\tfrac{1}{2}e^{-4j}) = 6(\tfrac{1}{2}e^{2j-4j}) = 3e^{-2j}$$

and dividing as in

$$\frac{6e^{2j}}{\tfrac{1}{2}e^{-4j}} = 6e^{2j}(\tfrac{1}{2})^{-1}(e^{-4j})^{-1}$$

$$= 6e^{2j}(2e^{4j}) = 12e^{6j}$$

On the other hand, addition is performed using the real and imaginary parts:

$$6e^{2j} + \tfrac{1}{2}e^{-4j} = 6\cos 2 + j6\sin 2 + \tfrac{1}{2}\cos(-4) + j\tfrac{1}{2}\sin(-4)$$

$$= (6\cos 2 + \tfrac{1}{2}\cos 4) + j(6\sin 2 - \tfrac{1}{2}\sin 4)$$

If necessary, the last expression can be reduced to polar coordinates using trigonometric identities. In particular, the polar-coordinate representation of $-2e^{j\pi/3}$ may be obtained this way:

$$-2e^{j\pi/3} = 2e^{j\pi/3}(-1)$$

$$= 2e^{j\pi/3}(e^{j\pi})$$

$$= 2e^{j(\pi/3 + \pi)}$$

$$= 2e^{j(4\pi/3)}$$

(see Figure 9), since $e^{j\pi} = -1$; in fact $e^{j\pi}$ is the polar-coordinate representation of -1.

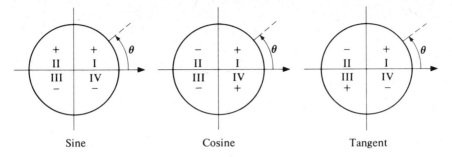

Sine   Cosine   Tangent

Figure 10 Sign versus quadrant for sine, cosine, and tangent.

10. Finding Arguments from the Real and Imaginary Parts

From x and y, the real and imaginary parts of a complex number $z = x + yj$, you can find $\cos \theta$ and $\sin \theta$ by dividing by the modulus:

$$\cos \theta = \frac{\text{Re } z}{|z|} = \frac{x}{(x^2 + y^2)^{1/2}}$$

$$\sin \theta = \frac{\text{Im } z}{|z|} = \frac{y}{(x^2 + y^2)^{1/2}}$$

In order to find θ itself from $\cos \theta$ and $\sin \theta$, you will need to find an inverse sine (Arcsin), inverse cosine (Arccos), or inverse tangent (Arctan). Except in the case of a few familiar angles, you will need some sort of table, slide rule, or computer to approximate the values. Such devices may be necessary, but they will not be sufficient, for none of them can tell you conclusively to which quadrant in the

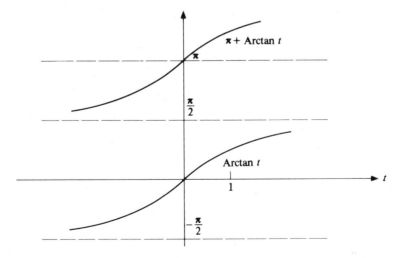

Figure 11 Graph of Arctan t and $\pi +$ Arctan t.

plane z belongs. That quadrant must be determined by inspecting the sign of $\sin \theta$ and $\cos \theta$. The sign of $\sin \theta$ is $+$ in the first and second quadrants, that of the cosine in the first and fourth quadrants. See Figure 10.

You should take this occasion to review (from trigonometry) the way in which tables of trigonometric functions are used to determine angles in the second, third, and fourth quadrants.

The most frequently used determination of θ uses the Arctan function (see Figure 11) in a sort of algorithm. First, determine whether $\cos \theta$ is zero.

1. If $\cos \theta = 0$ (that is, if $x = 0$), then

$$\theta = \begin{cases} \dfrac{\pi}{2} & \text{if } \sin \theta > 0 \\[2mm] \dfrac{\pi}{2} + \pi = \dfrac{3\pi}{2} & \text{if } \sin \theta < 0 \end{cases}$$

2. If $\cos \theta > 0$ (that is, if $x > 0$), then z is in the first or fourth quadrant, and

$$\theta = \text{Tan}^{-1} \frac{\sin \theta}{\cos \theta} = \text{Arctan} \frac{\sin \theta}{\cos \theta} \qquad \cos \theta > 0$$

$$= \text{Tan}^{-1} \frac{y}{x} \qquad x > 0$$

for Arctan takes values in the first and fourth quadrants. (See Figure 11.)

3. If $\cos \theta < 0$ (that is, if $x < 0$), then z is in the second or third quadrant, and

$$\theta = \pi + \text{Tan}^{-1} \frac{\sin \theta}{\cos \theta} = \pi + \text{Arctan} \frac{\sin \theta}{\cos \theta} \qquad \cos \theta < 0$$

$$= \pi + \text{Tan}^{-1} \frac{y}{x} \qquad x < 0$$

Thus, in the example $z = -1 - \sqrt{3}j$ given earlier, one could conclude from

$$\left. \begin{aligned} \cos \theta &= \frac{-1}{(1 + \sqrt{3}^2)^{1/2}} = -\frac{1}{2} \\[2mm] \sin \theta &= \frac{-\sqrt{3}}{(1 + \sqrt{3}^2)^{1/2}} = -\frac{\sqrt{3}}{2} \end{aligned} \right\} \tag{13}$$

that

$$\tan \theta = \frac{-1/2}{-\sqrt{3}/2} = \frac{1}{\sqrt{3}}$$

but it did not follow that $\theta = \pi/3$. (See Figure 9.) In fact, $\theta = 4\pi/3$. It is true that $\tan \pi/3 = 1/\sqrt{3}$, but

$$\cos \frac{\pi}{3} = \frac{1}{2} \qquad \sin \frac{\pi}{3} = \frac{\sqrt{3}}{2}$$

contrary to (13).

Thus, for example, if $\cos \theta = -\sqrt{3}/2$ and $\sin \theta = \frac{1}{2}$, then $e^{j\theta} = -\sqrt{3}/2 + \frac{1}{2}j$ lies in the second quadrant, and

$$\theta = \pi + \text{Arctan } \frac{\frac{1}{2}}{-\sqrt{3}/2}$$

$$= \pi + \text{Arctan } \frac{-1}{\sqrt{3}}$$

$$= \pi + \left(-\frac{\pi}{6}\right) = \frac{5\pi}{6}$$

If, instead, $\cos \theta = \sqrt{3}/2$ and $\sin \theta = -\frac{1}{2}$, then $e^{j\theta} = \sqrt{3}/2 + (-\frac{1}{2})j$ lies in the fourth quadrant, and

$$\theta = \text{Arctan } \frac{\frac{1}{2}}{-\sqrt{3}/2}$$

$$= \text{Arctan } \left(-\frac{1}{\sqrt{3}}\right) = -\frac{\pi}{6}$$

11. Sample Problems with Answers

Find the real and imaginary parts of each.

A. $-7 + \pi j$ *Answer:* Re $(-7 + \pi j) = -7$, Im $(-7 + \pi j) = \pi$

B. $(-7 + \pi j) - (2 + 3j)$
   Answer: Re $[(-7 + \pi j) - (2 + 3j)] = -9$; Im $[(-7 + \pi j) - (2 + 3j)] = \pi - 3$

C. $(-7 + \pi j)(2 + 3j)$
   Answer: Re $[(-7 + \pi j)(2 + 3j)] = -14 - 3\pi$; Im $[(-7 + \pi j)(2 + 3j)] = -21 + 2\pi$

D. $(-7 + \pi j)/(2 + 3j)$
   Answer: Re $[(-7 + \pi j)/(2 + 3j)] = (-14 + 3\pi)/13$; Im $[(-7 + \pi j)/(2 + 3j)] = (21 + 2\pi)/13$

E. $4e^{-(\pi/3)j}$ *Answer:* Re $4e^{-(\pi/3)j} = 2$; Im $4e^{-(\pi/3)j} = -2\sqrt{3}$

F. $4e^{-(\pi/3)j}e^{-1+(\pi/6)j}$
   Answer: Re $(4e^{-(\pi/3)j}e^{-1+(\pi/6)j}) = 2\sqrt{3}/e$; Im $(4e^{-(\pi/3)j}e^{-1+(\pi/6)j}) = -2/e$

G. $4e^{-(\pi/3)j}/e^{-1+(\pi/6)j}$
   Answer: Re $(4e^{-(\pi/3)j}/e^{-1+(\pi/6)j}) = 0$; Im $(4e^{-(\pi/3)j}/e^{-1+(\pi/6)j}) = -4e$

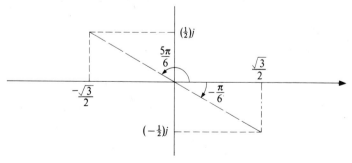

Figure 12 Arguments for $-\sqrt{3}/2 + (\frac{1}{2})j$ and $\sqrt{3}/2 + (-\frac{1}{2})j$.

H. $4e^{-(\pi/3)j} - e^{-1+(\pi/6)j}$

  Answer: Re $(4e^{-(\pi/3)j} - e^{-1+(\pi/6)j}) = 2 - \sqrt{3}/2e$; Im $(4e^{-(\pi/3)j} - e^{-1+(\pi/6)j}) = -2\sqrt{3} - \tfrac{1}{2}e$

Find the modulus, argument, and complex conjugate of each.

I. $-7 + \pi j$

  Answers: $|-7 + \pi j| = (49 + \pi^2)^{1/2}$; arg $(-7 + \pi j) = \pi + \mathrm{Tan}^{-1}(-\pi/7)$; $\overline{-7 + \pi j} = -7 - \pi j$

J. $(-7 + \pi j)/(2 + 3j)$

  Answers: $|(-7 + \pi j)/(2 + 3j)| = \sqrt{(-14 + 3\pi)^2 + (21 + 2\pi)^2}/13 = \sqrt{49 + \pi^2}/\sqrt{13}$;
  arg $[(-7 + \pi j)/(2 + 3j)] = \pi + \mathrm{Tan}^{-1}[(21 + 2\pi)/(14 - 3\pi)]$;
  $\overline{(-7 + \pi j)/(2 + 3j)} = (-14 + 3\pi)/13 - [(21 + 2\pi)/13]j$

K. $4e^{-(\pi/3)j}$ *Answers:* 4; $-\pi/3$; $4e^{(\pi/3)j}$

L. $4e^{-(\pi/3)j}/e^{-1+(\pi/6)j}$ *Answers:* $4e$; $-\pi/2$; $4e^{(\pi/2)j-1}$

12. Problems

Here is a list of complex numbers, to which the problems are to be applied.

$A = \sqrt{3} + j$	$B = 2 - 2j$	$C = (5 + 3j) - (1 - j)$
$D = -3$	$E = j^3$	$F = (3 + \pi j) + (-5 + 3j) - 13j$
$G = 4j - (2 + 7j)$	$H = (3 + 4j)(2j + 3j)$	$I = (-1 + 2j)(3 + j)$
$J = (-6 + 2j)(4 - 2j)$	$K = (1 + 2j)/(3 + j)$	$L = (3 + 2j)/(4 - j)$
$M = 5e^{-(\pi/4)j}$	$N = 2e^{(2\pi/3)j}$	$O = -3e^{\pi j}$
$P = e^{(\pi/2)j}e^{(\pi/3)j}$	$Q = e^{2+(\pi/2)j}$	$R = e^{-(\pi/3)j}$
$S = e^{-3-(\pi/3)j}$	$T = e^{(\pi/3)j} - e^{-(\pi/4)j}$	$U = e^{(\pi/4)j}/e^{(\pi/3)j}$
$V = -1 - 2j$	$W = 1 - 2j$	$X = -1 + 2j$
$Y = -(1 - 2j)$		

1. Find the real and imaginary parts of the complex numbers A through Y.
2. Find the modulus, argument, and complex conjugate of each of the complex numbers A through Y.
3. Plot, in the complex plane, the points A through Y.
4. Plot, in the complex plane, the complex conjugates of the points A through Y.
5. Write the complex numbers C through U in the form $a + bj$, where a and b are real.

13. Complex Numbers as Vectors

There is another way of writing a complex number, using its real and imaginary parts as the *components* of an *ordered pair:* $x + yj = (x, y)$ if x and y are real. When complex numbers are so written, we have

$$3 + 2j = (3, 2)$$

$$1 = (1, 0)$$

$$j = (0, 1)$$

In this notation, two complex numbers are added by adding their first components and then adding their second components. Thus, the calculation

$$(3 + 2j) + (7 - j) = 10 + j$$

becomes
$$(3, 2) + (7, -1) = (10, 1)$$

and
$$(3 + 2j) - (7 - j) = -4 + 3j$$

becomes
$$(3, 2) - (7, -1) = (-4, 3)$$

The ordered-pair notation emphasizes the idea of a complex number as a point in the plane, the real part being the first coordinate, and the imaginary part the second. The point $(x, y) = x + yj$ may also be thought of as a linear combination of $1 = (1, 0)$ and $j = (0, 1)$. Specifically,

$$x + yj = x(1, 0) + y(0, 1)$$
$$= (x, 0) + (0, y)$$
$$= (x, y)$$

The complex plane therefore consists of all those linear combinations of $1 = (1, 0)$ and $j = (0, 1)$ which have real coefficients. Expressed a little more succinctly, the complex plane is the collection of all real linear combinations of 1 and j. It is called a *two-dimensional real vector space*, and each complex number is called a *vector* in this space. The real coefficients are *scalars*.

If you have previously thought of vectors as being "arrows," then you may want to associate with the complex number (x, y) the arrow with its tail at $(0, 0)$ (the origin), and its head at (x, y). The vector (x, y) has both magnitude and direction. The magnitude is the modulus of (x, y), and the direction is specified by the argument. In other words, magnitude and direction are given directly by the polar-coordinate representation.

This vector space, the space of complex numbers, is denoted $\mathbf{C}$ and is distinguished from the *real* plane $\mathbf{R}^2$ by possessing the operation of complex multiplication:

$$(a, b)(x, y) = (ax - by, bx + ay)$$

which is the ordered-pair equivalent of the operation

$$(a + bj)(x + yj) = (ax - by) + (bx + ay)j$$

14. The Roots of Quadratic Polynomials with Real Coefficients

Polynomials which have no real roots must have complex roots. For instance, the polynomial $z^2 + 1$ has two roots, which are $\pm j$ as we can verify:

$$j^2 + 1 = -1 + 1 = 0 \qquad (-j)^2 + 1 = -1 + 1 = 0$$

For the quadratic polynomials, we can always find the roots using the quadratic formula (which is based on the technique of completing the square). The square roots of -16 are $\pm 4j$, and so the roots of the polynomial

$$5z^2 + 2z + 1 \tag{13}$$

are
$$\frac{-2 \pm j\sqrt{20 - 4}}{10} = \frac{-2 \pm 4j}{10} \tag{14}$$

or
$$\frac{-1 + 2j}{5} \quad \text{and} \quad \frac{-1 - 2j}{5}$$

One may verify that each of these is a root of (13) by substitution. Here is the calculation for the first root:

$$5\left(\frac{-1 + 2j}{5}\right)^2 + 2\frac{-1 + 2j}{5} + 1 = 5\left(-\tfrac{3}{25} - \tfrac{4}{25}j\right) - \tfrac{2}{5} + \tfrac{4}{5}j + 1 = 0$$

In general, the polynomial

$$Az^2 + Bz + C$$

will have two roots if $A \neq 0$. (If $A = 0$, it will have at most one root.) We shall consider the case in which A, B, and C are real, for it is this case which arises in the partial-fraction decompositions which occur when treating the differential equations of interest to us. For the same reason, we may assume that $A \neq 0$. (See Part 1 of Section C of this chapter.)

If $B^2 - 4AC > 0$, then the roots are real and distinct:

$$z = \frac{-B \pm \sqrt{B^2 - 4AC}}{2A} \qquad \text{if } B^2 - 4AC > 0$$

If $B^2 - 4AC = 0$, then the roots are real and equal:

$$z = \frac{-B}{2A} \quad \text{and} \quad \frac{-B}{2A} \qquad \text{if } B^2 - 4AC = 0$$

If $B^2 - 4AC < 0$, then the radicand is negative, its square roots are pure imaginary, i.e., real multiplies of j, and the roots of the polynomial are distinct complex conjugates of each other:

$$z = \frac{-B \pm j\sqrt{4AC - B^2}}{2A} \qquad \text{if } B^2 - 4AC < 0 \tag{15}$$

[Note that the roots of (13), as calculated in (14), are exactly those given by (15).]

If the roots of the polynomial $Az^2 + Bz + C$ are α_1 and α_2, then the polynomial may be factored as

$$Az^2 + Bz + C = A(z - \alpha_1)(z - \alpha_2)$$

15. The Partial-Fraction Decomposition with Complex Fractions

The partial-fraction decomposition can be carried a step further when the denominator is factored into linear factors with complex roots, if necessary. Thus,

$$\frac{1}{s^2 + 4} = \frac{1}{(s + 2j)(s - 2j)}$$

$$= \frac{A}{s + 2j} + \frac{B}{s - 2j}$$

$$= \frac{(A + B)s + (-2A + 2B)j}{s^2 + 4}$$

The equations determining A and B are these:

$$\text{Constant term:} \quad (-2A + 2B)j = 1 \quad \text{or} \quad A - B = \frac{j}{2} \bigg\}$$

$$\text{s term:} \qquad\qquad\qquad\qquad\qquad\qquad A + B = 0$$

The solution of the pair of equations is $A = j/4$, $B = -j/4$, and so

$$\frac{1}{s^2 + 4} = \frac{j}{4}\left(\frac{1}{s + 2j} - \frac{1}{s - 2j}\right) \tag{16}$$

16. The Laplace Transform of the Complex Exponential

The complex exponential e^{at} has a Laplace transform

$$\mathscr{L}[e^{at}](s) = \frac{1}{s - a} \qquad s > \text{Re } a$$

because the calculation in Section A by which $\mathscr{L}[e^{at}]$ is computed can be shown to be valid if $s > \text{Re } a$. Specifically,

$$\lim_{T \to \infty} \left(\frac{e^{(a-s)T}}{a - s} - \frac{1}{a - s}\right) = \lim_{T \to \infty} \left(\frac{e^{(\text{Re } a - s)T} e^{j(\text{Im } a)T}}{a - s} - \frac{1}{a - s}\right)$$

$$= \frac{1}{s - a} \qquad s > \text{Re } a$$

for the modulus of the factor $e^{j(\text{Im } a)T}$ is 1, while the other factor, $e^{(\text{Re } a - s)T}$, goes to zero as T increases, because $\text{Re } a - s < 0$.

Coupling the Laplace transform of the exponential with (16), the partial-fraction decomposition of $1/(s^2 + 4)$, gives this:

$$\mathscr{L}^{-1}\left[\frac{1}{s^2 + 4}\right](t) = \mathscr{L}^{-1}\left[\frac{j}{4}\left(\frac{1}{s + 2j} - \frac{1}{s - 2j}\right)\right](t) \tag{17}$$

$$= \frac{j}{4}(e^{-2jt} - e^{2jt}) \qquad t \geq 0 \tag{18}$$

It remains to see that (18), a curious-looking linear combination of complex exponentials, is in fact the familiar real function which we expected. To see this, we merely use the definition of the complex exponentials as follows:

$$\frac{j}{4}(e^{-2jt} - e^{2jt}) = \frac{j}{4}[\cos(-2t) + j\sin(-2t) - \cos 2t - j\sin 2t]$$

$$= \frac{j}{4}(-2j\sin 2t)$$

$$= \tfrac{1}{2}\sin 2t \tag{19}$$

Of course, $(\sin 2t)/2$ is indeed what we expected, the inverse Laplace transform of $1/(s^2 + 4)$. The surprise is that it can be calculated using the complex exponential. The relation between (18) and (19) has parallels involving the cosine, hyperbolic sine, and hyperbolic cosine, which you can work out for yourself. The most important ones we now give.

17. The Elementary Functions

The linear combinations of the complex exponential include the following familiar functions:

$$\frac{e^t + e^{-t}}{2} \equiv \cosh t \tag{20}$$

$$\frac{e^t - e^{-t}}{2} \equiv \sinh t \tag{21}$$

$$\frac{e^{jt} + e^{-jt}}{2} \equiv \tfrac{1}{2}(\cos t + j \sin t + \cos t - j \sin t)$$

$$\equiv \cos t \tag{22}$$

and
$$\frac{e^{jt} - e^{-jt}}{2j} \equiv \frac{1}{2j}(\cos t + j \sin t - \cos t + j \sin t)$$

$$\equiv \sin t \tag{23}$$

Each of these equations is an identity. Here are two more identities which are easy to observe from the previous four:

$$\cosh jt \equiv \frac{e^{jt} + e^{-jt}}{2}$$

$$\equiv \cos t \tag{24}$$

$$\sinh jt \equiv \frac{e^{jt} - e^{-jt}}{2} \equiv j\frac{e^{jt} - e^{-jt}}{2j}$$

$$\equiv j \sin t \tag{25}$$

Each of the functions listed above is a linear combination of exponentials, and each is a solution of a homogeneous linear differential equation with constant coefficients. Among the other such solutions are

$$e^t \sin t = e^t \frac{e^{jt} - e^{-jt}}{2j} = \frac{e^{t(1+j)} - e^{t(1-j)}}{2j}$$

$$e^t \cos t = \frac{e^{t(1+j)} + e^{t(1-j)}}{2}$$

and te^t, $t^2 e^t$, $t \cos t$, $t \sin t$, and so forth. The latter functions are not exponentials or linear combinations of them, but every function listed so far is a linear combination of functions of the form

$$t^k e^{zt} \tag{26}$$

where z is a complex number, and k is a nonnegative integer.

The collection of all linear combinations of all such functions does constitute exactly the set of all homogeneous solutions of linear differential equations with constant coefficients. You can see this fact for yourself if you consider the partial-fraction decomposition of rational functions whose denominator is of degree strictly greater than that of the numerator. We shall consider the matter again, from an entirely different point of view, when we discuss linear differential operators in Chapter IV. The functions in this collection are called the *elementary functions*, and they are the polynomials, the exponentials, their products, and their linear combinations, including the sinusoids.

18. More Problems

6. Write $\sin 10t$ as a linear combination of e^{j10t} and e^{-j10t} by writing

$$\sin 10t = Ae^{j10t} + Be^{-j10t}$$

using the definition of complex exponential, and then finding A and B.

7. Find the inverse Laplace transform of each of the following. Decompose the given rational function into fractions whose denominators are powers of first-degree polynomials; then take the inverse transforms to get linear combinations of functions of the form $t^k e^{zt}$.

(a) $\dfrac{1}{s-j}$

(b) $\dfrac{3}{(s-j)^2}$

(c) $\dfrac{s}{(s-j)^3}$

(d) $\dfrac{2+j}{(s-j)^2}$

(e) $\dfrac{1}{s^2+1}$

(f) $\dfrac{1}{s^2+s+1}$

(g) $\dfrac{1}{(s^2+s+1)^2}$

(h) $\dfrac{s}{(s^2+1)(s^2+s+1)}$

(i) $\dfrac{3+s}{(s-2)(s^2+s+1)}$

J. THE J_ω OPERATOR

One of the important facts which engineers need to determine about engineering systems is their response to oscillatory inputs. An electrical engineer designing an audio amplifier must pay careful attention to the frequency response of the system. He or she must make sure that the response of the amplifier is uniform; that is, that the response of the amplifier to a low-frequency input of given amplitude differs little from the response to a high-frequency input of the same amplitude. A mechanical engineer designing a shock-absorber system for an automobile must make sure that when the automobile is going over a road of prescribed bumpiness, the resulting vibrations transmitted through the shock-absorber system to the passengers are below a specified value.

The analysis of a system's response to oscillatory inputs is called *steady-state analysis*. In this analysis, we determine the sinusoidal part of the system output resulting from a sinusoidal input. Since sinusoidal functions play a central role in steady-state analysis, we shall discuss their properties. Then we shall define the J_ω operator and illustrate its use in determining the steady-state behavior of engineering systems giving rise to linear constant-coefficient differential equations.

1. Sinusoidal Functions

We begin with the sinusoidal function sin t whose graph is illustrated in Figure 1. (For additional discussion, see Section I of Chapter I.) The function sin t has a period of 2π seconds. The *frequency in cycles per second* (hertz) is the reciprocal of the period, so that its frequency is $1/2\pi$ hertz.

The peak amplitude of sin t is 1, since the maximum (over all t) of its absolute value is 1. If P_0 denotes peak amplitude, then

$$P_0 = \max_t |\sin t| = 1$$

(See also Section I of Chapter I.)

Next let us consider the function sin ωt which is illustrated in Figure 2. In the expression sin ωt, the term ωt is called the *argument* of the sine. If x is any function of t, then $x(t)$ is the *argument* of sin $x(t)$. The number ω multiplying t in the argument of the sine function is called the *frequency in radians per second* or *angular velocity*. The function sin t has a frequency of 1 radian/second.

Let T be the period of oscillation, f be the frequency in hertz, and ω be the frequency in radians per second. Then

$$f = \frac{1}{T} \quad \text{and} \quad \omega = 2\pi f$$

The function sin ωt has a period of $2\pi/\omega$ seconds, for when $t = 2\pi/\omega$ the argument of sin ωt is $\omega(2\pi/\omega) = 2\pi$. When t increases by $2\pi/\omega$, then ωt, the argument, increases from 0 to 2π, and the sine function goes through one oscillation. The frequency in radians per second of the function sin ωt is the number ω which multiplies t in the argument of the sine function. The frequency f in cycles per

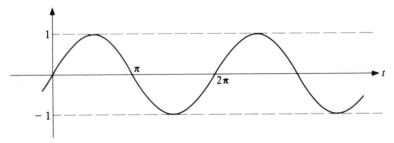

Figure 1 The graph of sin t.

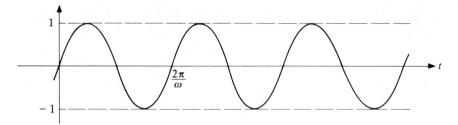

Figure 2 The graph of sin ωt.

second is the reciprocal of the period. Hence, $f = 1/(2\pi/\omega)$ cycles/second. The peak amplitude P_0 of sin ωt is 1, for

$$P_0 = \max_t |\sin \omega t| = 1$$

The graph of cos t is illustrated in Figure 3. Notice that cos t has a period of 2π seconds, a frequency of 1 radian/second, and a peak amplitude of 1, just as the function sin t has. What difference is there between sin t and cos t? If the graph of cos t were translated to the right by $\pi/2$ seconds, it would look like the graph of sin t. Or, equivalently, if the graph of sin t were translated to the left by $\pi/2$ seconds, it would look like cos t. This time difference of $\pi/2$ seconds is called a time *lead* or *lag*, depending on its sign: sin t lags cos t by $\pi/2$ seconds; cos t leads sin t by $\pi/2$ seconds. The time difference multiplied by the frequency in radians per second is called the *phase difference*. It is an angle. The *phase* of cos t *relative to* sin t is $\pi/2$ seconds $\times$ 1 radian/second $= \pi/2$ radians.

What is the phase of sin $(\omega t + \pi/3)$ relative to cos ωt?

$$\sin\left(\omega t + \frac{\pi}{3}\right) = \cos\left(\omega t - \frac{\pi}{2} + \frac{\pi}{3}\right) = \cos\left(\omega t - \frac{\pi}{6}\right)$$

The phase of sin $(\omega t + \pi/3)$ relative to cos ωt is $-\pi/6$ radian. The corresponding lead time is calculated this way:

$$\frac{-\pi/6 \text{ radian}}{\omega \text{ radians/second}} = \frac{-\pi}{6\omega} \text{ seconds}$$

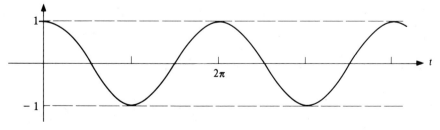

Figure 3 The graph of cos t.

Thus $\sin(\omega t + \pi/3)$ leads $\cos \omega t$ by $-\pi/6\omega$ seconds. Another way of saying this is that $\sin(\omega t + \pi/3)$ lags $\cos \omega t$ by $\pi/6\omega$ seconds.

Any sinusoidal function can be described by three parameters: period, peak amplitude, and phase relative to a sine or cosine. Since period, frequency in cycles per second, and frequency in radians per second are related, the first parameter can be specified by frequency in cycles per second, or frequency in radians per second, instead of period.

In the next part we shall discuss the sinusoidal steady-state response of engineering systems, using the J_ω operator. Note carefully how our study is based on frequency, peak amplitude, and phase.

2. Problems

1. Graph $\sin 3\pi t$.

2. Graph $\cos 3\pi t$.

3. By how much must $\cos 3\pi t$ be translated in order to look like $\sin 3\pi t$? What is the phase of $\sin 3\pi t$ relative to $\cos 3\pi t$?

4. What is the peak amplitude of $5 \cos 7t$?

5. What is the phase of $\sin 6\pi t$ relative to $\cos 6\pi t$?

6. What is the phase of $\cos 7\pi t$ relative to $\sin(7\pi t + \pi/4)$?

7. What is the phase of $\sin \omega t$ relative to $\cos(\omega t + \pi)$?

8. What is the phase of $\sin(\omega t + \phi)$ relative to $\cos \omega t$?

9. What is the phase of $\sin[\omega(t + t_0)]$ relative to $\cos \omega t$?

3. J_ω Operator Definition

We have seen that the behavior of many engineering systems can be described by linear differential equations with constant coefficients. In normal form, these linear differential equations have linear combinations of the unknown function and its derivatives on the left-hand side, and the known forcing function, which is determined from the inputs driving the engineering system, on the right-hand side. In this part we restrict our attention to those situations in which the forcing function is sinusoidal. Usually the general solution to such a differential equation is the sum of a particular sinusoid and the general solution to the corresponding homogeneous equation. The steady-state solution (sought using the J_ω operator) is only the particular-sinusoid part of the general solution.

The J_ω operator is a transformation which transforms sinusoidally forced differential equations to algebraic equations. We obtain the sinusoidal solution to the differential equation by solving the algebraic equation. We then use the inverse J_ω operator to transform our algebraic solution into the sinusoidal function we seek.

The J_ω operator assigns each and every function in its domain to some complex number. To specify the J_ω operator, we must define its domain and give a formula indicating how it converts a function in its domain to a complex number in its range.

The J_ω operator has for its domain the set of all functions having the form $A \cos (\omega t + \phi)$, where A and ϕ are real and A is nonnegative. Corresponding to each frequency ω, there is an operator J_ω. Thus there are many operators: $J_1, J_{2.5}, J_{1000}$ to name a few. The domain of the J_ω operator is the set

$$\{x(t) | x(t) = A \cos (\omega t + \phi) \text{ for some } A \geq 0 \text{ and } \phi \text{ real}\}$$

The range of the J_ω operator is the set of complex numbers. The J_ω operator is defined by

$$J_\omega[A \cos (\omega t + \phi)] = Ae^{j\phi} \tag{1}$$

To each sinusoidal function of the form $A \cos (\omega t + \phi)$, the J_ω operator assigns the complex number $Ae^{j\phi}$. $Ae^{j\phi}$ is called the *phasor* for the sinusoidal signal $A \cos (\omega t + \phi)$; A represents the peak value that the signal can attain, and ϕ represents the phase of the signal. Expressed in terms of real and imaginary parts, $Ae^{j\phi} = A \cos \phi + jA \sin \phi$. Using polar coordinates, we can plot the phasor in the complex plane as in Figure 4.

Let us practice using the J_ω operator.

$$J_\omega[\sin \omega t] = J_\omega \left[\cos \left(\omega t - \frac{\pi}{2} \right) \right] = 1e^{-j\pi/2} = -j$$

$$J_\omega[\cos \omega t] = 1e^{j0} = 1$$

$$J_\omega[-\sin \omega t] = J_\omega \left[\cos \left(\omega t + \frac{\pi}{2} \right) \right] = 1e^{j\pi/2} = j$$

$$J_\omega[A \sin \omega t] = J_\omega \left[A \cos \left(\omega t - \frac{\pi}{2} \right) \right] = Ae^{-j\pi/2} = -jA$$

$$J_\omega[B \cos \omega t] = Be^{j0} = B$$

$$J_2[B \cos 2t] = B$$

$$J_{7.2}[A \sin 7.2t] = -jA$$

$$J_{11.4}[-\sin 11.4t] = j$$

Having seen how the J_ω operator works, try to think of the complex number assigned by the J_6 operator to $A \cos 7t$. If you are thinking that $J_6[A \cos 7t] = A$, then think again. Is the sinusoid $A \cos 7t$ in the domain of J_6? It is not. Thus, J_6 cannot assign any number to $A \cos 7t$. The value of $J_6[A \cos 7t]$ is not defined, for J_6 refuses to work on any sinusoid unless it has a frequency of 6 radians/second.

Because of the strong link between the J_ω operator and the sinusoidal functions, we need to recall four trigonometric identities:

$$e^{j\phi} = \cos \phi + j \sin \phi \tag{2}$$

$$\cos (\alpha + \beta) = \cos \alpha \cos \beta - \sin \alpha \sin \beta \tag{3}$$

$$\sin (\alpha + \beta) = \sin \alpha \cos \beta + \cos \alpha \sin \beta \tag{4}$$

$$\sin^2 \phi + \cos^2 \phi = 1 \tag{5}$$

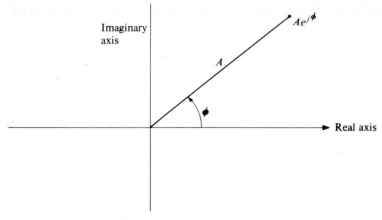

Figure 4 The phasor $Ae^{j\phi}$ for the signal $A \cos (\omega t + \phi)$.

These identities will prove quite useful as we inquire about what the J_ω operator does to linear combinations of sinusoids. The first fact we need to establish to do this inquiring is that linear combinations of sinusoids of a given frequency are sinusoids of the same frequency.

Consider the function $A \cos \omega t + B \sin \omega t$. If it is a sinusoid, it must be a sinusoid of frequency ω radians per second. Since a sinusoid of frequency ω can be represented by $C \cos (\omega t + \phi)$, we would like to know if there exists a positive value for C and a value for ϕ such that

$$C \cos (\omega t + \phi) = A \cos \omega t + B \sin \omega t \tag{6}$$

Using the trigonometric identity for the cosine of the sum of two angles, we obtain

$$C \cos (\omega t + \phi) = C(\cos \omega t \cos \phi - \sin \omega t \sin \phi) \tag{7}$$

Rearranging (7) gives

$$C \cos (\omega t + \phi) = (C \cos \phi) \cos \omega t + (-C \sin \phi) \sin \omega t \tag{8}$$

Identifying the coefficients of $\cos \omega t$ and $\sin \omega t$ of Equation (8) with those of Equation (6), we obtain

$$A = C \cos \phi \tag{9}$$

$$B = -C \sin \phi \tag{10}$$

Squaring and adding give

$$A^2 + B^2 = C^2 \cos \phi + C^2 \sin \phi = C^2(\cos^2 \phi + \sin^2 \phi) = C^2$$

so that
$$C = \sqrt{A^2 + B^2}$$

Using the expressions for C in (9) and (10), we see that ϕ is the angle whose cosine is $A/\sqrt{A^2 + B^2}$ and whose sine is $-B/\sqrt{A^2 + B^2}$:

$$\cos \phi = \frac{A}{\sqrt{A^2 + B^2}} \qquad \sin \phi = \frac{-B}{\sqrt{A^2 + B^2}}$$

When both A and B are positive, ϕ is an angle in the first quadrant. Table 1 shows the dependence of the quadrant of ϕ on the signs of A and B.

The immediate implication of Table 1 is that knowing the value of $A/\sqrt{A^2 + B^2}$ or knowing the value of $B/\sqrt{A^2 + B^2}$ is not enough to define the angle ϕ uniquely. For example, if $A/\sqrt{A^2 + B^2} = 1/\sqrt{2}$ and B is negative, then $\phi = +\pi/4$; but if B is positive, then $\phi = -\pi/4$.

Very often, the phase ϕ is expressed by the Arctangent function. We consider Tan^{-1} to provide principal values in quadrants I and IV. Hence,

$$\phi = \begin{cases} \mathrm{Tan}^{-1}\left(-\dfrac{B}{A}\right) & \text{when } A > 0 \\[2mm] \pi + \mathrm{Tan}^{-1}\left(-\dfrac{B}{A}\right) & \text{when } A < 0 \end{cases}$$

Figures 5 and 6 illustrate the geometry for ϕ. (Compare Part 10 of Section I.)
Now we illustrate that

$$J_\omega[A \cos \omega t + B \sin \omega t] = AJ_\omega[\cos \omega t] + BJ_\omega[\sin \omega t]$$

for the case $A > 0$. If $\phi = \mathrm{Tan}^{-1}(-B/A)$ and $C = \sqrt{A^2 + B^2}$, then

$$J_\omega[A \cos \omega t + B \sin \omega t] = J_\omega[C \cos (\omega t + \phi)]$$

$$= Ce^{j\phi}$$

$$= C(\cos \phi + j \sin \phi)$$

$$= C \cos \phi + jC \sin \phi$$

Table 1 The quadrant for the phase angle ϕ as a function of the signs of A and B for the equation

$C \cos (\omega t + \phi) = A \cos \omega t + B \sin \omega t$

	$A < 0$	$A > 0$
$B < 0$	II	I
$B > 0$	III	IV

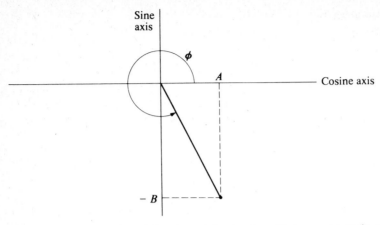

Figure 5 The computation of the phase ϕ for the sinusoid $A \cos \omega t + B \sin \omega t = C \cos (\omega t + \phi)$, where A and B are positive.

Since $A > 0$, the cosine of the angle whose tangent is $-B/A$ is $A/\sqrt{A^2 + B^2}$, and the sine of the angle whose tangent is $-B/A$ is $-B/\sqrt{A^2 + B^2}$. Hence,

$$J_\omega[A \cos \omega t + B \sin \omega t] = C \frac{A}{\sqrt{A^2 + B^2}} + jC \frac{-B}{\sqrt{A^2 + B^2}}$$

Since $C = \sqrt{A^2 + B^2}$,

$$J_\omega[A \cos \omega t + B \sin \omega t] = A - jB \tag{11}$$
$$= AJ_\omega[\cos \omega t] + BJ_\omega[\sin \omega t]$$

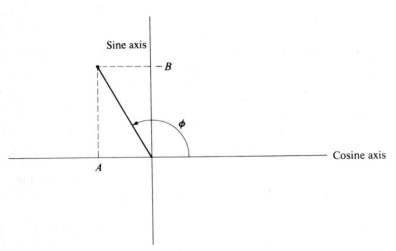

Figure 6 The computation of the phase ϕ for the sinusoid $A \cos \omega t + B \sin \omega t = C \cos (\omega t + \phi)$, where A and B are both negative.

More generally, it is true that

$$J_\omega[A_1 \cos (\omega t + \phi_1) + A_2 \cos (\omega t + \phi_2)]$$
$$= A_1 J_\omega[\cos (\omega t + \phi_1)] + A_2 J_\omega[\cos (\omega t + \phi_2)] \qquad (12)$$

For any two sinusoids of the same frequency ω, the action of J_ω on a linear combination is the same as the linear combination of the action of J_ω on the individual sinusoids. The J_ω operator is, in fact, a linear operator, and Equation (12) is exactly the property of linearity.

4. Finding Steady-State Solutions

We can use the J_ω operator to help us determine the steady-state solution to a nonhomogeneous linear differential equation with constant coefficients when the forcing function for the equation is a sinusoidal function $C \cos (\omega t + \phi)$. To use the J_ω operator on differential equations, we must determine what the operator does to derivatives of sinusoidal functions. Let us consider $J_\omega[(d/dt)[A \cos (\omega t + \phi)]]$.

$$J_\omega \left[\frac{d}{dt} A \cos (\omega t + \phi) \right] = J_\omega[-\omega A \sin (\omega t + \phi)]$$

$$= -\omega A J_\omega \left[\cos \left(\omega t + \phi - \frac{\pi}{2} \right) \right]$$

$$= -\omega A e^{j(\phi - \pi/2)} = -\omega A e^{j\phi} e^{-j\pi/2}$$

$$= -\omega A e^{j\phi}(-j)$$

$$= j\omega A e^{j\phi} = j\omega A J_\omega[\cos (\omega t + \phi)]$$

$$= j\omega\{J_\omega[A \cos (\omega t + \phi)]\} \qquad (13)$$

Thus, the J_ω operator acting on the derivative of a sinusoidal function yields $j\omega$ times the action of the operator on the sinusoidal function itself.

Consider the equation

$$\frac{dx}{dt}(t) + ax(t) = b \cos \omega t \qquad a > 0 \qquad (14)$$

We wish to know whether there is a function x which not only is a solution of this equation, but also is a sinusoid of frequency ω, the frequency of the nonhomogeneous term. If there is, then every term of Equation (14) becomes a sinusoid of the same frequency, and we may apply the J_ω operator to each term or to any sum of terms.

Let $X(\omega) = J_\omega[x(t)]$. Apply the J_ω operator to both sides of the equation.

$$J_\omega \left[\frac{dx}{dt}(t) + ax(t) \right] = J_\omega[b \cos \omega t]$$

We already know that J_ω is a linear operator [from Equation (12)]. Hence,

$$J_\omega \left[\frac{dx}{dt}(t) + ax(t) \right] = J_\omega \left[\frac{dx}{dt}(t) \right] + aJ_\omega[x(t)]$$

From Equation (13) we also know that for any function in its domain,

$$J_\omega \left[\frac{dx}{dt}(t) \right] = j\omega J_\omega[x(t)]$$

Thus, the equation $(dx/dt)(t) + ax(t) = \cos \omega t$ is transformed by the J_ω operator into a new equation,

$$j\omega X(\omega) + aX(\omega) = b \tag{15}$$

which is an algebraic equation, not a differential equation.

Solving this equation for $X(\omega)$ and recalling that $a > 0$, we obtain

$$X(\omega) = \frac{b}{a + \omega j} = \frac{b}{\sqrt{a^2 + \omega^2}\, e^{j\,\mathrm{Tan}^{-1}(\omega/a)}} = \frac{b}{\sqrt{a^2 + \omega^2}}\, e^{-j\,\mathrm{Tan}^{-1}(\omega/a)}$$

We now recognize $(b/\sqrt{a^2 + \omega^2})e^{-j\,\mathrm{Tan}^{-1}(\omega/a)}$ as the J_ω transform of a sinusoid having amplitude $b/\sqrt{a^2 + \omega^2}$ and phase $-\mathrm{Tan}^{-1}(\omega/a)$. That is, the inverse transform gives

$$x(t) = J_\omega^{-1}[X(\omega)]$$

$$= J_\omega^{-1} \left[\frac{b}{\sqrt{a^2 + \omega^2}}\, e^{-j\,\mathrm{Tan}^{-1}(\omega/a)} \right]$$

$$= \frac{b}{\sqrt{a^2 + \omega^2}} \cos \left(\omega t - \mathrm{Tan}^{-1} \frac{\omega}{a} \right) \tag{16}$$

This function is a sinusoid of frequency ω, and our analysis shows that it satisfies the given equation (14). Since (15), the transformed equation, has only one solution, the steady-state solution x must be unique. There are, however, other ways of writing it.

For instance, we could simplify $b/(a + j\omega)$ into its real and imaginary parts:

$$X(\omega) = \frac{b}{a + j\omega} = \frac{b}{a + j\omega} \frac{a - j\omega}{a - j\omega} = \frac{b(a - j\omega)}{a^2 + \omega^2} = \frac{ab}{a^2 + \omega^2} - j\frac{b\omega}{a^2 + \omega^2}$$

Recognizing $ab/(a^2 + \omega^2)$ as the J_ω transform of a cosine function having amplitude $ab/(a^2 + \omega^2)$, and $-j[b\omega/(a^2 + \omega^2)]$ as the J_ω transform of a sine function having amplitude $b\omega/(a^2 + \omega^2)$, we have

$$x(t) = bJ_\omega^{-1} \left[\frac{a}{a^2 + \omega^2} \right] + bJ_\omega^{-1} \left[-j\frac{\omega}{a^2 + \omega^2} \right]$$

$$= b \left(\frac{a}{a^2 + \omega^2} \cos \omega t + \frac{\omega}{a^2 + \omega^2} \sin \omega t \right) \tag{17}$$

Using trigonometric identities, it is easily shown that the two expressions (16) and (17) for $x(t)$ define in fact the same function.

5. Sample Problems with Answers

Find the sinusoidal steady-state solution to:

A. $\dfrac{dx}{dt}(t) + 4x(t) = 7 \cos 3t$ *Answer:* $x(t) = \frac{7}{5} \cos (3t - \text{Tan}^{-1} \frac{3}{4})$

B. $\dfrac{dx}{dt}(t) + 5x(t) = 3 \sin 12t$

      Answer: $x(t) = \frac{3}{13} \cos (12t + \phi)$, where $\cos \phi = -\frac{12}{13}$ and $\sin \phi = -\frac{5}{13}$

C. $\dfrac{dx}{dt}(t) + 7x(t) = 4 \cos (11t + \pi/3)$ *Answer:* $x(t) = \dfrac{4}{\sqrt{170}} \cos \left(11t + \dfrac{\pi}{3} - \text{Tan}^{-1} \frac{11}{7}\right)$

D. $\dfrac{dx}{dt}(t) + 10x(t) = 3 \cos \omega t$ *Answer:* $x(t) = \dfrac{30}{100 + \omega^2} \cos \omega t + \dfrac{3\omega}{100 + \omega^2} \sin \omega t$

E. $\dfrac{dx}{dt}(t) + 13x(t) = 5 \cos (\omega t + \phi)$

      Answer: $x(t) = \dfrac{65 \cos \phi + 5\omega \sin \phi}{169 + \omega^2} \cos \omega t + \dfrac{5\omega \cos \phi - 65 \sin \phi}{169 + \omega^2} \sin \omega t$

F. $\dfrac{d^2x}{dt^2}(t) + 2x(t) = 5 \cos 4t$ *Answer:* $x(t) = -\frac{5}{14} \cos 4t = \frac{5}{14} \cos (4t - \pi)$

G. $\dfrac{d^2x}{dt^2}(t) + 11x(t) = 7 \cos (\omega t + \phi)$ *Answer:* $x(t) = \dfrac{7}{11 - \omega^2} \cos (\omega t + \phi)$

H. $\dfrac{d^2x}{dt^2}(t) + 4\dfrac{dx}{dt}(t) + 3x(t) = \cos 5t$ *Answer:* $x(t) = \dfrac{1}{\sqrt{884}} \cos (5t - \pi + \text{Tan}^{-1} \frac{10}{11})$

I. $\dfrac{d^2x}{dt^2}(t) + a\dfrac{dx}{dt}(t) + bx(t) = c \cos \omega t$, where $b > \omega^2$

      Answer: $x(t) = \dfrac{c}{\sqrt{(b - \omega^2)^2 + (a\omega)^2}} \cos \left(\omega t - \text{Tan}^{-1} \dfrac{a\omega}{b - \omega^2}\right)$

J. $\dfrac{d^2x}{dt^2}(t) + a\dfrac{dx}{dt}(t) + bx(t) = c \cos \omega t$, where $b < \omega^2$

      Answer: $x(t) = \dfrac{c}{\sqrt{(b - \omega^2)^2 + (a\omega)^2}} \cos \left(\omega t - \pi - \text{Tan}^{-1} \dfrac{a\omega}{b - \omega^2}\right)$

6. Bode Plots

One of the useful ways to employ steady-state analysis is to consider what happens to the peak amplitude of the oscillations when the driving sinusoid keeps constant peak amplitude and the frequency ω increases. For example, if the behavior of a voltage in a system is described by

$$-\frac{d^2v}{dt^2}(t) + 11v(t) = \cos \omega t$$

where $\cos \omega t$ is the driving sinusoid, then the steady-state $v(t)$ is given by

$$v(t) = \frac{1}{11 + \omega^2}\cos \omega t$$

Notice how the peak amplitude of v changes with frequency. That is, as ω increases, the peak amplitude decreases. The graph of the peak amplitude of $v(t)$ as a function of ω is given in Figure 7.

In real engineering problems, the frequency ω can vary from zero to tens of thousands of radians per second. Thus, both the peak amplitude and ω can vary over large relative ranges. To allow graphing of these situations, it has become traditional to have both axes on log scales. The scaled peak amplitude is plotted in decibels (db), and its relationship to the peak amplitude $p(\omega)$ is defined by

$$db(\omega) = 20 \log p(\omega)$$

where log means the common logarithm, to the base 10.

Let us see how the db plot of peak amplitude is made for the function $1/(a^2 + \omega^2)$.

$$db(\omega) = 20 \log \frac{1}{a^2 + \omega^2}$$

When ω is very much smaller than a (written $\omega \ll a$), we have

$$\frac{1}{a^2 + \omega^2} \cong \frac{1}{a^2}$$

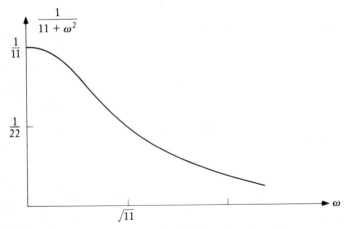

Figure 7 Graph of peak amplitude of $v(t)$ as a function of ω.

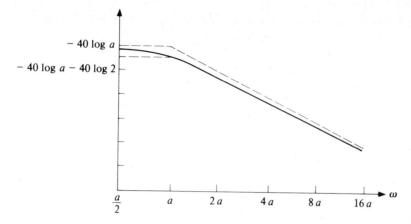

Figure 8 Graph of $20 \log [1/(a^2 + \omega^2)]$ versus $\log \omega$.

so that
$$db(\omega) \cong 20 \log \frac{1}{a^2} = -40 \log a$$

When $\omega = a$, we have
$$\frac{1}{a^2 + \omega^2} = \frac{1}{2a^2}$$

so that
$$db(\omega) = 20 \log \frac{1}{2a^2} = -40 \log a - 20 \log 2$$

Consider the relationship between $db(2\omega)$ and $db(\omega)$ for $\omega \gg a$.

$$db(2\omega) = 20 \log \frac{1}{a^2 + (2\omega)^2} \cong 20 \log \frac{1}{(2\omega)^2}$$

$$= 20 \log \frac{1}{\omega^2} + 20 \log \tfrac{1}{4}$$

$$\cong db(\omega) - 40 \log 2$$

Hence, for small ω we expect the graph to be rather flat, having a value just under $-40 \log a$. At $\omega = a$, we expect it to be $20 \log 2$ below $-40 \log a$, that is, approximately 6 decibels below $-40 \log a$. For large ω we expect the graph to decrease by $-40 \log 2$, or approximately 12 decibels, each time ω doubles. Figure 8 is a graph of the log-log plot of $1/(a^2 + \omega^2)$.

7. Examples

a. Steady-state RC problem Find the sinusoidal steady-state output voltage $v_o(t)$ produced by $v(t) = \cos \omega t$ in the system in Figure 9.

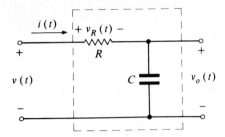

Figure 9 Simple RC circuit.

Using the electrical models, we have

$$\left. \begin{aligned} v(t) &= v_R(t) + v_o(t) \\ v_R(t) &= i(t)R \\ i(t) &= C\frac{dv_o}{dt}(t) \end{aligned} \right\}$$

Eliminating $v_R(t)$ and $i(t)$ gives the differential equation for $v_o(t)$:

$$v(t) = RC\frac{dv_o}{dt}(t) + v_o(t) \tag{18}$$

As before, we shall let the capital letter $V_o(\omega)$ denote $J_\omega[v_o]$. Take $v(t) = \cos \omega t$, and apply the J_ω operator to both sides of Equation (18):

$$1 = j\omega RCV_o(\omega) + V_o(\omega)$$

$$V_o(\omega) = \frac{1}{1 + j\omega RC} = \frac{1}{\sqrt{1 + (\omega RC)^2}\, e^{j\phi}} \qquad \text{where } \phi = \mathrm{Tan}^{-1}\,\omega RC \tag{19}$$

Apply the inverse J_ω operator to both sides of Equation (19):

$$v_o(t) = \frac{1}{\sqrt{1 + (\omega RC)^2}}\cos(\omega t - \phi) \qquad \text{where } \phi = \mathrm{Tan}^{-1}\,\omega RC$$

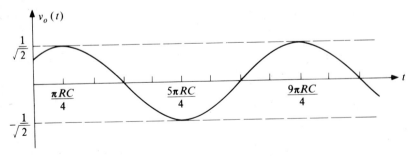

Figure 10 Graph of the sinusoid $1/\sqrt{2}\,\cos(t/RC - \pi/4)$.

The frequency of $v_o(t)$ is ω radians/second or $\omega/2\pi$ cycles/second. The period of $v_o(t)$ is $2\pi/\omega$ seconds. The peak amplitude of $v_o(t)$ is $1/\sqrt{1 + (\omega RC)^2}$. The phase of $v_o(t)$ is $-\mathrm{Tan}^{-1}\,\omega RC$, relative to $v(t)$.

When $\omega = 1/RC$, a sketch of $v_o(t)$ is easily made. (Figure 10; see Section I of Chapter I.) When $\omega \ll 1/RC$, the peak amplitude of $v_o(t)$ is approximately $1/\sqrt{1 + 0^2} = 1$. When $\omega \gg 1/RC$, the peak amplitude of $v_o(t)$ is approximately $1/\sqrt{0 + (\omega RC)^2} = 1/\omega RC$. (See Figure 11.)

b. Steady-state MBK problem Find the sinusoidal steady-state velocity $v_1(t)$ if $v(t) = \cos \omega t$ in the system illustrated in Figure 12.

Using the mechanical models, we obtain

$$f(t) = M\frac{dv_1}{dt}(t)$$

$$-f(t) + f_s(t) + f_d(t) = 0$$

$$v(t) - v_1(t) = \frac{1}{K}\frac{df_s}{dt}(t)$$

$$f_d(t) = B[v(t) - v_1(t)]$$

Eliminating the variables f, f_d, and f_s gives

$$M\frac{d^2 v_1}{dt^2}(t) + B\frac{dv_1}{dt}(t) + Kv_1(t) = Kv(t) + B\frac{dv}{dt}(t) \qquad (20)$$

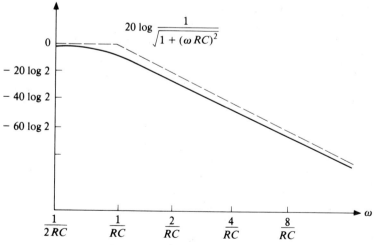

Figure 11 Graph of $20 \log 1/\sqrt{1 + (\omega RC)^2}$ versus $\log \omega$.

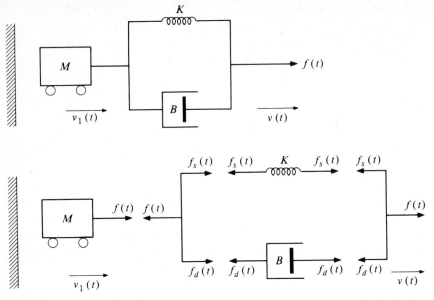

Figure 12 Spring-mass-dashpot mechanical system and the corresponding free-body diagrams of its elements.

Let $v(t) = \cos \omega t$, and apply the $J\omega$ operator to (20).

$$(j\omega)^2 M V_1(\omega) + j\omega B V_1(\omega) + K V_1(\omega) = K + j\omega B$$

$$V_1(\omega) = \frac{K + j\omega B}{-\omega^2 M + K + j\omega B} = \frac{1 + j(\omega B/K)}{1 - \omega^2(M/K) + j\omega(B/K)}$$

$$V_1(\omega) = \frac{[1 + j\omega(B/K)][1 - (M/K)\omega^2 - j\omega(B/K)]}{[1 - (M/K)\omega^2]^2 + [\omega(B/K)]^2}$$

$$V_1(\omega) = \frac{[1 - (M/K)\omega^2] + [\omega(B/K)]^2 + j\{\omega(B/K)[1 - (M/K)\omega^2] - \omega(B/K)\}}{[1 - (M/K)\omega^2]^2 + [\omega(B/K)]^2}$$

Applying the inverse J_ω operator, we obtain the steady-state solution

$$v_1(t)$$
$$= \frac{[1 - (M/K)\omega^2] + [\omega(B/K)]^2}{[1 - (M/K)\omega^2]^2 + [\omega(B/K)]^2} \cos \omega t + \frac{\omega(B/K)(M/K)\omega^2}{[1 - (M/K)\omega^2]^2 + [\omega(B/K)]^2} \sin \omega t$$

$$(21)$$

Alternatively, writing $V_1(\omega)$ in polar coordinates gives

$$V_1(\omega) = \frac{\sqrt{1 + [\omega(B/K)]^2}\; e^{j\phi_1}}{\sqrt{[1 - (M/K)\omega^2]^2 + [\omega(B/K)]^2}\; e^{j\phi_2}}$$

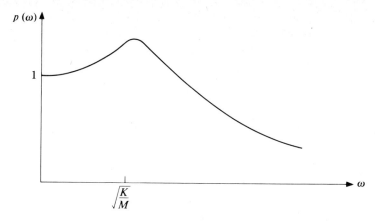

Figure 13 Peak velocity versus ω for Example b. As the frequency increases, the peak value of the velocity of the oscillation first increases to a maximum near $\sqrt{K/M}$ and then quickly decreases.

where

$$\phi_1 = \text{Tan}^{-1}\,\omega\frac{B}{K}$$

and

$$\phi_2 = \begin{cases} \text{Tan}^{-1}\dfrac{\omega(B/K)}{1-(M/K)\omega^2} & \text{if } \dfrac{M}{K}\omega^2 < 1 \\[3mm] \pi + \text{Tan}^{-1}\dfrac{\omega(B/K)}{1-(M/K)\omega^2} & \text{if } \dfrac{M}{K}\omega^2 > 1 \end{cases}$$

Thus,

$$v_1(t) = \sqrt{\frac{1+[\omega(B/K)]^2}{[1-(M/K)\omega^2]^2+[\omega(B/K)]^2}}\,\cos\,(\omega t + \phi_1 - \phi_2) \qquad (22)$$

Equations (21) and (22) give the same function, but (21) is written as a sum, while (22) is a phased cosine.

The radical in (22), the coefficient of the cosine, is called the *peak value of* $v_1(t)$ and designated $p(\omega)$.

c. MK steady state Given that the force $f(t) = A\cos(\omega t + \phi)$, determine the steady-state displacement $x(t)$ for the system shown in Figure 14. Assume that $x(t) = 0$ is the equilibrium position of the spring.

We have

$$\left.\begin{aligned} g(t) &= Kx(t) \\ f(t) - g(t) &= M\frac{d^2x}{dt^2}(t) \end{aligned}\right\}$$

Eliminating $g(t)$ gives

$$f(t) = M\frac{d^2x}{dt^2}(t) + Kx(t) \qquad (23)$$

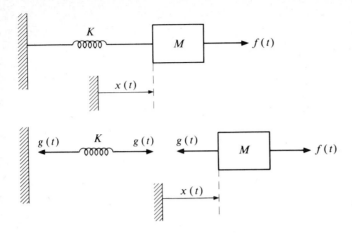

Figure 14 The mass-spring system and its free-body diagram.

Applying the J_ω operator to both sides of Equation (23), we obtain the trans-formed equation:

$$F(\omega) = (K - \omega^2 M)X(\omega)$$

Hence,
$$X(\omega) = \frac{1}{K - \omega^2 M} F(\omega)$$

If $f(t) = A(\cos \omega t + \phi)$, then $F(\omega) = Ae^{j\phi}$. Thus,

$$X(\omega) = \frac{1}{K - \omega^2 M} Ae^{j\phi}$$

Now apply the inverse J_ω operator, and obtain

$$x(t) = \frac{A}{K^2 - \omega^2 M} \cos (\omega t + \phi)$$

d. RLC steady state Given that $i(t) = \sin \omega t$, determine the steady-state voltage $v(t)$ for the system shown in Figure 15.
The element equations are

$$\left. \begin{array}{l} i_C(t) = C\dfrac{dv}{dt}(t) \\[2mm] i_R(t) = \dfrac{1}{R}v(t) \\[2mm] v(t) = L\dfrac{di_L}{dt} \end{array} \right\}$$

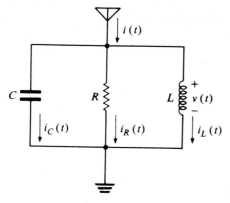

Figure 15 *RLC* parallel tuned circuit.

and the equation for either the upper or lower node is

$$i(t) = i_C(t) + i_R(t) + i_L(t)$$

Eliminating i_C, i_R, and i_L gives the equation

$$\frac{di}{dt}(t) = C\frac{d^2v}{dt^2}(t) + \frac{1}{R}\frac{dv}{dt}(t) + \frac{v}{L}(t)$$

Applying the J_ω operator to both sides of the equation gives

$$j\omega I(\omega) = (j\omega)^2 CV(\omega) + \frac{j\omega}{R}V(\omega) + \frac{V(\omega)}{L}$$

$$V(\omega) = \frac{j\omega I(\omega)}{1/L - \omega^2 C + j(\omega/R)} = \frac{j\omega L I(\omega)}{(1 - \omega^2 LC) + j\omega(L/R)}$$

Since $i(t) = \sin \omega t$, we obtain $I(\omega) = -j$ and

$$V(\omega) = \frac{\omega L}{\sqrt{(1 - \omega^2 LC)^2 + [\omega(L/R)]^2}\,e^{j\phi}}$$

where

$$\phi = \begin{cases} \text{Tan}^{-1}\dfrac{\omega L/R}{1 - \omega^2 LC} & \text{if } \omega^2 LC < 1 \\[3ex] \pi + \text{Tan}^{-1}\dfrac{\omega L/R}{1 - \omega^2 LC} & \text{if } \omega^2 LC > 1 \end{cases}$$

Applying the inverse J_ω operator yields

$$v(t) = \frac{\omega L}{\sqrt{(1 - \omega^2 LC)^2 + [\omega(L/R)]^2}}\cos(\omega t - \phi)$$

When $\omega(L/R)$ tends to be small, the peak voltage occurs when

$$\omega^2 LC = 1 \quad \text{or} \quad \omega = \sqrt{\frac{1}{LC}} \quad \text{or} \quad f = \frac{1}{2\pi\sqrt{LC}} \quad \text{cycles per second}$$

This frequency $1/2\pi\sqrt{LC}$ is called the *resonant frequency*, and the tuned circuit of Figure 15 forms the basis for radio and television tuners that select the station of interest.

8. Problems

10. Consider the series RLC tuned circuit of Figure 16. Suppose that $v(t) = \sin \omega t$. Determine the steady-state current $i(t)$ as a function of ω.

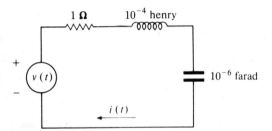

Figure 16 Simple series RLC tuned circuit.

11. Consider the RL circuit of Figure 17. Suppose that $v(t) = \cos \omega t$. Determine the steady-state current $i(t)$ as a function of ω.

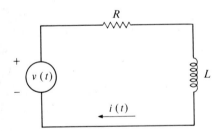

Figure 17 Simple RL series circuit.

12. Consider the RC filter circuit of Figure 18. Suppose that the input voltage is $v_i(t) = \sin \omega t$. Determine the output voltage $v_o(t)$.

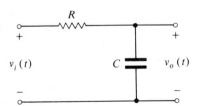

Figure 18 Low-pass-filter circuit.

13. Consider the RC filter circuit of Figure 19. Suppose that the input voltage is $v_i(t) = \cos \omega t$. Determine the output voltage $v_o(t)$.

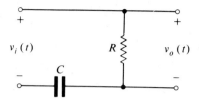

Figure 19 High-pass RC filter circuit.

14. Consider the mechanical system of Figure 20. Suppose that the driving force is $f(t) = 4 \cos \omega t$. Determine the steady-state velocity $v(t)$.

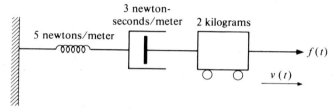

Figure 20 MKB mechanical system.

15. Consider the mechanical system of Figure 21. Suppose that the displacement is $x(t) = 4 \cos \omega t$ and that the tension in the spring is zero when $x(t) - x_1(t) = 0$. Determine the steady-state force $f(t)$.

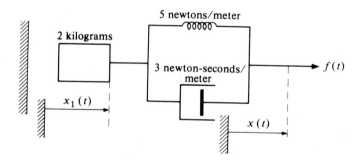

Figure 21 MKB mechanical system.

16. Consider the mechanical system of Figure 22. Determine the steady-state force $f(t)$ when $v(t) = 2 \cos \omega t$.

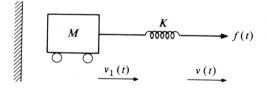

Figure 22 MK mechanical system.

17. Consider the mechanical system of Figure 23. Suppose that the tension in the spring is zero when $x(t) = 0$. Determine the steady-state displacement $x(t)$ when $f(t) = 2 \cos \omega t$.

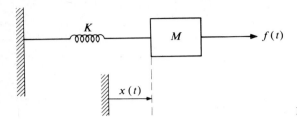

Figure 23 MK mechanical system.

LINEARITY EXPLAINED AND USED

The differential equations we have studied were said to be "linear." (See Section H of Chapter I.) Laplace transformation was shown to have a "linearity property," namely

$$\mathscr{L}[c_1 f_1 + c_2 f_2] = c_1 \mathscr{L}[f_1] + c_2 \mathscr{L}[f_2]$$

if c_1 and c_2 are constants. (See Section B of Chapter III.) The J_ω operator was shown to be "linear,"

$$J_\omega[A_1 \cos (\omega t + \phi_1) + A_2 \cos (\omega t + \phi_2)]$$
$$= A_1 J_\omega[\cos (\omega t + \phi_1)] + A_2 J_\omega[\cos (\omega t + \phi_2)]$$

(See Section J of Chapter III.) In many places the term "linear combination" was used to describe an expression like $af + bg$, where a and b are constants.

This chapter concerns linear operators, which play two different but related roles in our study. First, they can provide solution methods, as the Laplace transformation and the J_ω operator do, and they can provide formulas for solutions, as convolution did in Section G of Chapter III. Second, they can provide a convenient expression of general information about the collection of solutions (or general solution) of an equation. This information may not at once give us some particular solution we desire, but it so narrows the possibilities that very efficient solution methods may be used. Thus, the operators are used for developing (and critically scrutinizing) solution methods for differential-equation problems.

In Section A, the concept of linear operator is introduced, and many familiar things are shown to be linear operators. The most important feature of the section is the definition and explanation of the linear differential operators with constant coefficients. An application clarifies the elimination of unwanted variables as presented in Chapter II.

Thereafter, the chapter is divided into two subchapters. The first concerns two important solution methods for linear differential equations with constant coefficients. In explaining the methods, we also discuss the composition of operators.

The second subchapter concerns linear differential equations with nonconstant coefficients. The "equidimensional" equations are solved, and the methods of "variation of parameters" and "reduction of order" are explained. These are important solution methods. "Linear independence" is introduced to help describe the solutions.

A. LINEAR OPERATORS, FAMILIAR AND UNFAMILIAR

Look back for a moment at the three methods we have used for solving differential equations, and apply each of them to the equation

$$y' + 2y = f \tag{1}$$

where f is an unspecified continuous function, and y is unknown.

In Chapter I we developed the technique of separation of variables. In Section F we applied an extension of it to initial value problems such as

$$\left. \begin{array}{l} y' + 2y = f \\ y(0) = 0 \end{array} \right\} \tag{2}$$

There we derived a formula which tells us that

$$y(t) = e^{-2t} \int_0^t e^{2z} f(z) \, dz \tag{3}$$

The formula, which is actually a convolution, gives the solution y corresponding to each choice of the nonhomogeneous term, the function f.

In Chapter III we developed the technique of Laplace transformation. We apply it to (2) by writing the transformed problem

$$s\mathscr{L}[y] + 2\mathscr{L}[y] = \mathscr{L}[f] \tag{4}$$

solving that for the transformed solution

$$\mathscr{L}[y](s) = \mathscr{L}[f](s)\frac{1}{s+2} = \mathscr{L}[f](s)\mathscr{L}[e^{-2t}](s) \tag{5}$$

and then taking the inverse transform, by convolving f and e^{-2t}:

$$y(t) = f * e^{-2t} = \int_0^t f(z)e^{-2(t-z)} \, dz \tag{6}$$

In Chapter III we also developed the technique of using the J_ω operator. It can be applied to (1) if f is known to be a sinusoid of angular frequency ω, giving a transformed problem

$$j\omega J_\omega[y] + 2J_\omega[y] = J_\omega[f] \tag{7}$$

The solution of the transformed problem is

$$J_\omega[y] = \frac{J_\omega[f]}{2 + j\omega} = \frac{1}{\sqrt{\omega^2 + 4}} e^{-j\,\mathrm{Tan}^{-1}\,(\omega/2)} J_\omega[f] \tag{8}$$

Taking the inverse transform gives the steady-state solution of (1):

$$y(t) = \frac{1}{\sqrt{\omega^2 + 4}} f\left(t - \frac{1}{\omega}\,\mathrm{Tan}^{-1}\,\frac{\omega}{2}\right) \tag{9}$$

In this short review we have mentioned three different useful things, namely, the convolution formula, the Laplace transformation, and the J_ω operator. The Laplace-transform and J_ω-operator examples will be examined for the presence of "operators" in Part 1 below; the convolution formula will be the basis of Part 2. In Part 3 we shall make a provisional description of what a linear operator is, on the basis of these examples and some others. There the terminology will be introduced. In Parts 7 through 9 we define the linear differential operators with constant coefficients; in Parts 12 and 13 we apply them to the elimination of unwanted variables; and in Part 17 we show how the properties of linear operators provide information about the solutions of differential-equation problems.

1. The Familiar Linear Operators $\mathscr{L}$ and J_ω

The Laplace transform of a function f is defined by the formula

$$\mathscr{L}[f](s) = \int_0^\infty f(t)e^{-st}\, dt \tag{10}$$

We have seen that Laplace transformation replaces a whole initial value problem such as (2) with a new problem such as (4). In this replacement, every function in the original problem, whether known, unspecified, or unknown, is replaced with its transform.

The Laplace transformation is a linear operator. As is the case with nearly all our examples, the operator replaces each function with a different one. In this case f is replaced with its Laplace transform $\mathscr{L}[f]$. The symbol $\mathscr{L}$ is the name of the operator, $\mathscr{L}[f]$ is the transform of f, and $\mathscr{L}[f](s)$ is the value of $\mathscr{L}[f]$ at s. For instance,

$$\mathscr{L}[\cos 6t](s) = \frac{s}{s^2 + 36}$$

and for $s = 10$ we have

$$\mathscr{L}[\cos 6t](10) = \frac{10}{10^2 + 36}$$

In many engineering applications the functions f are functions of time. For this reason the collection of functions with Laplace transforms is known in applications as the *time domain*. In the terminology of operators it is called the *domain of the operator* $\mathscr{L}$. The formula (10), which tells exactly how to calculate the transform of f, is called the *assignment rule* of the operator $\mathscr{L}$. The collection of Laplace transforms, which we have called the *transform domain*, is the *range* of the operator $\mathscr{L}$.

The domain of an operator and the assignment rule of the operator will always completely characterize the operator, just as they do in Laplace transformation. To make things graphic and easy to remember, you may think of the domain and range of $\mathscr{L}$ as huge baskets full of functions, and of $\mathscr{L}$ as an arrow pointing from one to the other, as in Figure 1. $\mathscr{L}$ carries an entire problem from one basket to the other, transforming all the functions simultaneously.

The linearity property of $\mathscr{L}$ is the one that we have used very often. Namely, if c_1 and c_2 are constants, then

$$\mathscr{L}[c_1 f + c_2 g] = c_1 \mathscr{L}[f] + c_2 \mathscr{L}[g]$$

The J_ω operator also replaces whole problems with new ones. It assigns to each sinusoid $A \cos (\omega t + \phi)$ the complex number

$$J_\omega[A \cos (\omega t + \phi)] = Ae^{j\phi} \tag{11}$$

whether or not the constants A and ϕ have been specified in advance. The problem of finding a sinusoidal solution of an equation such as (1) is replaced with the problem of finding a complex number $J_\omega[y]$ satisfying (7).

The *domain* of the operator J_ω is the collection of all sinusoids with angular frequency ω; the *assignment rule* of J_ω is given by (11); the *range* of the operator J_ω is the collection $\mathbf{C}$ of all complex numbers. The domain and the assignment rule tell exactly what functions J_ω is to be used on, and exactly how it is to be applied to each element in its domain. See Figure 2.

We have used the linearity of the J_ω operator in Chapter III.

$$J_\omega[A_1 \cos (\omega t + \phi_1) + A_2 \cos (\omega t + \phi_2)]$$
$$= A_1 J_\omega[\cos (\omega t + \phi_1)] + A_2 J_\omega[\cos (\omega t + \phi_2)]$$
$$= A_1 e^{j\phi_1} + A_2 e^{j\phi_2}$$

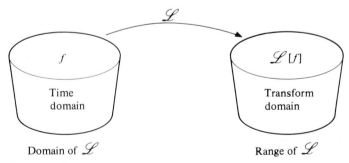

Domain of $\mathscr{L}$ Range of $\mathscr{L}$

Figure 1 Schematic diagram of the action of the Laplace transformation.

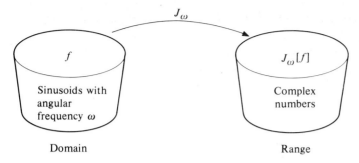

Figure 2 Schematic diagram of the action of the J_ω operator.

In particular, $\qquad J_\omega[A_1 \cos \omega t + A_2 \sin \omega t] = A_1 - jA_2$

The Laplace transformation and the J_ω operator are used to replace a given problem with a different and simpler one. In Part 2, which follows, we shall review solution techniques which utilize operators that do *not* transform the given problem into a problem in a transform domain.

2. More Examples of Linear Operators and Their Uses

In earlier chapters we studied some simple engineering systems and derived differential equations relating the system variables. We have often assumed that one of these variables represented something known, an input such as an applied voltage or force. We then derived an equation describing the effect that this input would have on another system variable, the output.

Now we shall emphasize the operator aspects of this situation. One operator will appear in the differential equation relating two system variables; another will appear in the formula for the output function as it is determined by the input function.

Consider the three systems shown in Figure 3. Each of the three systems can be modeled by an equation of the form

$$y' + \alpha y = f$$

Thus, anything we say about this equation and its solution has immediate application to the three systems. Taking $y(0) = 0$ gives the following initial value problem:

$$\left. \begin{array}{l} y' + \alpha y = f \\ y(0) = 0 \end{array} \right\} \tag{12}$$

The solution to the problem can be found using the techniques of Chapter I or the Laplace transformation. In the latter case, the transformed problem and its solution are these:

$$(s + \alpha)\mathscr{L}[y](s) = \mathscr{L}[f](s)$$

$$\mathscr{L}[y](s) = \mathscr{L}[f](s)\mathscr{L}[e^{-\alpha t}](s)$$

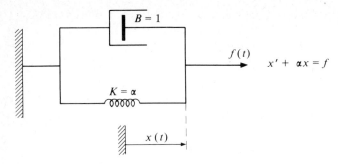

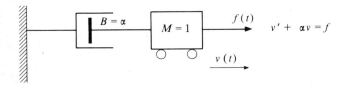

Figure 3a Spring-dashpot system (force-distance model).

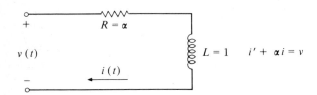

Figure 3b Mass-dashpot system (force-velocity model).

Figure 3c RL circuit (voltage-current model).

The solution may therefore be given as the convolution of f and $e^{-\alpha t}$.

$$y(t) = f * e^{-\alpha t} \qquad\qquad t \geq 0$$

$$= \int_0^t f(z) e^{-\alpha(t-z)}\, dz \qquad t \geq 0 \qquad (13)$$

$$= e^{-\alpha t} \int_0^t f(z) e^{\alpha z}\, dz \qquad t \geq 0$$

If f is defined on the whole real line, then the integration can be carried out for all t, and the resulting function is a solution on the whole real line. Since the derivative of (13) is

$$\frac{d}{dt}\left[e^{-\alpha t} \int_0^t f(z) e^{\alpha z}\, dz \right] = -\alpha e^{-\alpha t} \int_0^t f(z) e^{\alpha z}\, dz + e^{-\alpha t} f(t) e^{\alpha t}$$

$$= -\alpha e^{-\alpha t} \int_0^t f(z) e^{\alpha z}\, dz + f(t) \qquad \text{for all } t$$

substitution into (12) shows that the integral satisfies the equation *for all* t.

Therefore, in (13) the restriction $t \geq 0$ is unnecessary, and so the solution to (12) is

$$y(t) = e^{-\alpha t} \int_0^t f(z) e^{\alpha z} \, dz \qquad \text{for all } t$$

which we shall write as

$$y(t) = f * e^{-\alpha t} \qquad \text{for all } t \tag{14}$$

More generally, we now redefine the convolution $f * h$ of two functions f and h by the formula

$$(f * h)(t) = \int_0^t f(z) h(t - z) \, dz$$

This differs from what we did in systems analysis because we no longer define the convolution to be zero for $t < 0$. (The convolution as it appears in Section G of Chapter III is actually the special case in which both f and h are assumed to be zero on the half line $t < 0$, suitable for systems that are relaxed or at rest prior to $t = 0$. See the end of Part 12 of that section.)

Now look carefully at what (14) says. If f is any function continuous on all of the real line **R**, and if that function f is the nonhomogeneous term of (12), then $f * e^{-\alpha t}$ is the solution to (12). Think of it this way: In the system of Figure 3a, if the force is given by f (any continuous function of t), then the resulting displacement is given by

$$x(t) = f * e^{-\alpha t} \qquad \text{for all } t$$

If we think of the force as the *input* into the system, and of the displacement as the *output* of the system, then the system produces the output from the input as the operator produces x from f by convolution with $e^{-\alpha t}$.

Precisely similar things could be said of the other two systems. For Figure 3b, if f is the input, then the output is

$$v(t) = f * e^{-\alpha t}$$

For Figure 3c, if v is the input, then the output is

$$i(t) = v * e^{-\alpha t}$$

In each case, the output is calculated from the input by the operation of convolution with $e^{-\alpha t}$.

Since (14) gives the solution to (12) for every continuous function f, and since we have before us the way of interpreting every aspect of (12) and (14) in any of the three systems, let us concentrate our attention on (14). We shall see that it describes an operator, convolution with $e^{-\alpha t}$, which we shall call $\mathcal{T}$. The operator is to operate on every function continuous on the whole real line **R**. The collection of all such functions will be the *domain* of $\mathcal{T}$. The *assignment rule* for $\mathcal{T}$ is this:

$$\mathcal{T}[f](t) = f * e^{-\alpha t} \tag{15}$$

$$= \int_0^t f(z) e^{-\alpha(t-z)} \, dz \tag{16}$$

$$= e^{-\alpha t} \int_0^t f(z) e^{\alpha z} \, dz \tag{17}$$

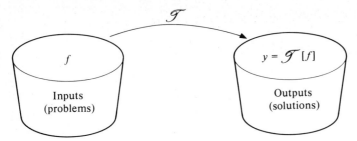

Figure 4 Schematic diagram of the action of the operator $\mathcal{T}$.

Each choice of f determines an initial value problem (12) for which $\mathcal{T}[f]$ is the solution. Thus, finding the operator $\mathcal{T}$ is tantamount to solving *all possible problems* of the form (12).

For contrast, observe the process of solving the problem without the operator, for only one choice of f, say $f(t) \equiv 3$.

$$y'(t) + \alpha y(t) = 3 \atop y(0) = 0 \Bigg| \qquad (18)$$

We shall assume that $\alpha \neq 0$. Then,

$$s\mathcal{L}[y](s) + \alpha\mathcal{L}[y](s) = \frac{3}{s}$$

$$\mathcal{L}[y](s) = \frac{3}{s}\frac{1}{s + \alpha}$$

$$= \frac{3}{\alpha}\left(\frac{1}{s} - \frac{1}{s + \alpha}\right)$$

Thus,
$$y(t) = \frac{3}{\alpha}(1 - e^{-\alpha t})$$

the formula actually holding for all t. This solution gives no clue as to what might be the solution of a problem which differs only in the nonhomogeneous terms, for instance,

$$y' + \alpha y(t) = t \atop y(0) = 0 \Bigg| \qquad (19)$$

By contrast, the operator $\mathcal{T}$ does give a formula for the solutions of both (18) and (19), but it requires a separate integration for each problem. When using the operator $\mathcal{T}$, we must find the solution of (18) this way:

$$y(t) = \mathcal{T}[3](t) = \int_0^t 3e^{-\alpha(t-z)} \, dz$$

$$= e^{-\alpha t} \int_0^t 3e^{\alpha z} \, dz$$

$$= e^{-\alpha t} \frac{3e^{\alpha z}}{\alpha} \Big|_0^t$$

$$= e^{-\alpha t} \left(\frac{3}{\alpha} e^{\alpha t} - \frac{3}{\alpha} \right)$$

$$= \frac{3}{\alpha} (1 - e^{-\alpha t}) \qquad \text{for all } t$$

For applications, it is important to observe that $\mathcal{T}$ has the property of *linearity*. That is, if c_1 and c_2 are constants, then

$$\mathcal{T}[c_1 f_1 + c_2 f_2](t) = e^{-\alpha t} \int_0^t [c_1 f_1(z) + c_2 f_2(z)]e^{\alpha z} \, dz$$

$$= e^{-\alpha t} \left[\int_0^t c_1 f_1(z)e^{\alpha z} \, dz + \int_0^t c_2 f_2(z)e^{\alpha z} \, dz \right]$$

$$= c_1 \mathcal{T}[f_1](t) + c_2 \mathcal{T}[f_2](t)$$

or $\qquad \mathcal{T}[c_1 f_1 + c_2 f_2] = c_1 \mathcal{T}[f_1] + c_2 \mathcal{T}[f_2] \qquad c_1, c_2$ constants

This linearity might be applied to the system of Figure 3a as follows. Let f_1 be one force, perhaps a constant force due to gravity, and let f_2 be another force, perhaps a sinusoidal one impressed on the system by a rotor. The displacement x of the system is then found using (17):

$$x = \mathcal{T}[f_1 + f_2]$$

$$= \mathcal{T}[f_1] + \mathcal{T}[f_2]$$

$$= f_1 * e^{-\alpha t} + f_2 * e^{-\alpha t}$$

That is, the output is simply the sum of $\mathcal{T}[f_1]$, which is the output that would result if f_1 alone were applied, and $\mathcal{T}[f_2]$, which would result if f_2 alone were applied. The model of the system implies this additivity property, and the linearity of the operator shows it.

Now let us leave the operator $\mathcal{T}$ and turn these systems front to back, interchanging the roles of input and output. This does not require any new differential equations, for the old ones give the relation between the variables, quite independently of any notion of input or output. Thus, in Figure 3a take x to be the input, and f to be the output; in Figure 3b take v to be the input, and f to be the output; and in Figure 3c take i to be the input, and v to be the output. Once again, each

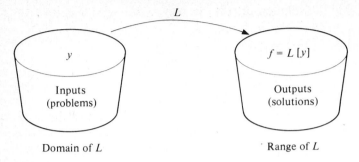

Domain of L Range of L

Figure 5 Schematic diagram of the action of L.

system can be modeled by an equation of the form

$$f = y' + \alpha y \tag{20}$$

For each possible input y we need to produce f, and (20) tells us just how to do it. In fact, (20) is the *assignment rule* for an operator we may call L:

$$L[y] = y' + \alpha y$$

The *domain* of L will be the collection of all functions which have a continuous derivative everywhere on the real line **R**. Thus, if

$$y(t) = \frac{3}{\alpha}(1 - e^{-\alpha t})$$

then
$$f(t) = L[y](t) = y'(t) + \alpha y(t)$$

$$= \frac{d}{dt}\left[\frac{3}{\alpha}(1 - e^{-\alpha t})\right] + \alpha \frac{3}{\alpha}(1 - e^{-\alpha t})$$

$$= 3e^{-\alpha t} + 3 - 3e^{-\alpha t}$$

$$= 3 \tag{21}$$

It is important to realize that L is also linear:

$$L[c_1 y_1 + c_2 y_2] = \frac{d}{dt}(c_1 y_1 + c_2 y_2) + \alpha(c_1 y_1 + c_2 y_2)$$

$$= c_1 y_1' + c_2 y_2' + \alpha c_1 y_1 + \alpha c_2 y_2$$

$$= c_1(y_1' + \alpha y_1) + c_2(y_2' + \alpha y_2)$$

$$= c_1 L[y_1] + c_2 L[y_2] \qquad \text{if } c_1 \text{ and } c_2 \text{ are constant}$$

Thus, from the sum of the inputs, the system produces the sum of the outputs; from the sum of two functions, the operator produces the sum of what it produces from the two individually.

The special case of L for $\alpha = 0$ gives the first "linear differential operator" of all, the *first-derivative operator D*, defined this way:

$$D[y] = y' \qquad \text{if } y \text{ is differentiable}$$

3. A Description of Operators

Some things are operators, and others are not. The full definition of the term *operator* should wait until "vector space" is defined, because an operator is simply a function whose domain and range are vector spaces. However, with few exceptions the operators to be studied now can be described as follows: each replaces a function with another, as Laplace transformation replaces a function with its transform. (The exceptional operators, such as the J_ω operator, may replace a function with a number, or vice versa.)

For some operators it is far more suggestive to use a word other than *replace* when describing the action of the operator. For instance, in discussing convolution, differentiation, and systems analysis, it was more descriptive to say that the operator *produced* one function from another, and there are several other descriptive phrases in use. The most common are the following, all of which have the *same mathematical meaning*:

T operates (or T acts) on y to produce $T[y]$.
T produces $T[y]$ from y.
T replaces y with $T[y]$.
T assigns $T[y]$ to y.
T carries y to $T[y]$.
T transforms y into $T[y]$.
T maps y to $T[y]$.

We use the first style simply to say that an operator operates. We use the second to emphasize the parallel between a physical system and a mathematical model of the system. We say that y is "replaced with" $T[y]$ when dealing with transforms where a whole problem is replaced with another. The expression "T assigns $T[y]$ to y" refers back to the "assignment rule" used in defining the action of T. This expression is nondescriptive, and it is usually preferred when no descriptive connotations are to be invoked. The expression "T carries y to $T[y]$" is used to emphasize the action of going from the domain to the range, from one set (or place) to another. The last two styles are also common in special circumstances, but we shall not use them.

The idea of "replacement," "production," or "assignment" should be growing familiar already. After all, the exponential function "assigns" to each number t another number e^t. The natural logarithm "replaces" each positive number t with its natural log, $\ln t$. The cosine function "produces" the number $\cos t$ from each number t.

Thus, operators such as $\mathscr{L}$ and convolution with $e^{\alpha t}$ will operate on functions, while the more familiar functions operate on numbers. The notation is chosen to emphasize this point. For instance, at various times we used these symbols:

$$\mathscr{L} \qquad \mathscr{L}[y] \qquad \mathscr{L}[y](s)$$

The first, $\mathscr{L}$, is the name of the Laplace transformation. Whenever we need to refer to this operator we can call it by name, saying, " $\mathscr{L}$ is the Laplace transformation," or " $\mathscr{L}$ is linear," or "Use $\mathscr{L}$ to transform the given problem."

The second symbol, $\mathscr{L}[y]$, represents the function produced when $\mathscr{L}$ operates on the function y. We call this function by name, saying, " $\mathscr{L}[y]$ is the Laplace transform of y," or " $\mathscr{L}[y]$ is the solution to the transformed problem," or "Multiply $\mathscr{L}[y]$ by 3." The third symbol, $\mathscr{L}[y](s)$, is the name of a number, the value of $\mathscr{L}[y]$ at s. For example, the value of $\mathscr{L}[\sin \pi t]$ at 7 is

$$\mathscr{L}[\sin \pi t](7) \qquad \text{or} \qquad \frac{\pi}{7^2 + \pi^2}$$

That is, $\mathscr{L}[y](s)$ is the number produced by the function $\mathscr{L}[y]$ at the point s.

Brackets are used to enclose the function on which $\mathscr{L}$ is operating, to remind readers whether they are reading about functions or numbers. The same notation may be used with some other operators to point out those parts of expressions that are functions. Thus, in (15), we used the brackets:

$$\mathscr{T}[f](t) = \int_0^t f(z)e^{-\alpha(t-z)}\, dz$$

and we could write

$$y = \mathscr{T}[f] = f * e^{-\alpha t}$$

as the solution of the problem (12).

However, it is customary to leave out the brackets when dealing with the first-derivative operator D. Thus, the definition of D becomes this:

$$Dy = \frac{dy}{dt} = y' \qquad \text{if } y \text{ is differentiable}$$

The value of the function is written as usual:

$$Dy(t) = \frac{dy}{dt}(t) = y'(t)$$

With this very informal description of "operator," you will know what to look for in the examples and discussion which follow. Each operator will be identified for you. You will want to look at it yourself, and ask: What functions does it act on, and what is its action on them? These are the important questions. Since the collection of things an operator operates on is called the *domain* (or *domain of definition*) of the operator, the first question to be asked about an operator should be this:

1. What is the domain of the operator?

The description of the action of an operator is called its *assignment rule*. The assignment rule must tell what each thing in the domain is replaced with. Very often the assignment rule can be stated in a formula or otherwise abbreviated. Since it is considered "known" how to take a derivative, the assignment rule for D can be written simply as

$$Dy = y'$$

but if it were not considered known, then the rest of the rule would also need to be written out, this way:

$$\text{where } y'(t) = \lim_{\Delta t \to 0} \frac{y(t + \Delta t) - y(t)}{\Delta t}$$

When $\mathscr{L}$ was defined at the beginning of Chapter III, a formula was needed by which one could calculate the value of the function $\mathscr{L}[y]$ at each point s:

$$\mathscr{L}[y](s) = \int_0^\infty y(t)e^{-st}\, dt$$

This sufficed to give the assignment rule for $\mathscr{L}$. Thus, the second question to be asked about an operator is

2. What is the assignment rule of the operator?

The domain and the assignment rule together characterize the operator.

Notice that a function f also has a domain and an assignment rule which tells how to calculate the function value $f(t)$ at each point t of the domain of f. A function also has a *range*, the collection of all function values. An operator also has a range. The *range* of an operator T is the collection of all elements which can be written as $T[x]$ for some x in the domain of T. Thus, the range of the Laplace transformation is exactly the transform domain, the set of all possible Laplace transforms $\mathscr{L}[y]$, one $\mathscr{L}[y]$ for each y in the domain of $\mathscr{L}$.

There is, therefore, a third question to ask about an operator, although it is subsidiary to the first two. That is,

3. What is the range of the operator?

To make all this terminology easier to remember, you may want to think of some picture such as Figure 6. The domain of the operator T is depicted as a kind of basket, full of functions. The range is also depicted as a basket, and the action of the operator is depicted as an arrow from the domain basket to the range basket. If T were the Laplace transformation, the left basket would be labeled "time domain," and the right basket would be labeled "transform domain."

As stated at the very beginning of this chapter, we have two uses for operators. First, we use some operators to give formulas for calculating particular solutions of certain differential equations. In earlier chapters we sometimes wrote such a formula without mentioning the operator explicitly, but in this chapter we name the operators. Second, we shall use other operators to give general properties of solutions. From these we can deduce the general solutions of equations, or identify

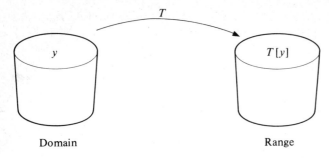

Domain   Range

Figure 6 Schematic representation of an operator.

the general solution when attempting to produce it piecemeal. That kind of descriptive understanding can help us remember the mechanics of solution methods already studied, and help us construct more efficient methods for important special cases.

4. Inverse Operators and the Solutions of Differential Equations

Previously, in Part 2, we solved the initial value problem (12),

$$\left.\begin{array}{l} y' + \alpha y = f \\ y(0) = 0 \end{array}\right| \tag{12}$$

by means of an operator $\mathcal{T}$ whose assignment rule is

$$\mathcal{T}[f] = f * e^{-\alpha t}$$

Further, we defined an operator L whose assignment rule is

$$L[y] = y' + \alpha y$$

in terms of which the initial value problem could be written

$$\left.\begin{array}{l} L[y] = f \\ y(0) = 0 \end{array}\right| \tag{22}$$

Now it is time to observe the fact that L exactly undoes the action of $\mathcal{T}$, and that $\mathcal{T}$ exactly undoes the action of L on functions y such that $y(0) = 0$. Such operators will be called "inverses" of one another.

An operator T_2 is said to be an *inverse* of the operator T_1 if the domain of T_2 contains the range of T_1 and if

$$T_2[T_1[f]] = f \qquad \text{for every } f \text{ in the domain of } T_1$$

The two operators are *mutually inverse* if each is an inverse of the other. (See Figure 7.)

In the example, $\mathcal{T}$ and L are mutually inverse. In fact, L states the problem and $\mathcal{T}$ solves it. The student's job is this: Given an L, either find the inverse $\mathcal{T}$ and

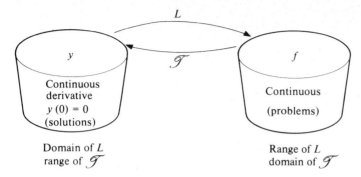

Figure 7 The operators L and $\mathcal{T}$ are mutually inverse.

apply it to the solution of the problem, or apply some other technique such as Laplace transformation or separation of variables, in which the inverse operator is only implicit.

In (12) and (22) the case $\alpha = 0$ shows the most important and familiar mutually inverse linear operators in all of calculus. They are

$$L[y](t) = y'(t) = \frac{dy}{dt}(t)$$

and

$$\mathcal{T}[f](t) = \int_0^t f(z)e^{0(t-z)}\, dz$$

$$= \int_0^t f(z)\, dz$$

differentiation and definite integration, respectively. Their linearity properties and inverse properties are well known.

Of these, the first will be used extensively. It has the special name D, the *first-derivative operator*, and its assignment rule is $Dy = y'$. We shall provide it with a domain as needed.

5. Sample Problems with Answers

A. The assignment rule for an operator T is given by $T[y](t) = 2y'(t) + 3$. Calculate the function produced by the action of T on:

(a) e^{-4t} *Answer:* $T[e^{-4t}](t) = -8e^{-4t} + 3$
(b) $2t^2$ *Answer:* $T[2t^2](t) = 8t + 3$
(c) $t \cos 5t$ *Answer:* $T[t \cos 5t](t) = -10t \sin 5t + 2 \cos 5t + 3$
(d) $4t^2 + 6 \ln t$ *Answer:* $T[4t^2 + 6 \ln t](t) = 16t + 12/t + 3$
(e) $e^{-4t} + 2t^2$ *Answer:* $T[e^{-4t} + 2t^2] = -8e^{-4t} + 8t + 3$

B. The assignment rule for an operator $\mathcal{T}$ is given by $\mathcal{T}[y] = e^{-4t} * y$. Calculate the function produced by the action of $\mathcal{T}$ on:

(a) $3t$ *Answer:* $\mathcal{T}[3t](t) = 3t/4 + \frac{3}{16}(e^{-4t} - 1)$
(b) 8 *Answer:* $\mathcal{T}[8](t) = 2(1 - e^{-4t})$
(c) e^{-2t} *Answer:* $\mathcal{T}[e^{-2t}](t) = \frac{1}{2}(e^{-2t} - e^{-4t})$

(d) $\sin 4t$ Answer: $\mathscr{T}[\sin 4t](t) = \frac{1}{8}(e^{-4t} - \cos 4t + \sin 4t)$
(e) $3e^{-2t} - \sin 4t$ Answer: $\mathscr{T}[3e^{-2t} - \sin 4t](t) = \frac{3}{2}e^{-2t} - \frac{13}{8}e^{-4t} + \frac{1}{8}(\cos 4t - \sin 4t)$
(f) $8 + 3t$ Answer: $\mathscr{T}[8 + 3t](t) = \frac{29}{16}(1 - e^{-4t}) + 3t/4$

C. Consider the operator T discussed in Problem A. From $T[e^{-4t}]$, $T[2t^2]$, and $T[e^{-4t} + 2t^2]$, calculated already, conclude whether or not T is linear.

Answer: Since $T[e^{-4t} + 2t^2] = -8e^{-4t} + 8t + 3$, whereas $T[e^{-4t}] + T[2t^2] = -8e^{-4t} + 8t + 6$, it follows that $T[e^{-4t} + 2t^2] \neq T[e^{-4t}] + T[2t^2]$, and therefore T is not linear.

D. Consider the operator $\mathscr{T}$ discussed in Problem B. Is it true that $\mathscr{T}[8 + 3t] = \mathscr{T}[8] + \mathscr{T}[3t]$? Do you suspect that $\mathscr{T}$ is linear?

Answer: Yes (actually, $\mathscr{T}$ is linear).

E. For the system in Figure 8, assume that $v(0) = 0$. Find the assignment rule for the operator L which acts on v to produce i. Also, find the assignment rule of the operator $\mathscr{T}$ which acts on i to produce v. Are L and $\mathscr{T}$ mutually inverse?

Answers: $L[v] = Cv' + (1/R)v$; $\mathscr{T}[i](t) = (1/C)i * e^{-(1/RC)t} = \int_0^t (1/C)i(z)e^{-(1/RC)(t-z)} \, dz$; yes, they are mutually inverse.

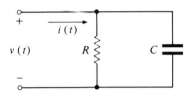

+

$i(t)$

$v(t)$ R C

−

Figure 8 RC parallel electric circuit.

F. For the system in Figure 9, assume that $f(0) = 0$ and $v(0) = 0$. Find the assignment rule for the operator $\mathscr{T}$ which acts on f to produce v. Also find the assignment rule for the operator L which acts on v to produce f. Are $\mathscr{T}$ and L mutually inverse?

Answers: $\mathscr{T}[f](t) = (1/B)f' * e^{-(K/B)t} = \int_0^t (1/B)f'(z)e^{-(K/B)(t-z)} \, dz$; $L[v](t) = (Kv + Bv') * e^{0t}$
$= \int_0^t [Kv(z) + Bv'(z)] \, dz$; yes, they are mutually inverse.

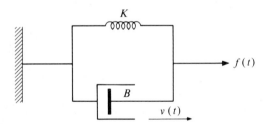

K

$f(t)$

B

$v(t)$

Figure 9 Spring-dashpot mechanical system.

G. For the system in Figure 10, assume that $f(0) = 0$ and $x(0) = 0$. Find the assignment rule for the operator which acts on x to produce f. Also find the assignment rule for the operator L which acts on f to produce x. Are $\mathscr{T}$ and L mutually inverse?

Answers: $\mathscr{T}[x](t) = Kx' * e^{-(K/B)t} = \int_0^t Kx'(z)e^{-(K/B)(t-z)} \, dz$; $L[f](t) = [(1/K)f' + (1/B)f] * e^{0t}$
$= \int_0^t [(1/K)f'(z) + (1/B)f(z)] \, dz$; yes, they are mutually inverse.

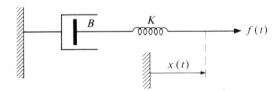

B K $f(t)$

$x(t)$

Figure 10 Spring-dashpot mechanical system.

6. Problems

1. The assignment rule for the operator L is given by $L[y] = 3y' + 7y$. Compute the function produced by the action of the operator L on:

    (a) $4t$ (b) $t \sin 3t$ (c) e^{7t}

    (d) t^2 (e) $e^{-2t} \sin 3t$ (f) $5e^{7t} + 6t^2$

    (g) $\ln 8t$ (h) $4t + \ln 8t$

2. The assignment rule for the operator $\mathcal{T}$ is given by $\mathcal{T}[y] = y * e^{-5t}$. Compute the function produced by the action of the operator $\mathcal{T}$ on:

    (a) $4t$ (b) $t \sin 3t$ (c) e^{7t}

    (d) t^2 (e) $e^{-2t} \sin 3t$ (f) $5e^{7t} + 6t^2$

    (g) 8 (h) e^{-5t}

3. Suppose that $L[y] = 3y' + 7y$, as in Problem 1. Is it true that

$$L[5e^{7t} + 6t^2] = 5L[e^{7t}] + 6L[t^2]$$

Do you suspect that L is linear?

4. Suppose that $\mathcal{T}[y] = y * e^{-5t}$, as in Problem 2. Is it true that

$$\mathcal{T}[5e^{7t} + 6t^2] = 5\mathcal{T}[e^{7t}] + 6\mathcal{T}[t^2]$$

Do you suspect that $\mathcal{T}$ is linear?

5. Suppose that $\mathcal{K}[y](t) = 4y(t) + 1$. Is it true that

$$\mathcal{K}[5e^{7t} + 6t^2] = 5\mathcal{K}[e^{7t}] + 6\mathcal{K}[t^2]$$

Do you suspect that $\mathcal{K}$ is linear?

6. For the system in Figure 11, assume that $f(0) = 0$. The differential equation describing the system is

$$f' + \frac{K}{B}f = Kv$$

Find the assignment rule for the operator L which acts on v to produce f. Also find the assignment rule for the operator $\mathcal{T}$ which acts on f to produce v. Are L and $\mathcal{T}$ mutually inverse?

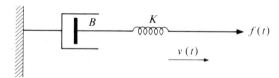

Figure 11 Spring-dashpot mechanical system.

7. For the system in Figure 12, assume that $f(0) = 0$ and $x(0) = 0$. The differential equation describing the system is

$$f' + \frac{K}{B}f = Kx'$$

    (a) Find the assignment rule for the operator Q which acts on x to produce f. Also find the assignment rule for the operator T which acts on f to produce x'. Are Q and T mutually inverse?

    (b) Find the assignment rule for the operator R which acts on f to produce x. Are Q and R mutually inverse?

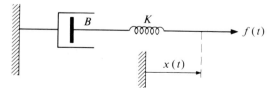

Figure 12 Spring-dashpot mechanical system.

8. For the system in Figure 13, assume that $v(0) = 0$ and $i(0) = 0$. Find the assignment rule for the operator K which acts on v to produce i. Also find the assignment rule for the operator T which acts on i to produce v. Are K and T mutually inverse?

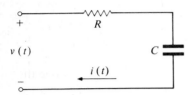

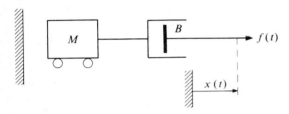

Figure 13 *RC* series circuit.

9. For the system in Figure 14, assume that $f(0) = 0$ and that $x(0) = x'(0) = 0$. Find the assignment rule for the operator L which acts on x to produce f. Also find the assignment rule for the operator K which acts on f to produce x. Are K and L mutually inverse?

Figure 14 Mass-dashpot mechanical system.

7. The Operators D^n

We have already mentioned an operator D, defined this way:
$$Dy = y' \qquad \text{if } y \text{ is differentiable}$$
If y is twice differentiable, then we may use D twice in succession:
$$y'' = D[y']$$
$$= D[D[y]]$$
$$= DDy$$
or, as it is usually abbreviated,
$$y'' = D^2 y \qquad \text{if } y \text{ is twice differentiable} \tag{23}$$

Equation (23) really defines an operator called D^2. Its domain is the collection of twice-differentiable functions; its assignment rule is $D^2 y = y''$. For instance,
$$D^2 e^{-3t} = D[De^{-3t}]$$
$$= D[-3e^{-3t}]$$
$$= 9e^{-3t}$$

Similarly, for each integer $n \geq 0$, an operator called D^n is defined this way:
$$D^n y = \frac{d^n y}{dt^n} = y^{(n)} \qquad \text{if } y \text{ is } n \text{ times differentiable} \tag{24}$$

D^2 as defined in (23) is a special case of D^n as defined in (24).

Similarly, D^3 has for its domain the collection of three-times-differentiable functions, and it assigns to each function its third derivative. The operator D^n, the nth-*derivative operator*, has for its domain the collection of n-times-differentiable functions.

Two of these operators, D^1 and D^0, have other names or representations. D^1 is, of course, D, and it is written either way. D^0 is the operator that does not take the derivative, but leaves the function unchanged. It is called the *identity* operator, and it is represented in many ways, of which three, I, 1, and D^0, will be used in the sections that follow. The first symbol is by far the most common, for I stands for "identity" when dealing with matrices and integral operators, too. The symbol D^0 we are almost through with. We shall use it only where we have to. We shall use the symbol 1 often because of its relation to multiplication. That is, the equation

$$3I[y] = 3y$$

may also be written

$$3 \cdot 1[y] = 3y$$

Clearly, it is easier to write 3 than $3I$ or $3 \cdot 1$, and so 3 *will also stand for the operation of multiplication* by 3. Thus,

$$3[y] = 3y \tag{25}$$

From here on, through the remainder of this section and most of the next two sections, we shall take the domain of D^n to be the collection of functions with n derivatives continuous on all of the real line. This collection is called $C^n(\mathbf{R})$, where C stands for continuity, $\mathbf{R}$ is the symbol for the real line, and n refers to the number of derivatives. For simplicity, the collection $C^0(\mathbf{R})$ of continuous functions on $\mathbf{R}$ is usually denoted $C(\mathbf{R})$. Thus, D^n may be depicted as in Figure 15.

For example, if a and b are constants, then

$$D^n[e^{at}] = a^n e^{at}$$

$$D^3[\sin bt] = -b^3 \cos bt$$

and $$D^2[at^3 + 4t^2 + 7t + 2b] = 6at + 8$$

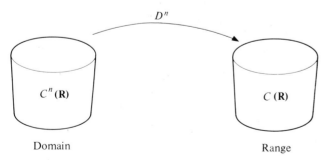

Domain   Range

Figure 15 Schematic representation of the operator D^n.

Since all the functions in $C^n(\mathbf{R})$ are defined on the same interval, the whole line $\mathbf{R}$, we can add them and make linear combinations. That is, if f_1 and f_2 are in $C^n(\mathbf{R})$, and if a_1 and a_2 are constants, then $a_1 f_1 + a_2 f_2$ is also a function in $C^n(\mathbf{R})$. The continuity of the nth derivatives of f_1 and f_2 at each point assures the continuity of the nth derivative of $a_1 f_1 + a_2 f_2$. Further,

$$\frac{d^n}{dt^n}[a_1 f_1(t) + a_2 f_2(t)] = \frac{d^n}{dt^n}[a_1 f_1(t)] + \frac{d^n}{dt^n}[a_2 f_2(t)]$$

$$= a_1 \frac{d^n}{dt^n}f_1(t) + a_2 \frac{d^n}{dt^n}f_2(t)$$

so that D^n is seen to be linear:

$$D^n[a_1 f_1 + a_2 f_2] = a_1 D^n f_1 + a_2 D^n f_2 \tag{26}$$

8. Linear Operators and Differential Equations with Constant Coefficients

Consider the differential equation

$$y'' + 3y' + 2y = 0 \tag{27}$$

and observe that the function e^{5t} is *not a solution*. How do we know? How do we show this? The long way is to solve the equation and then see whether e^{5t} is among the solutions. That method works but is unnecessarily laborious. A much easier way is simply to test whether e^{5t} fits the definition of "solution of the differential equation," this way: e^t is a solution if and only if

1. It is twice differentiable.
2. When it is substituted into the equation, it satisfies the equation at every point t.

If it fails to satisfy the equation at even a single point then it is not a solution.
    Since e^{5t} is twice differentiable, it is only necessary to attempt the substitution. The first and second derivatives of e^{5t} are $5e^{5t}$ and $25e^{5t}$, respectively. Thus, the result of substituting $y(t) = e^{5t}$ into the left-hand side of Equation (27) is

$$y''(t) + 3y'(t) + 2y(t) = 25e^{5t} + 3(5e^{5t}) + 2e^{5t} = 42e^{5t}$$

Now $42e^{5t}$ is not the same as the zero function, the right-hand side of (27), and so $y(t) = e^{5t}$ is *not* a solution of the equation.
    In the example just given, the critical step was calculating $y'' + 3y' + 2y$ for a given function y in $C^2(\mathbf{R})$. For any given function in $C^2(\mathbf{R})$, we can test whether or not it is a solution by evaluating the expression $y'' + 3y' + 2y$ and comparing the result with the zero function. In effect we have defined an operator T whose domain is $C^2(\mathbf{R})$ and whose assignment rule is

$$T[y] = y'' + 3y' + 2y$$

When we compare the zero function with the result of substituting y into the left-hand side of the equation, we are just comparing it with $T[y]$. Thus, the study of the solution of the equation

$$y'' + 3y' + 2y = 0$$

is equivalent to the study of answers to the question:
     What functions in $C^2(\mathbf{R})$ does T carry to 0?
The study of the equation

$$y'' + 3y' + 2y = f$$

is equivalent to the study of answers to the question:
     What functions in $C^2(\mathbf{R})$ does T carry to f?
These questions put the study of the equations into a somewhat different perspective and allow us new ways of looking at the subject.

     Now, at last, we are ready to define the operators which, like T, will focus our attention on new aspects of linear differential equations with constant coefficients.

9. Linear Differential Operators with Constant Coefficients

The operator T which we considered above has $C^2(\mathbf{R})$ for its domain and

$$T[y] = y'' + 3y' + 2y \tag{28}$$

for its assignment rule. Does T have a linearity property? We shall begin by seeing that it has. Second, we shall define linear differential operators with constant coefficients, of which T is an example, and observe that they all are linear. Then we shall introduce notation for such operators to make calculations with them as simple, efficient, and easily remembered as possible.

     First let us see that T itself is linear. Let y_1 and y_2 be any functions whatsoever in $C^2(\mathbf{R})$, and let c_1 and c_2 be constants.

$$T[c_1 y_1 + c_2 y_2] = \frac{d^2}{dt^2}(c_1 y_1 + c_2 y_2) + 3\frac{d}{dt}(c_1 y_1 + c_2 y_2) + 2(c_1 y_1 + c_2 y_2)$$

$$= (c_1 y_1'' + c_2 y_2'') + 3(c_1 y_1' + c_2 y_2') + 2(c_1 y_1 + c_2 y_2) \tag{29}$$

by the linearity of the first and second differentiations. The right-hand side of (29) may be rearranged by collecting in parentheses all the terms with y_1, and again all the terms with y_2. The result is this:

$$T[c_1 y_1 + c_2 y_2] = (c_1 y_1'' + 3c_1 y_1' + 2c_1 y_1) + (c_2 y_2'' + 3c_2 y_2' + 2c_2 y_2)$$

$$= c_1(y_1'' + 3y_1' + 2y_1) + c_2(y_2'' + 3y_2' + 2y_2) \tag{30}$$

The expression $y_1'' + 3y_1' + 2y_1$ in the first set of parentheses of (30) is $T[y_1]$, as can be seen by looking at the assignment rule for T, Equation (28). In the second set of parentheses is the expression for $T[y_2]$. Therefore,

$$T[c_1 y_1 + c_2 y_2] = c_1 T[y_1] + c_2 T[y_2]$$

and therefore T has the property which we call *linearity*.

Now for the definition: A *linear differential operator with constant coefficients and order n* is an operator whose assignment rule is of the form

$$L[y] = a_n y^{(n)} + a_{n-1} y^{(n-1)} + \cdots + a_3 y^{(3)} + a_2 y'' + a_1 y' + a_0 y$$

where $a_n, a_{n-1}, \ldots, a_1, a_0$ are constants, called the *coefficients* of the operator, and $a_n \neq 0$. Except when misunderstanding is likely, we shall simply call such a thing a *constant-coefficient operator*.

The domain of these operators may be selected to meet special needs, but for now the domain of such an operator of order n will be $C^n(\mathbf{R})$, the set of all functions with n continuous derivatives on the whole real line $\mathbf{R}$.

The proof that such an operator always has a linearity property is entirely analogous to the proof we gave for the special case, the operator T.

$L[c_1 y_1 + c_2 y_2]$

$$= a_n \frac{d^n}{dt^n}(c_1 y_1 + c_2 y_2) + \cdots + a_1 \frac{d}{dt}(c_1 y_1 + c_2 y_2) + a_0(c_1 y_1 + c_2 y_2)$$

$$= a_n(c_1 y_1^{(n)} + c_2 y_2^{(n)}) + \cdots + a_1(c_1 y_1' + c_2 y_2') + a_0(c_1 y_1 + c_2 y_2)$$

$$= c_1(a_n y_1^{(n)} + \cdots + a_1 y_1' + a_0 y_1) + c_2(a_n y_2^{(n)} + \cdots + a_1 y_2' + a_0 y_2)$$

$$= c_1 L[y_1] + c_2 L[y_2] \tag{31}$$

Thus, the name of this kind of operator is actually descriptive of its properties.

Finally, we must settle the notation. Typically the letter L is reserved for operators which are known to be linear, as these are. (For nonlinear operators, see Part 17 below. Any operator could be called T, $\mathcal{T}$, or K; L is reserved for linear operators. The operator which we used for an example could as well have been called L.) The operator whose assignment rule is

$$T[y] = y'' + 3y' + 2y$$

is often written as a "sum of operators." You are familiar with sums of functions. For instance, $3f_1 + 5f_2$ is a function whose assignment rule is

$$(3f_1 + 5f_2)(t) = 3f_1(t) + 5f_2(t)$$

Similarly, we define sums and *linear combinations of operators*. T would be $D^2 + 3D^1 + 2D^0$ or, more succinctly,

$$T = D^2 + 3D + 2$$

In this formula you will recognize D^1, which is D, and $2D^0$, which is $2I$, twice the identity operator, often written just 2. Thus, the action of T on y may be written this way:

$$T[y] = (D^2 + 3D + 2)[y]$$

Very often, even the brackets are omitted, as was done with the operators D^n alone:

$$Ty = (D^2 + 3D + 2)y$$

Warning: The right side of this equation is *not a product*. It represents the action of T on y, as described above, and the expression as a whole tells the function to which T carries y, namely $y'' + 3y' + 2y$.

10. Sample Problems with Answers

In the following problems, each operator L has $C^2(\mathbf{R})$ or $C^1(\mathbf{R})$ as its domain, and a function f is given. Find $L[f]$.

H. $L[y] = 4y'' + 2y' + y; f(t) = 3e^{-t}$ *Answer:* $L[f] = 9e^{-t}$

I. $L[y] = y' + 3y; f(t) = te^{-3t}$ *Answer:* $L[f] = e^{-3t}$

J. $L = D + 2; f(t) = \cos 5t$ *Answer:* $L[f] = -5 \sin 5t + 2 \cos 5t$

K. $L = 2D^2 + 3D + 1; f(t) = e^{-t} \sin 2t$ *Answer:* $L[f] = -8e^{-t} \sin 2t - 2e^{-t} \cos 2t$

To the left-hand side of each of the following differential equations there corresponds exactly one linear operator L with domain $C^2(\mathbf{R})$, such that the given equation may be written in the form $L[y] = f$, where f is the given right-hand side. For each, give the assignment rule for L.

L. $y'' + 4y' + 4y = e^t$ *Answer:* $L = D^2 + 4D + 4$

M. $6y'' + 2y = 7$ *Answer:* $L = 6D^2 + 2$

N. $3y''(t) + 13y'(t) = \cos 2\pi t$ *Answer:* $L = 3D^2 + 13D$

In each of the following problems an assignment rule is given. For each, choose from $C(\mathbf{R})$, $C^2(\mathbf{R})$, $C^3(\mathbf{R})$, etc., the largest collection to which the assignment rule may be applied, thereby completing the definition of the operator L. List the domain, the order of L, and the coefficients of L.

O. $Ly = D^3y$ *Answer:* $C^3(\mathbf{R})$; order $n = 3$; coefficients $a_3 = 1, a_2 = a_1 = a_0 = 0$

P. $Ly = 3D^2y + 4y$ *Answer:* $C^2(\mathbf{R})$; order $n = 2$; coefficients $a_2 = 3, a_1 = 0, a_0 = 4$

Q. $Ly = 2y + 3y' + 7y''$ *Answer:* $C^2(\mathbf{R})$; order 2; coefficients $a_2 = 7, a_1 = 3, a_0 = 2$

11. Problems

10. L has domain $C^2(\mathbf{R})$, and $L[y] = 3y'' + 6y' + 3y$.
   (a) What is $L[e^t]$?   (b) What is $L[e^{-t}]$?
   (c) What is $L[e^{3t}]$?   (d) What is $L[te^{-t}]$?
From your answers to parts *a* through *d*, can you determine:
   (e) Any solutions of the equation $3y'' + 6y' + 3y = 0$?
   (f) Any solutions of the equation $3y'' + 6y' + 3y = e^t$?

11. The solutions of the equation $y''' + 3y'' + 3y' + y = 0$ all have continuous third derivatives.
   (a) What is the operator L with domain $C^3(\mathbf{R})$ such that y satisfies the equation if and only if $Ly = 0$?
   (b) Which of the following are solutions of the given equation?

$$e^{-t} \qquad te^{-t} \qquad e^t \qquad te^t \qquad e^{-t} + 2te^{-t} \qquad 3e^t - te^t$$

   (c) Can the operator L be applied to functions in $C^4(\mathbf{R})$?
   (d) Give the order and the coefficients of L.

12. Using Operators to Eliminate Variables: An Example

Consider the problem of eliminating the variable x from the system of equations

$$x'' + 4x + y'' + 3y' + 2y = e^{2t} \tag{32a}$$

$$x' + 7x + y'' - y' - 2y = 3 \tag{32b}$$

We shall improve on the elimination method given in Chapter II. If we define some operators by

$$L_1 = D^2 + 4 \qquad L_3 = D^2 + 3D + 2 \atop L_2 = D + 7 \qquad L_4 = D^2 - D - 2 \Big\} \tag{33}$$

then the system may be written more succinctly as

$$L_1 x + L_3 y = e^{2t} \tag{34a}$$

$$L_2 x + L_4 y = 3 \tag{34b}$$

In this form the system reminds us of a pair of linear algebraic equations, which it is not because the operators do not simply multiply the variables. To eliminate x we shall apply L_2 to both sides of the first equation, and L_1 to both sides of the second, and then subtract. Here is what happens to the first equation:

$$L_2[L_1 x] + L_2[L_3 y] = L_2[e^{2t}] \tag{35}$$

or

$$(D + 7)[x'' + 4x] + (D + 7)[y'' + 3y' + 2y] = (D + 7)[e^{2t}]$$

$$(x''' + 7x'' + 4x' + 28x) + (y''' + 10y'' + 23y' + 14y) = 9e^{2t} \tag{36}$$

Here is the action of L_1 on the second equation:

$$L_1[L_2 x] + L_1[L_4 y] = L_1[3] \tag{37}$$

or

$$(D^2 + 4)[x' + 7x] + (D^2 + 4)[y'' - y' - 2y] = (D^2 + 4)[3]$$

$$(x''' + 7x'' + 4x' + 28x) + (y^{(4)} - y''' + 2y'' - 4y' - 8y) = 12 \tag{38}$$

Since the first set of parentheses is the same in both (36) and (38), the result of subtracting the two equations will be that x *and all its derivatives are eliminated.* That is, (36) minus (38) gives

$$(y''' + 10y'' + 23y' + 14y) - (y^{(4)} - y''' + 2y'' - 4y' - 8y) = 9e^{2t} - 12 \tag{39}$$

Combining like terms and dividing by -1 puts the equation in normal form:

$$y^{(4)} - 2y''' - 8y'' - 27y' - 22y = 12 - 9e^{2t} \tag{40}$$

The equation can be solved with Laplace transforms.

If we take the system of equations in operator form, as in (34a) and (34b), we have

$$L_1 x + L_3 y = e^{2t} \tag{41}$$

$$L_2 x + L_4 y = 3 \tag{42}$$

Operating with L_2 on the first and L_1 on the second gives (35) and (37), as follows:

$$L_2 L_1 x + L_2 L_3 y = L_2[e^{2t}] \tag{43}$$

$$L_1 L_2 x + L_1 L_4 y = L_1[3] \tag{44}$$

What we observed in (36) and (38) is that the first terms are equal; that is,

$$L_2 L_1 x = L_1 L_2 x \tag{45}$$

in this example, so that subtracting (37) from (35) or (44) from (43) results in the elimination of the terms in x, as follows:

$$L_2 L_3 y - L_1 L_4 y = L_2[e^{2t}] - L_1[3] \tag{46}$$

which is the same as (39). Of course, to obtain (39) from (46), one must use (33), the assignment rules for the operators.

On the other hand, the passage from (41) and (42) to (46) did not depend on the assignment rules (33) giving L_1, L_2, L_3, and L_4. Thus, the calculation is possible for any four linear operators, *so long as* (45) *holds*, so long as L_1 and L_2 *commute*, that is, $L_2 L_1 = L_1 L_2$.

That is the key to the method of elimination. Since we are using linear differential operators with constant coefficients, we may be sure that *all* the operators will commute and the method will work.

13. Elimination Method and Example

If $L_1, L_2, L_3, L_4, L_5, L_6, \ldots$ are linear differential operators, if $L_2 L_1 = L_1 L_2$, and if f_1 and f_2 are sufficiently differentiable, then the equations

$$L_1 x + L_3 y + L_5 z + \cdots = f_1 \tag{47}$$
$$L_2 x + L_4 y + L_6 z + \cdots = f_2 \tag{48}$$

can have x eliminated by the action of L_2 on the first and L_1 on the second, followed by subtraction. The result is the single equation

$$L_2 L_3 y - L_1 L_4 y + L_2 L_5 z - L_1 L_6 z \cdots = L_2 f_1 - L_1 f_2 \tag{49}$$

An example will show that (49) tells how the calculation is performed in practice. Let us eliminate x and its derivatives from the equations

$$x'' + 2x + 2y'' - y' = \cos 2t \tag{50}$$
$$x'' + x' + y' + 2y = 5 \sin 2t$$

Here we have

$$L_1 = D^2 + 2 \qquad L_3 = 2D^2 - D \qquad f_1 = \cos 2t$$
$$L_2 = D^2 + D \qquad L_4 = D + 2 \qquad f_2 = 5 \sin 2t \tag{51}$$

and the equations are

$$L_1 x + L_3 y = f_1$$
$$L_2 x + L_4 y = f_2$$

Operating with L_2 on the first and L_1 on the second gives

$$L_2 L_1 x + L_2 L_3 y = L_2 f_1$$
$$L_1 L_2 x + L_1 L_4 y = L_1 f_2$$

Subtracting the second from the first now gives

$$L_2 L_3 y - L_1 L_4 y = L_2 f_1 - L_1 f_2$$

which is just (49), derived on the spot rather than memorized. Substitution of (51) into this gives

$$(D^2 + D)(2D^2 - D)y - (D^2 + 2)(D + 2)y = (D^2 + D)\cos 2t - (D^2 + 2)5 \sin 2t$$

whose evaluation gives the desired equation,

$$2y^{(4)} - 3y'' - 2y' - 4y = -14 \cos 2t + 18 \sin 2t$$

14. Sample Problems with Answers

R. Consider the equations

$$\left. \begin{array}{r} 2x' + x + y'' + 3y = \sin 3t \\ x'' + 4y' + y = t \end{array} \right|$$

(a) Eliminate y, obtaining an equation in x alone.
Answer: $x^{(4)} - 5x'' - 6x' - x = 3t - 12 \cos 3t - \sin 3t$
(b) Eliminate x, obtaining an equation in y alone.
Answer: $y^{(4)} - 5y'' - 6y' - y = -2 - t - 9 \sin 3t$

15. Problems

12. Consider the system

$$\left. \begin{array}{r} x'' - x' + 3y'' + y = 4 \\ x'' + x - y' + 3y = t \end{array} \right|$$

(a) Eliminate x, obtaining an equation in y alone.
(b) Eliminate y, obtaining an equation in x alone.

13. Eliminate x from the system

$$\left. \begin{array}{r} x'' - 4x + 4y'' - 4y' + y = e^t \\ x'' - 4x' + 4x + 2y' - y = e^{3t} \end{array} \right|$$

16. Linear Differential Equations

As far back as Section H of Chapter I, we introduced terminology for *linear differential equations with constant coefficients*

$$a_n y^{(n)}(t) + \cdots + a_1 y'(t) + a_0 y(t) = f(t) \tag{52}$$

where $a_0, a_1, \ldots, a_n$ are constants. Equation (52) has *order* n if $a_n \neq 0$.

The other *linear differential equations* are those in which the coefficients are not constants but functions of t. Specifically, a differential equation is *linear* if it can be written in the form

$$a_n(t)y^{(n)}(t) + \cdots + a_1(t)y'(t) + a_0(t)y(t) = f(t) \tag{53}$$

where $a_0, a_1, \ldots, a_n$, and f are functions of t. If $a_n(t) \neq 0$ for every t on an interval (α, β), then the equation is of *order n on the interval* (α, β).

It is usually very easy to tell whether or not a differential equation is linear, for the form of (53) is easy to remember. On one side (the right), there is a function of t alone, called the *nonhomogeneous term*. On the other side, there is a sum of products, each product having one factor which is the unknown function y or one of its derivatives, the remaining factor being a function of t alone.

If the equation is of order n on an interval (α, β), and if the functions a_0, $a_1, \ldots, a_n$, and f are continuous there, then Theorem 2 from Section H of Chapter I asserts that there do exist functions which satisfy (53) at each point of the interval. These functions are called *solutions* of the equation, and they will have n continuous derivatives on the interval. The collection of all such functions is called the *general solution set* of the equation. (For examples and a discussion of applications, see Part 16 of Section B of Chapter III.)

If $y_0, y_1, \ldots, y_{n-1}$ are any n constants whatsoever, and if t_0 is any point of the interval (α, β), then among the solutions of the equation there is exactly one solution which satisfies not only (53) but also the n *initial conditions*

$$
\left.
\begin{aligned}
y(t_0) &= y_0 \\
y'(t_0) &= y_1 \\
y^{(n-1)}(t_0) &= y_{n-1}
\end{aligned}
\right\}
$$

A set of solutions is the general solution set when it contains the solutions of all the initial value problems.

We have seen numerous applications of such equations in cases in which all the coefficients were constant, and we have solved such equations.

In the case of constant coefficients we found a constant-coefficient operator L such that (52) could be written as $Ly = f$. Now we shall be able to write (53) in the same way, using an operator with nonconstant coefficients. Specifically, let L be the operator whose domain is the collection $C^n(\alpha, \beta)$ of all functions which have n continuous derivatives on the interval (α, β) and whose assignment rule is

$$
Ly(t) = a_n(t)y^{(n)}(t) + \cdots + a_1(t)y'(t) + a_0(t)y(t) \tag{54}
$$

In other words,

$$
L = a_n D^n + \cdots + a_1 D + a_0
$$

An operator whose assignment rule is of this form is called a *linear differential operator*, whether the coefficients $a_0, \ldots, a_n$ are constant or nonconstant functions of t. If the *lead coefficient* $a_n(t)$ never takes the value zero on the interval (α, β), then L has *order n* on (α, β). The proof that such an operator is actually linear is just like Equation (31).

It is clear that (53) may now be written

$$
Ly = f
$$

No universally effective method is known for solving linear differential equations with nonconstant coefficients, but some general properties of the solutions of equations with constant coefficients hold in the general case as well.

If we are given the coefficient functions $a_0, \ldots, a_n$, and if we hope to solve Equation (53) for every choice of the nonhomogeneous term f, then we certainly shall have to have a solution for it in the case $f \equiv 0$, for the zero function is one possible nonhomogeneous term. For instance, if (53) represents an engineering system for which f is the input, and if we wish to see the response y to every input and every choice of initial state, then among the responses we seek are the solutions of

$$a_n(t)y^{(n)}(t) + \cdots + a_1(t)y'(t) + a_0(t)y(t) = 0 \qquad (55)$$

or

$$Ly = 0$$

which is called the *homogeneous equation* associated with (53). Because the phrase "solutions of the associated homogeneous equation" is so long, these functions are usually called by the shorter name *homogeneous solutions* of (53). The terminology is almost universally used, but it is a little misleading, since a solution of (55) is not a solution of (53) at all. In fact, the function produced by substituting a homogeneous solution into the left-hand side of (53) is the same as that produced by substituting it into the left-hand side of (55), namely the zero function and not f, which is the right-hand side of (53).

Once the homogeneous solutions are found, a generally applicable method can be used to find the solutions of the given equation. (See Section F. A method applicable to the most important special case is given in Section C.)

There is, however, a fact of fundamental importance relating the general solution of (53) with that of the associated homogeneous equation (55). It is this: *If y_p is some solution of (53), no matter which one, and if h is the general solution of (55), then*

$$y = y_p + h \qquad (56)$$

is the general solution of (53).

For instance we have seen that the problem

$$\left. \begin{aligned} y'(t) + 2y(t) &= 3 \\ y(0) &= y_0 \end{aligned} \right\} \qquad (57)$$

has solution

$$y(t) = \tfrac{3}{2}(1 - e^{-2t}) + y_0 e^{-2t} \qquad (58)$$

Since this is so for each value of y_0, it follows that (58) is the general solution of (57). We may think of the first term of (58) as y_p:

$$y_p(t) = \tfrac{3}{2}(1 - e^{-2t})$$

for it is a solution of (57), the solution satisfying the initial condition $y_p(0) = 0$.

The second term of (58),

$$h(t) = y_0 e^{-2t}$$

is the general solution of the associated homogeneous equation

$$y'(t) + 2y(t) = 0$$

Now we must show that the property (56) holds for the general solution of *every* linear differential equation (53). First we show that every function of the required form is a solution of (53). Specifically, consider $y_p + y_h$, where y_h is some homogeneous solution, and substitute into the *left-hand* side of (53).

$$L[y_p + y_h] = a_n \frac{d^n}{dt^n}(y_p + y_h) + \cdots + a_0(y_p + y_h) \tag{59}$$

$$= (a_n y_p^{(n)} + a_n y_h^{(n)}) + \cdots + (a_0 y_p + a_0 y_h)$$

$$= (a_n y_p^{(n)} + \cdots + a_0 y_p) + (a_n y_h^{(n)} + \cdots + a_0 y_h)$$

$$= Ly_p + Ly_h \tag{60}$$

Now y_p already satisfies (53), for that was its property. Therefore, $Ly_p = f$. On the other hand, y_h was a homogeneous solution, so that $Ly_h = 0$. Hence (60) continues:

$$L[y_p + y_h] = Ly_p + Ly_h = f + 0 = f \tag{61}$$

That is, every function of the form $y_p + y_h$ is a solution of Equation (53).

Next we must show that every solution of (53) is of the form (56). In particular, let y_q be some solution of (53). We wish to know if it can be written in the form $y_p + y_h$. That is, if

$$g = y_q - y_p \qquad \text{or} \qquad y_q = y_p + g$$

then we must find out whether g is a homogeneous solution. There is one thing to try, namely substituting g into (55), the homogeneous equation associated with (53), and seeing whether it is a solution. We shall not do the substitution in detail, re-proving part of linearity as we did at (60), but rather use the linearity of the operator to shorten the calculation.

$$L[y_q - y_p] = Ly_q - Ly_p$$

$$= f - f = 0 \tag{62}$$

The equality $Ly_q = f = Ly_p$ is exactly the fact that y_q and y_p both satisfy (53).

Thus, from (62), it follows that $g = y_q - y_p$ is itself a homogeneous solution, which we may call y_h, and therefore

$$y_q = y_p + (y_q - y_p) = y_p + y_h$$

is of exactly the form required, (56). That completes the characterization: *The solutions are the functions of the form* $y_p + y_h$.

Here is an example of the way in which the characterization may be used.

Consider the differential equation

$$y'' + 3y' + 2y = 2 \qquad (63)$$

It is easy to see that $y_p(t) \equiv 1$ is a solution of (63), by substitution. The characteristic polynomial (Section C of Chapter III) shows that the corresponding homogeneous equation,

$$y'' + 3y' + 2y = 0 \qquad (64)$$

has solutions of the form e^{-t}, e^{-2t}, and $A_1 e^{-t} + A_2 e^{-2t}$. (We shall return to this in the next section.) The property (56) says, therefore, that

$$y(t) = 1 + A_1 e^{-t} + A_2 e^{-2t} \qquad (65)$$

is a solution of (63) for every choice of A_1 and A_2. (Try that out by substitution, to see that it is so.) In fact, this function y is the general solution of (63) because

$$y_h(t) = A_1 e^{-t} + A_2 e^{-2t}$$

is the general solution of (64).

We can also solve initial value problems. For instance, if

$$\left.\begin{array}{r} y''(t) + 3y'(t) + 2y(t) = 2 \\ y(0) = 0 \\ y'(0) = 1 \end{array}\right\} \qquad (66)$$

and if

$$y(t) = 1 + A_1 e^{-t} + A_2 e^{-2t} \qquad (67)$$

is the general solution, then

$$y(0) = 1 + A_1 e^0 + A_2 e^0$$
$$= 1 + A_1 + A_2 \qquad (68)$$

and

$$y'(t) = -A_1 e^{-t} - 2A_2 e^{-2t}$$

so that

$$y'(0) = -A_1 - 2A_2 \qquad (69)$$

Using (68) and (69) together with the initial values given in (66) gives these equations:

$$\begin{array}{ll} y(0): & 0 = 1 + A_1 + A_2 \\ y'(0): & 1 = -A_1 - 2A_2 \end{array}\Bigg\}$$

Solving these equations simultaneously gives

$$A_1 = -1 \qquad \text{and} \qquad A_2 = 0$$

Substitution into (67) gives the solution of the initial value problem, namely

$$y(t) = 1 - e^{-t}$$

In this part we have defined the linear differential operators, with assignment rules of the form

$$L[y](t) = a_n(t)y^{(n)}(t) + \cdots + a_1(t)y'(t) + a_0(t)y(t)$$

They have the property we call *linearity*, as we have shown, even if the coefficients are not constant. Let us consider a special case in which

$$E[y](t) = ty'(t) = tDy(t)$$

In this case, $a_1(t) = t$ and all the other coefficients are zero. Just to stress the linearity, we shall try out E on a linear combination, say $y(t) = c_1 y_1(t) + c_2 y_2(t)$.

$$E[c_1 y_1 + c_2 y_2](t) = tD[c_1 y_1(t) + c_2 y_2(t)]$$
$$= tc_1 Dy_1(t) + tc_2 Dy_2(t)$$
$$= c_1 tDy_1(t) + c_2 tDy_2(t)$$
$$= c_1 E[y_1](t) + c_2 E[y_2](t)$$

Therefore E is another example of a linear operator.

The letter E stands for "equidimensional." The equidimensional operators are those linear differential operators in which

$$a_k(t) = b_k t^k \qquad \text{where } b_k \text{ is constant}$$

That is, they have assignment rules of the form

$$L[y](t) = b_n t^n y^{(n)}(t) + \cdots + b_1 ty'(t) + b_0 y(t)$$

They will be studied further in Section D of this chapter.

17. Linearity, Nonlinearity, and Homogeneity

An operator T is *linear* if it has the following property: If y_1 and y_2 are any elements in the domain of T, and if c_1 and c_2 are constants such that $c_1 y_1 + c_2 y_2$ is also in the domain of T, then the identity

$$T[c_1 y_1 + c_2 y_2] = c_1 T[y_1] + c_2 T[y_2]$$

is satisfied.

All other operators are *nonlinear*. Since you are studying linear operators, you will want to be able to distinguish them from the nonlinear ones.

The property of linearity is composed of two parts, "homogeneity" and "additivity." An operator T is *homogeneous* if it satisfies the identity

$$T[cy] = cT[y]$$

for all y in the domain of T and all constants c such that cy is also in the domain of T. An operator T is *additive* if T satisfies the identity

$$T[y_1 + y_2] = T[y_1] + T[y_2]$$

for all y_1 and y_2 in the domain of T such that $y_1 + y_2$ is also in the domain of T.

It follows that T is linear if and only if T is both homogeneous and additive. Thus, a nonlinear operator may be recognized by either a failure to be additive or a failure to be homogeneous. Typically, it is easier to recognize the failure of homogeneity.

The homogeneity failure that is easiest to recognize is the failure to carry zero to zero. Every homogeneous operator must satisfy

$$0 = 0T[y] = T[0y] = T[0]$$

so that an operator lacking the property

$$T[0] = 0$$

is nonhomogeneous and therefore nonlinear. For example, if T has domain $C(\mathbf{R})$ and

$$T[y](t) = y(t) - 2 \qquad \text{for all } t$$

then $T[0](t) = -2 \neq 0$, and T is nonhomogeneous and therefore nonlinear.

An operator T may satisfy this "zero property" and still be nonhomogeneous. For instance, if T has $C(\mathbf{R})$ for its domain, and if $T[y] = y^3$ (that is, if T cubes each continuous function), then T carries zero to zero but is nonhomogeneous all the same. This fact may be shown with an example. If $y(t) \equiv 1$ and $c = 2$, then

$$T[y](t) \equiv 1$$

$$T[2y](t) \equiv [2y(t)]^3 = 8$$

although
$$2T[y](t) \equiv 2$$

Since $2 \neq 8$, it follows that $T[2y] \neq 2T[y]$, and therefore T is not homogeneous and thus not linear.

The technique demonstrated is typical of those employed for showing that an operator is not homogeneous. It consists of finding a specific pair, an element y in the domain and a constant c such that cy is also in the domain, but such that

$$cT[y] \neq T[cy]$$

For any given operator T, one such example suffices to show that T is not homogeneous. Such an example is called a *counterexample* to the possible homogeneity of T.

Counterexamples to additivity may also be found. Very often they are disguised forms of counterexamples to homogeneity. Consider the operator T of the previous example. We shall show that it is not additive by its action on y_1 and y_2, where each is identically 1. Thus,

$$T[y_1](t) \equiv 1^3 \equiv T[y_2](t)$$

$$T[y_1 + y_2](t) = [y_1(t) + y_2(t)]^3 \equiv (1 + 1)^3 = 8$$

although
$$T[y_1](t) + T[y_2](t) \equiv 1^3 + 1^3 = 2$$

Since $2 \neq 8$, it follows that

$$T[y_1](t) + T[y_2](t) \neq T[y_1 + y_2](t)$$

and therefore T is not additive.

There are operators which are homogeneous but not linear. Such an operator would pass every test for homogeneity, yet would fail some tests for additivity and therefore would be nonlinear. Here is an example of such an operator. Let the domain of T be $C^1(\mathbf{R})$, the collection of functions with a continuous derivative on the whole real line $\mathbf{R}$, and let

$$T[y](t) = \begin{cases} 0 & \text{if } y(t) = 0 \text{ and } y'(t) = 0 \\[2mm] \dfrac{[y(t)]^2 y'(t)}{[y(t)]^2 + [y'(t)]^2} & \text{otherwise} \end{cases}$$

By substitution and simplification, we can see that if c is any constant then $T[cy] = cT[y]$, so that T is homogeneous. However, $T[3] = 0$ and $T[t] = t^2/(t^2 + 1)$, while

$$T[t + 3] = \frac{(t + 3)^2}{(t + 3)^2 + 1} \neq \frac{t^2}{t^2 + 1} + 0 = T[t] + T[3]$$

Therefore T is not additive, even though it is homogeneous.

If one allows complex constants, then there are operators which are additive but not homogeneous. One such is Re, which carries each complex function f to its real part Re f.

The properties of linear operators can be expected to have important parallels in the properties of linear differential equations, which are of the form

$$Ly = f$$

where L is linear.

One such property was obtained in the previous part: *If y_p is some solution of a given linear differential equation, and if h is the general solution of the associated homogeneous equation, then the general solution of the given equation is*

$$y = y_p + h$$

Now let us consider homogeneous equations only. They are of the form

$$Ly = 0 \tag{70}$$

where L is some linear differential operator. A function y is a solution of (70) if and only if y satisfies the equation, that is,

$$Ly = 0$$

and this is so if and only if L carries y to the zero function. The function y is carried to zero if and only if every constant multiple is also carried to zero, for the homogeneity of L,

$$L[cy] = cLy \tag{71}$$

assures that the left-hand side of (71) is zero if and only if the right-hand side is zero. Thus, y is a solution of a homogeneous equation if and only if every constant multiple is also a solution. The property is more succinctly put this way: If a linear differential equation is homogeneous, then the collection of all solutions is "closed under taking constant multiples." Actually, much more is apparent, for the collection of all solutions is also *closed under taking linear combinations.* That is, if y_1 and y_2 are solutions, so that

$$Ly_1 = 0 \quad \text{and} \quad Ly_2 = 0$$

and if c_1 and c_2 are constants, then $c_1 y_1 + c_2 y_2$ is also a solution, for the linearity of L gives this:

$$L[c_1 y_1 + c_2 y_2] = c_1 Ly_1 + c_2 Ly_2$$
$$= c_1 \cdot 0 + c_2 \cdot 0 = 0$$

That is, $y = c_1 y_1 + c_2 y_2$ satisfies the equation $Ly = 0$.

For example, since the differential equation

$$y'' + 3y' + 2y = 0$$

is both linear and homogeneous, and since both e^{-t} and e^{-2t} are solutions, it follows from the previous remarks that

$$c_1 e^{-t} + c_2 e^{-2t}$$

is also a solution, no matter what may be the values of the constants c_1 and c_2.

Nonhomogeneous equations do not share this property. If y_p is a solution of the equation

$$Ly = f \tag{72}$$

where f is not the zero function, then $2y_p$ has the property

$$L[2y_p] = 2Ly_p = 2f$$
$$L[2y_p] \neq f$$

since $2f \neq f$. It follows that $2y_p$ is not a solution of (72), even though y_p is a solution. Thus, the collection of solutions of (72) is *not* closed under taking constant multiples nor under taking linear combinations.

All this discussion may now be summarized for convenient reference, as follows.

Theorem *If L is a linear differential operator, then the collection H of all solutions of the homogeneous equation*

$$Ly = 0 \tag{73}$$

is closed under taking linear combinations. On the other hand, if f is not the zero function, then the collection of all solutions of the nonhomogeneous equation

$$Ly = f \tag{74}$$

is not closed under taking constant multiples of linear combinations. Instead, if y_p *is some one solution of* (74), *then the solutions of* (74) *are exactly the functions of the form*

$$y = y_p + y_h$$

where y_h *is a solution of* (73), *a member of H.*

18. Sample Problems with Answers

Each of the following is the assignment rule for an operator defined on $C^1(\mathbf{R})$. Test each operator for linearity, and either show directly that it is linear or produce a counterexample and show that it is not.

S. $T[y] = 3y' + 4y$ *Answer:* Linear

T. $T[y](t) = 3y'(t) + 4$ *Answer:* Nonlinear; try $y(t) \equiv 0$

U. $T[y] = 3y' + 4(y)^2$ *Answer:* Nonlinear; try $y(t) \equiv 1$, $c = -1$

V. $T[y] = 3(y')^2 + 4y$ *Answer:* Nonlinear; try $y(t) = t$, $c = 2$

W. $T[y](t) = t^2 y'(t) + e^t y(t)$ *Answer:* Linear

Which of the following are linear differential equations and which are not?

X. $3y' + 4y = 10$ *Answer:* Linear

Y. $3y'(t) + 4 = 10t$ *Answer:* Linear; rewrite the equation as $3y'(t) = 10t - 4$

Z. $3y' + 4(y)^2 = 0$ *Answer:* Nonlinear

AA. $3[y'(t)]^2 + 4y(t) = 7e^t$ *Answer:* Nonlinear

BB. $t^2 y'(t) + e^t y(t) = \cos 3t$ *Answer:* Linear

CC. For the initial value problem

$$\left. \begin{array}{r} 3y' + 4y = 10 \\ y(0) = 2 \end{array} \right\}$$

find the general solution to the associated homogeneous equation, and a constant function which solves the given equation. From these, determine both the general solution of the given equation and the solution of the initial value problem.

    Answer: $y(t) = \frac{1}{2}(5 - e^{-4t/3})$

19. Problems

Test each of the following equations for linearity. For the linear ones, list the coefficients in order. For the nonlinear ones, provide a counterexample and show that the corresponding operator is nonlinear.

14. $y' + 3y = 4$ **15.** $ty'(t) + 3y(t) = 4$

16. $e^t y''(t) = 4$ **17.** $e^t y''(t) + t[y(t)]^2 = t^3$

18. $y(t)y'(t) + 3y(t) = t$ **19.** $y'(t) + e^{y(t)} = 0$

20. $y''(t) + e^t y'(t) + (\cos 2t)y(t) = \sin 2t$

21. Find the unique solution to the initial value problem

$$\left. \begin{array}{r} y' + 4y = 10t \\ y(0) = 1 \end{array} \right\}$$

by first determining the general solution for $y' + 4y = 10t$.

LINEAR DIFFERENTIAL OPERATORS WITH CONSTANT COEFFICIENTS

These operators were defined in the previous section. In this section and the next they will be used to find solutions of the most common differential equations of all. The solution methods are very efficient.

Each section begins with the procedure for the solution method, then progresses to its justification, and ends by showing the relation of the method to general information about the solutions of differential equations.

B. THE CHARACTERISTIC POLYNOMIAL; SOLUTIONS OF HOMOGENEOUS EQUATIONS

1. Example

The homogeneous equation

$$y'' + 3y' + 2y = 0 \tag{1}$$

has for its general solution the functions

$$y(t) = A_1 e^{-t} + A_2 e^{-2t} \tag{2}$$

where A_1 and A_2 are arbitrary constants. The calculation by which one goes quickly from (1) to (2) is this:

First, write the characteristic polynomial for (1), $s^2 + 3s + 2$. (This has already been defined in our discussion of the Laplace transform. It will be reviewed in the next part.) Writing the polynomial does not require any calculation, since its coefficients are the coefficients of the given differential equation.

Second, find the roots $\alpha_1 = -1$ and $\alpha_2 = -2$ of the characteristic polynomial. (These are named with the Greek letter *alpha*. Finding them is an algebraic calculation. It is exactly the same calculation as factoring the denominator of $\mathscr{L}[y]$, the Laplace transform of the solution y.)

Third, write down the two functions $e^{\alpha_1 t} = e^{-t}$ and $e^{\alpha_2 t} = e^{-2t}$, which are determined by the rule to be given in Part 3 below. No calculation is required.

Fourth, write down the general linear combination of these two functions:

$$A_1 e^{-t} + A_2 e^{-2t} \qquad A_1 \text{ and } A_2 \text{ arbitrary constants}$$

That is all there is to it. The only calculation is the finding of the roots of the polynomial (which also has to be done if one uses Laplace transforms). Therefore, the method is the quickest and most efficient way of finding a general solution. It does require memorizing the rule, which may be easier to remember if you bear this in mind: The most prominent solutions are exponentials, and at the very least you want to know which exponentials are solutions. They must satisfy the differential equation. Therefore, substitution of e^{st} into the equation should lead to a determination of the correct values of s.

Since

$$\frac{de^{st}}{dt} = se^{st} \quad \text{and} \quad \frac{d^2e^{st}}{dt^2} = s^2e^{st}$$

substitution into the left-hand side of the differential equation (1) must give

$$s^2e^{st} + 3se^{st} + 2e^{st} = 0$$

or

$$(s^2 + 3s + 2)e^{st} = 0$$

Since the factor e^{st} in this equation never takes the value zero, it follows that the values of s determined by the substitution are exactly the roots of the polynomial $s^2 + 3s + 2$, the first factor on the left-hand side, the characteristic polynomial.

2. The Characteristic Polynomial, Its Roots and Factors

The linear differential equation with constant coefficients

$$a_n y^{(n)} + a_{n-1} y^{(n-1)} + \cdots + a_1 y' + a_0 y = f$$

the corresponding operator L, and the corresponding homogeneous differential equation $Ly = 0$ or

$$a_n y^{(n)} + a_{n-1} y^{(n-1)} + \cdots + a_1 y' + a_0 y = 0$$

all have the same *characteristic polynomial*

$$p(s) = a_n s^n + a_{n-1} s^{n-1} + \cdots + a_1 s + a_0$$

The coefficient a_k multiplying $y^{(k)}$, the kth derivative of y, becomes the coefficient of s^k in the polynomial. The roots of the polynomial determine the general solution of the differential equation.

A number α is a *root with multiplicity m* if $(s - \alpha)^m$ is a factor of the polynomial but $(s - \alpha)^{m+1}$ is not. Thus, 3 is a root with multiplicity 2 of the polynomial

$$2s^3 - 12s^2 + 18s = 2s(s - 3)^2$$

Its multiplicity is determined by the fact that $(s - 3)^2$ is a factor of the polynomial but $(s - 3)^3$ is not a factor of it. The other root is zero, with multiplicity 1.

The roots of a polynomial should always be counted according to their multiplicities. In this example we would give the roots as 0, 3, 3, so that the multiplicities can be read off directly from the list of roots.

When so counted, the number of roots is the same as the degree of the polynomial (the power of the highest power of s with nonzero coefficient), which is the same as the order of the differential equation.

3. The Rule for Real Solutions

Consider a given homogeneous linear differential equation with constant coefficients and the corresponding characteristic polynomial $p(s)$. If the number α is a root of $p(s)$ with multiplicity m, then the m functions

$$e^{\alpha t} \qquad te^{\alpha t} \qquad t^2 e^{\alpha t} \qquad \cdots \qquad t^{m-2} e^{\alpha t} \qquad t^{m-1} e^{\alpha t}$$

are solutions of the differential equation, and so is every linear combination of them,

$$A_1 e^{\alpha t} + A_2 te^{\alpha t} + \cdots + A_m t^{m-1} e^{\alpha t} = e^{\alpha t}(A_1 + A_2 t + \cdots + A_m t^{m-1})$$

If $\alpha + j\beta$ and $\alpha - j\beta$ are a pair of complex roots of $p(s)$, conjugates of each other with $\beta \neq 0$, and if each of them has multiplicity m, then the $2m$ functions

$$e^{\alpha t} \cos \beta t \qquad te^{\alpha t} \cos \beta t \qquad \cdots \qquad t^{m-1} e^{\alpha t} \cos \beta t$$

and $\qquad\quad e^{\alpha t} \sin \beta t \qquad te^{\alpha t} \sin \beta t \qquad \cdots \qquad t^{m-1} e^{\alpha t} \sin \beta t$

are solutions of the differential equation, and so is every linear combination of them,

$$B_1 e^{\alpha t} \cos \beta t + B_2 te^{\alpha t} \cos \beta t + \cdots + B_m t^{m-1} e^{\alpha t} \cos \beta t$$

$$+ B_{m+1} e^{\alpha t} \sin \beta t + B_{m+2} te^{\alpha t} \sin \beta t + \cdots + B_{2m} t^{m-1} e^{\alpha t} \sin \beta t$$

$$= (B_1 + B_2 t + \cdots + B_m t^{m-1}) e^{\alpha t} \cos \beta t$$

$$+ (B_{m+1} + B_{m+2} t + \cdots + B_{2m} t^{m-1}) e^{\alpha t} \sin \beta t$$

The general solution of the differential equation is the sum of all the linear combinations for all the distinct roots.

In the examples which follow, notice that the coefficients are renamed, so that no two of the linear combinations have any coefficients in common and no two functions appear to be multiplied by the same coefficient.

4. Example: $p(s)$ Has Real, Equal Roots

The equation

$$y'' + 6y' + 9y = 0 \tag{3}$$

has characteristic polynomial

$$p(s) = s^2 + 6s + 9 = (s + 3)^2$$

The roots of $p(s)$ are -3, -3; that is, -3 with multiplicity $m = 2$. The rule therefore says that e^{-3t} and te^{-3t} are solutions of Equation (3) and that

$$y(t) = A_1 e^{-3t} + A_2 te^{-3t} \tag{4}$$

is the general solution of Equation (3).

That is all there is to the application of the rule. Let us verify the conclusion. First, let L be the operator corresponding to (3), with domain $C^2(\mathbf{R})$ and assignment rule

$$Ly = y'' + 6y' + 9y$$

Verifying the first conclusion from the rule consists in observing that

$$L[e^{-3t}] = 9e^{-3t} + 6(-3e^{-3t}) + 9e^{-3t}$$

$$= (9 - 18 + 9)e^{-3t} = 0 \tag{5}$$

and $L[te^{-3t}] = 9te^{-3t} - 6e^{-3t} + 6(-3te^{-3t} + e^{-3t}) + 9te^{-3t}$

$$= (9 - 18 + 9)te^{-3t} + (-6 + 6)e^{-3t} = 0 \tag{6}$$

That is, e^{-3t} and te^{-3t} are solutions of (3). The next conclusion is that $A_1 e^{-3t} + A_2 te^{-3t}$ is also a solution. The verification of this follows easily from the linearity of L, together with (5) and (6).

Let us now compare the results so far with the results of using Laplace transforms on the problem.

$$s^2 \mathscr{L}[y](s) - sy(0) - y'(0) + 6[s\mathscr{L}[y](s) - y(0)] + 9\mathscr{L}[y](s) = 0$$

$$\mathscr{L}[y](s) = \frac{sy(0) + y'(0) + 6y(0)}{(s + 3)^2}$$

$$= \frac{A_1}{s + 3} + \frac{A_2}{(s + 3)^2}$$

$$y(t) = A_1 e^{-3t} + A_2 te^{-3t} \qquad t \geq 0 \tag{7}$$

where A_1 and A_2 are some constants to be determined, as is usual in the partial-fraction decomposition. [In fact, $A_1 = y(0)$ and $A_2 = y'(0) + 3y(0)$.]

From our study of Laplace transforms, we know that (7) is the general solution. Comparison with (4) shows that the new method also gives the general solution, and more efficiently than Laplace transforms do. Everything since Equation (4) has been verification and comparison with other methods.

The satisfying of initial conditions must be done separately.

5. Example: $p(s)$ Has Complex Roots

The equation

$$y'' + 2y' + 5y = 0 \tag{8}$$

has characteristic polynomial $s^2 + 2s + 5$, whose roots are found using the quadratic formula as follows:

$$s = \frac{-2 \pm \sqrt{4 - 20}}{2} = -1 \pm 2j$$

each root with multiplicity 1. By the rule, $e^{-t} \cos 2t$ and $e^{-t} \sin 2t$ are solutions of (8), and the general solution is

$$y(t) = B_1 e^{-t} \cos 2t + B_2 e^{-t} \sin 2t$$
$$= e^{-t}(B_1 \cos 2t + B_2 \sin 2t) \tag{9}$$

Each choice of the constants B_1 and B_2 gives a solution of (8), a fact which is verified, as in the previous example, using the operator with domain $C^2(\mathbf{R})$ and assignment rule

$$Ly = y'' + 2y' + 5y$$

This result can also be compared with the Laplace-transform solution:

$$\mathscr{L}[y](s) = \frac{sy(0) + y'(0) + 2y(0)}{s^2 + 2s + 5}$$

$$y(t) = e^{-t}\left[y(0) \cos 2t + \frac{y'(0) + y(0)}{2} \sin 2t\right] \qquad t \geq 0$$

When comparing this formula with (9), notice that

$$B_1 = y(0)$$

$$B_2 = \frac{y'(0) + y(0)}{2}$$

Whether one takes $y(0)$ and $y'(0)$ as arbitrary (using the Laplace transform) or B_1 and B_2 as arbitrary (using the characteristic polynomial), there will be two arbitrary constants in the general solution of this second-order equation. The order of the equation, the degree of the characteristic polynomial $p(s)$, the number of roots (counting multiplicities) of $p(s)$, the number of functions given by the rule, the number of arbitrary coefficients required, and the number of arbitrary initial conditions which may be satisfied at a given point—all these numbers are equal, and in this problem each is the number 2.

6. Example: $p(s)$ Has Real, Distinct Roots

The equation

$$y'' + 3y' + 2y = 0 \tag{10}$$

has the characteristic polynomial

$$s^2 + 3s + 2 = (s + 1)(s + 2) \tag{11}$$

whose roots are -1 and -2, each with multiplicity 1. According to the rule, e^{-t} is a solution, along with every linear combination, which in this case is every multiple $A_1 e^{-t}$. Also according to the rule, e^{-2t} is a solution, along with every multiple. In fact, the rule appears to suggest that the multiples be called $A_1 e^{-2t}$. However, A_1 is a name already in use so that a different name must be chosen to distinguish it from the first. Let us choose A_2, for the multiples of e^{-2t} can certainly be called $A_2 e^{-2t}$. Now, according to the rule, we add $A_1 e^{-t}$ and $A_2 e^{-2t}$ to get the general solution:

$$y(t) = A_1 e^{-t} + A_2 e^{-2t} \qquad A_1 \text{ and } A_2 \text{ arbitrary} \qquad (12)$$

Let us verify that solution by direct substitution. The corresponding linear operator has domain $C^2(\mathbf{R})$ and assignment rule

$$L = D^2 + 3D + 2$$

The substitution goes this way, with linearity used to simplify the calculation:

$$L[A_1 e^{-t} + A_2 e^{-2t}] = A_1 L[e^{-t}] + A_2 L[e^{-2t}]$$
$$= A_1(e^{-t} - 3e^{-t} + 2e^{-t}) + A_2(4e^{-2t} - 6e^{-2t} + 2e^{-2t})$$
$$= A_1 \cdot 0 + A_2 \cdot 0 = 0$$

Thus, every function of the form (12) is a solution of (10).

Let us also compare (12) with the result of the Laplace-transform solution method:

$$s^2 \mathscr{L}[y](s) - sy(0) - y'(0) + 3[s\mathscr{L}[y](s) - y(0)] + 2\mathscr{L}[y](s) = 0$$

$$\mathscr{L}[y](s) = \frac{sy(0) + y'(0) + 3y(0)}{(s+1)(s+2)} \qquad (13)$$

$$= \frac{A_1}{s+1} + \frac{A_2}{s+2}$$

where A_1 and A_2 are to be determined, as is usual in the partial-fraction decomposition. Taking the inverse transform gives

$$y(t) = A_1 e^{-t} + A_2 e^{-2t} \qquad t \geq 0$$

and substitution shows that the solution is valid on the whole line. This agrees with (12) and shows that (12) is the general solution. Notice that the denominator of $\mathscr{L}[y](s)$ is exactly the characteristic polynomial. Determination of A_1 and A_2 from (13) shows that

$$A_1 = y'(0) + 2y(0)$$
$$A_2 = -y'(0) - y(0)$$

7. More Examples of the Use of the Characteristic Polynomial

The equation

$$y''' = 0 \qquad (14)$$

has the characteristic polynomial $p(s) = s^3$ whose root is 0 with multiplicity 3.

Hence,

$$A_1 e^{0t} + A_2 t e^{0t} + A_3 t^2 e^{0t} = A_1 + A_2 t + A_2 t^2$$

is the general solution.

The equation

$$y''' + 14y'' + 49y' = 0 \tag{15}$$

has the characteristic polynomial

$$p(s) = s^3 + 14s^2 + 49s = s(s + 7)^2 \tag{16}$$

whose roots are $-7, -7, 0$ (or -7 with multiplicity 2, and 0 with multiplicity 1). According to the rule, the general solution is

$$y(t) = A_1 e^{-7t} + A_2 t e^{-7t} + A_3 e^{0t}$$
$$= e^{-7t}(A_1 + A_2 t) + A_3$$

The equation

$$y''' + 4y' = 0 \tag{17}$$

has characteristic polynomial

$$p(s) = s^3 + 4s = s(s^2 + 4) \tag{18}$$

The roots are $0, +2j, -2j$, each with multiplicity 1, and so the general solution of the equation is

$$y(t) = A_1 + A_2 \cos 2t + A_3 \sin 2t$$

The equation

$$y^{(4)} + 18y'' + 81y = 0 \tag{19}$$

has characteristic polynomial

$$p(s) = s^4 + 18s^2 + 81 = (s^2 + 9)^2$$

The roots are $+3j$ and $-3j$, each with multiplicity 2, and so the general solution of the equation is

$$y(t) = B_1 \cos 3t + B_2 t \cos 3t + B_3 \sin 3t + B_4 t \sin 3t$$

8. Remarks on the Practical Application of the Rule

The four steps of the method were outlined in Part 1. In the first and second steps, one writes $p(s)$ and then finds all its roots and their multiplicities.

The third step is to write the list of functions corresponding to each real root and the list corresponding to each pair of complex conjugate roots. In each case the number of functions is the same as the number of multiplicities. Thus, a real root with multiplicity m contributes m functions. In (15), the real root -7 with multiplicity 2 contributed two functions, e^{-7t} and te^{-7t}, while the real root 0 with multiplicity 1 contributed only one function, the constant function $e^{0t} = 1$. For a pair of complex roots, be sure to count the multiplicities of both members of the pair. Thus, in (19), the root $3j$ has multiplicity 2, as does the root $-3j$, and so four functions are to be written. It is *not* the case that the cosines go with the root $3j$ and the sines with the other root; nor is the opposite the case. Instead, use the Rule for Real Solutions just as it is stated in Part 3: *to the pair* of complex roots, each with multiplicity 2, correspond the four functions given in the example.

The last step in the method consists in providing each of these functions with a coefficient, making sure that each function gets a coefficient which is different from all the others. If the order of the equation is n, then the degree of the polynomial will also be n, and therefore the number of roots (counting multiplicities) will also be n. Since there is one function in the list for each root (counting multiplicities), and one coefficient is required for each function, the number of coefficients will also be n, the order of the given differential equation.

Be sure to check, as the last thing, that you have provided the same number of coefficients as the order of the equation. If you have a different number, then you have made at least one error somewhere.

9. Sample Problems with Answers

In each problem, write the characteristic polynomial for the given homogeneous equation with constant coefficients. Then write all the roots together with their multiplicities, the functions given by the rule, and the general solution of the given equation.

A. $y''' + 2y'' + 5y' = 0$

Answers: $p(s) = s^3 + 2s^2 + 5s = s(s^2 + 2s + 5)$; roots: 0 with multiplicity 1, $-1 + 2j$ with multiplicity 1, $-1 - 2j$ with multiplicity 1; functions: $e^{0t} = 1$, $e^{-t}\cos 2t$, $e^{-t}\sin 2t$; $y(t) = A_1 + A_2 e^{-t}\cos 2t + A_3 e^{-t}\sin 2t$

B. $y'' - 6y = 0$

Answers: $p(s) = s^2 - 6 = (s - \sqrt{6})(s + \sqrt{6})$; roots: $\sqrt{6}$ with multiplicity 1, $-\sqrt{6}$ with multiplicity 1; functions: $e^{\sqrt{6}t}$, $e^{-\sqrt{6}t}$; $y(t) = A_1 e^{\sqrt{6}t} + A_2 e^{-\sqrt{6}t}$

C. $y^{(4)} + 8y''' + 16y'' = 0$

Answers: $p(s) = s^4 + 8s^3 + 16s^2 = s^2(s + 4)^2$; roots: 0 with multiplicity 2, -4 with multiplicity 2; functions: $e^{0t} = 1$, $te^{0t} = t$, e^{-4t}, te^{-4t}; $y(t) = A_1 + A_2 t + A_3 e^{-4t} + A_4 te^{-4t}$

10. Problems

For each of the following, find the characteristic polynomial, all its roots with multiplicities, and the general solution of the differential equation.

1. $y' = 0$	**2.** $y'' = 0$
3. $y' + 3y = 0$	**4.** $y'' + 8y = 0$
5. $y'' + 4y' + 5y = 0$	**6.** $y'' - 4y' + 4y = 0$
7. $y'' + 4y' + 3y = 0$	**8.** $y''' + 4y'' + 5y' = 0$
9. $y''' - 4y'' + 4y' = 0$	**10.** $y'' + 3y' = 0$
11. $y''' + 3y'' + 3y' + y = 0$	**12.** $y^{(4)} + 2y'' + y = 0$

For each of the following, write a homogeneous linear differential equation whose characteristic polynomial is the given polynomial, and write the general solution of the equation.

13. $p(s) = (s - 2)^4$ **14.** $p(s) = (s - 2)^3(s - 1)^2$ **15.** $p(s) = (s^3 + 4s)^2$

11. Determination of Damped and Undamped Sinusoidal Solutions

The determination of the sinusoidal solutions follows from that of the exponential ones, with the use of the complex exponential function, as developed in Section I of Chapter III. Consider the equation

$$y'' + 9y = 0 \tag{20}$$

and the operator $L = D^2 + 9$ and the polynomial $p(s) = s^2 + 9$. By direct computation,
$$L[e^{st}] = (s^2 + 9)e^{st} = p(s)e^{st}$$

Since the roots of $p(s)$ are $3j$ and $-3j$, it should follow that e^{3jt} and e^{-3jt} are carried to zero by L:

$$L[e^{3jt}] = p(3j)e^{3jt}$$
$$= [(3j)^2 + 9]e^{3jt}$$
$$= 0e^{3jt} = 0$$

and, similarly, $\qquad L[e^{-3jt}] = 0$

In fact, that is exactly what happens, for we showed (in Section I of Chapter III) that
$$D[e^{ct}] = ce^{ct} \qquad \text{if } c \text{ is complex}$$

By linearity, it then follows that

$$L[A_1 e^{3jt} + A_2 e^{-3jt}] = A_1 L[e^{3jt}] + A_2 L[e^{-3jt}]$$
$$= A_1 \cdot 0 + A_2 \cdot 0 = 0$$

so that the function

$$y(t) = A_1 e^{3jt} + A_2 e^{-3jt} \tag{21}$$

is a solution of (20), no matter what the constants A_1 and A_2 may be. That is all in accord with what the rule says about exponentials. We must now see what (21) has to say about sines and cosines.

In Section I of Chapter III, the complex exponential was defined by the formula

$$e^{x+jy} = e^x(\cos y + j \sin y)$$

If $x = 0$ and $y = 3t$ or $-3t$, we have

$$e^{3jt} = \cos 3t + j \sin 3t \qquad \text{and} \qquad e^{-3jt} = \cos 3t - j \sin 3t$$

Using this fact in (21) puts the solution of (20) into these forms:

$$y(t) = A_1 e^{3jt} + A_2 e^{-3jt}$$
$$= A_1(\cos 3t + j \sin 3t) + A_2(\cos 3t - j \sin 3t)$$
$$= (A_1 + A_2) \cos 3t + (A_1 - A_2)j \sin 3t$$
$$= B_1 \cos 3t + B_2 \sin 3t \tag{22}$$

where $$B_1 = A_1 + A_2 \quad \text{and} \quad B_2 = (A_1 - A_2)j \tag{23}$$

Now (22) is exactly what the rule said was the general solution of (20), but we have derived it by applying the *exponential* part of the rule to the complex roots of the polynomial.

Because of the relations between the (complex) exponential and the damped and undamped sinusoids (as given in Section I of Chapter III), the Rule for Real Solutions may be replaced by a new rule which follows.

12. The Rule for Complex Solutions

If $p(s)$ is the characteristic polynomial of a homogeneous linear differential equation with constant coefficients, and if α is a (real or complex) root of $p(s)$ with multiplicity m, then the functions

$$e^{\alpha t} \qquad te^{\alpha t} \qquad \cdots \qquad t^{m-1}e^{\alpha t} \tag{24}$$

are solutions of the differential equation. The general solution of the differential equation is obtained by taking the linear combination of all such functions, corresponding to *all* the roots of $p(s)$, where the coefficients in the linear combination are distinct, unspecified, arbitrary constants.

13. Example and Remarks on the Practical Application of the Rule

The equation

$$y'' + 2y' + 10y = 0$$

has the characteristic polynomial $s^2 + 2s + 10$ whose roots are

$$s = -1 \pm 3j$$

By the Rule for Real Solutions, the solution is

$$y(t) = e^{-t}(B_1 \cos 3t + B_2 \sin 3t)$$

and by the Rule for Complex Solutions,

$$y(t) = A_1 e^{(-1+3j)t} + A_2 e^{(-1-3j)t}$$
$$= e^{-t}(A_1 e^{3jt} + A_2 e^{-3jt})$$

As we have seen, the two forms of the solution are equivalent, since the B's can be calculated from the A's as in (22) and (23), and the A's can be calculated from the B's by solving (23) for the A's in terms of the B's, as follows:

$$A_1 = \frac{B_1 - jB_2}{2} \qquad A_2 = \frac{B_1 + jB_2}{2}$$

From the point of view of engineering applications, there is a definite advantage to using the original rule and the form of solution which contains only real functions, without any complex functions at all. Real-world phenomena, such as distance, force, voltage, and current, are all measured in real numbers. The solution in terms of damped sinusoids, the *real* functions, may be obtained from the complex rule and the identities relating them to the complex exponential, or they may be obtained from the original rule, which is ordinarily to be preferred.

The complex rule, however, is simpler to remember and easier to use in situations in which handling complex functions is no disadvantage.

14. Factoring Constant-Coefficient Operators

The factoring of operators will be used in the refined method for eliminating variables, with which the section ends. In Part 15 it is shown how the factors obtained determine solutions of equations.

To begin, observe that the action of $D^2 + 3D + 2$ on a function y can be described as the action of $D + 2$ on $(D + 1)[y]$:

$$(D + 2)[(D + 1)[y]] = (D + 2)[2y' + y]$$
$$= (y'' + y') + (2y' + 2y)$$
$$= y'' + 3y' + 2y$$
$$= (D^2 + 3D + 2)[y] \tag{25}$$

Thus we may write

$$D^2 + 3D + 2 = (D + 2)(D + 1) \tag{26}$$

where the right side of (25) is the *composition* of $D + 2$ on $D + 1$, that is, the result of following the action of $D + 1$ with that of $D + 2$. Note that (26) strongly suggests a multiplication, but it is *not multiplication* (because D is *not a number* and Dy is *not a number*). It is a composition, as was calculated at (25). Nevertheless, the two operators $D + 1$ and $D + 2$ are called the *factors of the composition* $(D + 2)(D + 1)$. In the factored form, the action of the operator is analyzed as a composition of first-order operators, which are very easy to describe, use, and remember.

We shall see that *every higher-order linear differential operator with constant coefficients can always be factored as the composition of first-order operators*. That is, the more complicated action of the higher-order operators can always be described in terms of a succession of simple actions.

The factoring of the operators is exactly parallel to factoring the characteristic polynomials and finding the roots. To continue the example, observe that the factorization of the characteristic polynomial

$$s^2 + 3s + 2 = (s + 2)(s + 1) \tag{27}$$

looks remarkably like (26). That is, $s + 2$ is the characteristic polynomial of the operator $D + 2$, $s + 1$ is that for $D + 1$, and $(s + 2)(s + 1)$ is the characteristic polynomial of the composition of $D + 2$ on $D + 1$. It will be apparent that, in general, *if L_1 and L_2 are constant-coefficient operators with characteristic polynomials $p_1(s)$ and $p_2(s)$, then the product $p_2(s)p_1(s)$ is the characteristic polynomial of the composition $L_2 L_1$.*

The technique of factoring now follows immediately. If

$$L = a_n D^n + a_{n-1} D^{n-1} + \cdots + a_0$$

is a linear differential operator with constant coefficients and order n, then its characteristic polynomial may be factored as follows:

$$p(s) = a_n(s - \alpha_n)(s - \alpha_{n-1}) \cdots (s - \alpha_1) \qquad (28)$$

where $\alpha_1, \alpha_2, \ldots, \alpha_n$ are the roots of the polynomial, each repeated according to its multiplicity. In a given problem, these roots and multiplicities would have to be found, as in the example. Having found them, one may at once write the factors of L and their characteristic polynomials:

$$L_1 = D - \alpha_1 \qquad\qquad p_1(s) = s - \alpha_1$$
$$L_2 = D - \alpha_2 \qquad\qquad p_2(s) = s - \alpha_2$$
$$\cdots\cdots\cdots\cdots\cdots\cdots \qquad\qquad \cdots\cdots\cdots\cdots\cdots\cdots$$
$$L_{n-1} = D - \alpha_{n-1} \qquad\qquad p_{n-1}(s) = s - \alpha_{n-1}$$
$$L_n = a_n D - a_n \alpha_n \qquad\qquad p_n(s) = a_n(s - \alpha_n)$$

Since the composition of operators corresponds to the product of the characteristic polynomials, it follows that L and $L_n L_{n-1} \cdots L_2 L_1$ have the same characteristic polynomial $p(s)$ as given in (28). Therefore they have the same coefficients and the same assignment rule. Since they have been taken to have the same domain $C^\infty(\mathbf{R})$ as well, it follows that they are the same operator.

For an example, let $Ly = 4y'' - 4y' + 5y$. The corresponding characteristic polynomial is

$$p(s) = 4s^2 - 4s + 5$$

Its roots, which may be found using the quadratic formula, are

$$\alpha_1 = \tfrac{1}{2} + j \qquad \alpha_2 = \tfrac{1}{2} - j$$

Therefore, $$p(s) = 4[s - (\tfrac{1}{2} - j)][s - (\tfrac{1}{2} + j)]$$
$$= p_2(s)p_1(s)$$

where $\quad p_1(s) = s - (\tfrac{1}{2} + j) \quad$ and $\quad p_2(s) = 4[s - (\tfrac{1}{2} - j)] = 4s - (2 - 4j)$

Now that we have the factors of the polynomial, we can write the factors of L:

$$L_1 = D - (\tfrac{1}{2} + j) \qquad \text{or} \qquad L_1[y] = y' - (\tfrac{1}{2} + j)y$$

and $\qquad L_2 = 4D - (2 - 4j) \qquad \text{or} \qquad L_2[y] = 4y' - (2 - 4j)y$

Thus $$4y'' - 4y' + 5y = (4D - (2 - 4j))(D - (\tfrac{1}{2} + j))[y]$$

Sometimes it is more convenient to use all real factors, in which case one must accept factors of order 2. For example, $3D^4 - 48$ has characteristic polynomial

$$p(s) = 3(s^4 - 16) = 3(s^2 - 4)(s^2 + 4)$$
$$= 3(s - 2)(s + 2)(s^2 + 4)$$

and so

$$(3D^4 - 48)[y] = 3(D - 2)(D + 2)(D^2 + 4)[y] \qquad (29)$$

It is to be emphasized that the right side of (29) is not multiplication but the composition of operators. In general, even for operators which do not have constant coefficients, the result of following the action of one operator L_1 with that of another L_2 is called the *composition* of L_2 on L_1, and it is written $L_2 L_1$. The assignment rule for $L_2 L_1$ is therefore

$$L_2 L_1[y] = L_2[L_1[y]]$$

Composition of L_2 on L_1 is only possible if $L_1[y]$ always lies in the domain of L_2, that is, if the domain of L_2 includes all of the range of L_1. This is always true if L_1 and L_2 are constant-coefficient operators because each has $C^\infty(\mathbf{R})$ as both domain and range.

In Part 12 of Section A it was pointed out that the "commutativity" property,

$$L_2 L_1 = L_1 L_2 \qquad (30)$$

is important for variable elimination methods. This property is possessed by any pair of constant-coefficient operators L_1 and L_2 because the product of the corresponding characteristic polynomials is just the product of numbers:

$$p_2(s)p_1(s) = p_1(s)p_2(s) \qquad \text{for all } s$$

However, examples in the Sample Problems will show that outside the class of constant-coefficient operators one cannot expect commutativity.

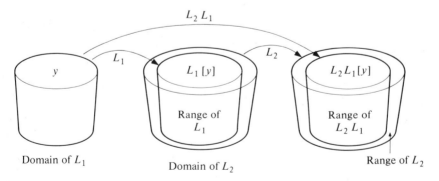

Figure 1 Schematic representation of the composition of L_2 on L_1.

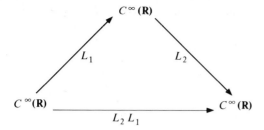

Figure 2 Diagrammatic representation of the composition of L_2 on L_1.

15. Annihilation: Which Equations Have a Given Solution?

Because of the fact that $(D - \alpha)[e^{\alpha t}]$ is the zero function, it follows that if L_1 is any constant-coefficient operator whatsoever, then

$$L_1(D - \alpha)y = 0 \tag{31}$$

is an equation for which $e^{\alpha t}$ is a solution. This is so because

$$L_1(D - \alpha)[e^{\alpha t}] = L_1[(D - \alpha)[e^{\alpha t}]] = L_1[0] = 0$$

In order to use this fact to help solve equations, we have only to observe that an exponential $e^{\alpha t}$ is a solution of a homogeneous constant-coefficient equation if and only if $s - \alpha$ is a factor of the characteristic polynomial, in which case the equation can be written in the form (31), according to the previous part. In this part we generalize this observation, taking as our tool the concept of annihilation.

A function y is said to be *annihilated* by a linear operator L if L carries y to 0. Thus, y is annihilated by L if and only if y is a solution of the homogeneous equation $L[y] = 0$. Asking what annihilates $t^k e^{\alpha t}$ is therefore equivalent to asking for the equations to which the function is a solution. A portion of the answer will be given here, and more will be given in the next part of the section.

If $(D - \alpha)^{k+1}$ is the operator resulting from applying $D - \alpha$ in succession $k + 1$ times, then $(D - \alpha)^{k+1}$ annihilates $t^k e^{\alpha t}$, for the differential equation $(D - \alpha)^{k+1}y = 0$ has characteristic polynomial $(s - \alpha)^{k+1}$. Therefore, we shall use an example to show how the annihilation is accomplished.

Applying $(D + 3)^2$ to te^{-3t} gives

$$(D + 3)^2 te^{-3t} = (D + 3)[(D + 3)te^{-3t}]$$
$$= (D + 3)e^{-3t} = 0$$

In this example, $\alpha = -3$, $k = 1$, and

$$(D + 3)te^{-3t} = e^{-3t}$$
$$(D + 3)^2 te^{-3t} = 0$$

In the general case,

$$(D - \alpha)t^k e^{\alpha t} = kt^{k-1}e^{\alpha t} + \alpha t^k e^{\alpha t} - \alpha t^k e^{\alpha t}$$

$$= kt^{k-1}e^{\alpha t}$$

and the action of $D - \alpha$ reduces the power of t by 1. After k successive applications of $D - \alpha$, one has

$$(D - \alpha)^k t^k e^{\alpha t} = k! \, t^0 e^{\alpha t}$$

$$= k! \, e^{\alpha t}$$

which is annihilated by one further application of $D - \alpha$. In fact, it follows that if f_k is any polynomial of degree k or less, then $(D - \alpha)^{k+1}$ annihilates every term of $f_k(t)e^{\alpha t}$.

Since $t^k e^{at} \cos bt$ and $t^k e^{at} \sin bt$ are linear combinations of $t^k e^{(a+bj)t}$ and $t^k e^{(a-bj)t}$, there are similar results for these functions. For instance, $D^2 + b^2$ annihilates $\cos bt$, $\sin bt$, and any linear combination of the two. In fact, performing the indicated differentiations shows that

$$(D^2 + b^2)[A_1 t^k \cos bt + A_2 t^k \sin bt] = [A_2 2bkt^{k-1} + A_1 k(k-1)t^{k-2}] \cos bt$$

$$+ [-A_1 2bkt^{k-1} + A_2 k(k-1)t^{k-2}] \sin bt$$

Continuing to apply $(D^2 + b^2)$ shows that if f_k and g_k are polynomials of degree at most k, then

$$(D^2 + b^2)^m[f_k(t) \cos bt + g_k(t) \sin bt] = f_{k-m}(t) \cos bt + g_{k-m}(t) \sin bt$$

which is zero if $m = k + 1$.

An argument which is exactly similar shows that if f_k and g_k are polynomials of degree at most k, then $[(D - a)^2 + b^2]^{k+1}$ annihilates the function

$$f_k(t)e^{at} \cos bt + g_k(t)e^{at} \sin bt$$

The results given above are summarized in Table 1.

Table 1 Functions annihilated by linear differential operators with constant coefficients

Operator	Functions annihilated*
$D - \alpha$	$Ae^{\alpha t}$
$(D - \alpha)^2$	$A_1 te^{\alpha t} + A_0 e^{\alpha t} = (A_1 t + A_0)e^{\alpha t}$
$(D - \alpha)^{k+1}$	$f_k(t)e^{\alpha t}$
$D^2 + b^2$	$A \cos bt + B \sin bt$
$(D^2 + b^2)^2$	$(A_1 t + A_0) \cos bt + (B_1 t + B_0) \sin bt$
$(D^2 + b^2)^{k+1}$	$f_k(t) \cos bt + g_k(t) \sin bt$
$(D - a)^2 + b^2$	$Ae^{at} \cos bt + Be^{at} \sin bt$
$[(D - a)^2 + b^2]^2$	$(A_1 t + A_0)e^{at} \cos bt + (B_1 t + B_0)e^{at} \sin bt$
$[(D - a)^2 + b^2]^{k+1}$	$f_k(t)e^{at} \cos bt + g_k(t)e^{at} \sin bt$

* f_k and g_k are polynomials of degree at most k.

If we want the homogeneous constant-coefficient equation of lowest order to which $t^k e^{\alpha t}$ is a solution, then we let $L_1 = (D - \alpha)^{k+1}$ and obtain the equation

$$L_1 y = 0$$

If we want an equation to which $t^k e^{\alpha t}$ and $t^m e^{\beta t}$ are solutions, with $\beta \neq \alpha$, we also let $L_2 = (D - \beta)^{m+1}$ and take

$$L_2 L_1 y = 0$$

We may be sure that $t^k e^{\alpha t}$ is a solution of the equation, but since it was shown in the previous part that $L_2 L_1 = L_1 L_2$ for constant-coefficient operators, it follows that

$$L_2 L_1 [t^m e^{\beta t}] = L_1 L_2 [t^m e^{\beta t}] = L_1 [0] = 0$$

and so the equation has the required solutions.

This gives us, at last, our strongest idea of the solutions of a constant-coefficient equation. If $Ly = 0$ is such an equation, and if L has the factors $a_n (D - \alpha_1)^{k_1}, \ldots,$ and $(D - \alpha_m)^{k_m}$, with $\alpha_1, \ldots, \alpha_m$ all different, then $t^k e^{\alpha_1 t}$ is a solution if k is an integer with $0 \leq k \leq k_1 - 1$, and it is annihilated by the factor $(D - \alpha_1)^{k+1}$ in the manner described above.

16. More Sample Problems with Answers

Find all functions of the form $t^k e^{\alpha t}$ which are annihilated by the following operators.

D. $(D - 6)^4$ *Answer:* $e^{6t}, te^{6t}, t^2 e^{6t}, t^3 e^{6t}$

E. $(D - 6)(D + 2)$ *Answer:* e^{6t}, e^{-2t}

F. $(D - 6)^2 (D + 2)$ *Answer:* e^{6t}, te^{6t}, e^{-2t}

G. $(D^2 - 4D - 12)^2$ *Answer:* $e^{6t}, te^{6t}, e^{-2t}, te^{-2t}$

H. $(D^2 + 1)^2$ *Answer:* $e^{jt}, te^{jt}, e^{-jt}, te^{-jt}$

Find an operator which annihilates each of the following.

I. te^{5t} *Answer:* $(D - 5)^2 = D^2 - 10D + 25$

J. $t^3 e^{5t}$ *Answer:* $(D - 5)^4$

K. $\cos 5t$ *Answer:* $D^2 + 25$

L. $t \cos 5t$ *Answer:* $(D^2 + 25)^2$

M. $e^{-t} + te^{5t}$ *Answer:* $(D + 1)(D - 5)^2$

N. $e^{-t} + t \cos 5t$ *Answer:* $(D + 1)(D^2 + 25)^2$

Write the general solution of each of the following equations.

O. $y'' - 4y' - 12y = 0$ *Answer:* $A_1 e^{6t} + A_2 e^{-2t}$

P. $y^{(4)} - 8y''' - 8y'' + 96y' + 144y = 0$ *Answer:* $A_1 e^{6t} + A_2 te^{6t} + A_3 e^{-2t} + A_4 te^{-2t}$

Q. $y^{(4)} + 2y'' + y = 0$ *Answer:* $A_1 e^{jt} + A_2 te^{jt} + A_3 e^{-jt} + A_4 te^{-jt}$

R. Show that the linear operators D and E with assignment rules $Dy = y'$ and $Ey = tDy$ are such that DE and ED are not equal.

    Answer: $DEy = D[ty'] = y' + ty''$, but $EDy = Ey' = ty''$

17. Problems

Find all functions of the form $t^k e^{\alpha t}$ which are annihilated by the following operators.

16. $(D + 4)^2$ **17.** $D^2 + 4$ **18.** $D^2 + 4D + 4$

19. $(D + 4)^2(D + 3)$ **20.** $(D^2 + 4)(D^2 + 3)$ **21.** $D^3 + 4D^2 + 4D$
22. $(D^2 + 4)^2(D + 3)$ **23.** $(D^2 + 4)^2(D^2 + 3)$ **24.** $(D^2 + 4)(D + 4)^2(D + 3)$

Find an operator which annihilates each of the following.

25. te^{-2t} **26.** $t^2 e^{-2t}$ **27.** $t^4 e^{8t}$

28. $\sin 2\pi t$ **29.** $3 \cos 2\pi t - 2 \sin 2\pi t$ **30.** $t \sin 2\pi t$

31. $t^2 \cos t$ **32.** $\cos 2\pi t + t \sin 2\pi t$

Find the general solution of each of the following equations:

33. $y'' + 2y' + y = 0$ **34.** $y'' + 6y' + 8y = 0$
35. $y''' + 6y'' + 8y' = 0$ **36.** $y'' + 4y' + 5y = 0$
37. $y''' + 4y'' + 5y' = 0$ **38.** $y^{(4)} + 2y'' + y = 0$
39. $(D^2 + 2)(D^2 + 4)[y] = 0$ **40.** $(D^2 + 4D + 5)(D + 3)[y] = 0$

In each of the following two problems, the assignment rules for linear operators L_1 and L_2 are given. Show that $L_1 L_2$ and $L_2 L_1$ are not equal.

41. $L_1 y = Dy$ and $L_2 y = t^2 D^2 y$ **42.** $L_1 y = D^2 y$ and $L_2 = tDy$

18. Last Words on Eliminating Variables

We are now ready to conclude our study of variable elimination. The equations considered in Part 12 of Section A,

$$x'' + 4x + y'' + 3y' + 2y = e^{2t} \tag{32a}$$
$$x' + 7x + y'' - y' - 2y = 3 \tag{32b}$$

may have y eliminated. The operators

$$L_1 = D^2 + 4 \qquad L_3 = D^2 + 3D + 2 = (D + 1)(D + 2)$$
$$L_2 = D + 7 \qquad L_4 = D^2 - D - 2 = (D + 1)(D - 2)$$

contain some common factors. Therefore we shall operate on (32a) with $D - 2$ rather than L_3, and on (32b) with $D + 2$ rather than L_4, because

$$(D - 2)L_3 = (D + 2)L_4 = (D + 2)(D + 1)(D - 2)$$

Thus,

$$(D - 2)[x'' + 4x] + (D - 2)[y'' + 3y' + 2y] = (D - 2)[e^{2t}] \tag{33}$$
$$(D + 2)[x' + 7x] + (D + 2)[y'' - y' - 2y] = (D + 2)[3] \tag{34}$$
$$(x''' - 2x'' + 4x' - 8x) + (y''' + y'' - 4y' - 4y) = 0$$
$$(x'' + 9x' + 14x) + (y''' + y'' - 4y' - 4y) = 6 \tag{35}$$

Subtraction gives an equation in x alone:

$$x''' - 3x'' - 5x' - 22x = -6 \tag{36}$$

Now let us see how the calculation looks if we use only the names of the operators. Let

$$F_0 = D + 1 \qquad F_3 = D + 2 \qquad F_4 = D - 2 \tag{37}$$

so that

$$L_3 = F_0 F_3 \qquad \text{and} \qquad L_4 = F_0 F_4 \tag{38}$$

Written with these names, Equations (32a) and (32b) are

$$L_1 x + F_0 F_3 y = e^{2t} \tag{39}$$

$$L_2 x + F_0 F_4 y = 3 \tag{40}$$

The action of F_4 on the first equation and F_3 on the second give

$$F_4 L_1 x + F_4 F_0 F_3 y = F_4[e^{2t}] \tag{41}$$

$$F_3 L_2 x + F_3 F_0 F_4 y = F_3[3] \tag{42}$$

Since $\qquad F_4 F_0 F_3 y = F_3 F_0 F_4 y = y''' + y'' - 4y' - 4y \tag{43}$

subtracting (42) from (41) gives

$$F_4 L_1 x - F_3 L_2 x = F_4[e^{2t}] - F_3[3] \tag{44}$$

or $\qquad x''' - 3x'' - 5x' - 22x = -6 \tag{45}$

The important points are these. First, the subtraction eliminates the terms containing y because of the equalities in (43). Those equalities may be verified using (37), the formulas for F_0, F_3, and F_4; but the equality of $F_4 F_0 F_3$ and $F_3 F_0 F_4$ is a consequence of the fact that *all three operators are linear differential operators with constant coefficients*.

The second important point is that the degree of (45) is lower than that of the equation we would have obtained using L_3 and L_4 instead of F_3 and F_4. Having lower order is definitely a practical advantage in using any solution method. The result of using L_3 and L_4 would have been

$$L_4 L_1 x - L_3 L_2 x = L_4[e^{2t}] - L_3[3]$$

or $\qquad F_0[F_4 L_1 x - F_3 L_2 x] = F_0[F_4[e^{2t}] - F_3[3]]$

the result of using F_0 on both sides of (44). Hence the equation showing all the derivatives would have been the result of using $F_0 = D + 1$ on both sides of (45), which is

$$x^{(4)} - 2x''' - 8x'' - 27x' - 22x = -6$$

19. Statement of the Refined Method

If $L_1 = F_1 F_0$, $L_2 = F_2 F_0$, L_3, L_4, L_5, L_6, ... are linear differential operators, and L_1 and L_2 have a common factor F_0, if the factors F_0, F_1, and F_2 commute, so that

$$F_2 F_0 = F_0 F_2 \qquad F_1 F_0 = F_0 F_1 \qquad F_2 F_1 = F_1 F_2$$

and if f_1 and f_2 are sufficiently differentiable, then the equations

$$L_1 x + L_3 y + L_5 z + \cdots = f_1$$

$$L_2 x + L_4 y + L_6 z + \cdots = f_2$$

can have x eliminated by the action of F_2 on the first, F_1 on the second, and subtraction. The result is the single equation

$$F_2 L_3 y - F_1 L_4 y + F_2 L_5 z - F_1 L_6 z + \cdots = F_2[f_1] - F_1[f_2]$$

20. Systems Containing More than Two Unknowns

When faced with a system such as

$$L_1 x + L_4 y + L_7 z = f_1 \tag{46}$$
$$L_2 x + L_5 y + L_8 z = f_2 \tag{47}$$
$$L_3 x + L_6 y + L_9 z = f_3 \tag{48}$$

we use systematic elimination to eliminate one variable completely, leaving one fewer variable and one fewer equation. (See Section D of Chapter II.) The operators involved must be linear and they must commute, or the method may fail. If the operators are linear differential operators with constant coefficients, then the elimination will always work.

The use of Equation (46) to eliminate x from (47) requires operating on both sides of (46) with L_2 and on (47) with L_1. Using (46) to eliminate x from (48) requires operating on (46) with L_3 and on (48) with L_1. Using (46) to eliminate x throughout the system of three equations therefore gives

$$L_2 L_4 y - L_1 L_5 y + L_2 L_7 z - L_1 L_8 z = L_2[f_1] - L_1[f_2]$$
$$L_3 L_4 y - L_1 L_6 y + L_3 L_7 z - L_1 L_9 z = L_3[f_1] - L_1[f_3]$$

a system of two equations in two variables, to which the method must be applied again to eliminate either y or z as desired.

21. Sample Problems with Answers

S. Consider the equations

$$-x''' + x' + y'' + 4y' + 4y = \cos t$$
$$x' + y'' - 4y = e^t$$

(a) Eliminate x, obtaining an equation in y alone.
Answer: $y^{(4)} - 4y''' + 4y' + 8y = \cos t$
(b) Eliminate y, obtaining an equation in x alone.
Answer: $x^{(4)} - 2x''' + 4x' = 3e^t + \sin t + 2 \cos t$

22. Problems

43. Consider the system
$$x'' - 4x + 4y'' - 4y' + y = e^t$$
$$x'' - 4x' + 4x + 2y' - y = e^{3t}$$

(a) Eliminate x, obtaining an equation in y alone.
(b) Eliminate y, obtaining an equation in x alone.

44. Consider the system

$$\left.\begin{array}{r} x'' + y'' + z'' = 0 \\ x' + x + 3y' + y + z' = t \\ x'' + 2x' + x + 9y'' - y + z''' + z' = e^{4t} \end{array}\right\}$$

Eliminate x and y to find an equation in z alone.

C. THE METHOD OF UNDETERMINED COEFFICIENTS FOR SOLVING NONHOMOGENEOUS EQUATIONS

The previous section contained rules for solving homogeneous equations

$$Ly = 0$$

In this section we use an extension of the same technique, giving a procedure for solving nonhomogeneous equations

$$Ly = f$$

provided that the equation has constant coefficients and the nonhomogeneous term f is a linear combination of terms of the form

$$t^m e^{at} \cos bt \qquad \text{and} \qquad t^m e^{at} \sin bt$$

As before, we begin with examples. Then we give the method and end with a statement of the result for the simplest cases.

1. Example

Consider the equation

$$y'' + 5y' + 6y = 7e^{-3t} \tag{1}$$

and imagine y to be some function which satisfies the equation.

Observe that $D + 3$ annihilates the right-hand side of (1). Thus, $D + 3$ applied to both sides of (1) gives

$$(D + 3)(D^2 + 5D + 6)y = (D + 3)7e^{-3t} = 0$$

or $$(D + 3)(D^2 + 5D + 6)y = 0 \tag{2}$$

an equation which y and all other solutions of (1) must satisfy. Now (2) is homogeneous with constant coefficients. It is of the form $Ly = 0$, where L has characteristic polynomial

$$p(s) = (s + 3)(s^2 + 5s + 6) = (s + 2)(s + 3)^2$$

Thus y lies in the general solution set of $Ly = 0$, or

$$y = A_1 e^{-2t} + A_2 e^{-3t} + A_3 t e^{-3t} \tag{3}$$

for some choice of A_1, A_2, and A_3. In this way, the problem of finding y has been reduced to that of selecting three numbers to make (3) satisfy (1). Some of these numbers will be arbitrary. The others will be completely determined by substitution into (1), as follows:

$$(D^2 + 5D + 6)(A_1 e^{-2t} + A_2 e^{-3t} + A_3 te^{-3t})$$

$$= A_1(D^2 + 5D + 6)e^{-2t} + A_2(D^2 + 5D + 6)e^{-3t} + A_3(D^2 + 5D + 6)te^{-3t}$$

$$= A_1(0) + A_2(0) + A_3(-1 + 0t)e^{-3t}$$

$$= -A_3 e^{-3t} \tag{4}$$

Thus, $$(D^2 + 5D + 6)y = -A_3 e^{-3t}$$

and so the substitution into (1) gives

$$-A_3 e^{-3t} = 7e^{-3t} \qquad \text{or} \qquad A_3 = -7$$

Putting this number back into (3), the form of y, gives

$$y(t) = A_1 e^{-2t} + A_2 e^{-3t} - 7te^{-3t} \tag{5}$$

Substitution verifies that $-7te^{-3t}$ is a solution of (1). The first two terms of (5) constitute the general homogeneous solution of (1), and so (5) is the general solution of (1).

In this example, A_3 was the "undetermined coefficient."

2. The Method

The method is useful for problems which can be written in the following form:

Solve the equation $$L_1 y = f \tag{6}$$

where L_1 is a linear differential operator with constant coefficients, and f can be annihilated by another such operator.

Step 1. Find an operator (of the lowest order) which annihilates f, and call it L_2. Use L_2 on both sides of (6) to obtain

$$L_2 L_1 y = L_2 f = 0 \tag{7}$$

Step 2. Since the desired function y is annihilated by $L_2 L_1$, write the characteristic polynomial for $L_2 L_1$, and write the general solution of (7).

Step 3. Acting with L_1 on that general solution allows substitution into the given equation (6) and permits determination of the values for some of the unknown coefficients. The coefficients not so determined must be arbitrary. As a check on your work you should verify that the corresponding terms constitute the general homogeneous solution.

If n_1 and n_2 are the orders of L_1 and L_2, respectively, then n_2 coefficients will be determined by step 3, and there will be n_1 terms in the homogeneous solution.

3. Example

Solve
$$y'' + 9y = 7 \cos 2t \tag{8}$$

Here $L_1 = D^2 + 9$, while $L_2 = D^2 + 4$ annihilates $7 \cos 2t$. Thus,
$$L_2 L_1 = (D^2 + 4)(D^2 + 9)$$

annihilates any solution y. The characteristic polynomial for $L_2 L_1$ is
$$p(s) = (s^2 + 4)(s^2 + 9)$$

and the general solution of
$$L_2 L_1 y = 0$$

is therefore
$$y(t) = A_1 \cos 3t + A_2 \sin 3t + A_3 \cos 2t + A_4 \sin 2t$$

The action of L_1 on this expression annihilates the first two terms, which are homogeneous solutions of the given equation (8). Specifically,
$$L_1 y = y'' + 9y = 5A_3 \cos 2t + 5A_4 \sin 2t$$

and substitution of y into (8) shows that
$$7 \cos 2t = 5A_3 \cos 2t + 5A_4 \sin 2t$$

or
$$A_3 = \tfrac{7}{5} \qquad A_4 = 0$$

Thus, the general solution of (8) is
$$y(t) = A_1 \cos 3t + A_2 \sin 3t + \tfrac{7}{5} \cos 2t$$

4. Sample Problems with Answers

Use the method of undetermined coefficients to find the general solution of each of the following.

A. $y'(t) + 3y(t) = 4e^{-3t}$ Answer: $y(t) = 4te^{-3t} + A_1 e^{-3t}$

B. $y''(t) + 4y(t) + 5y(t) = te^t$ Answer: $y(t) = \tfrac{1}{10}te^t - \tfrac{3}{50}e^t + e^{-2t}(A_1 \cos t + A_2 \sin t)$

C. $y''(t) = 2t^2 + 3$ Answer: $y(t) = \tfrac{1}{6}t^4 + \tfrac{3}{2}t^2 + A_1 t + A_2$

D. $y''(t) + 4y(t) = e^t \cos t$ Answer: $y(t) = e^t(\tfrac{2}{5} \cos t + \tfrac{1}{10} \sin t) + A_1 \cos 2t + A_2 \sin 2t$

E. $y'(t) + 3y(t) = 4e^{-3t} + \cos 4t$ Answer: $y(t) = 4te^{-3t} + \tfrac{3}{25} \cos 4t + \tfrac{4}{25} \sin 4t + A_1 e^{-3t}$

5. Problems

Use the method of undetermined coefficients to find the general solution of each of the following.

1. $y'(t) + 6y(t) = 2$

2. $y''(t) + 6y'(t) = 2$

3. $y'(t) + 6y(t) = t$

4. $y''(t) + 6y'(t) = -t^2$

5. $y'(t) + 6y(t) = e^{2t} \sin 4t$

6. $y'(t) + 6y(t) = te^{2t} \sin 4t$

7. $y'(t) + 6y(t) = 2 + t$

8. $y'(t) + 6y(t) = t + e^{-6t}$

9. $y''(t) + 4y'(t) + 5y(t) = 3 \cos t + \sin t$

10. $y''(t) + 4y'(t) + 4y(t) = e^{-2t}$

11. $y''(t) + 4y'(t) + 4y(t) = e^{-2t} + 7t$

6. Summary for the Simplest Case: The Simple Rule

Consider a linear differential equation with constant coefficients,

$$a_n y^{(n)} + \cdots + a_1 y' + a_0 y = f \quad \text{or} \quad L[y] = f \tag{9}$$

1. If the nonhomogeneous term f is a constant multiple of an exponential $e^{\alpha t}$ which is not a homogeneous solution of (9), then (9) has a solution $y_p(t) = C_1 e^{\alpha t}$.

2. If the nonhomogeneous term f is a sinusoid of angular frequency β, that is, if f is a linear combination of $\cos \beta t$ and $\sin \beta t$, and if f is not a homogeneous solution of (9), then (9) has a solution $y_p(t) = C_1 \cos \beta t + C_2 \sin \beta t$.

3. If the nonhomogeneous term f is a polynomial of degree m, and if the constant function 1 is not a homogeneous solution of (9), then (9) has a solution which is a polynomial of degree m,

$$y_p(t) = C_1 + C_2 t + \cdots + C_{m+1} t^m$$

4. If y_p is a solution of (9), and if the general homogeneous solution is h, then the general solution of (9) is $y = y_p + h$.

7. Examples of the Simplest Case

The equation

$$y''(t) + 3y'(t) + 2y(t) = 3e^{4t} \tag{10}$$

has characteristic polynomial $s^2 + 3s + 2 = (s + 1)(s + 2)$, and homogeneous solution

$$h(t) = A_1 e^{-t} + A_2 e^{-2t}$$

Since $3e^{4t}$ is an exponential function which is not a homogeneous solution, part 1 of the rule says that there exists a function

$$y_p(t) = C_1 e^{4t}$$

which is a solution of (10). It remains to find the value of C_1. Substitution of $C_1 e^{4t}$ into the left-hand side of (10) gives

$$4^2 C_1 e^{4t} + 3(4C_1 e^{4t}) + 2C_1 e^{4t} \quad \text{or} \quad 30C_1 e^{4t}$$

Since $C_1 e^{4t}$ is to satisfy (10), the function $30C_1 e^{4t}$ must be f itself. That is,

$$30C_1 e^{4t} = 3e^{4t}$$

from which it follows that $C_1 = \frac{1}{10}$, or

$$y_p(t) = \frac{1}{10} e^{4t}$$

The calculation is easily described using the operator L determined by (10) whose assignment rule is

$$L[y] = y'' + 3y' + 2y$$

Thus, y_p satisfies the equation if and only if

$$L[y_p] = 3e^{4t}$$

Since
$$L[y_p] = L[C_1 e^{4t}] = 30C_1 e^{4t}$$

it must follow that

$$30C_1 e^{4t} = 3e^{4t}$$

or
$$C_1 = \tfrac{1}{10}$$

The general solution of (10) is obtained by adding $h(t)$, in this way:

$$y(t) = y_p(t) + h(t)$$
$$= \tfrac{1}{10}e^{4t} + A_1 e^{-t} + A_2 e^{-2t} \tag{11}$$

Now consider an initial value problem with the same differential equation, say

$$\left.\begin{array}{c} y''(t) + 3y'(t) + 2y(t) = 3e^{4t} \\ y(1) = 0 \\ y'(1) = 0 \end{array}\right\}$$

The solution of the problem may be obtained from the general solution (11) by determining the values of the remaining constants, A_1 and A_2, so as to make (11) satisfy the initial conditions.

$$\left.\begin{array}{l} 0 = y(1) = \tfrac{1}{10}e^4 + A_1 e^{-1} + A_2 e^{-2} \\ 0 = y'(1) = \tfrac{4}{10}e^4 - A_1 e^{-1} - 2A_2 e^{-2} \end{array}\right\} \tag{12}$$

The values of A_1 and A_2 which satisfy these conditions are

$$A_1 = -\tfrac{6}{10}e^5 \quad \text{and} \quad A_2 = \tfrac{5}{10}e^6$$

so that
$$y(t) = \tfrac{1}{10}e^{4t} - \tfrac{6}{10}e^5 e^{-t} + \tfrac{5}{10}e^6 e^{-2t}$$
$$= \tfrac{1}{10}(e^{4t} - 6e^{5-t} + 5e^{6-2t})$$

Because of the fact that the initial conditions are specified at $t = 1$, it is convenient to write y in such a way that $t - 1$ plays a prominent part, namely

$$y(t) = \tfrac{1}{10}(e^4 e^{4(t-1)} - 6e^4 e^{-(t-1)} + 5e^4 e^{-2(t-1)})$$

$$= \frac{e^4}{10}(e^{4(t-1)} - 6e^{-(t-1)} + 5e^{-2(t-1)})$$

Anticipating this, one could write the general solution (11) in the following equivalent form:

$$y(t) = \tfrac{1}{10}e^{4t} + B_1 e^{-(t-1)} + B_2 e^{-2(t-1)}$$

The equations for determining B_1 and B_2 from the initial conditions are

$$
\left.
\begin{aligned}
0 = y(1) &= \frac{e^4}{10} + B_1 + B_2 \\[2mm]
0 = y'(1) &= \frac{4e^4}{10} - B_1 - 2B_2
\end{aligned}
\right\}
$$

which are much nicer to solve than Equations (12), because the coefficients are nicer. The values are

$$B_1 = -\tfrac{6}{10}e^4 = A_1 e^{-1} \qquad \text{and} \qquad B_2 = \tfrac{5}{10}e^4 = A_2 e^{-2}$$

8. Examples in Which the Simplest-Case Results Cannot Be Used

The equation

$$y''(t) + 3y'(t) + 2y(t) = 3te^{4t} \tag{13}$$

cannot be treated with the simple rule, for $3te^{4t}$ is neither an exponential nor a sinusoid nor a polynomial, but a product of a polynomial and an exponential. It can be treated with the general method given in Part 2. The general solution is

$$\tfrac{1}{10}te^{4t} - \tfrac{11}{300}e^{4t} + A_1 e^{-t} + A_2 e^{-2t}$$

The equation

$$y''(t) + 3y'(t) + 2y(t) = 3e^{-2t} \tag{14}$$

has characteristic polynomial $s^2 + 3s + 2 = (s + 2)(s + 1)$ and general homogeneous solution

$$h(t) = A_1 e^{-t} + A_2 e^{-2t}$$

Since $3e^{-2t}$, the nonhomogeneous term of (14), is a homogeneous solution, it follows that (14) cannot be treated with the simple rule. It can be treated with the general method. The general solution is

$$y(t) = -3te^{-2t} + A_1 e^{-t} + A_2 e^{-2t}$$

The equation

$$y'''(t) + 3y''(t) = 120t^3 - 24t^2 \tag{15}$$

has characteristic polynomial $s^3 + 3s^2 = s^2(s + 3)$ and general homogeneous solution

$$h(t) = A_1 + A_2 t + A_3 e^{-3t}$$

Since the nonhomogeneous term of (15) is a polynomial but the constant function $t^0 = 1$ is a homogeneous solution of (15), it follows that the equation cannot be treated with the simple rule. With the general method, it can be shown that the general solution of (15) is

$$y(t) = 2t^5 - 4t^4 + \tfrac{16}{3}t^3 - \tfrac{16}{3}t^2 + A_1 + A_2 t + A_3 e^{-3t}$$

In each of these three examples it should be observed that the simple rule could not give the functions which appear in the general solution of the equation, even if one ignored the hypotheses.

9. Sample Problems with Answers

For each of the following equations, either show why the simple rule may not be applied, or else apply the rule and find both the solution which the rule describes and the general solution of the equation. In the answers, only the general solution is given.

F. $y' + 6y = -t^2$ *Answer:* $y(t) = -\tfrac{1}{6}t^2 + \tfrac{1}{18}t - \tfrac{1}{108} + A_1 e^{-6t}$

G. $y'' + 4y' + 5y = \pi e^{-3t}$ *Answer:* $y(t) = (\pi/3)e^{-3t} + e^{-2t}(A_1 \cos t + A_2 \sin t)$

H. $y'' + 3y' + 2y = 7 \cos 2t$ *Answer:* $y(t) = -\tfrac{7}{20} \cos 2t + \tfrac{21}{20} \sin 2t + A_1 e^{-t} + A_2 e^{-2t}$

I. $y' + 6y = 4e^{-6t}$

     Answer: The simple rule is not applicable, because the nonhomogeneous term is a homogeneous solution.

J. $y' + 6y = \ln t$

     Answer: The simple rule is not applicable, because the nonhomogeneous term is not of the required form.

K. $y' + 6y = e^t \cos t$

     Answer: The simple rule is not applicable, because the nonhomogeneous term is not of the required form.

L. $y'' + 6y' = -t^2$

     Answer: The simple rule is not applicable, because the nonhomogeneous term is a polynomial but the constant function $t^0 = 1$ is a homogeneous solution.

M. Solve the initial value problem

$$\left. \begin{aligned} y' + 6y &= -t^2 \\ y(3) &= 0 \end{aligned} \right\}$$

given that the general solution of the differential equation is

$$y(t) = -\tfrac{1}{6}t^2 + \tfrac{1}{18}t - \tfrac{1}{108} + A_1 e^{-6t}$$

Answer: $y(t) = -\tfrac{1}{108}t^2 + \tfrac{1}{18}t - \tfrac{1}{6} + \tfrac{145}{108}e^{-6(t-3)}$

N. Solve the initial value problem

$$\left. \begin{aligned} y'' + 4y' + 5y &= \pi e^{-3t} \\ y(2\pi) &= 0 \\ y'(2\pi) &= 1 \end{aligned} \right\}$$

given that the general solution of the differential equation is

$$y(t) = \frac{\pi}{2}e^{-3t} + e^{-2t}(A_1 \cos t + A_2 \sin t)$$

Answer: $y(t) = (\pi/2)e^{-3t} + e^{-2(t-2\pi)}\{-(\pi/2)e^{-6\pi} \cos t + [1 + (\pi/2)e^{-6\pi}] \sin t\}$

10. Problems

Use the simple rule to find the general solution of each of the following differential equations.

12. $y' + 3y = 1$ **13.** $y' + 3y = 2t$ **14.** $y' + 3y = 4e^{-2t}$

15. $y' + 3y = 2 \cos \pi t$ **16.** $y' + 3y = \cos \pi t + 2 \sin \pi t$

17. $y'' + 4y' + 4y = -e^t$ **18.** $y'' + 4y' + 4y = t^2$

19. $y'' + 4y' + 4y = -3 \sin t$ **20.** $y'' + 4y' + 4y = t + 2$

Solve the following initial value problems, using the simple rule.

21. $\left.\begin{array}{l} y' + 3y = 4e^{-2t} \\ \quad y(1) = 0 \end{array}\right\}$ **22.** $\left.\begin{array}{l} y' + 3y = 2 \cos \pi t \\ \quad y(2) = 1 \end{array}\right\}$

23. $\left.\begin{array}{l} y'' + 4y' + 4y = -e^t \\ \quad y(1) = 1 \\ \quad y'(1) = 0 \end{array}\right\}$ **24.** $\left.\begin{array}{l} y'' + 4y' + 4y = t \\ \quad y(1) = 1 \\ \quad y'(1) = 0 \end{array}\right\}$

25. The simple rule says that there is a polynomial of degree 1 which is a solution of the equation $y' + 3y = 2t$. Is there more than one polynomial solution of the equation? Find all polynomials of degree 0 or degree 2 which are solutions of the equation.

26. The simple rule says that there is a multiple of e^t which is a solution of the equation $y'' + 4y' + 4y = -e^t$. Is there a multiple of e^{-2t} which satisfies the equation? Are there any other multiples of exponentials which are solutions of the equation?

LINEAR DIFFERENTIAL OPERATORS
WITH GENERAL COEFFICIENTS

In Part 16 of Section A, the linear differential operators were defined. Each has an assignment rule of the form

$$L = a_n D^n + a_{n-1} D^{n-1} + \cdots + a_1 D + a_0$$

The coefficients a_n, ..., a_0 are functions of t alone—perhaps constant, perhaps not. In the remainder of the chapter we allow nonconstant coefficients.

Equations with general coefficients are not usually suitable for treatment with Laplace transforms. Instead, the process of solution is typically broken into two parts: finding the general homogeneous solution h and then finding a particular solution y_p for the given equation. The general solution of the given equation is then $y_p + h$. Typically, the trouble in solving an equation arises because one does not know how to find the general *homogeneous* solution. An exception is the equidimensional equation, which may be solved in terms of functions already known. The procedure is given in Section D. In Section E we discuss the way in which the general solution can be written as linear combinations of a few solutions. Section F concerns ways of making use of parts of a desired solution to either simplify a problem or solve it outright.

D. THE EQUIDIMENSIONAL EQUATION

The *equidimensional operators* are linear combinations of the linear operators

$$E_n = t^n D^n$$

For example, the formula

$$E_2 + 3E_1 + E_0 = t^2 D^2 + 3tD + 1$$

defines an equidimensional operator with assignment rule

$$Ly = t^2 y'' + 3ty' + y$$

These operators lead to *equidimensional differential equations*, $Ly = f$, in which every term is of the form $b_n t^n D^n$ for some constant b_n. For example,

$$t^2 y'' + 3ty' + y = f$$

is an equidimensional equation.

While the operator D^n carries exponentials to themselves in this way:

$$D^n[e^{\alpha t}] = \alpha^n e^{\alpha t}$$

the equidimensional operators carry monomials to themselves:

$$E_n[t^r] = t^n D^n[t^r]$$

$$= t^n r(r-1) \cdots (r-n+1)t^{r-n}$$

$$= r(r-1) \cdots (r-n+1)t^r$$

For this reason, equidimensional equations tend to have solutions of the form t^r, and solving one is largely a question of finding the correct values of r.

1. Examples

The differential equation in the initial value problem

$$\left. \begin{array}{r} 3ty'(t) + 2y(t) = 0 \\ y(1) = 5 \end{array} \right\} \tag{1}$$

is equidimensional with $b_0 = 2$ and $b_1 = 3$. Note that it is not an equation with constant coefficients, because if it is written in the form

$$a_1(t)y'(t) + a_0(t)y(t) = 0$$

then $a_1(t) = 3t$. However, it is linear.

The operator $L = 3E_1 + 2E_0$ carries monomials to themselves, and so we may try it on a general monomial t^r.

$$L[t^r] = 3tDt^r + 2t^r$$

$$= 3trt^{r-1} + 2t^r$$

$$= (3r + 2)t^r \tag{2}$$

This last function is the zero function if and only if $3r + 2 = 0$, or $r = -\frac{2}{3}$. Thus $t^{-2/3}$ is a solution of the equation, and so is any multiple

$$y(t) = At^{-2/3}$$

but no other monomial is a solution.

To satisfy the initial condition $y(1) = 5$, we take $A = 5$.

$$y(t) = 5t^{-2/3}$$

Actually, this problem could also have been solved using separation of variables, because the equidimensional equation is first-order, but substitution of the monomial t^r is effective with an equidimensional equation of *any* order. To illustrate, consider

$$t^3 y''' + 3t^2 y'' = 0 \qquad (3)$$

For this, $\quad L[t^r] = t^3 r(r-1)(r-2)t^{r-3} + 3t^2 r(r-1)t^{r-2}$

$$= r(r-1)(r+1)t^r$$

which is the zero function if and only if r has one of the values 0, 1, and -1. Thus, $t^0 = 1$, t, and $t^{-1} = 1/t$ are solutions, and so is any linear combination. The general solution is

$$y(t) = A_0 + A_1 t + \frac{A_2}{t}$$

2. Equidimensional Operators and Their Characteristic Polynomials

The expression $t^3 y'''(t)$ which appears on the left-hand side of (3) describes the action of the operator E_3. In general,

$$E_n[y](t) = t^n D^n y(t)$$

$$= t^n \frac{d^n y}{dt^n}(t) \qquad \text{if } y \text{ is } n \text{ times differentiable} \qquad (4)$$

The assignment rule for the general equidimensional operator is

$$L = b_n E_n + \cdots + b_1 E_1 + b_0$$

$$= b_n t^n D^n + \cdots + b_1 tD + b_0$$

where $b_0, \ldots, b_n$ are constants. As the examples illustrate, an equidimensional operator L carries the monomial t^r to the product of t^r and a polynomial $q(r)$ which is independent of t:

$$L[t^r] = q(r)t^r$$

$$= [b_n r(r-1) \cdots (r-n+1) + \cdots + b_1 r + b_0]t^r \qquad (5)$$

It follows that an equidimensional equation, $Ly = f$, has a characteristic polynomial $q(r)$ which is entirely different from the polynomial of the same name possessed by a linear differential equation with constant coefficients.

3. The Rule for Homogeneous Equidimensional Equations

If $q(r)$ is the characteristic polynomial of the homogeneous equidimensional equation

$$L[y] = 0 \qquad (6)$$

and if $\overset{\circ}{\alpha}$ is a root of $q(r)$ having multiplicity m, then

$$A_1 t^\alpha + A_2 t^\alpha \ln t + A_3 t^\alpha (\ln t)^2 + \cdots + A_m t^\alpha (\ln t)^{m-1}$$

$$= t^\alpha [A_1 + A_2 \ln t + A_3 (\ln t)^2 + \cdots + A_m (\ln t)^{m-1}] \qquad (7a)$$

is a solution of (6), valid on the right half line, $t > 0$. It will also be valid on the left half line, $t < 0$, if α is an integer and $m = 1$. If (7a) is not valid on the left half line, then the solution there which corresponds to the root α with multiplicity m is obtained from (7a) by replacing t with $-t$ or $|t|$ throughout, giving

$$|t|^\alpha [A_1 + A_2 \ln |t| + A_3 (\ln |t|)^2 + \cdots + A_m (\ln |t|)^{m-1}] \qquad (7b)$$

The general solution on the right half line is the sum of all expressions of the form (7a); that on the left half line is the sum of those of the form (7b).

Since $t = e^{\ln t}$ and $t^r = e^{r \ln t}$, it follows that if each of $\alpha + \beta j$ and $\alpha - \beta j$ is a root of $q(r)$ with multiplicity m, and if $\beta \neq 0$, then

$$t^{\alpha + \beta j} = t^\alpha e^{j\beta \ln t} = t^\alpha [\cos (\beta \ln t) + j \sin (\beta \ln t)]$$

and $t^{\alpha - \beta j} = t^\alpha [\cos (\beta \ln t) - j \sin (\beta \ln t)]$. Adding or subtracting shows that

$$\frac{t^{\alpha + \beta j} + t^{\alpha - \beta j}}{2} = t^\alpha \cos (\beta \ln t) \qquad \frac{t^{\alpha + \beta j} - t^{\alpha - \beta j}}{2j} = t^\alpha \sin (\beta \ln t)$$

and the solution for $t > 0$ can be written this way:

$$t^\alpha [B_1 \cos (\beta \ln t) + \cdots + B_m (\ln t)^{m-1} \cos (\beta \ln t)$$

$$+ B_{m+1} \sin (\beta \ln t) + \cdots + B_{2m} (\ln t)^{m-1} \sin (\beta \ln t)] \qquad (8)$$

There is also a solution for $t < 0$, which can be obtained from (8) by replacing t with $-t$ or $|t|$ throughout, just the way (7b) was obtained from (7a).

4. Examples and Comments

The equation

$$t^3 y'''(t) + 6t^2 y''(t) + 6ty'(t) = 0$$

has characteristic polynomial

$$q(r) = r(r - 1)(r - 2) + 6r(r - 1) + 6r$$

$$= r(r^2 - 3r + 2 + 6r - 6 + 6)$$

$$= r(r^2 + 3r + 2)$$

$$= r(r + 1)(r + 2)$$

The roots are $0, -1, -2$, each with multiplicity 1, and

$$y(t) = A_1 + A_2 t^{-1} + A_3 t^{-2}$$

$$= A_1 + \frac{A_2}{t} + \frac{A_3}{t^2} \qquad t \neq 0$$

The solution is valid on the half line $t > 0$ and on the half line $t < 0$, but it is not defined at $t = 0$, where both $1/t$ and $1/t^2$ are not defined.

The equation

$$ty'(t) + \alpha y(t) = 0 \tag{9}$$

has characteristic polynomial $r + \alpha$. The rule gives the solution

$$y(t) = At^{-\alpha}$$

If α is a nonpositive integer, then $t^{-\alpha}$ is defined for all t. If α is a positive integer, then $t^{-\alpha}$ is defined except at $t = 0$. If α is not an integer, then $t^{-\alpha}$ may be a real number only for $t > 0$. To be on the safe side, we may write the general solution as

$$y(t) = A|t|^{-\alpha} = \frac{A}{|t|^\alpha} \qquad t \neq 0$$

or, better yet, as

$$y(t) = \begin{cases} A_1(-t)^{-\alpha} & t < 0 \\ A_2 t^{-\alpha} & t > 0 \end{cases}$$

where A_1 and A_2 may be *different* constants.

Theorem 2 in Section H of Chapter I says that Equation (9) has solutions on every interval where the lead coefficient is not zero. Here the lead coefficient is $a_1(t) = b_1 t = t$, which is not zero on every interval which does not contain the point $t = 0$. The largest such intervals are the half lines $t > 0$ and $t < 0$, the intervals mentioned above.

The theorem also says that initial conditions can be set arbitrarily at interior points of such an interval, that is, at all points other than the end point of the interval, $t = 0$. For equidimensional equations in general, and for Equation (9) in particular, initial conditions usually *cannot* be set arbitrarily at $t = 0$. For example,

$$ty'(t) - 2y(t) = 0$$

has general solution

$$y(t) = At^2$$

For each value of A, $y(t)$ will have value zero at $t = 0$.

The uniqueness of the solution of an initial value problem is assured on the interval containing the initial point, by the same theorem. That is, the initial value problem

$$\left. \begin{aligned} ty'(t) - 2t &= 0 \\ y(1) &= 1 \end{aligned} \right\} \tag{10}$$

has the unique solution

$$y(t) = t^2$$

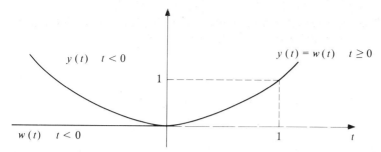

Figure 1 Solutions of (10)

on the interval $t > 0$. However, on the whole line the function $y(t) = t^2$ is not the only solution of (10), for

$$w(t) = u(t)t^2$$

is differentiable everywhere, even at $t = 0$:

$$w'(t) = 2tu(t)$$

and w not only satisfies the differential equation on the whole line, but also satisfies the initial condition.

5. Sample Problems with Answers

Find the general solution of the equation valid on the right half line, $t > 0$, and solve the initial value problem there.

A. $ty'(t) + 3y(t) = 0$
$\qquad\qquad y(2) = 1$

Answers: $y(t) = A_1/t^3$, $t > 0$
$\qquad\quad y(t) = (2/t)^3$, $t > 0$

B. $4t^2y''(t) + 4ty'(t) - y(t) = 0$
$\qquad\qquad y(3) = 0$
$\qquad\qquad y'(3) = 5$

Answers: $y(t) = A_1 t^{-1/2} + A_2 t^{1/2}$, $t > 0$
$\qquad\quad y(t) = -15[(3/t)^{1/2} - (t/3)^{1/2}]$, $t > 0$

C. $t^3y'''(t) + 5t^2y''(t) + 4ty'(t) = 0$
$\qquad\qquad y(1) = 0$
$\qquad\qquad y'(1) = 1$
$\qquad\qquad y''(1) = 2$

Answers: $y(t) = A_1 + A_2 t^{-1} + A_3 t^{-1} \ln t$, $t > 0$
$\qquad\quad y(t) = 5 - 5t^{-1} - 4t^{-1} \ln t$, $t > 0$

D. $4t^2y''(t) + 4ty'(t) + y(t) = 0$
$\qquad\qquad y(5) = 0$
$\qquad\qquad y'(5) = \frac{3}{4}$

$\qquad$ *Answers:* $y(t) = A_1 \cos\left(\tfrac{1}{2} \ln t\right) + A_2 \sin\left(\tfrac{1}{2} \ln t\right)$, $t > 0$

$$y(t) = \tfrac{15}{2} \sin\left(\tfrac{1}{2} \ln \tfrac{t}{5}\right) = \tfrac{15}{2}[\sin\left(\tfrac{1}{2} \ln t\right) \cos\left(\tfrac{1}{2} \ln 5\right) - \cos\left(\tfrac{1}{2} \ln t\right) \sin\left(\tfrac{1}{2} \ln 5\right)], \ t > 0$$

Find the general solution of the equation valid on the left half line, $t < 0$, and solve the initial value problem there.

E. $ty'(t) + 3y(t) = 0$ *Answer:* $y(t) = A_1/(-t)^3 = A_2/t^3, \; t < 0$

     $y(-1) = 2$ $y(t) = -2/t^3, \; t < 0$

F. $4t^2y''(t) + 4ty'(t) - y(t) = 0$

     $y(-1) = 0$ *Answer:* $y(t) = A_1(-t)^{-1/2} + A_2(-t)^{1/2}, \; t < 0$

     $y'(-1) = 1$ $y(t) = (-t)^{-1/2} - (-t)^{1/2}, \; t < 0$

6. Problems

For each differential equation, find the general solution valid on the half line containing the given initial point; *also* solve the initial value problem.

1. (a) $2ty'(t) + y(t) = 0$ (b) $2ty'(t) + y(t) = 0$ (c) $2ty'(t) + y(t) = 0$

        $y(4) = 1$ $y(-1) = 7$ $y(-4) = -3$

2. (a) $9t^2y''(t) + 9ty'(t) - y(t) = 0$ (b) $9t^2y''(t) + 9ty'(t) - y(t) = 0$

             $y(1) = 0$ $y(-1) = 0$

             $y'(1) = 1$ $y'(-1) = 1$

3. (a) $t^2y''(t) + 5ty'(t) + 4y(t) = 0$ (b) $t^2y''(t) + 5ty'(t) + 4y(t) = 0$

             $y(2) = 1$ $y(-1) = 1$

             $y'(2) = 0$ $y'(-1) = 0$

4. (a) $t^2y''(t) + 7ty'(t) + 13y(t) = 0$ (b) $t^2y''(t) + 7ty'(t) + 13y(t) = 0$

             $y(1) = 1$ $y(-1) = 1$

             $y'(1) = 0$ $y'(-1) = 0$

7. Relation to Operators with Constant Coefficients

Let $z = \ln t$. If y and x are differentiable functions such that $x(z) = y(t)$, that is, if $x(\ln t) = y(t)$ for each $t > 0$, then x' and y' are related, as can be shown using the following application of the chain rule:

$$x'(z) = \frac{dx}{dz}(z)$$

$$= \frac{dx}{dz}(\ln t)$$

$$= \frac{dt}{dz}\frac{dx}{dt}(\ln t)$$

$$= e^z \frac{dx}{dt}(\ln t)$$

$$x'(z) = e^z \frac{dy}{dt}(t)$$

$$= ty'(t)$$

For instance, if $x(z) = e^{4z}$, then $x'(z) = 4e^{4z} = 4t^4$. On the other hand, $y(t) = x(z) = e^{4z} = t^4$, and

$$ty'(t) = t4t^3 = 4t^4$$

Thus, $x(z) = e^{\alpha z}$ satisfies

$$x'(z) - \alpha x(z) = 0 \qquad \text{or} \qquad (D - \alpha)x = 0$$

if and only if $y(t) = x(z) = e^{\alpha z} = e^{\alpha \ln t} = t^\alpha$ satisfies

$$ty'(t) - \alpha y(t) = 0 \qquad \text{or} \qquad (E - \alpha)y = 0$$

Similarly, x satisfies

$$(D - \alpha_1)^{m_1} \cdots (D - \alpha_n)^{m_n}[x] = 0$$

if and only if y satisfies

$$(E - \alpha_1)^{m_1} \cdots (E - \alpha_n)^{m_n}[y] = 0$$

Therefore, where t and $e^{\alpha t}$ appear in the rules for constant-coefficient operators, they are replaced in the rule for equidimensional operators by $\ln t$ and t^α on the right half line, and by $\ln |t|$ and $|t|^\alpha$ on the left half line.

The characteristic polynomials of $E - \alpha_1$ and $E - \alpha_2$ are $r - \alpha_1$ and $r - \alpha_2$, and

$$q(r) = (r - \alpha_1)(r - \alpha_2)$$

is the characteristic polynomial of the composition

$$(E - \alpha_1)(E - \alpha_2) = EE - (\alpha_1 + \alpha_2)E + \alpha_1 \alpha_2$$

Notice particularly that $EE \neq E_2$. In fact, $EE[t^r] = r^2 t^r$, which has characteristic polynomial r^2, while

$$E_2[t^r] = t^2 \frac{d^2}{dt^2}(t^r) = r(r-1)t^r$$

has characteristic polynomial $r(r-1)$. Thus,

$$(E - \alpha_1)(E - \alpha_2) = EE - (\alpha_1 + \alpha_2)E + \alpha_1 \alpha_2$$

$$= E_2 + (1 - \alpha_1 - \alpha_2)E_1 + \alpha_1 \alpha_2$$

To factor an equidimensional operator, for instance,

$$L = E_2 + 4E + 2$$

factor the characteristic polynomial

$$q(r) = r(r - 1) + 4r + 2$$
$$= r^2 + 3r + 2$$
$$= (r + 1)(r + 2)$$

and write the corresponding factors for L, namely,

$$L = (E + 1)(E + 2)$$

To annihilate $(\ln t)^k t^\alpha$, one may use the operator $(E - \alpha)^{k+1}$; and to annihilate either

$$(\ln t)^k t^\alpha \cos (b \ln t) \qquad \text{or} \qquad (\ln t)^k t^\alpha \sin (b \ln t) \qquad (11)$$

one may use

$$\{[E - (a + bj)][E - (a - bj)]\}^{k+1} = [E_2 + (1 - 2a)E + a^2 + b^2]^{k+1}$$

There is also a method of undetermined coefficients for nonhomogeneous equidimensional equations. The most important special case is this: If L is an equidimensional operator, and if t^α is not a homogeneous solution of

$$L[y] = At^\alpha \qquad (12)$$

then (12) has a solution of the form

$$y_p(t) = Ct^\alpha$$

This is, of course, part 1 of the simple rule for constant coefficients, transformed by replacing t and e^z with $\ln t$ and t^α.

The general procedure for the method of undetermined coefficients is rarely used in practice, because the functions (11) are not very commonly encountered. However, it may be carried out as described in Section C, being applied to equations of the form

$$L_1 y = f$$

where L_1 is equidimensional and f may be annihilated by some equidimensional operator L_2. Do not forget, though, to use E in place of D, and $q(r)$ in place of $p(s)$.

E. FORMS OF A HOMOGENEOUS SOLUTION; LINEAR INDEPENDENCE

A solution of a linear differential equation can always be analyzed into the form

$$y = y_p + y_h$$

where y_p is another solution of the equation, and y_h is a homogeneous solution. Adding a homogeneous solution y_0 to y_p and subtracting the same from y_h gives the same function y, but analyzed differently, as

$$y = y_q + y_k$$

where $y_q = y_p + y_0$ and $y_k = y_h - y_0$. Thus, a given solution may be written in different ways, some convenient for some purposes and some for others.

When writing the solutions of initial value problems the "basic solution system" is often useful, and so we begin with that, observing forms of solution which are useful for special purposes. Then we work up to the concept of "linear independence," which shows what it is that all the forms of solution have in common, and which is of practical value in dealing with more difficult cases, in which the whole general solution cannot be found by a single method.

1. Example: Rewriting the Solution of a Differential Equation

Consider an initial value problem with unspecified initial values

$$y''(t) - 4y(t) = 7e^{-t} \tag{1a}$$

$$y(0) \text{ unspecified}$$
$$y'(0) \text{ unspecified} \tag{1b}$$

We can use Laplace transforms to find the general solution of Equation (1a):

$$\mathcal{L}[y](s) = \frac{7 + y(0)s(s+1) + y'(0)(s+1)}{(s+1)(s^2-4)}$$

$$= -\frac{7}{3}\frac{1}{s+1} + \frac{7 + 6y(0) + 3y'(0)}{12}\frac{1}{s-2}$$

$$+ \frac{7 + 2y(0) + y'(0)}{4}\frac{1}{s+2}$$

$$y(t) = -\frac{7}{3}e^{-t} + \frac{7 + 6y(0) + 3y'(0)}{12}e^{2t} + \frac{7 + 2y(0) - y'(0)}{4}e^{-2t} \tag{2}$$

Equation (2) looks rather complicated, but it shows explicitly how the initial values enter into the general solution of (1a). It is very convenient for graphing when $y(0)$ and $y'(0)$ are known, for then the coefficients can be simplified (calculated), and the solution simplifies into the form

$$y(t) = -\tfrac{7}{3}e^{-t} + A_1 e^{2t} + A_2 e^{-2t} \tag{3}$$

Formula (3) is also that obtained by the use of the characteristic polynomial and the method of undetermined coefficients, which is easier to remember and use than the technique of Laplace transformation.

There are other ways of rewriting (2). For instance, recombining terms gives the formulas

$$y(t) = (-\tfrac{7}{3} + \tfrac{7}{12}e^{2t} + \tfrac{7}{4}e^{-2t}) + y(0)\frac{e^{2t} + e^{-2t}}{2} + y'(0)\frac{e^{2t} - e^{-2t}}{4}$$

$$= y_p(t) + y(0)\cosh 2t + \frac{y'(0)}{2}\sinh 2t \tag{4}$$

The latter shows very clearly the dependence upon the initial conditions. The parentheses enclose a function y_p, itself a solution of the differential equation, which satisfies the initial conditions

$$y_p(0) = 0 \qquad y'_p(0) = 0$$

The last two terms of (4) are multiples of the hyperbolic functions

$$\cosh 2t = \frac{e^{2t} + e^{-2t}}{2} \qquad \text{and} \qquad \sinh 2t = \frac{e^{2t} - e^{-2t}}{2} \tag{5}$$

These functions are homogeneous solutions—in fact, solutions of very special initial value problems. That is, $\cosh 2t$ is the solution of the initial value problem

$$\left.\begin{aligned} y'' - 4y &= 0 \\ y(0) &= 1 \\ y'(0) &= 0 \end{aligned}\right\} \tag{6}$$

and $(\sinh 2t)/2$ is the solution of the problem

$$\left.\begin{aligned} y'' - 4y &= 0 \\ y(0) &= 0 \\ y'(0) &= 1 \end{aligned}\right\} \tag{7}$$

Together, these two functions $\cosh 2t$ and $(\sinh 2t)/2$ constitute what we shall come to call a "basic solution system" for the homogeneous equation $y'' - 4y = 0$. The term will be defined in Part 3, after more examples have been given. Such systems of functions will be used throughout the section.

The functions e^{2t} and e^{-2t} are linear combinations of $\cosh 2t$ and $\sinh 2t$. In fact, it is apparent from (5) that

$$e^{2t} = \cosh 2t + \sinh 2t \qquad \text{and} \qquad e^{-2t} = \cosh 2t - \sinh 2t \tag{8}$$

Thus the solution given in (3) could be rewritten as

$$\begin{aligned} y(t) &= -\tfrac{7}{3}e^{-t} + A_1(\cosh 2t + \sinh 2t) + A_2(\cosh 2t - \sinh 2t) \\ &= -\tfrac{7}{3}e^{-t} + (A_1 + A_2)\cosh 2t + (A_1 - A_2)\sinh 2t \\ &= -\tfrac{7}{3}e^{-t} + B_1 \cosh 2t + B_2 \sinh 2t \end{aligned} \tag{9}$$

This is a form alternative to (3). We shall now derive the form alternative to (2). Since $\sinh 0 = 0$ and $\cosh 0 = 1$ [see (6) and (7)], it follows from (9) that

$$y(0) = -\tfrac{7}{3} + B_1 \qquad \text{or} \qquad B_1 = y(0) + \tfrac{7}{3}$$

Differentiating (9) and using the conditions on the derivatives in (6) and (7) shows that

$$y'(0) = \tfrac{7}{3} + 2B_2 \qquad \text{or} \qquad B_2 = \tfrac{1}{2}[y'(0) - \tfrac{7}{3}]$$

Substituting these expressions for B_1 and B_2 into (9) gives

$$y(t) = -\tfrac{7}{3}e^{-t} + [y(0) + \tfrac{7}{3}] \cosh 2t + [y'(0) - \tfrac{7}{3}] \frac{\sinh 2t}{2} \tag{10}$$

This formula for y should be compared with the others that we have already derived, namely:

$$y(t) = -\tfrac{7}{3}e^{-t} + B_1 \cosh 2t + B_2 \sinh 2t \tag{11}$$

$$y(t) = (-\tfrac{7}{3}e^{-t} + \tfrac{7}{12}e^{2t} + \tfrac{7}{4}e^{-2t}) + y(0) \cosh 2t + y'(0) \frac{\sinh 2t}{2} \tag{4}$$

$$y(t) = -\tfrac{7}{3}e^{-t} + A_1 e^{2t} + A_2 e^{-2t} \tag{3}$$

$$y(t) = -\tfrac{7}{3}e^{-t} + \frac{7 + 6y(0) + 3y'(0)}{12} e^{2t} + \frac{7 + 2y(0) - y'(0)}{4} e^{-2t} \tag{2}$$

Any one of these forms may be the most useful in one application or another. Forms (3) and (11) are the quickest to derive, the easiest to remember, and the most succinct to write out. They lack explicit information about the role of the initial values, as given in the other forms. Formula (2) gives the quickest route for calculating values and graphing the solution of an initial value problem in which $y(0)$ and $y'(0)$ are specified. Formulas (4) and (10) are compromises using the basic solution system and showing explicitly the dependence on the initial values. Of the two, (10) shows that particular solution (of the given equation) which is simplest to write, whereas (4) shows that particular solution which satisfies the conditions

$$y(0) = 0 \qquad y'(0) = 0$$

2. Example: The Effect of Changing the Initial Point

According to Theorem 2 in Section H of Chapter I, initial conditions for the differential equation (1a)

$$y''(t) - 4y(t) = 7e^{-t}$$

might be set at any point at all. In the previous part we set them at $t = 0$. For comparison, we now set them at $t = 5$ and derive the forms of solution corresponding to those we have studied. Of course, (3) and (11) are still useful; the only difficulty with the others is that they mention the numbers $y(0)$ and $y'(0)$ which would have to be calculated somehow.

Here, however, are calculations that can be done directly and efficiently, using initial values set at $t = 5$. First, the general homogeneous solution may be rewritten

$$A_1 e^{2t} + A_2 e^{-2t} = A_1 e^{10} e^{2(t-5)} + A_2 e^{-10} e^{-2(t-5)}$$

$$= C_1 e^{2(t-5)} + C_2 e^{-2(t-5)}$$

and substituted into (3), giving a form much easier to use:

$$y(t) = -\tfrac{7}{3}e^{-t} + C_1 e^{2(t-5)} + C_2 e^{-2(t-5)} \tag{12}$$

To evaluate the coefficients in terms of $y(5)$ and $y'(5)$, evaluate (12) and its derivative at $t = 5$, as follows

$$y(5) = -\tfrac{7}{3}e^{-5} + C_1 + C_2$$
$$y'(5) = \tfrac{7}{3} + 2C_1 - 2C_2$$

and solve the resulting equations for C_1 and C_2.

$$C_1 = \tfrac{1}{12}[6y(5) + 3y'(5) + 7e^{-5}]$$
$$C_2 = \tfrac{1}{4}[2y(5) - y'(5) + 7e^{-5}]$$

Substituting these values into (12) gives the following formula, analogous to (2):

$$y(t) = -\tfrac{7}{3}e^{-t} + \frac{6y(5) + 3y'(5) + 7e^{-5}}{12}e^{2(t-5)} + \frac{2y(5) - y'(5) + 7e^{-5}}{4}e^{-2(t-5)}$$

Collecting multiples of $y(5)$ and $y'(5)$ gives this:

$$y(t) = -\tfrac{7}{3}e^{-t} + \tfrac{7}{12}e^{-5}e^{2(t-5)} + \tfrac{7}{4}e^{-5}e^{-2(t-5)}$$

$$+ y(5)\frac{e^{2(t-5)} + e^{-2(t-5)}}{2} + y'(5)\frac{e^{2(t-5)} - e^{-2(t-5)}}{4}$$

$$= 7e^{-5}\left(-\frac{e^{-(t-5)}}{3} + \frac{e^{2(t-5)}}{12} + \frac{e^{-2(t-5)}}{4}\right)$$

$$+ y(5)\cosh 2(t-5) + y'(5)\frac{\sinh 2(t-5)}{2}$$

Obtaining the remaining forms is a process that starts with the replacement of the exponentials in (12) with the corresponding hyperbolic functions, as in (8) and (9):

$$y(t) = -\tfrac{7}{3}e^{-t} + D_1 \cosh 2(t-5) + D_2 \sinh 2(t-5)$$

Evaluation of the coefficients in terms of $y(5)$ and $y'(5)$ gives the formula

$$y(t) = -\tfrac{7}{3}e^{-t} + [y(5) + \tfrac{7}{3}e^{-5}] \cosh 2(t-5) + [y'(5) - \tfrac{7}{3}e^{-5}]\frac{\sinh 2(t-5)}{2}$$

The functions $\cosh 2(t-5)$ and $[\sinh 2(t-5)]/2$ which appear here constitute the " basic solution system with initial point $t = 5$," because they respectively satisfy the following initial value problems:

$$\left.\begin{array}{c} y'' - 4y = 0 \\ y(5) = 1 \\ y'(5) = 0 \end{array}\right\} \quad \text{and} \quad \left.\begin{array}{c} y'' - 4y = 0 \\ y(5) = 0 \\ y'(5) = 1 \end{array}\right\} \tag{13}$$

What you may have observed in this example actually holds for every linear differential equation with *constant* coefficients: If

$$v_0(t), \ \ldots, \ v_{n-1}(t)$$

is the basic solution system with initial point zero, then

$$v_0(t - t_0), \ldots, v_{n-1}(t - t_0)$$

is the system with initial point t_0. However, this translation property is not true in the general case, defined below. For an example, see Part 4.

3. The Basic Solution System with Initial Point $t = t_0$

If L is an arbitrary linear differential operator of order n on an interval (α, β), and if t_0 is a point with $\alpha < t_0 < \beta$, then the homogeneous equation $Ly = 0$ has a unique solution v_0 satisfying

$$\left.\begin{aligned} v_0(t_0) &= 1 \\ v_0'(t_0) &= 0 \\ \cdots\cdots\cdots \\ v_0^{(n-1)}(t_0) &= 0 \end{aligned}\right\}$$

In the example in Part 1, we were given $t_0 = 0$, and we found $v_0(t) = \cosh 2t$, which solves the initial value problem

$$\left.\begin{aligned} v_0'' - 4v_0 &= 0 \\ v_0(0) &= 1 \\ v_0'(0) &= 0 \end{aligned}\right\}$$

In the example of Part 2, we were given $t_0 = 5$, and we found $v_0(t) = \cosh 2(t - 5)$.

The same equation, $Ly = 0$, also has a unique solution v_1 satisfying the initial conditions

$$\left.\begin{aligned} v_1(t_0) &= 0 \\ v_1'(t_0) &= 1 \\ v_1''(t_0) &= 0 \\ \cdots\cdots\cdots \\ v_1^{(n-1)}(t_0) &= 0 \end{aligned}\right\}$$

In the examples of Parts 1 and 2 we had $v_1(t)$ equal to $(\sinh 2t)/2$ and $[\sinh 2(t - 5)]/2$, respectively, as can be seen from (7) and (13). These pairs of functions,

$$v_0(t) = \cosh 2t \qquad v_1(t) = \tfrac{1}{2} \sinh 2t$$

in the first example, and

$$v_0(t) = \cosh 2(t - 5) \qquad v_1(t) = \tfrac{1}{2} \sinh 2(t - 5)$$

in the second example, constitute basic solution systems for the equation $y'' - 4y = 0$ with initial points $t = 0$ and $t = 5$, respectively. Here is the definition for the general situation:

If the homogeneous linear differential equation

$$Ly = 0$$

is of order n on an interval $\alpha < t < \beta$ containing the point t_0, $\alpha < t_0 < \beta$, then the equation has as its *basic solution system with initial point* $t = t_0$ the n functions v_0, $v_1, v_2, \ldots, v_{n-1}$ which are those solutions of the equation satisfying, respectively, the following sets of initial conditions:

$$
\left.\begin{aligned}
v_0(t_0) &= 1 \\
v_0'(t_0) &= 0 \\
v_0''(t_0) &= 0 \\
\cdots\cdots\cdots \\
v_0^{(n-1)}(t_0) &= 0
\end{aligned}\right\}
\quad
\left.\begin{aligned}
v_1(t_0) &= 0 \\
v_1'(t_0) &= 1 \\
v_1''(t_0) &= 0 \\
\cdots\cdots\cdots \\
v_1^{(n-1)}(t_0) &= 0
\end{aligned}\right\}
\quad \cdots \quad
\left.\begin{aligned}
v_{n-1}(t_0) &= 0 \\
v_{n-1}'(t_0) &= 0 \\
v_{n-1}''(t_0) &= 0 \\
\cdots\cdots\cdots \\
v_{n-1}^{(n-1)}(t_0) &= 1
\end{aligned}\right\}
\tag{14}
$$

In the previous parts we saw two examples of basic solution systems; in the next part we shall see more examples. The point to notice is this: If we can solve *the n initial value problems indicated in (14), then we can solve* all *the initial value problems with initial point t_0, for the formula*

$$
y(t) = y(t_0)v_0(t) + y'(t_0)v_1(t) + y''(t_0)v_2(t) + \cdots + y^{(n-1)}(t_0)v_{n-1}(t)
$$

then gives the unique function y satisfying the differential equation and having the specified values $y(t_0)$, $y'(t_0)$, $\ldots$, $y^{(n-1)}(t_0)$.

The corresponding formula for a nonhomogeneous equation

$$
Ly = f
$$

for which some particular solution y_p is known, is

$$
y(t) = y_p(t) + [y(t_0) - y_p(t_0)]v_0(t) + [y'(t_0) - y_p'(t_0)]v_1(t) + \cdots
$$
$$
+ [y^{(n-1)}(t_0) - y_p^{(n-1)}(t_0)]v_{n-1}(t) \tag{15}
$$

In the previous example, Equation (10) was of this form.

The practical use of the basic solution systems and Equation (15) occurs when it is required to predict the response of a system to many different inputs and initial conditions. Then the basic solution system is found, and a new prediction can be made by finding y_p and then substituting into (15).

4. More Examples of Basic Solution Systems

Here is an example in which these functions are required. The general solution of

$$
y''' = 0 \tag{16}
$$

is

$$
y(t) = C_1 + C_2 t + C_2 t^2
$$

The basic solution system with initial point at $t = 0$ consists of the three functions which satisfy the differential equation (16) and the following three sets of initial conditions:

$$\left.\begin{aligned} v_0(0) &= 1 \\ v_0'(0) &= 0 \\ v_0''(0) &= 0 \end{aligned}\right\} \quad \text{or} \quad v_0(t) \equiv 1 \qquad\qquad \left.\begin{aligned} v_1(0) &= 0 \\ v_1'(0) &= 1 \\ v_1''(0) &= 0 \end{aligned}\right\} \quad \text{or} \quad v_1(t) = t$$

and
$$\left.\begin{aligned} v_2(0) &= 0 \\ v_2'(0) &= 0 \\ v_2''(0) &= 1 \end{aligned}\right\} \quad \text{or} \quad v_2(t) = \frac{t^2}{2}$$

In terms of these three functions, the general solution of (16) is

$$y(t) = y(0)v_0(t) + y'(0)v_1(t) + y''(0)v_2(t)$$

$$= y(0) + y'(0)t + \frac{y''(0)}{2}t^2$$

For the same differential equation, the basic solution system with initial point $t = -7$ consists of the three functions which satisfy the following three sets of initial conditions:

$$\left.\begin{aligned} v_0(-7) &= 1 \\ v_0'(-7) &= 0 \\ v_0''(-7) &= 0 \end{aligned}\right\} \quad \text{or} \quad v_0(t) \equiv 1 \tag{17a}$$

$$\left.\begin{aligned} v_1(-7) &= 0 \\ v_1'(-7) &= 1 \\ v_1''(-7) &= 0 \end{aligned}\right\} \quad \text{or} \quad v_1(t) = t + 7 \tag{17b}$$

and
$$\left.\begin{aligned} v_2(-7) &= 0 \\ v_2'(-7) &= 0 \\ v_2''(-7) &= 1 \end{aligned}\right\} \quad \text{or} \quad v_2(t) = \frac{(t + 7)^2}{2} \tag{17c}$$

In terms of these three functions, the general solution of (16) may be written as

$$y(t) = y(-7)v_0(t) + y'(-7)v_1(t) + y''(-7)v_2(t)$$

$$= y(-7) + y'(-7)(t + 7) + \frac{y''(-7)}{2}(t + 7)^2$$

In terms of the same functions, the general solution of the nonhomogeneous equation

$$y''' = 3$$

may be written as in (15):

$$y(t) = \tfrac{1}{2}t^3 + \left[y(-7) + \frac{7^3}{2}\right]v_0(t) + \left[y'(-7) - \frac{3 \times 7^2}{2}\right]y_1(t)$$

$$+ [y''(-7) + 21]y_2(t)$$

$$= \tfrac{1}{2}t^3 + \left[y(-7) + \frac{7^3}{2}\right] + \left[y'(-7) - \frac{3 \times 7^2}{2}\right](t + 7)$$

$$+ [y''(-7) + 21]\frac{(t + 7)^2}{2}$$

Now let us look at basic solution systems for differential equations with nonconstant coefficients, for example,

$$t^2 y''(t) - t y'(t) + y(t) = 0$$

for which the general solution on either $t > 0$ or $t < 0$ may be written as

$$y(t) = C_1 t + C_2 t \ln |t|$$

The basic solution system with initial point $t = 1$ consists of the two functions defined by the two initial value problems

$$\begin{aligned} v_0(1) &= 1 \\ v_0'(1) &= 0 \end{aligned} \quad \text{or} \quad v_0(t) = t - t \ln t$$

and

$$\begin{aligned} v_1(1) &= 0 \\ v_1'(1) &= 1 \end{aligned} \quad \text{or} \quad v_1(t) = t \ln t$$

The general solution on the half line $t > 0$ may therefore be written

$$y(t) = y(1)v_0(t) + y'(1)v_1(t)$$
$$= y(1)(t - t \ln t) + y'(1)t \ln t$$

In terms of the same functions, the general solution of the nonhomogeneous equation

$$t^2 y''(t) - t y'(t) + y(t) = t^{1/2}$$

may be written as in (15):

$$y(t) = 4t^{1/2} + [y(1) - 4]v_0(t) + [y'(1) - 2]v_1(t)$$
$$= 4t^{1/2} + [y(1) - 4](t - t \ln t) + [y'(1) - 2]t \ln t$$

For comparison, the basic solution system with initial point $t_0 = e$ consists of $v_0(t) = (2t - t \ln t)/2$ and $v_1(t) = -t + t \ln t$ which is not the result of translating the system with initial point at $t_0 = 1$, given above.

5. Sample Problems with Answers

A. The general solution of the equation $y'' + y = 0$ may be written

$$y(t) = A_1 \cos t + A_2 \sin t \qquad (18)$$

(a) Find the basic solution system with initial point $t = 0$ for the equation $y'' + y = 0$, and write the general solution in terms of those functions.

Answer: $y(t) = y(0) \cos t + y'(0) \sin t$

(b) Show by substitution that $\cos(t - 5)$ and $\sin(t - 5)$ are solutions of the equation.

(c) Find the basic solution system with initial point $t = 5$ for the equation $y'' + y = 0$, and write the general solution in terms of those functions.

Answer: $y(t) = y(5) \cos(t - 5) + y'(5) \sin(t - 5)$

(d) One solution of the equation is $\cos t$. Use trigonometric identities to find coefficients C_1 and C_2 such that $\cos t$ may be written as a linear combination of $\cos(t - 5)$ and $\sin(t - 5)$, namely $\cos t = C_1 \cos(t - 5) + C_2 \sin(t - 5)$.

Answer: $\cos t = \cos 5 \cos(t - 5) - \sin 5 \sin(t - 5)$

B. (a) Find the basic solution system with initial point $t = 0$ for the equation $y'' + 4y' + 4y = 0$, and write the general solution of the equation in terms of those functions.

Answer: $y(t) = y(0)(e^{-2t} + 3te^{-2t}) + y'(0)te^{-2t}$

(b) Do the same for the initial point $t = 3$.

Answer: $y(t) = y(3)[e^{-2(t-3)} + 3(t-3)e^{-2(t-3)}] + y'(3)(t-3)e^{-2(t-3)}$

C. Write each of the following functions as a linear combination of e^{2t} and $\sinh 2t$:

(a) $3e^{2t}$ (b) $7e^{-2t}$ (c) $4 \cosh 2t$

Answers: (a) $3e^{2t} = 3e^{2t} + 0 \sinh 2t$; (b) $7e^{-2t} = -14 \sinh 2t + 7e^{2t}$; (c) $4 \cosh 2t = -4 \sinh 2t + 4e^{2t}$

D. Write each of the functions 1, t, and t^2 as a linear combination of the functions 1, $t - 1$, and $(t - 1)^2$.

Answers: $1 = 1 + 0(t - 1) + 0(t - 1)^2 = 1$; $t = (t - 1) + 1$; $t^2 = (t - 1)^2 + 2(t - 1) + 1$

6. Problems

For each of the following problems, find the basic solution system with the given initial point t_0 for the given differential equation. Also write the general solution of the equation in terms of the given basic solution system.

1. (a) $y'' + 9y = 0$; $t_0 = 0$ (b) $y'' + 9y = 0$; $t_0 = 2$

2. (a) $y'' + 3y' + 2y = 0$; $t_0 = 0$ (b) $y'' + 3y' + 2y = 0$; $t_0 = -3$

3. (a) $y'' + 8y' + 16y = 0$; $t_0 = 0$ (b) $y'' + 8y' + 16y = 0$; $t_0 = 5$

4. (a) $y'' + 8y' + 20y = 0$; $t_0 = 0$ (b) $y'' + 8y' + 20y = 0$; $t_0 = -1$

5. (a) $t^2 y'' + ty' - 25y = 0$; $t_0 = 1$ (b) $t^2 y'' + ty' - 25y = 0$; $t_0 = -2$

6. (a) $t^2 y'' + 3ty' + y = 0$; $t_0 = 1$ (b) $t^2 y'' + 3ty' + y = 0$; $t_0 = -4$

7. (a) $4t^2 y'' + 4ty' + y = 0$; $t_0 = 3$ (b) $4t^2 y'' + 4ty' + y = 0$; $t_0 = -5$

8. The following three functions constitute a basic solution system for a certain differential equation. Identify the initial point, and determine which function is v_0, which is v_1, and which is v_2:

$$\tfrac{1}{2}(e^{t-1} - e^{-t+1}) \qquad \tfrac{1}{2}(e^{t-1} - 2 + e^{-t+1}) \qquad 1$$

7. Linear Combinations and Linear Independence

Recall that the *linear combinations* of $f_1, f_2, \ldots, f_n$ are the things that can be written

$$c_1 f_1 + c_2 f_2 + \cdots + c_n f_n \qquad \text{where } c_1, c_2, \ldots, c_n \text{ are constants}$$

It is clear that for each linear homogeneous differential equation, each basic solution system contains exactly as many functions as the order of the equation. Now it should also be noticed that no one of these functions is a linear combination of the others. For instance, v_0 can never be written as

$$c_1 v_1 + \cdots + c_{n-1} v_{n-1}$$

Let us observe that fact in the solution set for the equation $y''' = 0$, with initial point $t = -7$, as given in (17a) through (17c) in Part 4:

$$v_0(t) \equiv 1 \qquad v_1(t) = t + 7 \qquad v_2(t) = \frac{(t+7)^2}{2} \tag{19}$$

First, $v_0(-7) = 1$, while every linear combination of v_1 and v_2 takes value zero at $t = -7$:

$$c_1 v_1(-7) + c_2 v_2(-7) = c_1 \cdot 0 + c_2 \cdot 0 = 0$$

Second, $v_1'(-7) = 1$, while the first derivative of every linear combination of v_0 and v_2 takes value zero at $t = -7$:

$$c_0 v_0'(-7) + c_2 v_2'(-7) = c_0 \cdot 0 + c_2 \cdot 0 = 0$$

Finally, $v_2''(-7) = 1$, while the second derivative of every linear combination of v_0 and v_1 takes value zero at $t = -7$:

$$c_0 v_0''(-7) + c_1 v_1''(-7) = c_0 \cdot 0 + c_1 \cdot 0 = 0$$

Thus, not one of the three functions is a linear combination of the other two.

Notice that the argument just given did not depend at all on the formulas for v_0, v_1, and v_2 as given in (19), but only on the initial values taken at the initial point. Neither did it depend on the location of the initial point, nor on the differential equation. Thus, we can give the same argument for *every* basic solution system for *every* linear homogeneous differential equation with *any* initial point, varying the argument only to accommodate the different orders of the equations. The conclusion will be stated in such a way that we shall be able to refer to it easily later on:

Theorem 3 *If $Ly = 0$ is any linear homogeneous equation of order n on an interval $\alpha < t < \beta$ containing some point t_0, then the basic solution system with initial point t_0 consists of exactly n functions, no one of which is a linear combination of the other $n - 1$ functions.*

If the expression "is a linear combination of" seems cumbersome to you, then try the following expression, which means the same thing: f is *spanned by* $f_1, \ldots, f_n$ if and only if f is a linear combination of $f_1, \ldots, f_n$, and the collection of all things spanned by $f_1, \ldots, f_n$ is called the *span of* $f_1, \ldots, f_n$. In these terms we would say, for instance, that a basic solution system spans the general solution set of a homogeneous linear differential equation, but that no function in the system is spanned by the others.

If not one of a collection $f_1, \ldots, f_n$ is spanned by the others (that is, if none is a linear combination of the others), then we say that the collection $f_1, \ldots, f_n$ is a *linearly independent collection*. We have therefore seen that a basic solution system is linearly independent. A collection which is not linearly independent is called a *linearly dependent collection*. Thus, a collection is linearly dependent if and only if at least one member is a linear combination of the other members.

Our first use of linear independence is to help us simplify expressions for solutions of differential equations. For instance, the equation

$$y'' + 4y' + 4y = 0 \tag{20}$$

has many solutions, including these:

$$e^{-2t} \qquad 3e^{-2t} \qquad (t-1)e^{-2t} \qquad -te^{-2t} \qquad te^{-2t} + 10^{-2t} \tag{21}$$

The general solution is actually a linear combination of these,

$$A_1 e^{-2t} + A_2(3e^{-2t}) + A_3(t-1)e^{-2t} + A_4(-te^{-2t}) + A_5(te^{-2t} + 10e^{-2t}) \tag{22}$$

It is easy to see that the collection (21) is linearly dependent, and thus it is easy to simplify the general solution from (21) to the simpler form

$$y(t) = (A_1 + 3A_2 - A_3 + 10A_5)e^{-2t} + (A_3 - A_4 + A_5)te^{-2t}$$
$$= B_1 e^{-2t} + B_2 te^{-2t}$$

Of course, we could also span the general solution with a different pair of members of (21), for instance, e^{-2t} and $(t-1)e^{-2t}$:

$$y(t) = (A_1 + 3A_2 - A_4 + 11A_5)e^{-2t} + (A_3 - A_4 + A_5)(t-1)e^{-2t}$$
$$= C_1 e^{-2t} + C_2(t-1)e^{-2t}$$

Thus, if a function is written as a linear combination of a linearly dependent collection, then it may be simplified by being written in terms of fewer functions. The concept of linear dependence reminds us to simplify; it also reminds us *how* to simplify.

If we "simplify" a solution this way, do we retain all parts of the solution? That is, if the general solution is written as the linear combination

$$y(t) = A_1 f_1(t) + A_2 f_2(t) + A_3 f_3(t) + \cdots + A_{m-1} f_{m-1}(t) + A_m f_m(t) \tag{23}$$

and if f_m is itself a linear combination of $f_1, f_2, \ldots, f_{m-1}$, then can the general solution be spanned by $f_1, f_2, \ldots, f_{m-1}$? Yes it can, for if

$$f_m = C_1 f_1 + C_2 f_2 + \cdots + C_{m-1} f_{m-1}$$

then substitution into the right-hand side of (23) shows that y can be written

$$y = A_1 f_1 + \cdots + A_{m-1} f_{m-1} + A_m(C_1 f_1 + \cdots + C_{m-1} f_{m-1})$$
$$= (A_1 + A_m C_1)f_1 + \cdots + (A_{m-1} + A_m C_{m-1})f_{m-1}$$

and is therefore spanned by $f_1, \ldots, f_{m-1}$.

Of course, it may be possible to simplify still further. For instance, if f_{m-1} is spanned by $f_1, \ldots, f_{m-2}$, then so is the general solution. If we continue in this way, each time extracting one function which is spanned by those remaining, where do we end? We end when we reach a collection of which none is a linear combination of the others, and which nevertheless has the same span as the original collection. This fact will be very useful later on. Therefore it, too, will be written out for convenient reference.

Theorem 4 (simplification theorem) *If $f_1, \ldots, f_m$ is a linearly dependent collection, then elimination of one member which is spanned by the others leaves a smaller collection with the same span. Continuing by successive elimination, one can always obtain a smaller collection which has the same span as $f_1, \ldots, f_m$ and which is, in addition, linearly independent.*

There clearly remains the question of how far we must simplify in order to extract this linearly independent collection of solutions which spans the general solution. Perhaps you have guessed. If the order of the equation is n, then the number of solutions left at the end of the simplification will also be n.

This brings us to the most important single use of linear independence in all of the study of differential equations:

Theorem 5 *If $Ly = 0$ is a linear differential equation with order n on some interval, and if n functions $y_1, \ldots, y_n$ are solutions on that interval, then they span the general solution of the equation if and only if they constitute a linearly independent collection.*

(A demonstration of this fact will be given in Part 14 of this section.)

It is because of this fact that the rules for the use of the characteristic polynomial give the general solution of a homogeneous equation with constant coefficients. For example, the equation

$$y'' + 2y' + y = 0$$

has characteristic polynomial $p(s) = (s + 1)^2$, so that e^{-t} and te^{-t} are solutions. Also, every function of the form $A_1 e^{-t} + A_2 te^{-t}$ is a solution. Since te^{-t} is not a constant multiple of e^{-t}, it is clear that e^{-t} and te^{-t} form a linearly independent collection; since there are two of them, they span the general solution of the given second-order equation.

Here is an example of a second-order equation with nonconstant coefficients:

$$t^2 y'' + 4ty' + 2y = 0$$

for which two solutions, t^{-1} and t^{-2}, may be found. It follows from the linearity of the equation that $A_1 t^{-1} + A_2 t^{-2}$ is also a solution of the equation for each choice of the constants A_1 and A_2. Since t^{-2} is not a constant multiple of t^{-1}, and vice versa, it must be that the collection consisting of t^{-1} and t^{-2} is linearly independent. It follows from Theorem 5, then, that

$$y(t) = A_1 t^{-1} + A_2 t^{-2}$$

is the general solution of the equation.

Our last example is the third-order equation

$$(2t + t^2)y''' + (6 - t^2)y'' - (6 + 2t)y' = 0 \tag{24}$$

Nothing that we have studied so far tells us how to solve the equation, but it is easy to check by substitution that 1, e^t, and t^{-1} are solutions of the equation. Since no one of those three functions could possibly be a linear combination of the other two, the collection must be linearly independent. By Theorem 5 they span the general solution. Hence,

$$y(t) = A_1 + \frac{A_2}{t} + A_3 e^t$$

is the general solution of (24).

Linear-independence arguments are superfluous in cases in which one has an algorithm or formula for generating the desired solution or the general solution of a differential equation. Thus, the correct use of the Laplace-transform calculations of the previous chapter and the separation of variables technique in Chapter I may obviate the explicit use of linear independence. However, the example given above begins to show its importance in other cases, for Theorem 5 places a definite bound on the number of functions one needs to find. Equation (24) is a homogeneous third-order equation, and so three linearly independent solutions will span all the solutions.

Furthermore, the Laplace transform is a rather complicated tool. Unless one uses it frequently, one is likely to forget the transforms of even the most common functions. Moreover, even the use of the characteristic polynomial can become uncertain in the memory. However, if one remembers that n linearly independent solutions span the general solution set of a homogeneous nth-order equation, then one can remember or discover solutions one at a time and still know for sure when one has obtained them all. (It is usually advantageous to postpone the effort of finding the basic solution system until one is ready to begin satisfying initial conditions.)

When one has all the homogeneous solutions in hand, then one can always obtain as well the solutions of the given equation, using the method of variation of parameters to be described in the next section. (This method works even when the method of undetermined coefficients is not applicable.) Linear independence is important in the use of this method, as it is on every occasion when one needs to verify that one has all the homogeneous solutions.

The next part of this section concerns techniques for showing whether a collection is linearly independent or not. The section concludes with a derivation (proof) of Theorem 5 and an application to the use of the characteristic polynomial.

8. Techniques for Showing Linear Dependence or Independence

In the previous part, linear dependence was defined this way: The collection $f_1, \ldots, f_m$ is linearly dependent if some one member is a linear combination of the others. To show linear dependence using the definition directly, we may have to test *each* member of the collection to see whether it is spanned by the others.

For an example, consider the functions 1, t, and t^2. It is easy to see that the constant function 1 is not a linear combination of t and t^2, for every such combination,

$$C_1 t + C_2 t^2$$

takes the value zero at $t = 0$, whereas 1 does not. We have not yet finished showing that 1, t, and t^2 are linearly independent, however, for it might happen that one of the other members, perhaps t, is a linear combination of the rest. Let us see whether it is or not. A linear combination

$$C_0 + C_2 t^2 \tag{25}$$

takes value C_0 at $t = 0$. If (25) is really going to be the function t, then (25) must take value zero at $t = 0$, and hence $C_0 = 0$. If (25) is going to be t, then the remaining constant C_2 must be chosen so that $C_2 t^2$ is t for each value of t. Naturally, this is impossible, and so we conclude that t is not a linear combination of 1 and t^2.

Can it still happen that t^2 is a linear combination of 1 and t? The same kind of argument can be used to show that this, too, is impossible.

Of course, the same kind of argument can be used to show that no one of the monomials

$$1 \quad t \quad t^2 \quad \cdots \quad t^m$$

is a linear combination of the others, and therefore the collection is linearly independent according to the definition given at the beginning of the previous section. This fact is very useful to remember, but the calculations used to derive it turn out to be unnecessarily lengthy. In practice, a demonstration of linear independence is nearly always either a reference to a previously investigated example, such as the monomials given above, or a use of the following criterion, which is equivalent to the definition but suggests a more efficient technique of calculation.

Linear-independence criterion *If $f_1, f_2, \ldots, f_n$ is a given collection, then:*
1. *If numbers $c_1, c_2, \ldots, c_m$ can be found, not all of them zero, such that*

$$c_1 f_1 + c_2 f_2 + \cdots + c_m f_m = 0 \tag{26}$$

   then the collection is linearly dependent.
2. *On the other hand, if Equation (26) implies that $c_1 = 0$, $c_2 = 0$, $\ldots$, and $c_m = 0$, then the collection is linearly independent.*

For example, 1, t, $t^2, \ldots, t^m$ are linearly independent because the general linear combination

$$c_m t^m + \cdots + c_2 t^2 + c_1 t + c_0$$

can be the zero function only if all $m + 1$ of the coefficients are zero. Beware, however, of the fact that the monomials and polynomials are being considered as functions, and that a function is the *zero function* if and only if it is *identically zero*. Thus, the fact that

$$t^2 - 3t + 2 = 0 \qquad \text{if and only if} \qquad t = 1, 2$$

is entirely consistent with the fact that

$$f(t) = t^2 - 3t + 2 \text{ is not the zero function}$$

More examples will be given in the next part.

The proof of the criterion consists only of observing that

$$f_1 = c_2 f_2 + c_3 f_3 + \cdots + c_m f_m \tag{27}$$

if and only if

$$-f_1 + c_2 f_2 + c_3 f_3 + \cdots + c_m f_m = 0 \tag{28}$$

which is (26) with $c_1 = -1$ and then noticing that the corresponding statements about $f_2, f_3, \ldots, f_m$ are also true.

9. Examples of the Use of the Linear-Independence Criterion

Let us see whether e^{-t}, e^{2t}, and e^{3t} constitute a linearly independent collection by setting

$$c_1 e^{-t} + c_2 e^{2t} + c_3 e^{3t} = 0 \tag{29}$$

and drawing conclusions about c_1, c_2, and c_3. Start by dividing out the most rapidly growing exponential, giving

$$c_1 e^{-4t} + c_2 e^{-t} + c_3 = 0$$

and take the limit as t increases, which is

$$c_1 \cdot 0 + c_2 \cdot 0 + c_3 = 0 \qquad \text{or} \qquad c_3 = 0$$

Substituting this value into (29) gives

$$c_1 e^{-t} + c_2 e^{2t} = 0$$

to which the same argument can be applied, showing that $c_2 = 0$. It necessarily follows that $c_1 = 0$ also. Thus,

$$\text{If} \qquad c_1 e^{-t} + c_2 e^{2t} + c_3 e^{3t} = 0 \qquad \text{then} \qquad c_1 = c_2 = c_3 = 0$$

and the linear-independence criterion says that therefore e^{-t}, e^{2t}, and e^{3t} form a linearly independent collection.

A similar argument shows that if $\alpha_1, \alpha_2, \ldots$, and α_m are distinct real numbers, then $e^{\alpha_1 t}, e^{\alpha_2 t}, \ldots$, and $e^{\alpha_m t}$ form a linearly independent set.

For another example, consider sinusoids of differing frequencies. Specifically, suppose that $\omega_1 > \omega_2 > 0$, and consider the collection consisting of $\cos \omega_1 t$, $\sin \omega_1 t$, $\cos \omega_2 t$, and $\sin \omega_2 t$. First set

$$c_1 \cos \omega_1 t + c_2 \sin \omega_1 t + c_3 \cos \omega_2 t + c_4 \sin \omega_2 t = 0 \tag{30}$$

and then see whether or not c_1, c_2, c_3, and c_4 need all be zero. Evaluation of (30) at $t = 0$ gives

$$c_1 + c_3 = 0 \qquad (31)$$

while evaluation at $t = \pi/\omega_1$ gives

$$-c_1 + c_3 \cos \pi \frac{\omega_2}{\omega_1} + c_4 \sin \pi \frac{\omega_2}{\omega_1} = 0 \qquad (32)$$

Differentiation of (30), followed by evaluation at $t = 0$ and $t = \pi/\omega_1$, respectively, gives

$$\omega_1 c_2 + \omega_2 c_4 = 0 \qquad (33)$$

and

$$-\omega_1 c_2 - \omega_2 c_3 \sin \pi \frac{\omega_2}{\omega_1} + \omega_2 c_4 \cos \pi \frac{\omega_2}{\omega_1} = 0 \qquad (34)$$

Elimination of c_1 from (31) and (32), and of c_2 from (33) and (34), gives the simultaneous equations

$$\left. \begin{aligned} c_3 \left(1 + \cos \pi \frac{\omega_2}{\omega_1} \right) + c_4 \sin \pi \frac{\omega_2}{\omega_1} = 0 \\[2mm] -c_3 \sin \pi \frac{\omega_2}{\omega_1} + c_4 \left(1 + \cos \pi \frac{\omega_2}{\omega_1} \right) = 0 \end{aligned} \right\}$$

whose only solution is $c_3 = c_4 = 0$, since the inequalities $\omega_1 > \omega_2 > 0$ imply that $0 < \pi(\omega_2/\omega_1) < \pi$ and hence that $\cos \pi(\omega_2/\omega_1) \neq -1$. Substitution into (31) and (33) shows that c_1 and c_2 are also zero, and that the four sinusoids constitute a linearly independent set.

To illustrate a slightly different method of calculation, consider the same collection and suppose that $\omega_1 > \omega_2 > 0$. Evaluation of (30) at $t = 0$ leads to

$$c_1 + c_3 = 0 \qquad (31)$$

while evaluation of the second derivative at $t = 0$ gives

$$\omega_1^2 c_1 + \omega_2^2 c_3 = 0 \qquad (35)$$

Solution of (31) and (35) simultaneously shows that

$$c_3 = -c_1 \quad \text{and} \quad c_1(\omega_1^2 - \omega_2^2) = 0$$

Since $\omega_1 > \omega_2 > 0$ we have $c_1 = c_3 = 0$.

Evaluation of the first and third derivatives of (30) at $t = 0$ gives the pair of equations

$$\left. \begin{aligned} \omega_1 c_2 + \omega_2 c_4 = 0 \\ \omega_1^3 c_2 + \omega_2^3 c_4 = 0 \end{aligned} \right\}$$

from which we conclude that

$$\omega_2 c_4 = -\omega_1 c_2 \quad \text{and} \quad \omega_1 c_2(\omega_1^2 - \omega_2^2) = 0$$

Since $\omega_1 > \omega_2 > 0$ we have $c_2 = c_4 = 0$.

More generally, any collection of sinusoids with nonzero frequencies is linearly independent if the squares of the frequencies are all distinct, or, still more generally, if the frequencies of no more than two have the same squares, and of those two sinusoids neither is a multiple of the other. Thus, $\sin t$, $\cos t$, and $\sin (t + \frac{1}{10})$ form a linearly dependent set, while $\sin t$, $\cos (t + \frac{1}{10})$, $\cos 3t$, $\sin 4t$, and $\cos 4t$ form a linearly independent set.

10. Example of Reference to a Known Independent Collection

Finally, there is a very general technique, that of writing all elements of the collection in terms of a different collection which is *known* to be linearly independent. Thus, the linear independence of t, $t + 1$, $t + 2$, and $(t + 1)^2$ can be tested by writing the equation

$$c_1 t + c_2(t + 1) + c_3(t + 2) + c_4(t + 1)^2 = 0 \tag{36}$$

and then writing the left-hand side as a linear combination of 1, t, and t^2, in this way:

$$0 = c_1 t + c_2 t + c_2 + c_3 t + 2c_3 + c_4 t^2 + 2c_4 t + c_4$$
$$= c_4 t^2 + (c_1 + c_2 + c_3 + 2c_4)t + (c_2 + 2c_3 + c_4)1$$

Since 1, t, and t^2 form a linearly independent set, it must follow that

$$\begin{aligned} t^2: & & c_4 = 0 \\ t: & & c_1 + c_2 + c_3 + 2c_4 = 0 \\ 1: & & c_2 + 2c_3 + c_4 = 0 \end{aligned} \Big\}$$

from which we see that

$$\begin{aligned} c_1 + c_2 + c_3 &= 0 \\ c_2 + 2c_3 &= 0 \end{aligned} \Big\}$$

For any value of c_3, then, (36) can be satisfied as long as

$$\begin{aligned} c_2 &= -2c_3 \\ c_1 &= -c_2 - c_3 = c_3 \\ c_4 &= 0 \end{aligned} \Bigg\}$$

Thus, t, $t + 1$, $t + 2$, and $(t + 1)^2$ constitute a linearly dependent set.

This technique is the basis for setting various coefficients equal, in the partial-fraction decomposition and in the method of undetermined coefficients.

11. Sample Problems with Answers

In each of the following problems, express the first of the functions as a linear combination of the others, or else show that it is not spanned by them.

E. e^{-2t}; $\sinh 2t$, $(\cosh 2t)/4$, e^{-3t}, 1, t^2
  Answer: $e^{-2t} = -\sinh 2t + 4(\cosh 2t)/4 + 0e^{-3t} + 0 \cdot 1 + 0t^2$

F. $\cos(t + 3)$; $\cos t$, $\sin t$, $\cos 2t$
  Answer: $\cos(t + 3) = \cos 3 \cos t - \sin 3 \sin t$

G. $t^2 + 1$; $(t + 2)^2$, $(t + 1)$, t
  Answer: $t^2 + 1 = (t + 2)^2 - 3(t + 1) + t$

H. $t^2 + 1$; $(t + 2)^3$, $t + 1$, t
  Answer: $t^2 + 1$ is not spanned by $(t + 2)^3$, $t + 1$, and t

In each of the next three problems, discover whether or not the given collection is linearly independent.

I. $\cos 2t$, $\sin^2 t$, 1 *Answer:* Dependent, since $\cos 2t = 1 - 2\sin^2 t$

J. $\cos t$, $t \cos t$, $t^2 \cos t$
  Answer: Independent, since

$$c_1 \cos t + c_2 t \cos t + c_3 t^2 \cos t = (c_1 + c_2 t + c_3 t^2) \cos t = 0$$

implies that $c_1 + c_2 t + c_3 t^2 = 0$

K. $(t + 1)^2$, $(t + 2)^2$, t^2 *Answer:* Independent

L. Use linear independence to show that if

$$1 + A_1 e^t + A_2 e^{5t} = 3e^t - 2e^{5t} + B$$

then $A_1 = 3$, $A_2 = -2$, and $B = 1$.
  Answer: Rewrite the equation as $(A_1 - 3)e^t + (A_2 + 2)e^{5t} + (1 - B)e^{0t} = 0$. Since e^t, e^{5t}, and e^{0t} are linearly independent, in the last equation the coefficient of each must be zero.

M. Reduce the following linearly dependent collection to a linearly independent collection with the same span by successively removing functions which are linear combinations of those which remain: $(t + 1)^2$, $t + 1$, t^3, t^2, t, 1.
  Answer: Any collection of four which contains t^3 and either $(t + 1)^2$ or t^2 [for instance t^3, $(t + 1)^2$, t^2, and $t + 1$] will be linearly independent.

12. Problems

In each of the following problems, express the first of the functions as a linear combination of the others, or else show that it is not spanned by them.

9. e^t; $\cosh t$, $3 \sinh t$, $\sin t$ **10.** e^t; $\cosh t$, $2e^{-t}$, -1

11. $\cosh 3t$; $\sinh 3t$, $3t^2$, $5e^{-3t}$ **12.** 1; t^2, $(t + 1)^2$, $(t + 2)^2$

13. $\frac{1}{2}\sin(2t + 3)$; $\cos 2t$, $\sin 2t$, 1 **14.** $\sin 2t$; $\cos 2t$, $\sin t$

15. 1; $(t + 1)^2$, t **16.** $\cosh t$; $\sinh t$, $\sinh 2t$, $\cosh 2t$

In each of the next four problems, discover whether or not the given collection is linearly independent.

17. t^2, $(t + 1)^2$, $(t + 2)^2$ **18.** t^2, $(t + 1)^2$, $(t + 2)^2$, $(t + 3)^2$

19. (a) $\cos \pi(t - 1)$, $\cos \pi t$ (b) $\cos \pi(t - \frac{1}{2})$, $\cos \pi t$
  (c) $\cos \pi(t - \frac{1}{2})$, $\cos \pi t$, $\cos \pi(t + \frac{1}{2})$ (d) $\cos \pi(t - \frac{1}{2})$, $\cos \pi t$, $\sin \pi t$
  (e) $\cos \pi(t - \frac{1}{4})$, $\cos \pi t$, $\cos \pi(t + \frac{1}{4})$

20. (a) e^t, e^{2t}, te^t (b) e^t, e^{-t}, te^t (c) e^t, e^{-t}, te^{2t}

21. Use linear independence to show that if

$$c_1 \sin t + c_2 \cos t = 3 \sin t + 7 \cos t$$

then $c_1 = 3$ and $c_2 = 7$.

22. Find a function in the following list which is spanned by the others, remove it, and test the remaining collection of four functions for linear independence:

$$(t+1)^3 \qquad (t+1)^2 \qquad t+1 \qquad t+7 \qquad t-1$$

23. From the following collection of functions, select a (possibly smaller) collection with the same span but which is linearly independent:

$$(t+1)^3 \qquad (t+1)^2 \qquad t^2 \qquad 2t+1 \qquad t^3 - t - 1$$

13. Remarks on the Rules for Using the Characteristic Polynomial

Consider a homogeneous linear differential equation with constant coefficients

$$Ly = 0 \tag{37}$$

whose characteristic polynomial may be factored as

$$p(s) = C(s - \alpha_1)^{m_1}(s - \alpha_2)^{m_2} \cdots (s - \alpha_n)^{m_n}$$

where C is constant, and $\alpha_1, \alpha_2, \ldots, \alpha_n$ are distinct numbers, real or complex. The rules give solutions of (37) of the form

$$g_1(t)e^{\alpha_1 t} + g_2(t)e^{\alpha_2 t} + \cdots + g_n(t)e^{\alpha_n t} \tag{38}$$

where $g_1, g_2, \ldots, g_n$ are polynomials of degree $m_1 - 1, m_2 - 1, \ldots, m_n - 1$, respectively. We shall show that (38) gives the general solution of (37), as follows. First we shall show the linear independence of the collection of all the individual terms of the form $t^k e^{\alpha t}$, of which (38) is the general linear combination, and of which there are $m_1 + m_2 + \cdots + m_n$. Since this number is the same as the degree of $p(s)$ and the order of Equation (37), it must follow from Theorem 5 that (38) is the general solution of (37).

To show linear independence, we must show that if $g_1, g_2, \ldots,$ and g_n are given polynomials, and if

$$g_1(t)e^{\alpha_1 t} + g_2(t)e^{\alpha_2 t} + \cdots + g_n(t)e^{\alpha_n t} \equiv 0 \tag{39}$$

then $\qquad g_1(t) \equiv 0 \qquad g_2(t) \equiv 0 \qquad \cdots \qquad g_n(t) \equiv 0$

First consider the numbers $\alpha_1, \alpha_2, \ldots, \alpha_n$ as complex numbers. Choose one with the largest modulus, and take it to be α_1, since the order in which the terms are listed is immaterial. Throughout (39), evaluate not at t but at $\bar{\alpha}_1 t$, obtaining

$$g_1(\bar{\alpha}_1 t)e^{|\alpha_1|^2 t} + g_2(\bar{\alpha}_1 t)e^{\alpha_2 \bar{\alpha}_1 t} + \cdots + g_n(\bar{\alpha}_1 t)e^{\alpha_n \bar{\alpha}_1 t} \equiv 0$$

Division by $e^{|\alpha_1|^2 t}$ gives

$$g_1(\bar{\alpha}_1 t) + g_2(\bar{\alpha}_1 t)e^{(\alpha_2 \bar{\alpha}_1 - |\alpha_1|^2)t} + \cdots + g_n(\bar{\alpha}_1 t)e^{(\alpha_n \bar{\alpha}_1 - |\alpha_1|^2)t} \equiv 0 \tag{40}$$

The next step will be to take the limit of the left-hand side of (40). We shall see that in each of the last $n - 1$ terms the exponential assures that the limit is zero. Since only the zero polynomial has limit zero as t increases, it must follow that g_1 is the zero polynomial; that is, $g_1 = 0$.

Consider the exponent $(\alpha_2 \alpha_1 - |\alpha_1|^2)t$. Since

$$|\alpha_2 \bar{\alpha}_1| = |\alpha_2||\alpha_1| \le |\alpha_1|^2$$

it follows that $\alpha_2 \bar{\alpha}_1$ falls inside a circle of radius $|\alpha_1|^2$, and therefore $\alpha_2 \bar{\alpha}_1 - |\alpha_1|^2$ falls in the left half plane and has negative real part. (See Figure 1.) Hence, $\alpha_2 \alpha_1 - |\alpha_1|^2$ can be written as $\alpha + \beta j$, where α is negative and *not* zero, so that

$$\lim_{t \to \infty} g_2(\bar{\alpha}_1 t)e^{\alpha t} = 0$$

Since $e^{\beta j t}$ is just a sinusoid, and therefore bounded, it follows that

$$\lim_{t \to \infty} g_2(\bar{\alpha}_1 t)e^{\alpha t + \beta j t} = 0 \qquad \text{or} \qquad \lim_{t \to \infty} g_2(\bar{\alpha}_1 t)e^{\alpha_2 \bar{\alpha}_1 t - |\alpha_1|^2 t} = 0$$

Of course the same is true of the other exponential terms in (40), and so we may conclude that

$$\lim_{t \to \infty} g_1(\bar{\alpha}_1 t) = 0$$

Since all nonconstant polynomials increase unboundedly as t increases, it follows that g_1 is constant. Since g_1 has limit zero, that constant must be zero, so that

$$g_1(t) \equiv 0$$

Thus, (39) becomes

$$g_2(t)e^{\alpha_2 t} + \cdots + g_n(t)e^{\alpha_n t} \equiv 0 \tag{41}$$

Equation (41) is simpler than (39) and can be treated in the same way, to show that another of the polynomials is identically zero. After $n - 1$ instances of this treatment, we get $g_1 = 0$, $g_2 = 0$, ..., $g_{n-1} = 0$, and so $g_n(t)e^{\alpha_n t} \equiv 0$, from which it follows that $g_n = 0$ too. This completes the argument that all the coefficients of all the terms $t^k e^{\alpha t}$ appearing in (39) are zero. Therefore, those terms constitute a

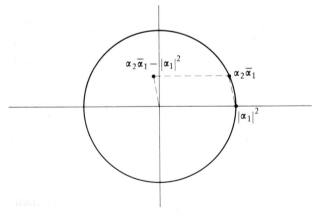

Figure 1 Showing that $\operatorname{Re}(\alpha_2 \bar{\alpha}_1 - |\alpha_1|^2) < 0$ if $|\alpha_2| \le |\alpha_1|$.

linearly independent set of $m_1 + m_2 + \cdots + m_n$ solutions of a linear differential equation of order $m_1 + m_2 + \cdots + m_n$, and Theorem 5 says that therefore they give the general solution. Thus, the rules for the use of the characteristic polynomial are correct.

14. Remarks: Linearly Independent Sets Are Minimal

This last part of the section is devoted to one of the most important properties of linearly independent sets. In one respect this property marks the beginning of linear algebra. We shall state and prove the property. Then we shall use it to give a proof of Theorem 5 and discuss its applications to differential equations.

> **Replacement lemma, or minimality lemma** *Among all the collections with a given span, those which are linearly independent have the fewest members. Specifically, if $f_1, \ldots, f_n$ and $g_1, \ldots, g_m$ have the same span, and if $g_1, \ldots, g_m$ is a linearly independent set, then $m \leq n$. If, in addition, $f_1, \ldots, f_n$ is linearly independent also, then $n = m$. Conversely, if $g_1, \ldots, g_n$ and $f_1, \ldots, f_n$ have the same span as well as the same number of elements, then one set is linearly independent if and only if the other is also.*

In other words, the minimum number of elements required to span a given collection is independent of the way one chooses to do the spanning. This fact is quite significant, and one of its immediate consequences will be a proof of Theorem 5, which says that a linear differential equation of order n always requires exactly n linearly independent functions to span the general homogeneous solution.

Here we begin the proof of the lemma. Let $f_1, \ldots, f_n$ and $g_1, \ldots, g_m$ be two collections having the same span, and let $g_1, \ldots, g_m$ be linearly independent. Since g_1 lies in the span of the f's, it follows that

$$g_1, f_1, \ldots, f_n$$

is a dependent collection. Some element of this collection, perhaps f_n, is therefore a linear combination of the elements written before it. We may remove this element without affecting the span. We may suppose the element to be removed is f_n, and conclude that

$$g_1, f_1, \ldots, f_{n-1}$$

has the same span.

Now continue. Since g_2 lies in the span, it follows that

$$g_1, g_2, f_1, \ldots, f_{n-1}$$

is dependent, and we remove one element which is a linear combination of its predecessors. The element to be removed cannot possibly be g_1 or g_2, since their predecessors are from $g_1, \ldots, g_n$, a linearly independent set. Thus the element to be removed is from among the f's, and we may assume it is f_{n-1}. Therefore,

$$g_1, g_2, f_1, \ldots, f_{n-2}$$

has the same span.

Continue further: In $m - 1$ steps we come to the collection

$$g_1, \ldots, g_{m-1}, f_1, \ldots, f_{n-m+1}$$

since $g_1, \ldots, g_{m-1}$ does not by itself have the same span, for it lacks g_m. Thus $n - m + 1 \geq 1$, or $n \geq m$, which is the first of our desired conclusions.

If $f_1, \ldots, f_n$ is *also* a linearly independent set, then the f's and g's may be interchanged in our argument, showing that $m \geq n$, and therefore that $m = n$. That is the second of our desired conclusions.

There remains the converse to show. Suppose that $f_1, \ldots, f_n$ is not a linearly independent collection. We can extract from it a collection with the same span but which *is* linearly independent (Theorem 4). Let us say that this smaller collection has m elements, and observe that $m < n$. Since every linearly independent collection with the same span as $f_1, \ldots, f_n$ must have exactly m elements, it must follow that no collection with the same span but containing n elements can be linearly independent. Thus $g_1, \ldots, g_n$ is a dependent set.

This concludes the proof of the lemma.

Our conclusion may be phrased another way. If $f_1, \ldots, f_n$ is any finite collection, then there exists a number m with the following property: A collection with the same span as $f_1, \ldots, f_n$ has exactly m members if and only if it is linearly independent, for otherwise it has at least $m + 1$ members. There is the essence of the minimality. The number m is called the *dimension* of the span of $f_1, \ldots, f_n$.

Theorem 5 is now virtually proved. If $Ly = 0$ is a linear equation with order n on some interval, then it has a basic solution system $v_1, \ldots, v_n$ which spans the general solution, by Theorem 2 in Section H of Chapter I. The basic solution system is a linearly independent set of exactly n functions. By the minimality lemma, other collections of functions which span the general solution are linearly independent if and only if they consist of exactly n solutions. Smaller collections cannot span; larger collections cannot be linearly independent.

In terms of operators, one might express Theorem 5 this way: If L is a linear differential operator of order n on a given interval, then the set of elements annihilated by that operator is of dimension n and is spanned by any collection of n functions annihilated by L, just so long as the collection is linearly independent.

F. VARIATION OF PARAMETERS AND REDUCTION OF ORDER

As we saw earlier, all the solutions of a linear differential equation

$$Ly = f \tag{1}$$

may always be written in the form

$$y = y_p + y_h$$

where y_p is some other solution of the given equation, and y_h is some homogeneous solution, a function satisfying the equation $Ly_h = 0$.

This section rises from the fact that if the homogeneous solutions are all known, then a particular solution y_p may always be found, and with it the general solution. The general idea is this: If $y_1, y_2, \ldots, y_n$ are known homogeneous solutions, so that the general homogeneous solution

$$y_h(t) = C_1 y_1(t) + \cdots + C_n y_n(t)$$

is also known, then one asks if the n arbitrary constants (parameters) $C_1, \ldots, C_n$ may not be replaced with functions

$$c_1(t), \ldots, c_n(t)$$

such that
$$y(t) = c_1(t)y_1(t) + \cdots + c_n(t)y_n(t) \tag{2}$$

is a solution of the given (nonhomogeneous) equation. That is, $c_1(t), \ldots, c_n(t)$ become unknown functions in a new problem, which can be solved by a technique called *variation of parameters* which we shall show. Once they are found, y itself may be written down at once, in the form (2).

The section ends with the use of the same technique to replace a problem involving an nth-order linear equation with one of order $n - 1$ if only one homogeneous solution is already known.

1. Example and Method: A First-Order Equation

The equidimensional equation

$$ty'(t) + y(t) = \ln t \qquad t > 0 \tag{3}$$

has a homogeneous solution

$$y_1(t) = t^{-1} \tag{4}$$

and general homogeneous solution

$$Cy_1(t) = Ct^{-1}$$

We shall set

$$y(t) = c_1(t)y_1(t) = \frac{c_1(t)}{t} \tag{5}$$

where c_1 is a function—an unknown function—and see what information we can discover about c_1. Since

$$y'(t) = c_1'(t)y_1(t) + c_1(t)y_1'(t) \tag{6}$$

the substitution of y and y' into (3) uses (5) and (6), giving a new equation

$$t[c_1'(t)y_1(t) + c_1(t)y_1'(t)] + c_1(t)y_1(t) = \ln t$$

or
$$tc_1'(t)t^{-1} + tc_1(t)(-t^{-2}) + c_1(t)t^{-1} = \ln t \tag{7}$$

$$c_1'(t) = \ln t \tag{8}$$

Integration by parts gives the unknown function:

$$c_1(t) = \int 1 \ln t \, dt$$

$$= t \ln t - \int tt^{-1} \, dt$$

$$= t \ln t - t + C \tag{9}$$

Now we can write a formula for y, using (5), as follows:

$$y(t) = c_1(t)y_1(t)$$
$$= (t \ln t - t + C)t^{-1}$$
$$= \ln t - 1 + Ct^{-1}$$

In fact, y is thus given in the form $y_p + y_h$, where $y_p(t) = \ln t - 1$ and $y_h(t) = Ct^{-1}$. One may verify by substitution that y_p is in fact a solution of the given equation:

$$ty_p'(t) + y_p(t) = t(t^{-1}) + \ln t - 1 = \ln t$$

Now let us see how to use the method on *any* first-order linear equation

$$a_1(t)y'(t) + a_0(t)y(t) = f(t) \tag{10}$$

or
$$Ly = f$$

for which a nonzero homogeneous solution y_1 is known. Set

$$y(t) = c_1(t)y_1(t) \tag{11}$$

where c_1 is a new unknown function. [As soon as c_1 is calculated, y may also be calculated using (11), since y_1 is known.] Differentiating (11) gives

$$y' = c_1' y_1 + c_1 y_1'$$

and substitution into (10) gives

$$a_1 c_1' y_1 + a_1 c_1 y_1' + a_0 c_1 y_1 = f$$
$$a_1 c_1' y_1 + c_1(a_1 y_1' + a_0 y_1) = f \tag{12}$$

In (12), the term in parenthesis is $L[y_1]$, which is zero since y_1 is a homogeneous solution of (10). This is the first important point about the method, for in every case it gives the simplification seen in the example when going from (7) to (8). The second important point is that the result of solving (12) is a formula for c_1', formula (13).

$$a_1 c_1' y_1 + c_1 L[y_1] = f$$

$$a_1 c_1' y_1 = f$$

$$c_1'(t) = \frac{f(t)}{a_1(t)y_1(t)} \tag{13}$$

Every factor on the right in (13) is known, and integration gives $c_1(t)$ itself. In the example, we had

$$a_1(t) = t \qquad y_1(t) = t^{-1} \qquad f(t) = \ln t$$

so that

$$c_1'(t) = \frac{\ln t}{t t^{-1}} = \ln t$$

and

$$c_1(t) = t \ln t - t + C$$

In general, definite integration of (13) leads to this:

$$c_1(t) = \int_{t_0}^{t} \frac{f(z)}{a_1(z)y_1(z)} \, dz$$

From (11),

$$y(t) = c_1(t)y_1(t) = y_1(t) \int_{t_0}^{t} \frac{f(z)}{a_1(z)y_1(z)} \, dz \tag{14}$$

This equation gives a particular solution, in fact the one such that $y(t_0) = 0$. To obtain the general solution we need only add the general homogeneous solution. The definite integral is given in (14) for use in the formula for the general solution in terms of an initial value at t_0:

$$y(t) = y_1(t) \frac{y(t_0)}{y_1(t_0)} + y_1(t) \int_{t_0}^{t} \frac{f(z)}{a_1(z)y_1(z)} \, dz \tag{15}$$

However, the usual formula is written in terms of two indefinite integrals and under the assumption that the equation is in *normal form*, that is, $a_1(t) \equiv 1$. The general solution of the equation

$$y'(t) + a(t)y(t) = f(t)$$

is

$$y(t) = e^{-\int a(t)\,dt} \left[C + \int f(t) e^{\int a(t)\,dt} \, dt \right] \tag{16}$$

If you are going to use formula (16) from memory, then you should remember that two integrations are required. First, $\int a(t) \, dt$ is the integral of the coefficient, and the reciprocal of the exponential of that integral is $y_1(t) = e^{-\int a(t)\,dt}$ which is a homogeneous solution. After y_1 is found, there remains the other integration,

$$\int f(t) e^{\int a(t)\,dt} \, dt = \int \frac{f(t)}{y_1(t)} \, dt$$

The constant C appearing in (16) is actually redundant, for an arbitrary constant of integration is inherent in each indefinite integral.

If the coefficient a is constant, then (15) and (16) will give the same solution as the formula derived using convolution in Equation (98) in Section G of Chapter III. In all cases it is the same as that given in Section F of Chapter I.

2. Example: A Second-Order Equation

Consider the equidimensional equation

$$t^2 y'' - 2ty' + 2y = t \ln t \tag{17}$$

The corresponding homogeneous equation is

$$t^2 y'' - 2ty' + 2y = 0$$

and the general homogeneous solution is

$$C_1 t + C_2 t^2 \tag{18}$$

We seek a solution of the given equation (17) of the form

$$y(t) = c_1(t)t + c_2(t)t^2 \tag{19}$$

$$= c_1(t)y_1(t) + c_2(t)y_2(t)$$

where $y_1(t) = t$ and $y_2(t) = t^2$.

It is usual to begin the calculation by dividing out the lead coefficient of the given equation, putting the equation into normal form with lead coefficient identically 1:

$$y'' - 2t^{-1}y' + 2t^{-2}y = t^{-1} \ln t \tag{20}$$

Into (20) we shall have to substitute y and its derivatives, insofar as we know them. Differentiating (19) gives

$$y'(t) = c_1(t) + c_2(t)2t + c_1'(t)t + c_2'(t)t^2 \tag{21}$$

We shall set the sum of the c' terms equal to zero, giving an equation which will surprise you if you have never seen the method before, but which will be explained (and justified) in the last paragraph of this part of the section. Thus,

$$c_1'(t)t + c_2'(t)t^2 = 0 \quad \text{or} \quad c_1'(t) + c_2'(t)t = 0 \tag{22}$$

Of course, (22) now allows us to simplify (21) in this way:

$$y'(t) = c_1(t) + c_2(t)2t \tag{23}$$

Differentiation gives

$$y''(t) = c_1'(t) + c_2'(t)2t + 2c_2(t) \tag{24}$$

Substitution of (24), (23), and (19) into the given equation (20) gives

$$c_1'(t) + c_2'(t)2t + 2c_2(t) - 2t^{-1}[c_1(t) + c_2(t)2t] + 2t^{-2}[c_1(t)t + c_2(t)t^2] = t^{-1} \ln t$$

Collecting multiples of c_1', c_2', c_1, and c_2 gives

$$c_1'(t) + c_2'(t)2t + c_1(t)(-2t^{-1} + 2t^{-1}) + c_2(t)(2 - 4 + 2) = t^{-1} \ln t$$

or
$$c_1'(t) + c_2'(t)2t = t^{-1} \ln t \qquad (25)$$

Equations (22) and (25) may now be solved for c_1' and c_2':

$$\left. \begin{aligned} c_1'(t) + c_2'(t)t &= 0 \\ c_1'(t) + c_2'(t)2t &= t^{-1} \ln t \end{aligned} \right| \qquad (26)$$

Subtracting the first equation from the second shows that $c_2'(t)t = t^{-1} \ln t$ and so

$$\left. \begin{aligned} c_2'(t) &= t^{-2} \ln t \\ c_1'(t) &= -t^{-1} \ln t \end{aligned} \right|$$

With c_1' and c_2' in hand, it only remains to integrate them:

$$c_1(t) = \int c_1'(t) \, dt = -\int (\ln t) \frac{1}{t} \, dt$$

$$= -\frac{(\ln t)^2}{2} + C_1$$

$$c_2(t) = \int c_2'(t) \, dt = \int \frac{1}{t^2} \ln t \, dt$$

$$= -\frac{1}{t} \ln t - \int -\frac{1}{t} \frac{1}{t} \, dt$$

$$= -\frac{1}{t} (\ln t + 1) + C_2$$

Finally, substitution into (19) gives the solution:

$$y(t) = c_1(t)t + c_2(t)t^2$$

$$= \frac{-t(\ln t)^2}{2} - t(\ln t + 1) + C_1 t + C_2 t^2$$

This is the general solution of (20), and of (17).

Now the unexplained step must be explained. If (21) had been differentiated as it stood, and then substitution were made into (20), the result would have been a single equation of the second order in both unknowns, c_1 and c_2. There are very many pairs of functions satisfying this equation. What we did in (22) was to specify one pair by setting an additional equation that the pair must satisfy. The result of so specifying is a pair of equations which *can always be solved* for a *unique* pair of derivatives c_1' and c_2'. We now give the general technique of solution.

3. Method of Variation of Parameters: The Second-Order Equation

Consider a second-order linear differential equation, in normal form in that the lead coefficient is identically 1,

$$y''(t) + a_1(t)y'(t) + a_0(t)y(t) = f(t) \tag{27}$$

or
$$Ly = f$$

If the general homogeneous solution

$$C_1 y_1(t) + C_2 y_2(t)$$

is known, then a particular solution may be found of the form

$$y(t) = c_1(t)y_1(t) + c_2(t)y_2(t)$$

or
$$y = c_1 y_1 + c_2 y_2 \tag{28}$$

It is important to remember that y_1 and y_2 must be known and must generate the general homogeneous solution. Any two linearly independent homogeneous solutions will suffice. It is the functions c_1 and c_2 which become the unknowns of the problem, from which y is recovered by use of (28).

Differentiating (28) gives

$$y' = c_1 y_1' + c_2 y_2' + c_1' y_1 + c_2' y_2$$

Setting the sum of the c' terms equal to zero gives

$$c_1' y_1 + c_2' y_2 = 0 \tag{29}$$

and
$$y' = c_1 y_1' + c_2 y_2' \tag{30}$$

Differentiating the latter gives

$$y'' = c_1' y_1' + c_2' y_2' + c_1 y_1'' + c_2 y_2''$$

Substituting this, together with (30) and (28), into the given equation (27) gives

$$c_1' y_1' + c_2' y_2' + c_1 y_1'' + c_2 y_2'' + a_1(c_1 y_1' + c_2 y_2') + a_0(c_1 y_1 + c_2 y_2) = f$$

Collecting the terms multiplied by c_1 and those multiplied by c_2 gives

$$c_1' y_1' + c_2' y_2' + c_1 Ly_1 + c_2 Ly_2 = f$$

or
$$c_1' y_1' + c_2' y_2' = f \tag{31}$$

since y_1 and y_2, being homogeneous solutions, are annihilated by L. The equations satisfied by c_1' and c_2' are (29) and (31):

$$\left. \begin{array}{l} c_1' y_1 + c_2' y_2 = 0 \\ c_1' y_1' + c_2' y_2' = f \end{array} \right\} \tag{32}$$

These equations may be solved for the unknown c_1' by multiplying the first equation by y_2' and the second by y_2, and subtracting. The result is

$$c_1'(y_1 y_2' - y_1' y_2) = -y_2 f$$

$$c_1' = \frac{-y_2 f}{y_1 y_2' - y_1' y_2}$$

$$c_1(t) = \int_{t_0}^{t} \frac{-f(z)y_2(z)}{y_1(z)y_2'(z) - y_1'(z)y_2(z)} dz$$

Similarly,

$$c_2' = \frac{y_1 f}{y_1 y_2' - y_1' y_2}$$

$$c_2(t) = \int_{t_0}^{t} \frac{f(z)y_1(z)}{y_1(z)y_2'(z) - y_1'(z)y_2(z)} dz$$

The integrations have been written as definite integrals to emphasize the fact that the integration is carried out before substitution into (28), which gives

$$y(t) = y_1(t) \int_{t_0}^{t} \frac{-f(z)y_2(z)}{y_1(z)y_2'(z) - y_1'(z)y_2(z)} dz$$

$$+ y_2(t) \int_{t_0}^{t} \frac{f(z)y_1(z)}{y_1(z)y_2'(z) - y_1'(z)y_2(z)} dz + C_1 y_1(t) + C_2 y_2(t)$$

where C_1 and C_2 are the arbitrary constants necessary in the general solution. In terms of indefinite integrals, the solution is written

$$y(t) = y_1(t) \int \frac{-f(t)y_2(t)}{y_1(t)y_2'(t) - y_1'(t)y_2(t)} dt + y_2(t) \int \frac{f(t)y_1(t)}{y_1(t)y_2'(t) - y_1'(t)y_2(t)} dt$$

The required arbitrary constants C_1 and C_2 are inherent in the indefinite integrals.

Memorization of the method may be difficult unless done as follows. First remember that the given equation is in normal form, with lead coefficient 1, and that y is to be a sum of function multiples of known homogeneous solutions. Always write down "$y_1 = \cdots$" and "$y_2 = \cdots$." Next, remember the simultaneous equations (32), particularly that they are equations in the *first derivatives* of the unknown coefficient functions, namely c_1' and c_2'.

Those are the essential points to remember, for with that much you can easily solve the simultaneous equations and integrate, or derive whatever part of the method you may need.

4. Method of Variation of Parameters: The Equation of Order n

The given equation is of the form

$$a_n(t)y^{(n)}(t) + \cdots + a_0(t)y(t) = f(t)$$

First divide through by the lead coefficient a_n, putting the equation into normal form:

$$Ly(t) = y^{(n)}(t) + \frac{a_{n-1}(t)}{a_n(t)} y^{(n-1)}(t) + \cdots + \frac{a_0(t)}{a_n(t)} y(t) = \frac{f(t)}{a_n(t)} \tag{34}$$

The method may be used if a linearly independent collection of n homogeneous solutions

$$y_1, \ldots, y_n$$

is known. We seek a particular solution of the form

$$y(t) = c_1(t)y_1(t) + \cdots + c_n(t)y_n(t)$$

or

$$y = c_1 y_1 + \cdots + c_n y_n \tag{35}$$

where $c_1, \ldots, c_n$ are the new unknown functions. Differentiate (35), obtaining

$$y' = c_1 y_1' + \cdots + c_n y_n' + c_1' y_1 + \cdots + c_n' y_n$$

Set the sum of the c' terms equal to zero

$$c_1' y_1 + \cdots + c_n' y_n = 0 \tag{36}$$

and conclude that

$$y' = c_1 y_1' + \cdots + c_n y_n' \tag{37}$$

Differentiate (37) in turn, obtaining

$$y'' = c_1 y_1'' + \cdots + c_n y_n'' + c_1' y_1' + \cdots + c_n' y_n'$$

Set the sum of the c' terms equal to zero

$$c_1' y_1' + \cdots + c_n' y_n' = 0 \tag{38}$$

and conclude that

$$y'' = c_1 y_1'' + \cdots + c_n y_n'' \tag{39}$$

Continue to differentiate. Each of the first $n - 1$ differentiations is followed by setting an expression in $c_1', \ldots, c_n'$ equal to zero. For instance, the $(n - 1)$st differentiation is followed by setting the equation

$$c_1' y_1^{(n-2)} + \cdots + c_n' y_n^{(n-2)} = 0 \tag{40}$$

and concluding that

$$y^{(n-1)} = c_1 y_1^{(n-1)} + \cdots + c_n y_n^{(n-1)} \tag{41}$$

The nth differentiation, the last, is

$$y^{(n)} = c_1' y_1^{(n-1)} + \cdots + c_n' y_n^{(n-1)} + c_1 y_1^{(n)} + \cdots + c_n y_n^{(n)} \tag{42}$$

The nth and last equation cannot be set arbitrarily. Instead, y and its derivatives are replaced in (34) with (35), (37), (39), ..., (41), and (42). Combining terms gives

$$c_1' y_1^{(n-1)} + \cdots + c_n' y_n^{(n-1)} + c_1 L y_1 + \cdots + c_n L y_n = \frac{f}{a_n}$$

or

$$c_1' y_1^{(n-1)} + \cdots + c_n' y_n^{(n-1)} = \frac{f}{a_n} \tag{43}$$

because $L y_1 = 0, \ldots, L y_n = 0$.

Thus, the collection of n equations determining $c_1', \ldots, c_n'$ consists of the $n-1$ equations set during differentiation, plus Equation (43), which was obtained by substitution:

$$\left. \begin{array}{l} c_1' y_1 + \cdots + c_n' y_n = 0 \\[4pt] c_1' y_1' + \cdots + c_n' y_n' = 0 \\[4pt] \cdots\cdots\cdots\cdots\cdots\cdots\cdots \\[4pt] c_1' y_1^{(n-2)} + \cdots + c_n' y_n^{(n-2)} = 0 \\[4pt] c_1' y_1^{(n-1)} + \cdots + c_n' y_n^{(n-1)} = \dfrac{f}{a_n} \end{array} \right\} \tag{44}$$

These equations should now be solved for $c_1', \ldots, c_n'$, from which $c_1, \ldots, c_n$ are obtained by integration. The desired solution is then achieved by substituting the known functions $y_1, \ldots, y_n$ and $c_1, \ldots, c_n$ into Equation (35).

For ease of memorizing, notice that Equations (44) are equations in which $c_1', \ldots, c_n'$ are unknown, that the coefficients are the known functions $y_1, \ldots, y_n$ and their derivatives, each equation containing one higher-order derivative than the preceding, and that only the last equation has a nonzero right side, having f/a_n instead of zero. The n equations are linear algebraic equations in the n unknowns $c_1', \ldots, c_n'$. The first $n-1$ equations are arbitrary in that they do not depend on properties $c_1, \ldots, c_n$ must have in order that the function y satisfy (34). They are, however, exactly those equations which lead to the simplest form of the final equation, which is not at all arbitrary but obtained by substitution into the given equation.

5. Example: An Equation of Order 3

The functions t^{-1}, t, and t^2 are linearly independent homogeneous solutions of the equation

$$t^3 y''' + t^2 y'' - 2t y' + 2y = t \ln t \tag{45}$$

whose order is 3. We seek a particular solution of the form

$$y(t) = c_1(t) t^{-1} + c_2(t) t + c_3(t) t^2 \tag{46}$$

using the method of variation of parameters. Of course, we are taking

$$y_1(t) = t^{-1} \qquad y_2(t) = t \qquad y_3(t) = t^2$$

Having specified the form of solution (46) and identified y_1, y_2, and y_3, our next step is to write the given equation in normal form

$$y''' + t^{-1}y'' - 2t^{-2}y' + 2t^{-3}y = t^{-2} \ln t$$

and identify the proper form of the nonhomogeneous term in (34)

$$\frac{f(t)}{a_3(t)} = t^{-2} \ln t$$

Next we write down the three simultaneous equations, either by differentiating as in the previous part or by recalling Equations (44) so derived:

$$\left. \begin{aligned} c_1' t^{-1} + c_2' t + c_3' t^2 &= 0 \\ -c_1' t^{-2} + c_2' + 2c_3' t &= 0 \\ 2c_1' t^{-3} \qquad + 2c_3' &= t^{-2} \ln t \end{aligned} \right\}$$

For convenience, we multiply these equations by t, t^2, and t^3, respectively, obtaining

$$\left. \begin{aligned} c_1' + c_2' t^2 + c_3' t^3 &= 0 \\ -c_1' + c_2' t^2 + 2c_3' t^3 &= 0 \\ 2c_1' \qquad + 2c_3' t^3 &= t \ln t \end{aligned} \right\} \tag{47}$$

and solve for c_1', c_2', and c_3'. Eliminating c_2' from the first two equations by subtraction gives

$$\left. \begin{aligned} 2c_1' - c_3' t^3 &= 0 \\ 2c_1' + 2c_3' t^3 &= t \ln t \end{aligned} \right\} \tag{48}$$

Eliminating c_1' by subtraction gives

$$3c_3' t^3 = t \ln t \qquad \text{or} \qquad c_3'(t) = \tfrac{1}{3}t^{-2} \ln t$$

from which it follows from (48) that

$$c_1'(t) = \tfrac{1}{6}t \ln t$$

and from (47) that

$$c_2'(t) = -\tfrac{1}{2}t^{-1} \ln t$$

Indefinite integration of these three functions gives us

$$c_1(t) = \frac{1}{6} \int t \ln t \, dt = \frac{1}{6}\left(\frac{t^2}{2} \ln t - \int \frac{t^2}{2} \frac{1}{t} dt \right)$$

$$= \frac{1}{6}\left(\frac{t^2}{2} \ln t - \frac{t^2}{4} \right) + A_1$$

$$c_2(t) = -\frac{1}{2} \int t^{-1} \ln t \, dt = -\frac{1}{4}(\ln t)^2 + A_2$$

and $$c_3(t) = \frac{1}{3} \int t^{-2} \ln t \, dt = \frac{1}{3} \left(-t^{-1} \ln t - \int -t^{-1}t^{-1} \, dt \right)$$

$$= \tfrac{1}{3}(-t^{-1} \ln t - t^{-1}) + A_3$$

Thus, from (46),

$$y(t) = c_1(t)y_1(t) + c_2(t)y_2(t) + c_3(t)y_3(t)$$

$$= \frac{1}{6}\left(\frac{t \ln t}{2} - \frac{t}{4} \right) - \frac{t}{4}(\ln t)^2 + \frac{1}{3}(-t \ln t - t) + A_1 t^{-1} + A_2 t + A_3 t^2$$

$$= -\frac{t}{4}(\ln t)^2 - \frac{t}{4}\ln t + A_1 t^{-1} + \left(A_2 - \frac{3}{8} \right) t + A_3 t^2$$

$$= -\frac{t}{4}[(\ln t)^2 + \ln t] + B_1 t^{-1} + B_2 t + B_3 t^2$$

where B_1, B_2, and B_3 are arbitrary. This is the general solution of the given equation (45).

6. Reduction of Order

The same general technique will allow us to reduce the order of the equation to be solved whenever a homogeneous solution is known. Substitution shows at once that $y_1(t) = t$ is a homogeneous solution of the linear equation

$$t^2 y'' - t(t + 2)y' + (t + 2)y = 5t^3 \tag{49}$$

Since the equation is of second order, we do not have the general homogeneous solution and cannot use variation of parameters. However, the equation does have solutions of the form

$$y(t) = c(t)y_1(t) = c(t)t \tag{50}$$

In fact, every solution of the equation can be written in this form, and we shall show that the result of substituting (50) into (49) is a new linear differential equation whose solution is c' and whose order is 1 less than that of the given equation. That is true in general; therefore, the method is called *reduction of order*. In the example, the new equation will be of order 1, and hence solvable by the methods already given.

Differentiating (50) begins the method:

$$y'(t) = c'(t)t + c(t)$$

$$y''(t) = c''(t)t + 2c'(t)$$

Substituting into (49) and simplifying the result gives the equation

$$t^3 c'' + 2t^2 c' - t(t + 2)(c't + c) + (t + 2)ct = 5t^3$$

Collecting multiples of c'', c', and c we obtain

$$t^3 c'' + c'(2t^2 - t^3 - 2t^2) + c[-t(t+2) + t(t+2)] = 5t^3$$

$$t^3 c'' - t^3 c' = 5t^3 \tag{51}$$

$$c'' - c' = 5$$

This is, to be sure, a second-order equation, but it is of first order in c', for c itself does not appear:

$$\frac{d}{dt}(c') - c' = 5$$

Its general solution is

$$c'(t) = -5 + Ae^t$$

Hence

$$c(t) = -5t + Ae^t + B$$

Substitution into (50) gives

$$y(t) = -5t^2 + Ate^t + Bt$$

which is the general solution of (49). Of course, A and B are the arbitrary constants appearing in the general homogeneous solution $Ate^t + Bt$, while $-5t^2$ is a particular solution.

Notice that the result of the substitution is an equation in which c (undifferentiated) appears only when it multiplies exactly the terms of the original equation. That is, if

$$Ly = f$$

is the given equation of order n, then the result of the substitution is an equation of the form

$$cLy_1 + (\text{terms in } c', c'', \ldots, c^{(n)}) = f$$

Since $Ly_1 = 0$, the equation simplifies and becomes of order $n - 1$ in the unknown function c'. In the example, the term cLy_1 is seen in (51). The method may be used on linear equations of any order.

7. Sample Problems with Answers

In each of the first four problems, find a particular solution using variation of parameters; also find the general solution. (In two of the problems, some homogeneous solutions are given in parenthesis.)

A. $y' - 2y = e^{2t}$ Answer: $y(t) = te^{2t} + Ae^{2t}$

B. $y'' + y = 1/\cos t$, on the interval $-\pi/2 < t < \pi/2$
      Answer: $y(t) = (\cos t) \ln \cos t + t \sin t + A_1 \cos t + A_2 \sin t$

C. $t^2 y'' - ty' = t^3 e^t$ $(1, t^2)$ Answer: $y(t) = te^t - e^t + A_1 + A_2 t^2$

D. $t^3 y''' + 3t^2 y'' = 1$ $(t^{-1}, 1, t)$ Answer: $y(t) = -\ln t + A_1 t^{-1} + A_2 + A_3 t$

In each of the next two problems, reduce the order of the equation, using the homogeneous solution given in parenthesis.

E. $y''' - 8y = 4e^{2t}$ (e^{2t}) Answer: $c''' + 6c'' + 12c' = 4$

F. $t^2 y'' - 3ty' + 4y = t^3 \ln t$ (t^2) Answer: $c'' + (1/t)c' = (1/t) \ln t$

8. Problems

In each case, find a particular solution using variation of parameters; also find the general solution.

1. $y'' - 5y' + 6y = 2e^t$ 2. $y'' - y' - 2y = 2e^{-t}$

3. $y''' - 6y'' + 9y' = 4te^{3t}$ 4. $y'' + 2y' + y = 3e^{-t}$

5. $t^2 y'' + 7ty' + 5y = t^{-2}$ 6. $y'' + y = \sin t$

Use the formula for the solution of first-order linear equations to solve:

7. $y' + 4ty = 3t$ 8. $y' + (\cos 3t)y = 6 \cos 3t$

Use the homogeneous solution given in parenthesis to reduce the order, and then solve each of the following equations.

9. $t^2 y'' - 3ty' + 4y = 1$ (t^2) 10. $t^2 y'' - ty' = 0$ (t^2)

11. $t^2 y'' - (3t^2 + 2t)y' + (3t + 2)y = 0$ (t)

12. $t^2 y'' - (3t^2 + 2t)y' + (3t + 2)y = -2t^3$ (t)

9. Remarks on Variation of Parameters

Suppose that $y_1, \ldots, y_n$ constitute a linearly independent collection of homogeneous solutions of a linear differential equation

$$a_n(t)y^{(n)}(t) + \cdots + a_0(t)y(t) = f(t) \qquad (52)$$

of order n on an interval $\alpha < t < \beta$. By the method of variation of parameters, we seek a solution of the form

$$y_p(t) = c_1(t)y_1(t) + \cdots + c_n(t)y_n(t)$$

where $y_1, \ldots, y_n$ are known and the coefficients satisfy the equations

$$\begin{rcases} c_1'(t)y_1(t) + \cdots + c_n'(t)y_n(t) = 0 \\ c_1'(t)y_1'(t) + \cdots + c_n'(t)y_n'(t) = 0 \\ \cdots\cdots\cdots\cdots\cdots\cdots\cdots\cdots\cdots\cdots\cdots \\ c_1'(t)y_1^{(n-2)}(t) + \cdots + c_n'(t)y_n^{(n-2)}(t) = 0 \\ c_1'(t)y_1^{(n-1)}(t) + \cdots + c_n'(t)y_n^{(n-1)}(t) = \dfrac{f(t)}{a_n(t)} \end{rcases} \qquad (53)$$

We shall show that these equations are necessarily solvable for $c_1'(t), \ldots, c_n'(t)$. Specifically, we shall show that if t_0 is any point of the interval, then (53) is solvable at $t = t_0$.

Let v_{n-1} be the nth member of the basic solution system with initial point t_0 (see Section E). Specifically, v_{n-1} is that homogeneous solution of (52) which satisfies the initial conditions

$$\begin{rcases} v_{n-1}(t_0) = 0 \\ v_{n-1}'(t_0) = 0 \\ \cdots\cdots\cdots\cdots\cdots \\ v_{n-1}^{(n-2)}(t_0) = 0 \\ v_{n-1}^{(n-1)}(t_0) = 1 \end{rcases} \qquad (54)$$

Since v_{n-1} is a homogeneous solution, it may be written as a linear combination

$$v_{n-1}(t) = \alpha_1 y_1(t) + \cdots + \alpha_n y_n(t) \tag{55}$$

Differentiating (55) and substituting into (54) gives the n equations

$$\left.\begin{aligned} v_{n-1}(t_0) &= \alpha_1 y_1(t_0) + \cdots + \alpha_n y_n(t_0) = 0 \\ &\cdots\cdots\cdots\cdots\cdots\cdots\cdots\cdots\cdots\cdots\cdots\cdots\cdots\cdots \\ v_{n-1}^{(n-1)}(t_0) &= \alpha_1 y_1^{(n-1)}(t_0) + \cdots + \alpha_n y_n^{(n-1)}(t_0) = 1 \end{aligned}\right\} \tag{56}$$

These equations are solvable, and they determine $\alpha_1, \ldots, \alpha_n$, since v_{n-1} is a linear combination of $y_1, \ldots, y_n$ (see Theorem 5 in Part 7 of Section E). If we now set

$$c_1'(t_0) = \frac{\alpha_1 f(t_0)}{a_n(t_0)} \qquad \cdots \qquad c_n'(t_0) = \frac{\alpha_n f(t_0)}{a_n(t_0)}$$

then comparison of (56) with (53) shows that $c_1'(t_0), \ldots, c_n'(t_0)$ satisfies (53) at t_0. Of course, the numbers $\alpha_1, \ldots, \alpha_n$ depend on the point t_0.

Since t_0 is any point of the interval, what we have really accomplished is the solution of (53) at every point of the interval, which was what we needed.

POWER-SERIES SOLUTION METHODS

A *power series may be differentiated term by term on the interior of its interval of convergence, and the sum of the differentiated series will be the derivative of the sum of the original series.* On this fact is based a technique for finding a solution of a differential equation by representing it as the sum of an infinite series determined directly from substitution into the equation.

The equation of our first example below could be solved by other means. After discussing where and why the technique is used, we apply it to equations not readily solved by other means. A natural extension called the *method of Frobenius* is illustrated. The appendix ends with remarks on some well-known and useful classes of functions often defined by their series expansions or the differential equations which they solve.

1. Examples: Techniques of Power-Series Substitution

Solve the initial value problem

$$y' + 7y = 0 \tag{1}$$
$$y(0) = 2 \tag{2}$$

and find the general solution of (1).

To begin, write the unknown function y as the sum of a power series about the initial point $t_0 = 0$:

$$y(t) = \sum_{k=0}^{\infty} b_k t^k \tag{3}$$

The coefficients of the series, $b_0, b_1, \ldots, b_k, \ldots$, will be the new unknowns in the problem. Of these, the first is determined from the given initial condition, since

$$b_0 = y(t_0) = y(0) = 2$$

The other coefficients will be calculated from it.

The idea is to substitute this series into the left-hand side of (1), obtaining for $y' + 7y$ a power series whose coefficients are known because they are all zero. Since each is a linear combination of the unknowns $b_0, b_1, \ldots$, we may solve for the latter. From them, the solution y is determined by formula (3).

The important step, then, is the substitution of (3) into (1). This requires the power series for y', which is obtained by differentiating (3) term by term to obtain

$$y'(t) = \sum_{k=0}^{\infty} kb_k t^{k-1}$$

The power series for $y' + 7y$ is now calculated.

$$y'(t) + 7y(t) = \sum_{k=0}^{\infty} kb_k t^{k-1} + 7 \sum_{k=0}^{\infty} b_k t^k$$

The two series will be added by adding like powers of t. Since t^3 is the $k = 3$ term in the second series but the $k = 4$ term in the first, we shall change the dummy variable in the series for y', using a double change of variable:

$$y'(t) = \sum_{k=0}^{\infty} kb_k t^{k-1} \tag{4}$$

(Substitute $k - 1 = m$ or $k = m + 1$.)

$$y'(t) = \sum_{m+1=0}^{\infty} (m + 1)b_{m+1} t^m$$

(Substitute $k = m$.)

$$y'(t) = \sum_{k+1=0}^{\infty} (k + 1)b_{k+1} t^k$$

(Many people prefer to change the dummy variable in a single step, replacing k with $k + 1$ wherever it appears, but this can be confusing.) Finally, since the $k + 1 = 0$ term contains the factor $k + 1$ whose value is zero, the series may begin with $k = 0$. That is,

$$y'(t) = \sum_{k+1=0}^{\infty} (k + 1)b_{k+1} t^k$$

$$= \sum_{k=0}^{\infty} (k + 1)b_{k+1} t^k \tag{5}$$

Now the series for y' and y may be substituted into the left-hand side of the differential equation. The series for $y' + 7y$ is therefore

$$y'(t) + 7y(t) = \sum_{k=0}^{\infty} (k+1)b_{k+1}t^k + 7\sum_{k=0}^{\infty} b_k t^k \tag{6}$$

$$= \sum_{k=0}^{\infty} [(k+1)b_{k+1} + 7b_k]t^k \tag{7}$$

$$= \sum_{k=0}^{\infty} 0t^k \tag{8}$$

where the right-hand side of the last equality is the series expansion for the right-hand side of the differential equation. Since (7) and (8) are power series for a single function 0 about a single point $t_0 = 0$, the coefficients of (7) must be equal to those for (8), and so

$$(k+1)b_{k+1} + 7b_k = 0 \qquad k = 0, 1, 2, \ldots \tag{9}$$

Equation (9), called the *recurrence relation*, is equivalent to the sequence of equations

$$
\left.
\begin{aligned}
b_1 + 7b_0 &= 0 && \text{or} && b_1 = -7b_0 \\
2b_2 + 7b_1 &= 0 && \text{or} && b_2 = -7b_1 \cdot \tfrac{1}{2} \\
3b_3 + 7b_2 &= 0 && \text{or} && b_3 = -7b_2 \cdot \tfrac{1}{3} \\
4b_4 + 7b_3 &= 0 && \text{or} && b_4 = -7b_3 \cdot \tfrac{1}{4} \\
&\cdots\cdots\cdots\cdots\cdots\cdots\cdots\cdots\cdots
\end{aligned}
\right\} \tag{10}
$$

each of which determines one unknown coefficient from its predecessors. Since $b_0 = y(0) = 2$ is already known, Equations (10) show us how to calculate, one after another, each of the remaining coefficients $b_1 = -14$, $b_2 = 49$, ...

In many problems it is possible to find a general formula for b_k as a function of b_0. In this example we have

$$b_1 = -7b_0$$

and so forth. Thus,

$$b_2 = \frac{-7b_1}{2} = \frac{(-7)(-7b_0)}{2} = \frac{49b_0}{2}$$

$$b_1 = (-7)(2)$$

$$b_2 = (-7)(-7)(2)(\tfrac{1}{2}) = (-7)^2(2)(\tfrac{1}{2})$$

$$b_3 = (-7)(-7)^2(2)(\tfrac{1}{2})(\tfrac{1}{3}) = (-7)^3 \frac{2}{3!}$$

$$b_4 = (-7)(-7)^3(2)\left(\frac{1}{3!}\right)(\tfrac{1}{4}) = (-7)^4 \frac{2}{4!}$$

$$\cdots\cdots\cdots\cdots\cdots\cdots\cdots\cdots\cdots$$

$$b_k = (-7)^k \frac{2}{k!} \qquad k \geq 1 \tag{11}$$

From these coefficients, Taylor's series for the solution y is now constructed:

$$y(t) = \sum_{k=0}^{\infty} (-7)^k(2)\frac{1}{k!} t^k = 2 \sum_{k=0}^{\infty} \frac{(-7t)^k}{k!} \tag{12}$$

Ordinarily, a solution will have to be left in this form, but in this case we may recognize the series for the exponential and conclude that $y(t) = 2e^{-7t}$.

Had we left the initial value unspecified, we would have obtained

$$b_0 = y(0)$$
$$b_k = (-7)^k \frac{y(0)}{k!} \qquad k \geq 1 \tag{13}$$

instead of (11), and

$$y(t) = y(0) \sum_{k=0}^{\infty} \frac{(-7t)^k}{k!} = y(0)e^{-7t} \tag{14}$$

instead of (12). Here $b_0 = y(0)$ is the arbitrary constant to be expected in the general solution.

It is also possible to solve nonhomogeneous equations in the same way; for instance,

$$y' + 7y = 5t \tag{15}$$

The substitution proceeds just as before down to (8), where we have instead

$$5t = 0 + 5t + \sum_{k=2}^{\infty} 0t^k$$

as the series for the nonhomogeneous term. Thus,

$$\sum_{k=0}^{\infty} [(k+1)b_{k+1} + 7b_k]t^k = 0 + 5t + \sum_{k=2}^{\infty} 0t^k \tag{16}$$

Therefore the coefficients satisfy the recurrence relation

$$(k+1)b_{k+1} + 7b_k = 0 \qquad k \neq 1$$
$$2b_2 + 7b_1 = 5 \tag{17}$$

analogous to (9), equivalent to the sequence of equations

$$b_1 + 7b_0 = 0$$
$$2b_2 + 7b_1 = 5$$
$$3b_3 + 7b_2 = 0$$
$$4b_4 + 7b_3 = 0$$
$$\cdots\cdots\cdots$$

As before, b_0 is arbitrary (unless determined by some initial condition), and the other coefficients may be found in turn:

$$b_0 = y(0) \quad \text{is arbitrary}$$

$$b_1 = -7b_0$$

$$b_2 = (-7b_1 + 5)(\tfrac{1}{2}) = [(-7)^2 b_0 + 5](\tfrac{1}{2})$$

$$b_3 = (-7)[(-7)^2 b_0 + 5](\tfrac{1}{2})(\tfrac{1}{3})$$

$$b_4 = (-7)^2[(-7)^2 b_0 + 5](\tfrac{1}{2})(\tfrac{1}{3})(\tfrac{1}{4})$$

$$\cdots\cdots\cdots\cdots\cdots\cdots\cdots\cdots$$

$$b_k = (-7)^{k-2} \frac{(-7)^2 b_0 + 5}{k!}$$

$$= (-7)^k \frac{b_0 + 5(-7)^{-2}}{k!} \qquad k \ge 2$$

Thus,
$$y(t) = b_0 + (-7)b_0 t + [b_0 + 5(-7)^{-2}] \sum_{k=2}^{\infty} \frac{(-7t)^k}{k!}$$

$$= b_0 - 7b_0 t + (b_0 + \tfrac{5}{49}) \sum_{k=0}^{\infty} \frac{(-7t)^k}{k!} - (b_0 + \tfrac{5}{49})(1 - 7t)$$

$$= \left[y(0) + \frac{5}{49} \right] \sum_{k=0}^{\infty} \frac{(-7t)^k}{k!} + \frac{5t}{7} - \frac{5}{49}$$

$$= \left[y(0) + \frac{5}{49} \right] e^{-7t} + \frac{5t}{7} - \frac{5}{49}$$

is the general solution of the nonhomogeneous equation. Ordinarily, the solution would have to be left as a series.

2. Sample Problems with Answers

Use the substitution of power series about the given point t_0. Find the recurrence relation and then the series representation for the required solution.

A. $y' + 2y = 0$, about $t_0 = 0$. Find the general solution.
    Answer: $(k + 1)b_{k+1} + 2b_k = 0$ for $k = 0, 1, 2, \ldots$; $y(t) = y(0) \sum_{k=0}^{\infty} (-2t)^k/k!$

B. $y' + 2y = 0$, about $t_0 = 4$. Satisfy the initial condition $y(4) = 3$.
    Answer: $(k + 1)b_{k+1} + 2b_k = 0$ for $k = 0, 1, 2, \ldots$; $y(t) = 3 \sum_{k=0}^{\infty} [-2(t - 4)]^k/k!$

C. $y' + 2y = t^2$, about $t_0 = 4$. Find the general solution.
    Answer: First find the power-series expansion for t^2 about $t_0 = 4$, which is $t^2 = (t - 4)^2 + 8(t - 4) + 16$. Then find the recurrence relation: $(k + 1)b_{k+1} + 2b_k = 0$ for $k = 3, 4, \ldots$. The other equations defining coefficients are: $b_1 + 2b_0 = 16$; $2b_2 + 2b_1 = 8$; $3b_3 + 2b_2 = 1$.

$$y(t) = 3 + 10(t - 4) - 6(t - 4)^2 - \frac{13}{4} \sum_{k=3}^{\infty} \frac{[-2(t - 4)]^k}{k!}$$

$$= 3 + 10(t - 4) - 6(t - 4)^2 - \frac{13}{4} \sum_{k=0}^{\infty} \frac{[-2(t - 4)]^k}{k!} + \frac{13}{4}[1 - 2(t - 4) + 2(t - 4)^2]$$

$$= \tfrac{25}{4} + \tfrac{7}{2}(t - 4) + \tfrac{1}{2}(t - 4)^2 - \tfrac{13}{4}e^{-2(t-4)}$$

D. $y' + ty = 0$, about $t_0 = 0$. Find the general solution.

Answer: $b_1 = 0$; $(k + 1)b_{k+1} + b_{k-1} = 0$ for $k = 1, 2, \ldots$

$$y(t) = y(0)\left(1 - \frac{t^2}{2} + \frac{t^4}{4 \cdot 2} - \frac{t^6}{6 \cdot 4 \cdot 2} \cdots\right)$$

$$= y(0) \sum_{m=0}^{\infty} \frac{(-t^2/2)^m}{m!} = y(0)e^{-t^2/2}$$

3. Problems

Use the substitution of power series about the given point t_0. Find the recurrence relation and then the series representation for the required solution.

1. $y' + 5y = 0$
   (a) About $t_0 = 0$. Find the general solution.
   (b) About $t_0 = 0$. Satisfy the initial condition $y(0) = 7$.
   (c) About $t_0 = 2$. Find the general solution.
   (d) About $t_0 = 2$. Satisfy the initial condition $y(2) = \quad 3$.

2. $y' + 5y = -1$
   (a) About $t_0 = 0$. Satisfy the initial condition $y(0) = -2$.
   (b) About $t_0 = 2$. Satisfy the initial condition $y(2) = 0$.

3. $y' + 5y = t$
   (a) About $t_0 = 0$. Satisfy the initial condition $y(0) = 3$.
   (b) About $t_0 = 1$. Satisfy the initial condition $y(1) = 2$.

4. $y' + t^2 y = 0$
   (a) About $t_0 = 0$. Find the general solution.
   (b) About $t_0 = 1$. Satisfy the initial condition $y(1) = 5$.

4. The Use of the Technique: Examples

The method of variation of parameters can provide a particular solution of a linear differential equation whenever the complete homogeneous solution is known. It follows that the important gap in our collection of techniques has been our inability to find homogeneous solutions unless the equation (1) is first order, or (2) has constant coefficients, or (3) is equidimensional. For this reason we shall treat only homogeneous equations. A nonhomogeneous equation can be treated the same way, but in practice one would undertake it only in problems whose solution was sufficiently important to justify the complexity of the calculation.

If initial conditions are given at a point t_0, or if there is any other reason to obtain an expansion about some point t_0 other than the origin, then the known and unknown functions should all be expanded about that same point. In every other respect, the method proceeds as before. For instance, in (15) the expansion of $5t$ about t_0 would be

$$5t = 5t_0 + 5(t - t_0) + \sum_{k=2}^{\infty} 0(t - t_0)^k$$

since

$$f(t) = \sum_{k=0}^{\infty} \frac{f^{(k)}(t_0)}{k!} (t - t_0)^k \tag{18}$$

is the formula for the power series for f about t_0.

Thus, the solution of the initial value problem

$$\left.\begin{array}{r} y' + 7y = 5t \\ y(t_0) = A \end{array}\right\}$$

would have $b_0 = A$ and the recurrence relation

$$\left.\begin{array}{r} b_1 + 7b_0 = 5t_0 \\ 2b_2 + 7b_1 = 5 \\ (k + 1)b_{k+1} + 7b_k = 0 \qquad k \geq 2 \end{array}\right\}$$

instead of (17), but the steps of the method are exactly the same as before. For this reason, hereafter all our examples will use power-series expansions only about the origin $t_0 = 0$.

In the examples given, we have solved first-order equations, obtaining in each case a solution containing just one arbitrary constant. A solution in terms of power series about t_0 may be obtained whenever the equation may be written in the form

$$y' + a_0(t)y = f(t)$$

where the lead coefficient is 1 and both a_0 and f have power-series expansions about t_0. Given an equation of any order, *if it can be written in the form*

$$y^{(n)} + a_{n-1}(t)y^{(n-1)} + \cdots + a_1(t)y' + a_0(t)y = f(t)$$

where the lead coefficient is 1 and the coefficients $a_{n-1}, \ldots, a_1, a_0$ and nonhomogeneous term f all have power-series expansions about t_0 convergent on an interval $t_0 - \alpha < t < t_0 + \alpha$, then all solutions of the equation on that interval will have power-series expansions about t_0 convergent upon that interval, and the series substitution technique will find them all. Such a point t_0 is called a *regular point* for the equation.

For example, the equation

$$y'' + ty = 0 \tag{19}$$

has only regular points. We let $Ly = y'' + ty$, and we seek the general solution in the form

$$y(t) = \sum_{k=0}^{\infty} b_k t^k$$

Differentiating twice and substituting into the equation gives

$$0 = Ly = \sum_{k=0}^{\infty} k(k-1)b_k t^{k-2} + t \sum_{k=0}^{\infty} b_k t^k$$

$$= \sum_{k=2}^{\infty} k(k-1)b_k t^{k-2} + \sum_{k=0}^{\infty} b_k t^{k+1}$$

$$= \sum_{k=0}^{\infty} (k+2)(k+1)b_{k+2} t^k + \sum_{k=1}^{\infty} b_{k-1} t^k$$

$$= 2b_2 t^0 + \sum_{k=1}^{\infty} [(k+2)(k+1)b_{k+2} + b_{k-1}]t^k$$

The right-hand side is the power-series expansion for Ly, which is zero. We deduce that

$$(k + 2)(k + 1)b_{k+2} + b_{k-1} = 0 \qquad k \geq 1$$

Reindexed, with k replaced by $k - 2$ throughout, this is

$$k(k - 1)b_k + b_{k-3} = 0 \qquad k \geq 3$$

or

$$b_k = \frac{-b_{k-3}}{k(k - 1)} \qquad k \geq 3 \Bigg) \tag{20}$$

We also have

$$b_2 = 0 \Bigg\}$$

Each coefficient is determined by the one that is three coefficients before it. Thus $b_3, b_6, \ldots$ are multiples of b_0; and $b_4, b_7, \ldots$ are multiples of b_1; and $b_5, b_8, \ldots$ are multiples of b_2, which is zero. There is no restriction on b_0 or b_1, which are therefore the two arbitrary constants expected, and the general solution is

$$y(t) = b_0 \left(1 - \frac{1}{3 \cdot 2} t^3 + \frac{1}{6 \cdot 5 \cdot 3 \cdot 2} t^6 + \cdots \right)$$

$$+ b_1 \left(t - \frac{1}{4 \cdot 3} t^4 + \frac{1}{7 \cdot 6 \cdot 4 \cdot 3} t^7 + \cdots \right)$$

5. The Method of Frobenius

Bessel's equation of order $\frac{1}{2}$

$$t^2 y'' + t y' + (t^2 - \tfrac{1}{4}) y = 0 \tag{21}$$

does not have a regular point at $t_0 = 0$. This is because in normal form

$$y'' + \frac{1}{t} y' + \left(1 - \frac{1}{4t^2} \right) y = 0$$

its coefficients $a_1(t) = 1/t$ and $a_2(t) = 1 - 1/4t^2$ do not have power-series expansions about $t_0 = 0$. Nevertheless, the equation does have solutions readily found by the *method of Frobenius*, which generalizes the power-series substitution and which we illustrate using Equation (21). The method consists in substituting

$$y(t) = t^p \sum_{k=0}^{\infty} b_k t^k = \sum_{k=0}^{\infty} b_k t^{k+p} \tag{22}$$

and determining the exponent p and the coefficients b_k so as to satisfy the given differential equation. Taking $b_0 \neq 0$ fixes the first term at $b_0 t^p$ and determines the possible values of p, called the *exponents* of the equation at $t = 0$.

The differentiated series are

$$y'(t) = \sum_{k=0}^{\infty} (k + p)b_k t^{k+p-1}$$

and

$$y''(t) = \sum_{k=0}^{\infty} (k + p)(k + p - 1)b_k t^{k+p-2}$$

(Unlike the derivatives of a power series, their first terms need not vanish.) Substitution into the differential equation (21) gives

$$0 = Ly = t^2 y'' + ty' - \frac{y}{4} + t^2 y$$

$$= t^2 \sum_{k=0}^{\infty} (k + p)(k + p - 1)b_k t^{k+p-2} + t \sum_{k=0}^{\infty} (k + p)b_k t^{k+p-1}$$

$$- \tfrac{1}{4} \sum_{k=0}^{\infty} b_k t^{k+p} + t^2 \sum_{k=0}^{\infty} b_k t^{k+p}$$

$$= \sum_{k=0}^{\infty} [(k + p)^2 - \tfrac{1}{4}]b_k t^{k+p} + \sum_{k=0}^{\infty} b_k t^{k+p+2}$$

$$= \sum_{k=0}^{\infty} [(k + p)^2 - (\tfrac{1}{2})^2]b_k t^{k+p} + \sum_{k=2}^{\infty} b_{k-2} t^{k+p}$$

$$= [p^2 - (\tfrac{1}{2})^2]b_0 t^p + [(1 + p)^2 - (\tfrac{1}{2})^2]b_1 t^{1+p}$$

$$+ \sum_{k=2}^{\infty} \{[(k + p)^2 - (\tfrac{1}{2})^2]b_k + b_{k-2}\}t^{k+p}$$

$$= t^p \left([p^2 - (\tfrac{1}{2})^2]b_0 + [(1 + p)^2 - (\tfrac{1}{2})^2]b_1 t \right.$$

$$\left. + \sum_{k=2}^{\infty} \{[(k + p)^2 - (\tfrac{1}{2})^2]b_k + b_{k-2}\}t^k \right) \qquad (23)$$

Since t^p is zero only at $t = 0$, the other factor in (23), the series in parenthesis, must be identically zero, and every term of the series must vanish. Since $b_0 \neq 0$, the vanishing of the first term requires that the factor multiplying b_0 in the first term must vanish instead. The resulting equation is called the *indicial equation*, whose roots are the exponents. In this example the indicial equation is

$$p^2 - (\tfrac{1}{2})^2 = 0 \qquad (24)$$

and the exponents are

$$p = \pm\tfrac{1}{2}$$

The solution corresponding to the larger value is determined now. For $p = \frac{1}{2}$:

$$k = 1: \qquad [(\tfrac{3}{2})^2 - (\tfrac{1}{2})^2]b_1 \qquad\qquad = 0 \qquad \text{or} \qquad b_1 = 0$$

$$k \geq 2: \qquad [(k + \tfrac{1}{2})^2 - (\tfrac{1}{2})^2]b_k + b_{k-2} = 0 \qquad\qquad (25)$$

$$(k^2 + k)b_k = -b_{k-2}$$

$$b_k = \frac{-b_{k-2}}{k(k + 1)} \qquad\qquad (26)$$

Since $b_1 = 0$, it follows from the recurrence formula (25) or its equivalent (26) that b_3, b_5, and all odd terms are zero. Also, if k is even

$$b_k = \frac{(-1)^{k/2}b_0}{(k + 1)!}$$

$$= \frac{(-1)^m b_0}{(2m + 1)!} \qquad \text{where } k = 2m, \; m \geq 0$$

Thus, (22) gives

$$y_1(t) = t^{1/2} \sum_{k=0}^{\infty} b_k t^k$$

$$= b_0 \sum_{m=0}^{\infty} \frac{(-1)^m t^{2m + 1/2}}{(2m + 1)!}$$

Ordinarily, a series solution method would end at this point, but in this particular case the similarity to the sine series may be used to write

$$y_1(t) = b_0 t^{-1/2} \sum_{m=0}^{\infty} \frac{(-1)^m t^{2m + 1}}{(2m + 1)!}$$

$$= b_0 t^{-1/2} \sin t$$

The coefficients for the solution y_2 corresponding to $p = -\frac{1}{2}$ must now be found, using (23). These we shall call $c_0, c_1, \ldots$, to distinguish them from the coefficients of y_1. For $p = -\frac{1}{2}$:

$$k = 1: \qquad [(\tfrac{1}{2})^2 - (\tfrac{1}{2})^2]c_1 \qquad\qquad = 0 \qquad \text{or} \qquad 0c_1 = 0 \qquad (27)$$

$$k \geq 2: \qquad [(k - \tfrac{1}{2})^2 - (\tfrac{1}{2})^2]c_k + c_{k-2} = 0$$

$$c_k = \frac{-c_{k-2}}{k(k - 1)} \qquad\qquad (28)$$

Equation (28) shows that all the even terms are multiples of c_0. In fact, if $k = 2m$, then

$$c_k = c_{2m} = \frac{(-1)^m c_0}{(2m)!}$$

and

$$t^{-1/2} \sum_{m=0}^{\infty} \frac{(-1)^m c_0 t^{2m}}{(2m)!} = c_0 t^{-1/2} \cos t$$

is at least a portion of the desired series.

Equation (28) also shows that all the odd terms are multiples of c_1, and (27) indicates that c_1 is arbitrary. Can it happen that there are *three* arbitrary constants in the general solution of a second-order equation? It cannot. The same calculations as before will show that the sum of the odd terms is a constant multiple of y_1, the solution for $p = \frac{1}{2}$, and so the general solution of the equation is

$$y(t) = y_2(t) + y_1(t) = c_0 t^{-1/2} \sum_{m=0}^{\infty} \frac{(-1)^m t^{2m}}{(2m)!} + b_0 t^{-1/2} \sum_{m=0}^{\infty} \frac{(-1)^m t^{2m+1}}{(2m+1)!}$$

$$= t^{-1/2}(c_0 \cos t + b_0 \sin t) \tag{29}$$

6. Remarks on the Method of Frobenius and Its Use

When seeking a series solution near a point t_0, you do not have to use the method of Frobenius if t_0 is a regular point of the differential equation. Other points are called *singular points* of the equation, and they are often of interest in applications.

From now on we shall treat only second-order equations. A singular point t_0 of a second-order equation

$$a_2(t)y'' + a_1(t)y' + a_0(t)y = 0$$

is classified as a *regular singular point* if

$$(t - t_0)\frac{a_1(t)}{a_2(t)} \qquad \text{and} \qquad (t - t_0)^2 \frac{a_0(t)}{a_2(t)}$$

both have power-series expansions about t_0. Otherwise, the singular point t_0 is an *irregular singular* point. The treatment of irregular singular points will not be considered.

The method of Frobenius will always produce at least one solution about a regular singular point. The indicial equation will have two roots, which may be real or complex, equal or distinct. If their difference is not an integer, then the method may be used to produce the two linearly independent solutions, one for each exponent. If the difference is an integer, then the method will produce a solution corresponding to the root with the larger real part, call it p_1. We may call that solution $y_1(t)$. The first term of its series will be $b_0(t - t_0)^{p_1}$.

Now let $u_1 = y_1/b_0$. If the difference between the roots is an integer, then an attempt to use the method to find a solution y_2 corresponding to the root p_2 with smaller real part will fail to determine $c_{p_1 - p_2}$ in terms of c_0, for one of two reasons. On the one hand, the recurrence relation may assert that the coefficient $c_{p_1 - p_2}$ is a second arbitrary constant, just as occurred in the example. In this case the value of that coefficient should be taken to be zero, the effect of doing so being to change the sum of the series by a multiple of the sum of the other series, $c_{p_1 - p_2} u_1(t)$. In this case both solutions are obtained.

On the other hand, the recurrence relation may give an equation which contradicts the arbitrary choice of c_0 independent of b_0. This signals the fact that the second (linearly independent) solution cannot be obtained using the method of Frobenius. In this case there is a second solution

$$u_2(t) = u_1(t) \ln (t - t_0) + u_3(t) = u_1(t) \ln (t - t_0) + (t - t_0)^{p_2} \sum_{k=0}^{\infty} c_k(t - t_0)^k \quad (30)$$

where u_1 is the sum of the series corresponding to the exponent p_1. The coefficients c_k are determined by substituting u_2 into the given differential equation. It is worth noting that unless $p_2 = p_1$, c_0 is not arbitrary but determined by the coefficients of u_1. The general solution is

$$y(t) = Au_1(t) + Bu_2(t)$$

A simple example of the latter case may be obtained by using the method of Frobenius on the equidimensional equation

$$Ly = t^2 y'' - ty' + y = 0 \qquad (31)$$

The substitution of

$$y(t) = t^p \sum_{k=0}^{\infty} b_k t^k$$

leads to the indicial equation

$$(p - 1)^2 = 0 \qquad \text{from which} \quad p = 1, 1$$

Continuing leads to the first solution, $u_1(t) = t$. Since $p_1 = p_2$, it is clear from the outset that the method of Frobenius will not produce a second solution independent of the first. The second solution may therefore be found in the form (30):

$$u_2(t) = u_1(t) \ln t + u_3(t) = u_1(t) \ln t + t^{p_2} \sum_{k=0}^{\infty} c_k t^k$$

$$= t \ln t + \sum_{k=0}^{\infty} c_k t^{k+1}$$

Substituting this into (31) gives

$$0 = L[u_2](t) = \sum_{k=1}^{\infty} k^2 c_k t^{k+1}$$

from which it follows that c_0 is arbitrary but $c_k = 0$ for $k \geq 1$. Since $p_2 = p_1$, the second series is in fact the same series as the first. The general solution is

$$y(t) = At + Bt \ln t$$

7. Sample Problems with Answers

E. Use power-series substitution to solve the initial value problem $y'' - 4y = 0$, $y(0) = 3$, $y'(0) = 0$. Give the recurrence relation as well.

Answer: $(k + 2)(k + 1)b_{k+2} - 4b_k = 0$ for $k \geq 0$;

$$y(t) = 3 \sum_{k \text{ even}} \frac{(2t)^k}{k!} + 0 \sum_{k \text{ odd}} \frac{(2t)^k}{k!} = 3 \sum_{m=0}^{\infty} \frac{(2t)^{2m}}{(2m!)} = 3 \cosh 2t$$

F. For the equation

$$t^2 y'' + \frac{t}{t-1} y' + \left[\frac{1}{t} + \frac{1}{(t-1)^2} \right] y = 0$$

classify the points $t_0 = 0, 1, -1$ as to whether they are regular points, regular singular points, or irregular singular points.

Answer: In normal form the equation is

$$y'' + \frac{1}{t(t-1)} y' + \left[\frac{1}{t^3} + \frac{1}{t^2(t-1)^2} \right] y = 0$$

At each t_0, $(t - t_0)/t(t - 1)$ has a power-series expansion. At each t_0 except $t_0 = 0$, $(t - t_0)^2 \times [1/t^3 + 1/t^2(t - 1)^2]$ has a power-series expansion. Thus, -1 is a regular point, $+1$ is a regular singular point, and 0 is an irregular singular point.

Use the method of Frobenius to find all those solutions of the following equations which have the form $t^p \sum_{k=0}^{\infty} b_k t^k$ with $b_0 \neq 0$. In each case give the indicial equation, and for each exponent of the equation either find the solution, showing the recurrence relation, or tell how the solution method fails to give a solution.

G. $t^2 y'' + ty' - y = 0$
 Answer: The indicial equation $p^2 - 1 = 0$ implies that $p = \pm 1$. For the exponent $p = 1$, the recurrence relation is $[(k + 1)^2 - 1]b_k = 0$ for $k \geq 1$, and so $y_1(t) = t(1 + 0t + 0t^2 + \cdots) = t$ is a solution. For the exponent $p = -1$, the recurrence relation is $[(k - 1)^2 - 1]c_k = 0$ for $k \geq 1$, and so $y_2(t) = t^{-1}(1 + 0t + 0t^2 + \cdots) = t^{-1}$ is a solution.

H. $t^2 y'' + ty' + t^2 y = 0$ (*Bessel's equation* of order 0)
 Answer: The indicial equation $p^2 = 0$ implies that $p = 0$ with multiplicity 2. Thus, the method of Frobenius can give only one solution. The recurrence relation is $k^2 b_k + b_{k-2} = 0$ for $k \geq 2$. Since b_1 is zero, it follows that

$$y_1 = \sum_{k \text{ even}} (-1)^{k/2} \frac{t^k}{[k(k-2) \cdots (1)]^2} = \sum_{m=0}^{\infty} (-1)^m \frac{t^{2m}}{2^{2m}(m!)^2} = \sum_{m=0}^{\infty} (-1)^m \frac{(t/2)^{2m}}{(m!)^2}$$

8. Problems

In the next four problems, use power-series substitution to solve the initial value problem. In each, give the recurrence relation as well.

5. $y'' + 4y = 0$, $y(0) = 0$, $y'(0) = 3$
6. $y'' + 3y' + 2y = 0$, $y(0) = 1$, $y'(0) = -1$ (give the first six terms of the series)
7. $y'' + 4y = 0$, $y(0) = 1$, $y'(0) = 2$ (give the first four terms of the series)
8. $y'' + (t + 1)y = 0$, $y(0) = 1$, $y'(0) = 0$ (give the first four terms of the series)

For the next three problems, classify the points $t = 0, 1, 2$ as to whether they are regular points, singular points, or irregular singular points for the equation given.

9. $(t - 1)y'' + ty' + t^2 y = 0$
10. $ty'' + (t - 2)y'/t + y/(t - 2) = 0$
11. $t^2 y'' + ty' + \sqrt{t - 1} \, y = 0$

For the following four problems, use the method of Frobenius to find all solutions of the form $t^p \sum_{k=0}^{\infty} b_k t^k$. Give the indicial equation, the exponents, and for each exponent the recurrence relation and either a solution or an explanation of how the solution method fails to give a solution.

12. $t^2 y'' + t y' + (t^2 - 1)y = 0$ (*Bessel's equation* of order 1)

13. $t^2 y'' + t y' - \frac{1}{9} y = 0$

14. $t^2 y'' + t y' - \frac{1}{4} y = 0$

15. $t y'' + y = 0$

16. Do the same, but use the series $(t - 1)^p \sum_{k=0}^{\infty} b_k (t - 1)^k$, for *Legendre's equation*:

    (a) $(1 - t^2)y'' - 2t y' + 6y = 0$

    (b) $(1 - t^2)y'' - 2t y' + 12y = 0$

17. For the equations of the indicated problems, find the second solution of the form $u_1(t) \ln t + u_3(t)$, finding three terms of the series for u_3.

    (a) Problem 12

    (b) Problem 15

    (c) Sample Problem H

    (d) Problem 16a [find $u_1(t) \ln (t - 1) + u_3(t)$]

18. Solve, finding three terms of each infinite series about $t = 1$: $(1 - t^2)y'' - 2t y' = 0$. (*Legendre's equation.*)

9. Bessel Functions and Legendre Polynomials

There are many ways of defining functions. Some functions are defined by algebraic formulas such as $3t^2 - 5t^{1/3} + 2$. Some, such as $\cos t$ and $\sin t$, are defined directly from geometric considerations. The exponential and the Arctangent are among many defined as inverses of other functions. Some functions are defined as integrals. One such function is the natural logarithm,

$$\ln t = \int_1^t z^{-1} \, dz \qquad t > 0$$

which also may be considered as the solution of the initial value problem

$$\left. \begin{array}{l} y'(t) = t^{-1} \\ y(1) = 0 \end{array} \right\}$$

    The functions to be discussed in this part also arise as solutions of certain differential equations, Bessel's equation and Legendre's equation. The formulas giving series representations for these functions can be derived using the series substitutions already described. Therefore, we shall present the series without deriving them. Other useful properties will also be mentioned without derivation. For comparison, we first list properties of the sinusoids in the same way that we shall list those of the Bessel functions and Legendre polynomials.

a. Sinusoids The functions $\cos t$ and $\sin t$ are linearly independent solutions of the equation

$$y'' + y = 0 \tag{32}$$

They are the sums of the power series

$$\cos t = \sum_{k=0}^{\infty} (-1)^k \frac{t^{2k}}{(2k)!} \qquad \text{and} \qquad \sin t = \sum_{k=0}^{\infty} (-1)^k \frac{t^{2k+1}}{(2k + 1)!}$$

on the whole real line. Near $t = 0$, the cosine and sine may be approximated as

$$\cos t \cong 1 - \frac{t^2}{2} \qquad \sin t \cong t$$

Both are periodic with period 2π:

$$\cos (t + 2\pi) = \cos t \qquad \text{and} \qquad \sin (t + 2\pi) = \sin t \qquad \text{for all } t$$

They satisfy other symmetry properties as well:

$$\cos (-t) = \cos t \qquad\qquad \sin (-t) = -\sin t$$

$$\cos \left(\frac{\pi}{2} + t \right) = -\cos \left(\frac{\pi}{2} - t \right) \qquad \sin \left(\frac{\pi}{2} + t \right) = \sin \left(\frac{\pi}{2} - t \right)$$

$$\cos (t + \pi) = -\cos t \qquad\qquad \sin (t + \pi) = -\sin t$$

and the cofunction (translation) property

$$\sin t = \cos \left(t - \frac{\pi}{2} \right)$$

and the Pythagorean theorem

$$\cos^2 t + \sin^2 t = 1$$

from which it follows that $|\cos t| \leq 1$ and $|\sin t| \leq 1$. They are the most commonly used periodic or oscillatory functions.

Among the other properties that relate them to each other are the derivative formulas

$$\frac{d}{dt} \cos t = -\sin t \qquad \text{and} \qquad \frac{d}{dt} \sin t = \cos t$$

and the orthogonality relations

$$\int_{-\pi}^{\pi} \sin mt \cos nt \, dt = 0$$

and

$$\int_{-\pi}^{\pi} \cos mt \cos nt \, dt = \int_{-\pi}^{\pi} \sin mt \sin nt \, dt = \begin{cases} 0 & m \neq n \\ \dfrac{\pi}{2} & m = n \neq 0 \end{cases}$$

where m and n are integers. The values are tabulated in tables of trigonometric functions, and their graphs are discussed in Section I of Chapter I and Section J of Chapter III.

b. Bessel functions Constant multiples of the *Bessel function* J_n are solutions of *Bessel's equation of order n*

$$t^2 y'' + ty + (t^2 - n^2)y = 0 \tag{33}$$

which has a regular singular point at $t = 0$. The method of Frobenius leads to the series representation

$$J_n(t) = \left(\frac{t}{2}\right)^n \sum_{k=0}^{\infty} (-1)^k \frac{(t/2)^{2k}}{k! \, (n+k)!} \tag{34}$$

on the whole real line if n is a nonnegative integer. Near $t = 0$, J_n may be approximated as

$$J_n(t) \cong \frac{(t/2)^n}{n!} \tag{35}$$

while for large positive values of t,

$$J_n(t) \cong \left(\frac{2}{\pi t}\right)^{1/2} \cos\left[t - \frac{\pi}{4}(1 - 2n)\right] \tag{36}$$

Thus, the behavior is oscillatory although not periodic, and the amplitude of oscillation decays as $x^{-1/2}$ does (not exponentially). The function J_n is an important description of wave phenomena. (See Problem 12 and Sample Problem H.) Graphs of J_0 and J_1 are given in Figure 1.

The functions J_n and their derivatives are related by the recurrence relations

$$\frac{J_{n-1}(t) + J_{n+1}(t)}{2} = \frac{n}{t} J_n(t) \tag{37}$$

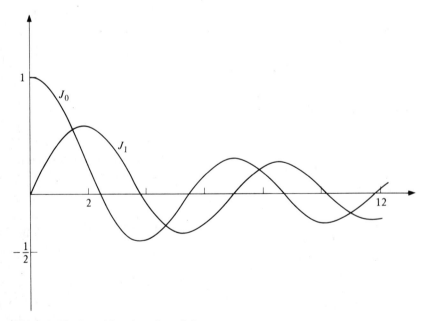

Figure 1 The Bessel functions J_0 and J_1.

and
$$\frac{J_{n-1}(t) - J_{n+1}(t)}{2} = J_n'(t) \tag{38}$$

If n is not an integer, then the factor $(n + k)!$ in the series (34) must be replaced with

$$\Gamma(n + k + 1) = \int_0^\infty z^{n+k} e^{-z} \, dz$$

a value of the gamma function, which generalizes the factorial and which we shall not discuss further. The sum of the series gives a solution of the equation on the right half line $t > 0$, and the subsequent properties (35) to (38) are valid also. If n is not an integer, then the general solution of Equation (33) is

$$y(t) = A_1 J_n(t) + A_2 J_{-n}(t)$$

If n is a positive integer, then the second solution has logarithmic behavior near $t = 0$ and may be found by the substitution described earlier. The more commonly used forms are given in standard reference works, along with tables of values.
    If $n = \frac{1}{2}$, then the functions have formulas in terms of the sinusoids,

$$J_{1/2}(t) = \left(\frac{2}{\pi t}\right)^{1/2} \sin t \quad \text{and} \quad J_{-1/2}(t) = \left(\frac{2}{\pi t}\right)^{1/2} \cos t$$

showing both the oscillatory character and the amplitude decay. [See Equations (21) and (29) of Part 5.]
    The Bessel functions also satisfy an orthogonality relation. If $n \geq 0$, and α_1 and α_2 are distinct roots of the equation $J_n(\alpha) = 0$, then

$$\int_0^1 t J_n(\alpha_1 t) J_n(\alpha_2 t) \, dt = 0$$

c. Legendre polynomials *Legendre's equation,*

$$(1 - t^2)y'' - 2ty' + n(n + 1)y = 0 \tag{39}$$

has regular singular points at $t = \pm 1$. If n is a nonnegative integer, then there is a solution which is a polynomial called the nth *Legendre polynomial* P_n. Power-series substitution gives the formula

$$P_n(t) = 2^{-n} \sum_{k=0}^{[n/2]} (-1)^k \frac{(2n - 2k)!}{(n - 2k)! \, (n - k)!} t^{n-2k} \tag{40}$$

where
$$\left[\frac{n}{2}\right] = \begin{cases} \dfrac{n}{2} & \text{if } n \text{ is even} \\[2mm] \dfrac{n-1}{2} & \text{if } n \text{ is odd} \end{cases}$$

P_n is of degree n, and all the powers of t appearing in it are even if n is even but odd if n is odd. Thus, P_n is an even or odd function accordingly as n is even or odd. (See Problems 16, 17d, and 18 in Part 8.)

Near $t = 0$, $P_n(t)$ may be approximated by

$$P_n(t) \cong \begin{cases} \dfrac{(-1)^{n/2}}{2^n} \dfrac{n!}{(n/2)!} t^0 & \text{if } n \text{ is even} \\[3mm] \dfrac{(-1)^{(n-1)/2}}{2^n} \dfrac{(n+1)!}{[(n+1)/2]!} t^1 & \text{if } n \text{ is odd} \end{cases}$$

while for large values of t (positive or negative)

$$P_n(t) \cong \frac{(2n)!}{(n!)^2} \left(\frac{t}{2}\right)^n$$

On the interval $-1 \leq t \leq 1$ they satisfy

$$|P_n(t)| \leq 1$$

with equality at $t = \pm 1$. They may also be given by *Rodrigue's formula*,

$$P_n(t) = \frac{(-1)^n}{2^n n!} \frac{d^n}{dt^n} (1 - t^2)^n$$

and they satisfy the recurrence relation

$$(n + 1)P_{n+1}(t) = (2n + 1)tP_n(t) - nP_{n-1}(t)$$

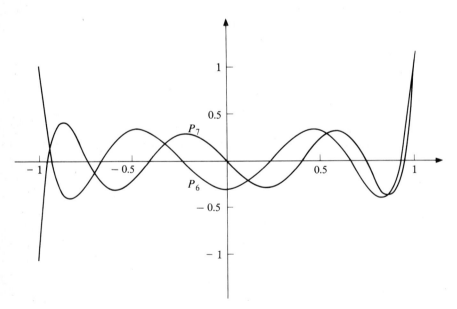

Figure 2 Typical Legendre polynomials: P_6 and P_7.

Another linear independent solution to the differential equation will show logarithmic behavior near $t = \pm 1$, and it may be given by

$$Q_n(t) = \frac{1}{2} P_n(t) \ln \left| \frac{1+t}{1-t} \right| - \sum_{k=1}^{n} \frac{1}{k} P_{k-1}(t) P_{n-k}(t) \qquad t \neq \pm 1$$

The Legendre polynomials satisfy the orthogonality relations

$$\int_{-1}^{1} P_n(t) P_m(t) \, dt = \begin{cases} 0 & \text{if } n \neq m \\ \dfrac{2}{2n+1} & \text{if } n = m \end{cases}$$

Graphs of P_6 and P_7 are given in Figure 2.

10. Sample Problem with Answer

I. For the Legendre polynomials P_0, P_1, P_2, and P_3, verify that (40) and Rodrigue's formula give the same formula.

Answer: $P_0(t) = 1$; $P_1(t) = t$; $P_2(t) = \frac{1}{2}(3t^2 - 1)$; $P_3(t) = \frac{1}{2}(5t^3 - 3t)$

ANSWERS TO SELECTED PROBLEMS

CHAPTER 1

Section I.B

Part 1

1. $y(t) = 2 - \cos t$

5. $y(t) = \frac{2}{3}e^{3t} - \frac{2}{3}e^3$

8. $y(t) = -7 + \ln|t|$

3. $y(t) = -2 - \cos t$

7. $y(t) = \ln|t|$

Part 6

11. $y(t) = \frac{1}{5}\cos 5t + 2t + C$

13. $y(t) = 2t^2 + 6t + C$

Part 9

14. $y(t) = \begin{cases} 1 & t \le 1 \\ t & t > 1 \end{cases}$

16. $y(t) = \begin{cases} t & t \le 0 \\ \frac{1}{3}e^{3t} - \frac{1}{3} & t > 0 \end{cases}$

Section I.D

Part 9

1. (a) $y(t) = 5e^{-3t}$ (c) $y(t) = 2e^{12}e^{-3t} = 2e^{-3(t-4)}$

2. (b) $y(t) = 7e^{-5(t-4)}$

3. $y(t) = 3e^{-(\sin 2t)/2}$

5. (a) $y(t) = 10\,\dfrac{e^{-(e^{3t})/3}}{e^{-(e^{-6})/3}} = 10e^{(e^{-6}-e^{3t})/3}$

7. $y(t) = -5e^{-3t}$

9. $y(t) = 0$ all t

11. $y(t) = \dfrac{15}{t+2}$ $t > -2$

13. (a) $y' + \dfrac{1}{2}\left(\dfrac{1}{t+10}\right) y(t) = 0$ $t \ge 0$

(e) $t_2 = 10e^2 - 10 = 10(e^2 - 1)$ seconds

$p(t_2) = \dfrac{e^{-2}}{200}$ cubic meters of pollutant per cubic meter of water

14. (a) $y(t) = Ce^{-\pi t}$ (b) $y(t) = Ce^{-(\sin t)/2}$

Part 14

15. $y(t) = 5e^{3(t-2)}$ all t

17. $y(t) = 4e^{-1/2(e^{2t} - e^{-6})}$ all t

19. $y(t) = -2e^{-\cos t + \cos 1}$ all t

22. $y(t) = 2e^{(3/t) + 3}$ $t < 0$

Section I.E

Part 8

1. $y(t) = 2 + e^{-5t}$

3. $y(t) = 2 - e^{-5t}$

7. $y(t) = 2 + Ce^{5t}$

9. (b) $y(t) = \begin{cases} -\frac{4}{5} + \frac{4}{5}e^{5t} & t \le 0 \\ 0 & t > 0 \end{cases}$ (d) $y(t) = \begin{cases} \frac{4}{5} - \frac{4}{5}e^{-5(t+1)} & t \le 0 \\ \frac{4}{5} - \frac{4}{5}e^{-5} & t > 0 \end{cases}$

Part 10

10. $N(t) = N(t_0)e^{\alpha(t - t_0)} = (1{,}000{,}000)e^{\alpha(t - t_0)}$ t in days. Since $2{,}000{,}000 = 1{,}000{,}000^\alpha$, it follows that $\alpha = \ln 2$. Monday noon: 4,000,000; previous Friday: 500,000.

11. $s(t) = s(t_0)e^{-\lambda(t - t_0)}$ and so $-s'(t) = \lambda s(t_0)e^{-\lambda(t - t_0)}$
and $-s'(t) = [-s'(t_0)]e^{-\lambda(t - t_0)}$

Measure $-s'(t_0)$ and $-s'(t_1)$ at chosen times t_0 and t_1, and solve the latter equation for λ:

$$\lambda = \frac{1}{t_1 - t_0} \ln \left(\frac{-s'(t_0)}{-s'(t_1)}\right)$$

15. Let $y(t)$ be the concentration at time t:

$$y(t) = \frac{1}{2t + 1}$$

17. Loss is $1 - e^{-1/6}$ oz.

Section I.F

Part 4

1. $y(t) = t - 1 + e^{-t}$

3. $y(t) = \dfrac{1}{2} e^t \left(t + \dfrac{1}{t}\right)$

4. (a) For v constant we have $v' = 0$ and : $up' + gp = M_0 g + Bv$.

(b) p satisfies the initial condition $p(0) = 0$.

$$p'(t) = \frac{M_0 g + Bv}{u} e^{-g(t - t_0)/u}$$

6. $v(t) = e^{B_0(1-\alpha t)^3/3\alpha M} \left[v(0) + \left(g - \dfrac{f}{M} \right) \displaystyle\int_0^t e^{-B_0(1-\alpha z)^3/3\alpha M} \, dz \right]$

Section I.J

Part 4

1. Linear; separable; $y(t) = 1 + Ct^{-1}$.   **2.** Nonlinear; separable; $y(t) = -\dfrac{1}{t+C}$, also $y \equiv 0$.

5. Nonlinear; separable; $y(t) = \pm\sqrt{2 \sin t + C}$.   **7.** Nonlinear; nonseparable.

Section I.K

Part 4

2.

n	t_n	y_n	y_n'
0	1	0.7500	−0.6614
1	1.1	0.6839	−0.7296
2	1.2	0.6109	−0.7917
3	1.3	0.5317	−0.8469
4	1.4	0.4470	−0.8945
5	1.5	0.3576	−0.9339
6	1.6	0.2642	−0.9645
7	1.7	0.1677	−0.9858
8	1.8	0.0692	−0.9976
9	1.9	−0.0306	−0.9995
10	2.0	−0.1306	−0.9914

Part 9

6. Nyström's method; $y'(1) = -\sqrt{7/4}$, $y''(1) = -\frac{3}{4}$; $y_1 = 0.6801$.

n	t_n	y_n	y_n'
0	1	0.7500	−0.6614
1	1.1	0.6801	−0.7332
2	1.2	0.6034	−0.7975
3	1.3	0.5206	−0.8538
4	1.4	0.4326	−0.9015
5	1.5	0.3403	−0.9403
6	1.6	0.2445	−0.9696
7	1.7	0.1464	−0.9892
8	1.8	0.0467	−0.9989
9	1.9	−0.0534	−0.9986
10	2.0	−0.1530	−0.9882

8. Milne's method; $y'(-2) = -\frac{1}{2}$, $y''(-2) = -\frac{1}{2}$; $y_1\{1\} = -3.0256$.

n	t_n	$y_n^{(0)}$	$y_n^{(0)'}$	$\dfrac{y_n^{(0)'} + y_{n-1}^{(1)'}}{2}$	$y_n^{(1)}$	$y_n^{(1)'}$	C_0
0	-2				-3	-0.5	
1	-1.95				-3.0256	-0.5516	
2	-1.90	-3.0552	-0.6080	-0.5798	-3.0580	-0.6095	-0.0028
3	-1.85	-3.0865	-0.6684	-0.6389	-3.0895	-0.6700	-0.0030
4	-1.80	-3.1250	-0.7361	-0.7030	-3.1283	-0.7379	-0.0033
5	-1.75	-3.1633	-0.8076	-0.7728	-3.1668	-0.8096	-0.0035
6	-1.70	-3.2092	-0.8878	-0.8487	-3.2132	-0.8901	-0.0040
7	-1.65	-3.2558	-0.9732	-0.9316	-3.2599	-0.9757	-0.0041
8	-1.60	-3.3107	-1.0692	-1.0225	-3.3154	-1.0721	-0.0047
9	-1.55	-3.3671	-1.1724	-1.1222	-3.3722	-1.1756	-0.0051
10	-1.50	-3.4330	-1.2886	-1.2321	-3.4386	-1.2924	-0.0056
11	-1.45	-3.5014	-1.4148	-1.3536	-3.5075	-1.4190	-0.0061
12	-1.40	-3.5805	-1.5575	-1.4882	-3.5874	-1.5625	-0.0069
13	-1.35	-3.6638	-1.7139	-1.6382	-3.6713	-1.7195	-0.0075
14	-1.30	-3.7593	-1.8918	-1.8057	-3.7680	-1.8984	-0.0087
15	-1.25	-3.8612	-2.0889	-1.9937	-3.8707	-2.0966	-0.0095
16	-1.20	-3.9777	-2.3147	-2.2056	-3.9886	-2.3238	-0.0109
17	-1.15	-4.1031	-2.5679	-2.4459	-4.1153	-2.5785	-0.0122
18	-1.10	-4.2464	-2.8604	-2.7194	-4.2605	-2.8732	-0.0141
19	-1.05	-4.4026	-3.1930	-3.0331	-4.4186	-3.2082	-0.0160
20	-1.00	-4.5813	-3.5813	-3.3948	-4.6000	-3.6000	-0.0187

CHAPTER II

Section A

Part 7

1. $v(t) = -5R$ volts; $p(t) = 25R$ watts   **2.** $i(t) = 2Ct$ amperes; $p(t) = 2Ct^3$ watts

5. $i(t) = \begin{cases} 0 \text{ amperes} & t \le t_0 \\ -\dfrac{2}{L}(t - t_0) \text{ amperes} & t > t_0 \end{cases}$

7. Draw a circuit diagram for the circuit. You may copy Figure 14, but you *must define* v and i on it.

$$i(10) = 1.5 \times 10^{-3} \text{ amperes}; \quad p(10) = 2.25 \times 10^{-2} \text{ watts}; \quad w(10) = 2.25 \times 10^{-1} \text{ joules}$$

10. On your circuit diagram you *must define* the currents through the two resistors.

$$i(t) = \left(\frac{1}{R_1} + \frac{1}{R_2}\right) v(t)$$

12. On your circuit diagram you *must define* the voltages across the inductors.

$$v(t) = (L_1 + L_2)i'(t)$$

13. The equation relating v_L and v is $v'_L + (R/L)V_L = v'$.

(c) $v_L(t) = e^{-(R/L)(t-t_0)} \left[4 + \dfrac{e^{(R/L)(t-t_0)}}{10^2 + (R/L)^2} \left(10 \cos 10t + \dfrac{R}{L} \cos 10t \right) \right.$

$$\left. - \dfrac{1}{10^2 + (R/L)^2} \left(10 \sin 10t_0 + \dfrac{R}{L} \cos 10t_0 \right) \right] \text{ volts}$$

(d) $v_L(t) = \left(4 + \dfrac{3L}{R} \right) e^{-(R/L)(t-t_0)} - \dfrac{3L}{R} \text{ volts}$

$$p(t) = \dfrac{1}{R} \left[\left(4 + \dfrac{3L}{R} \right) e^{-(R/L)(t-t_0)} - \dfrac{3L}{R} - 2 + 3t \right] \left[\left(4 + \dfrac{3L}{R} \right) e^{-(R/L)(t-t_0)} - \dfrac{3L}{R} \right] \text{ watts}$$

15. The equation relating i_L and i is $i'_L + (R/L)i_L = (R/L)i$. Determine v from i_L.

(b) $i_L(t) = 4 - 5e^{-(R/L)(t-t_0)}$ amperes; $v(t) = 5Re^{-(R/L)(t-t_0)}$ volts

(c) $i_L(t) = \dfrac{Re^{-6t}}{R - 6L} + \left[i_L(t_0) - \dfrac{Re^{-6t_0}}{R - 6L} \right] e^{-(R/L)(t-t_0)}$ amperes

$$v(t) = \dfrac{-6RLe^{-6t}}{R - 6L} - R \left[i_L(t_0) - \dfrac{Re^{-6t_0}}{R - 6L} \right] e^{-(R/L)(t-t_0)} \text{ volts}$$

17. (a) $v_C(t) = \frac{1}{3} \sin 10t + 10 - \frac{1}{3} \sin 10$ volts

(b) $v_C(t) = -\frac{1}{3}te^{-3t} - \frac{1}{9}e^{-3t} + 10 + \frac{4}{9}e^{-3}$ volts

(c) $v_C(t) = e^{-t} \left(\dfrac{-\sin 5t - 5 \cos 5t}{26C} \right) + e^{-1} \left(\dfrac{\sin 5t + 5 \cos 5}{26C} \right) + \dfrac{10}{C}$ volts

19. (b) $i_R(t) = t - RC + \left[\dfrac{4}{R} - (t_0 - RC) \right] e^{-(t-t_0)/RC}$ amperes

$$i_C(t) = -RC + \left[\dfrac{4}{R} - (t_0 - RC) \right] e^{-(t-t_0)/RC} \text{ amperes}$$

Section B

Part 7

1. (a) $x(t) = 10 + 2t + 10t^2$ meters; $p(t) = 40 + 400t$ watts

(c) $x(t) = 10t^2 - \dfrac{20}{(30\pi)^2} \cos 30\pi t + 2t + 10 + \dfrac{20}{(30\pi)^2}$ meters

$$p(t) = 20(1 + \cos 30\pi t) \left[10t^2 + 2t + 10 + \dfrac{20}{(30\pi)^2} (1 - \cos 30\pi t) \right] \text{ watts}$$

2. (c) $x(t) = \frac{5}{2}te^{-t}$ meters; $w(t) = \frac{25}{4} t^2 e^{-2t}$ joules

(d) $x(t) = \frac{1}{2}t^2$ meters; $w(t) = \frac{1}{4}t^4$ joules

3. (b) $f(t) = 200 \cos 1000t$ newtons; $p(t) = 20{,}000 \cos^2 1000t$ watts

(d) $f(t) = 10$ newtons; $p(t) = 25$ watts

5. On the free-body diagram you *must define* both the force and the distance.

$$w(t) = 20 \text{ joules}; f(t) = 1 \text{ newton}$$

6. (a) $f_1 - f_2 = M_1 x''_1$; (c) $x'_1 = x'_2$; $x''_1 = x''_2$; (d) $f_1 = (M_1 + M_2)x''_1$

Section C

Part 4

2. (b) $f(t) = -15t + \dfrac{2}{\pi}\cos 15\pi t - \dfrac{2}{\pi} + 2$ newtons;   (c) $f(t) = -15t + 9te^{-2t} + \frac{9}{2}e^{-2t} + 1$ newtons

3. (c) $f(t) = 200\sin 10t$ newtons

5. On the free-body diagram you *must define* the tensions through the elements (through each dashpot and the spring).

$$(B_1 + B_2)f' + Kf = KB_1v + B_1B_2v'$$

$$f(t) = \begin{cases} 10B_1 + (3 - 10B_1)e^{-K(t-t_0)/(B_1+B_2)} \text{ newtons} & t_0 \le t < 10 \\ 20B_1 + [(3 - 10B_1)e^{-K(10-t_0)/(B_1+B_2)} - 10B_1]e^{-K(t-t_0)/(B_1+B_2)} \text{ newtons} & 10 \le t \end{cases}$$

Section D

Part 3

1. $MBv'' + MKv' + KBv = Bf' + Kf^-$ **3.** $MBg' + (BK + MK)g = BKf$

5. $MBv_1'' + (MK + BK)v_1 = Bf'$ **7.** $Bg' + Kg = BKv$

9. $Bv_1' + Kv_1 = Bv'$ **11.** $g' = Kv_1$

Part 7

13. (a) $v_1(t_0) = v(t_0) + Ri_R(t_0);\ v_1'(t_0) = \dfrac{i_C(t_0)}{C}$

   (c) $i_R(t_0)$ is given; $i_R'(t_0) = \dfrac{1}{R}\left[\dfrac{1}{C}i_C(t_0) - v'(t_0)\right]$

   (e) $i_C(t_0)$ is given; $i_C'(t_0) = \dfrac{1}{RC}i_C(t_0) - \dfrac{1}{R}v'(t_0) - \dfrac{1}{L}v(t_0) - \dfrac{R}{L}i_R(t_0)$

CHAPTER III

Section A

Part 8

1. $\mathcal{L}[C](s) = \dfrac{C}{s}$; $\mathcal{L}[0](s) = s\left(\dfrac{C}{s}\right) - C = 0$

3. $\mathcal{L}[e^{7t}](s) = \dfrac{1}{s-7}$; $\mathcal{L}\left[\dfrac{d}{dt}e^{-7t}\right](s) = \dfrac{s}{s-7} - 1 = \dfrac{7}{s-7}$; $\mathcal{L}\left[\dfrac{d^2}{dt^2}e^{7t}\right](s) = \dfrac{49}{s-7}$

5. $\mathcal{L}[te^{-4t}](s) = \dfrac{1}{(s+4)^2}$; $\mathcal{L}\left[\dfrac{d}{dt}te^{-4t}\right](s) = \dfrac{s}{(s+4)^2}$; $\mathcal{L}\left[\dfrac{d^2}{dt^2}te^{-4t}\right](s) = \dfrac{-8s - 16}{(s+4)^2}$

7. $\mathcal{L}[\sin 4t](s) = \dfrac{4}{s^2 + 16}$; $\mathcal{L}\left[\dfrac{d}{dt}\sin 4t\right](s) = \dfrac{4s}{s^2 + 16}$; $\mathcal{L}\left[\dfrac{d^2}{dt^2}\sin 4t\right](s) = \dfrac{-64}{s^2 + 16}$

10. $\mathcal{L}[t^2 - 1](s) = \dfrac{2}{s^3} - \dfrac{1}{s}$; $\mathcal{L}\left[\dfrac{d}{dt}[t^2 - 1]\right](s) = \dfrac{2}{s^2}$; $\mathcal{L}\left[\dfrac{d^2}{dt^2}[t^2 - 1]\right](s) = \dfrac{2}{s}$

11. $\mathcal{L}[y](s) = \dfrac{2}{s - 3}$ 　　　　　　　　　　　 **13.** $\mathcal{L}[y](s) = \dfrac{4}{(s + 3)(3s + 1)}$

14. $\mathcal{L}[y](s) = \dfrac{12s + 27}{3s^2 + 6s + 2}$ 　　　　　 **15.** $\mathcal{L}[y](s) = \dfrac{-4}{s - 3}$; $y(t) = -4e^{3t}$

17. $\mathcal{L}[y](s) = \dfrac{6}{(s + 3)^2}$; $y(t) = 6te^{-3t}$ 　　 **19.** $\mathcal{L}[y](s) = \dfrac{1}{s^2 + 8s + 16}$; $y(t) = te^{-4t}$

21. $\mathcal{L}[y](s) = \dfrac{3}{s + 2}$; $y(t) = 3e^{-2t}$ 　　　 **25.** $f(t) = -3e^{-8t}$

27. $f(t) = 1 + e^{-t}$ 　　　　　　　　　　　　 **29.** $f(t) = -4te^{-2t}$

Section B

Part 6

1. $\mathcal{L}^{-1}\left[\dfrac{-3}{s}\right](t) = 3$ 　　 $t \geq 0$ 　　　　　 **4.** $\mathcal{L}^{-1}\left[\dfrac{3s}{2s^2 + 6}\right](t) = \dfrac{3}{2}\cos\sqrt{3}t$ 　　 $t \geq 0$

6. $\mathcal{L}^{-1}\left[\dfrac{4}{s^2 + 3}\right](t) = \dfrac{4}{\sqrt{3}}\sin\sqrt{3}t$ 　 $t \geq 0$ 　 **8.** $\mathcal{L}^{-1}\left[\dfrac{s + 3}{s^2 + 9}\right](t) = \cos 3t + \sin 3t$ 　 $t \geq 0$

11. $\mathcal{L}^{-1}\left[\dfrac{14}{3s^2 + 2}\right](t) = \dfrac{14}{\sqrt{6}}\sin\sqrt{\dfrac{2}{3}}t$ 　 $t \geq 0$

13. (c) $\mathcal{L}[y](s) = \dfrac{s + 1}{s^2 + 3}$; $y(t) = \cos\sqrt{3}t + \dfrac{1}{\sqrt{3}}\sin\sqrt{3}t$ 　 $t \geq 0$

14. (a) $\mathcal{L}[y](s) = \dfrac{3s}{s^2 + 4}$; $y(t) = 3\cos 2t$ 　 $t \geq 0$

15. $\mathcal{L}[y](s) = \dfrac{s + 3}{s^2 + 8}$; $y(t) = \cos 2\sqrt{2}t + \dfrac{3}{2\sqrt{2}}\sin 2\sqrt{2}t$ 　 $t \geq 0$

17. $\mathcal{L}[y](s) = \dfrac{2}{3s + 5}$; $y(t) = \dfrac{2}{3}e^{-(5/3)t}$ 　 $t \geq 0$

Part 14

23. $\mathcal{L}^{-1}\left[\dfrac{s - 2}{(s - 2)^2 + 16}\right](t) = e^{2t}\cos 4t$ 　 $t \geq 0$

25. $\mathcal{L}^{-1}\left[\dfrac{s - 3}{s^2 - 6s + 10}\right](t) = e^{3t}\cos t$ 　 $t \geq 0$

27. $\mathcal{L}^{-1}\left[\dfrac{1}{(s + 2)^2 + 9}\right](t) = \dfrac{1}{3}e^{-2t}\sin 3t$ 　 $t \geq 0$

29. $\mathcal{L}^{-1}\left[\dfrac{1}{s^2 + 6s + 11}\right](t) = \dfrac{1}{\sqrt{2}}e^{-3t}\sin\sqrt{2}t$ 　 $t \geq 0$

31. $\mathcal{L}^{-1}\left[\dfrac{2s + 3}{s^2 + 4s + 6}\right](t) = 2e^{-2t}\cos\sqrt{2}t - \dfrac{1}{\sqrt{2}}e^{-2t}\sin\sqrt{2}t$ 　 $t \geq 0$

33. $\mathcal{L}^{-1}\left[\dfrac{-7}{(s - 2)^3}\right](t) = -\dfrac{7}{2}t^2 e^{2t}$ 　 $t \geq 0$ 　　 **35.** $\mathcal{L}^{-1}\left[\dfrac{s}{(s - 2)^3}\right](t) = te^{2t} + e^{2t}$ 　 $t \geq 0$

37. $\mathcal{L}\left[\int_0^t \sin 2z \, dz\right](s) = \dfrac{2}{s(s^2 + 4)}$ **39.** $\mathcal{L}\left[\int_0^t (z^3 - \cos \frac{1}{2}z) \, dz\right](s) = \dfrac{6}{s^5} - \dfrac{4s}{4s^3 + s}$

41. $\mathcal{L}[y](s) = \dfrac{2}{s^2 - 2s + 5}$; $y(t) = e^t \sin 2t$ $t \geq 0$

43. $\mathcal{L}[y](s) = \dfrac{3}{s^2 + 2s + 5}$; $y(t) = \dfrac{3e^{-t}}{2} \sin 2t$ $t \geq 0$

45. $\mathcal{L}[y](s) = \dfrac{-2s - 1}{2s^2 + s + 3}$; $y(t) = -e^{-t/4} \cos \dfrac{\sqrt{47}}{4} t - \dfrac{1}{\sqrt{47}} e^{-t/4} \sin \dfrac{\sqrt{47}}{4} t$ $t \geq 0$

47. $\mathcal{L}[y](s) = \dfrac{2s + 1}{s(s^2 + 2s + 5)}$ **49.** $\mathcal{L}[y](s) = \dfrac{1}{(s^2 + 1)(s^2 + 2s + 5)}$

51. $\mathcal{L}[y](s) = \dfrac{-s^3 - 7s^2 - 16s - 12}{(s + 2)^2(3s^2 + 2s + 3)}$

Part 18

55. $y(t) = y(0)e^t \cos 2t + \dfrac{y(0) + y'(0)}{2} e^t \sin 2t$ $t \geq 0$

57. $y(t) = y(0)e^{t/3} \cos \dfrac{2\sqrt{2}}{3} t + \dfrac{3y'(0) - y(0)}{2\sqrt{2}} e^{t/3} \sin \dfrac{2\sqrt{2}}{3} t$ $t \geq 0$

Section C

Part 7

1. $\mathcal{L}^{-1}\left[\dfrac{2s - 3}{s^2 + 2s + 5}\right](t) = 2e^{-t} \cos \sqrt{5}t - \sqrt{5}e^{-t} \sin \sqrt{5}t$ $t \geq 0$

3. $\mathcal{L}^{-1}\left[\dfrac{-6}{s^2 + 2s - 8}\right](t) = e^{-4t} - e^{2t}$ $t \geq 0$ **5.** $\mathcal{L}^{-1}\left[\dfrac{s}{3s^2 + 18s + 27}\right](t) = \frac{1}{3}e^{-3t} - te^{-3t}$ $t \geq 0$

7. $\mathcal{L}^{-1}\left[\dfrac{s}{(s + 1)^2(s + 2)}\right](t) = -te^{-t} + 2e^{-t} - 2e^{-2t}$ $t \geq 0$

9. $\mathcal{L}^{-1}\left[\dfrac{1}{(s - 1)^2(s + 1)^2}\right](t) = \frac{1}{4}(te^t - e^t + te^{-t} + e^{-t})$ $t \geq 0$

12. $\mathcal{L}^{-1}\left[\dfrac{4s^2}{(s + 1)^2(s - 3)}\right](t) = -te^{-t} + \frac{7}{4}e^{-t} + \frac{9}{4}e^{3t}$ $t \geq 0$

14. $\mathcal{L}^{-1}\left[\dfrac{2s + 1}{s^2(2s^2 + 1)}\right](t) = t + 2 - 2 \cos \dfrac{t}{\sqrt{2}} - \sqrt{2} \sin \dfrac{t}{\sqrt{2}}$ $t \geq 0$

16. $\mathcal{L}^{-1}\left[\dfrac{-2s}{(s^2 + 4)(s^2 + 6s + 13)}\right](t) = \frac{1}{25}[-3 \cos 2t - 4 \sin 2t + 3e^{-3t} \cos 2t + 13e^{-3t} \sin 2t]$ $t \geq 0$

19. $y(t) = e^{-t} - e^{-2t}$ $t \geq 0$ **21.** $y(t) = te^{-t} - e^{-t} + e^{-2t}$ $t \geq 0$

23. $y(t) = \frac{6}{5} - \frac{1}{5}e^{-t/2} \cos t + \frac{19}{10}e^{-t/2} \sin t$ $t \geq 0$

25. $y(t) = te^{-t} + t^2e^{-t}$ $t \geq 0$ **27.** $y(t) = -2t^2 + 12t - 35 + 35e^{-t/3}$ $t \geq 0$

29. $x(t) = \dfrac{1}{2}\left(\sin t - \dfrac{1}{\sqrt{2}} \sin \sqrt{2}t\right)$ $t \geq 0$ **31.** $v_R(t) = 4e^{-t/3} - 4e^{-t/2} - \frac{2}{3}te^{-t/2}$ $t \geq 0$

34. $x(t) = \dfrac{1}{2}\left(\sin t - \dfrac{1}{\sqrt{2}}\sin\sqrt{2}\,t\right)$ $t \geq 0$

Part 10

37. $p(s) = (s+1)(s+3);\ y(t) = A_1 e^{-t} + A_2 e^{-3t}$ $t \geq 0$

39. $p(s) = [(s+2)^2 + 1]^1;\ y(t) = e^{-2t}(A_1 \cos t + A_2 \sin t)$ $t \geq 0$

41. $p(s) = s^3;\ y(t) = A_1 t^2 + A_2 t + A_3$ $t \geq 0$

43. $p(s) = (s^2 + 3)^2;\ y(t) = A_1 \cos\sqrt{3}\,t + A_2 \sin\sqrt{3}\,t + A_3 t \cos\sqrt{3}\,t + A_4 t \sin\sqrt{3}\,t$ $t \geq 0$

45. $y^{(6)} + 6y^{(5)} + 11y^{(4)} + 12y^{(3)} + 10y'' + 6y' + 9y = 0$

$y(t) = A_1 \cos 3t + A_2 \sin 3t + A_3 \cos t + A_4 \sin t + A_5 t \cos t + A_6 t \sin t$ $t \geq 0$

Section D

Part 8

1. $\mathscr{L}[f](s) = \dfrac{e^{-s} + 3e^{-2s}}{s};\ f(t) = u(t-1) + 3u(t-2)$

2. $\mathscr{L}[f](s) = \dfrac{1 - 3e^{-s}}{s};\ f(t) = 4 - 3u(t+1) - 3u(t-1)$

5. (a) $\mathscr{L}^{-1}\left[\dfrac{e^{-3s}}{s}\right](t) = u(t-3)$ $t \geq 0$

(b) $\mathscr{L}^{-1}\left[\dfrac{e^{-\pi s} + 2e^{-4s} - 4e^{-6s}}{s}\right](t) = u(t-\pi) + 2u(t-4) - 4u(t-6)$ $t \geq 0$

7. (b) $\mathscr{L}^{-1}\left[\dfrac{se^{-\pi s}}{(s^2+1)}\right](t) = u(t-\pi)\cos(t-\pi)$ $t \geq 0$

(d) $\mathscr{L}^{-1}\left[e^{-3s}\left(\dfrac{1}{s} - \dfrac{1}{s+2}\right)\right](t) = u(t-3)(1 - e^{-2(t-3)})$ $t \geq 0$

8. (c) $\mathscr{L}^{-1}\left[e^{-3s}\left(\dfrac{s}{s^2+3s+2}\right)\right](t) = u(t-3)(2e^{-2(t-3)} - e^{-(t-3)})$ $t \geq 0$

(d) $\mathscr{L}^{-1}\left[(1 - e^{-3s})\left(\dfrac{s}{s^2+3s+2}\right)\right](t) = 2e^{-2t} - e^{-t} - u(t-3)[2e^{-2(t-3)} - e^{-(t-3)}]$ $t \geq 0$

(f) $\mathscr{L}^{-1}\left[e^{-3s}\left(\dfrac{4s+3}{s^2+6s+13}\right)\right](t) = u(t-3)[4e^{-3(t-3)}\cos 2(t-3) - \tfrac{9}{2}e^{-3(t-2)}\sin 2(t-3)]$

$t \geq 0$

9. $y(t) = \tfrac{1}{3} - \tfrac{2}{3}e^{-3t} - u(t-2)(\tfrac{1}{3} - \tfrac{1}{3}e^{-3(t-2)})$ $t \geq 0$

12. $y(t) = \sin t + u(t-2\pi)(1 - \cos(t-2\pi))$ $t \geq 0$

14. $y(t) = \tfrac{1}{3}e^t + \tfrac{2}{3}e^{-t/2}\cos\dfrac{\sqrt{3}}{2}t + u(t-3)\left[-1 + \tfrac{1}{3}e^{(t-3)} + \tfrac{2}{3}e^{-(t-3)/2}\cos\dfrac{\sqrt{3}}{2}(t-3)\right]$ $t \geq 0$

15. $y(t) = u(t-2)\tfrac{1}{9}[-1 + 3(t-2) + e^{-3(t-2)}]$ $t \geq 0$

Part 13

17. $\mathcal{L}[f(t)](s) = \int_2^\infty (t - 2)^2 e^{-st}\, dt = e^{-2}\dfrac{2}{s^3}$

19. $\mathcal{L}[f(t)](s) = \int_\pi^{2\pi} e^{-st} \sin t\, dt = \dfrac{-e^{-\pi s} - e^{-2\pi s}}{s^2 + 1}$

21. $\mathcal{L}[u(t - 2)e^{-(t-2)}](s) = \dfrac{e^{-2s}}{s + 1}$

23. $\mathcal{L}[f](s) = \dfrac{e^{-3s}(2e^3)}{s + 1}$

25. $\mathcal{L}[f](s) = \dfrac{1 - e^{-2s}}{s^2}$

27. $\mathcal{L}[f](s) = \dfrac{e^{-\pi s} + e^{-2\pi s}}{s^2 + 1}$

29. $y(t) = e^{-t} + u(t - 1)(1 - e^{-(t-1)})$ $t \geq 0$
$\quad y'(t) = -e^{-t} + u(t - 1)e^{-(t-1)}$ $t \geq 0$

31. $y(t) = 1 - e^{-t} + (1 - t)u(t - 1)$ $t \geq 0$
$\quad y'(t) = e^{-t} - u(t - 1)$ $t \geq 0$

Section E

Part 7

1. (a) $y(t) = 2u(t - 1)e^{-4(t-1)}$ $t \geq 0$ (c) $y(t) = 2e^{-4t}$ $t \geq 0$
3. $y(t) = e^{-2t} + u(t - 1)\frac{3}{2}(1 - e^{-2(t-1)}) - u(t - 2)e^{-2(t-2)}$ $t \geq 0$
5. $y(t) = u(t - 1)e^{-2(t-1)} \sin (t - 1)$ $t \geq 0$

7. $\mathcal{L}^{-1}\left[\dfrac{2s + 1}{s + 2}\right](t) = 2\delta(t) - 3e^{-2t}$ $t \geq 0$

Part 11

11. (a) $x(t) = \dfrac{1 - \cos 2t}{8}$ $t \geq 0$ (b) $x(t) = \frac{1}{2} \sin 2t$ $t \geq 0$

12. $f = 2v' + 4v$; (a) $v(t) = \frac{1}{2} - \frac{1}{2}e^{-2t}$ $t \geq 0$; (b) $v(t) = \frac{1}{2}e^{-2t}$ $t \geq 0$

13. $v' + v = \dfrac{2i}{3}$; (b) $v(t) = \dfrac{2e^{-t}}{3}$ $t \geq 0$; (c) $v(t) = \frac{2}{3}$ $t \geq 0$

14. (b) $\int_0^1 \delta(t - \frac{1}{2})\, dt = 1$; (c) $\int_0^1 \delta(t - 1)\, dt = 0$ 17. $\int_0^\pi \sin \left(t - \dfrac{\pi}{3}\right) \delta \left(t - \dfrac{\pi}{4}\right) dt = -\sin \dfrac{\pi}{12}$

18. (a) $\int_0^{10} (\sin z)\delta(z - t)\, dz = \sin t$ if $0 \leq t < 10$ (b) $\int_0^{10} (\sin z)\delta(z - t)\, dz = 0$ if $10 \leq t$

Section F

Part 8

1. (b) $v(t) = -\sqrt{6}\, u(t - t_1) \sin \sqrt{6}\, (t - t_1)$ $t \geq 0$, if $t_1 \geq 0$
   (c) $v(t) = x'(t)$ $t \geq 0$
2. On the circuit diagram *you must define* the voltage across the resistor; (b) $v_0(t) = 2e^{-6t}$ $t \geq 0$

Section G

Part 11

2. (a) $u(t-1)\frac{1}{3}(1-e^{-3(t-1)})$ (b) $u(t-1)\frac{1}{3}(1-e^{-3(t-1)})$

5. $u(t-e)\frac{1}{3}(1-e^{-3(t-3)})$ **6.** (a) $\delta(t)_* \cos \pi t = 1$

9. $y(t) = \dfrac{e^{2t}-e^{-2t}\cos 3t - \frac{4}{3}e^{-2t}\sin 3t}{25}$ $t \geq 0$

10. $y(t) = \frac{1}{3}e^{-2(t-1)}\sin 3(t-1)$ $t \geq 0$

11. (d) $y(t) = \frac{1}{13} + \frac{25}{13}e^{-2t}\cos 3t + \frac{38}{39}e^{-2t}\sin 3t$ $t \geq 0$

14. $f' + \dfrac{B}{M}f = Bx''$ (a) $x(t) = \frac{1}{3}t^3 + \dfrac{B}{12M}t^4$ $t \geq 0$

15. (a) $H(s) = \dfrac{R}{LS+R}$ (b) $h(t) = u(t)\dfrac{R}{L}e^{-(R/L)t}$

(c) $v_0(t) = u(t)\dfrac{R/L}{400+(R/L)^2}\left(20e^{-Rt/L} - 20\cos 20t + \dfrac{R}{L}\sin 20t\right)$

(d) $v_0(t) = u(t)(1-e^{-Rt/L}) + u(t-3)(e^{-R(t-3)/L}-1)$

17. (a) $H(s) = \dfrac{s^2}{s+B/M}$ (b) $h(t) = \delta'(t) - \dfrac{B}{M}\delta(t) + u(t)\left(\dfrac{B}{M}\right)^2 e^{-Bt/M}$

(c) $x(t) = u(t)\left(\dfrac{1}{3}t^2 + \dfrac{B}{12M}t^4\right)$

20. (a) $H(s) = \dfrac{sB_2}{Ms^2 + (B_1+B_2)s + K}$

(c) If $B_1 > 1$, then $h(t) = \dfrac{u(t)}{\beta-\gamma}(e^{-\gamma t} - e^{-\beta t})$

where $\gamma = \dfrac{-(1+B_1) + \sqrt{(1+B_1)^2 - 4}}{2}$

$\beta = \dfrac{-(1+B_1) - \sqrt{(1+B_1)^2 - 4}}{2}$

If $B_1 = 1$, then $h(t) = u(t)(e^{-t} - te^{-t})$

If $B_1 < 1$, then $h(t) = u(t)e^{-\gamma t}\left(\cos\sqrt{1-\gamma^2}\,t - \dfrac{\gamma}{\sqrt{1-\gamma^2}}\sin\sqrt{1-\gamma^2}\,t\right)$

where $\gamma = (1+B_1)/2$

Section H

Part 4

1. The jump is zero; $\lim\limits_{t\to\infty} x(t) = -\frac{1}{2}$

4. $v'(t) + \dfrac{B}{M}v(t) = \dfrac{3}{M}u(t)$

$\qquad v(0) = -3 - \dfrac{5}{M}$

Section I

Part 12

1. (b) $\text{Re}(2 - 2j) = 2$; $\text{Im}(2 - 2j) = -2$ (d) $\text{Re}(-3) = -3$; $\text{Im}(-3) = 0$

(i) $\text{Re}[(-1 + 2j)(3 + j)] = -5$; $\text{Im}[(-1 + 2j)(3 + j)] = 5$

(k) $\text{Re}\left(\dfrac{1 + 2j}{3 + j}\right) = \dfrac{1}{2}$; $\text{Im}\left(\dfrac{1 + 2j}{3 + j}\right) = \dfrac{1}{2}$ (m) $\text{Re}(5e^{-(\pi/4)j}) = \dfrac{5}{\sqrt{2}}$; $\text{Im}(5e^{-(\pi/4)j}) = \dfrac{-5}{\sqrt{2}}$

(n) $\text{Re}(2e^{(2\pi/3)j}) = -1$; $\text{Im}(2e^{(2\pi/3)j}) = \sqrt{3}$ (q) $\text{Re}(e^{2 + (\pi/2)j}) = 0$; $\text{Im}(e^{2 + (\pi/2)j}) = e^2$

(t) $\text{Re}(e^{(\pi/3)j} - e^{-(\pi/4)j}) = \cos\dfrac{\pi}{3} - \cos\left(\dfrac{-\pi}{4}\right)$; $\text{Im}(e^{(\pi/3)j} - e^{-(\pi/4)j}) = \sin\dfrac{\pi}{3} + \sin\dfrac{\pi}{4}$

2. (b) $|2 - 2j| = \sqrt{8}$; $\text{Arg}(2 - 2j) = -\dfrac{\pi}{4}$; $\overline{(2 - 2j)} = 2 + 2j$

(d) $|-3| = 3$; $\text{Arg}(-3) = \pi$; $\overline{-3} = -3$

(i) $|(-1 + 2j)(3 + j)| = 5\sqrt{2}$; $\text{Arg}((-1 + 2j)(3 + j)) = \dfrac{3\pi}{4}$; $\overline{(-1 + 2j)(3 + j)} = -5 - 5j$

(k) $\left|\dfrac{1 + 2j}{3 + j}\right| = \dfrac{1}{\sqrt{2}}$; $\text{Arg}\left(\dfrac{1 + 2j}{3 + j}\right) = \dfrac{\pi}{4}$; $\overline{\left(\dfrac{1 + 2j}{3 + j}\right)} = \dfrac{1 - j}{2}$

(m) $|5e^{-(\pi/4)j}| = 5$; $\text{Arg}(5e^{-(\pi/4)j}) = \dfrac{-\pi}{4}$; $\overline{5e^{-(\pi/4)j}} = 5e^{(\pi/4)j}$

(n) $|2e^{(2\pi/3)j}| = 2$; $\text{Arg}(2e^{(2\pi/3)j}) = \dfrac{2\pi}{3}$; $\overline{2e^{(2\pi/3)j}} = 2e^{-(2\pi/3)j}$

(q) $|e^{2 + (\pi/2)j}| = e^2$; $\text{Arg}(e^{2 + (\pi/2)j}) = \dfrac{\pi}{2}$; $\overline{e^{2 + (\pi/2)j}} = e^{2 - (\pi/2)j}$

(t) $|e^{(\pi/3)j} - e^{-(\pi/4)j}| = \left[\left(\dfrac{1 - \sqrt{2}}{2}\right)^2 + \left(\dfrac{\sqrt{3} - \sqrt{2}}{2}\right)^2\right]^{1/2}$

$\text{Arg}(e^{(\pi/3)j} - e^{-(\pi/4)j}) = \pi + \text{Tan}^{-1}\left(\dfrac{\sqrt{3} - \sqrt{2}}{1 - \sqrt{2}}\right)$

$\overline{e^{(\pi/3)j} - e^{-(\pi/4)j}} = e^{-(\pi/3)j} - e^{(\pi/4)j}$

(w) $|1 - 2j| = \sqrt{5}$; $\text{Arg}(1 - 2j) = \text{Tan}^{-1}(-2)$; $\overline{(1 - 2j)} = 1 + 2j$

5. (d) $-3 = -3 + 0j$ (i) $(-1 + 2j)(3 + j) = -5 + 5j$

(k) $\dfrac{1 + 2j}{3 + j} = \dfrac{1}{2} + \dfrac{1}{2}j$ (m) $5e^{-(\pi/4)j} = 5\cos\dfrac{\pi}{4} + \left(-5\sin\dfrac{\pi}{4}\right)j$

(q) $e^{2 + (\pi/2)j} = 0 + e^2 j$ (t) $e^{(\pi/3)j} - e^{-(\pi/4)j} = \dfrac{1 - \sqrt{2}}{2} + \dfrac{\sqrt{3} - \sqrt{2}}{2}j$

Part 18

6. $A = \dfrac{1}{2j}$, $B = -\dfrac{1}{2j}$; $\sin 10t = \dfrac{1}{2j}(e^{j10t} - e^{-j10t})$

7. (b) $\mathscr{L}^{-1}\left[\dfrac{3}{(s - j)^2}\right](t) = 3te^{jt}$ $t \geq 0$

(e) $\mathscr{L}^{-1}\left[\dfrac{1}{s^2 + 1}\right](t) = \mathscr{L}^{-1}\left[\dfrac{1}{(s - j)(s + j)}\right](t) = \dfrac{1}{2j}(e^{jt} - e^{-jt})$ $t \geq 0$

(g) $\mathcal{L}^{-1} \left[\dfrac{1}{s^2 + s + 1} \right] (t) = -\dfrac{1}{3} te^{\alpha t} - \dfrac{2j}{3\sqrt{3}} e^{\alpha t} - \dfrac{1}{3} te^{\beta t} + \dfrac{2j}{3\sqrt{3}} e^{\beta t}$ $t \geq 0$

where $\alpha = (-1 + \sqrt{3}\,j)/2$ and $\beta = (-1 + \sqrt{3}\,j)/2$

Section J

Part 2

3. Translate by $t_0 = \frac{1}{6}$. The phase of $\sin 3\pi t$ relative to $\cos 3\pi t$ is $\pi/2$.

5. The phase of $\sin 6\pi t$ relative to $\cos 6\pi t$ is $\pi/2$.

7. The phase of $\sin \omega t$ relative to $\cos (\omega t + \pi)$ is $3\pi/2$.

9. The phase of $\sin (\omega(t + t_0))$ relative to $\cos \omega t$ is $\omega t_0 + \pi/2$.

Part 8

11. On the circuit diagram you *must define* the voltages across the circuit elements.

$$i(t) = \frac{1}{(R^2 + \omega^2 L^2)^{1/2}} \cos \left(\omega t - \mathrm{Tan}^{-1} \frac{\omega L}{R} \right)$$

$$= \frac{R}{R^2 + \omega^2 L^2} \cos \omega t + \frac{\omega L}{R^2 + \omega^2 L^2} \sin \omega t$$

13. On the circuit diagram you *must define* the voltage across the capacitor and the current

$$v_0(t) = \frac{CR\omega}{(1 + C^2 R^2 \omega^2)^{1/2}} \cos \left(\omega t + \mathrm{Tan}^{-1} \frac{1}{CR\omega} \right)$$

$$= \frac{(CR\omega)^2 \cos \omega t - CR\omega \sin \omega t}{1 + C^2 R^2 \omega^2}$$

15. On the free-body diagram you *must define* the tension in spring, dashpot, and connector to the mass.

$$f(t) = \frac{2\omega^2 [(\omega^2 - 25)^2 + (6\omega^3)^2]^{1/2}}{(\frac{5}{2} - \omega^2)^2 + (\frac{3}{2}\omega)^2} \cos (\omega t + \phi)$$

where $\phi = \begin{cases} \mathrm{Tan}^{-1} \dfrac{6\omega^2}{\omega^2 - 25} & \omega^2 - 25 > 0 \\[3mm] \dfrac{\pi}{2} & \omega^2 - 25 = 0 \\[3mm] \pi + \mathrm{Tan}^{-1} \dfrac{6\omega^3}{\omega^2 - 25} & \omega^2 - 25 < 0 \end{cases}$

17. On the free-body diagram you *must define* the tension in the spring.

$$x(t) = \begin{cases} \dfrac{2}{K - M\omega} \cos \omega t & \text{if } K - M\omega > 0 \\[3mm] \dfrac{2}{M\omega - K} \cos (\omega t - \pi) & \text{if } K - M\omega < 0 \end{cases}$$

CHAPTER IV

Section A

Part 6

1. (c) $L[e^{7t}] = 28e^{7t}$ (d) $L[t^2] = 7t^2 + 6t$
   (f) $L[5e^{7t} + 6t^2] = 140e^{7t} + 42t^2 + 36t$

2. (b) $\mathcal{T}[t \sin 3t] = t(\frac{5}{34} \sin 3t - \frac{3}{34} \cos 3t) - \frac{16}{1156} \sin 3t + \frac{30}{1156} \cos 3t - \frac{30}{1156}e^{-5t}$
   (c) $\mathcal{T}[e^{7t}] = \frac{1}{12}(e^{7t} - e^{-5t})$ (e) $\mathcal{T}[e^{-2t} \sin 3t] = \frac{1}{6}(e^{-2t} \sin 3t - e^{-2t} \cos 3t + e^{-5t})$

3. It is true; L is linear.

7. (a) $Q[x](t) = Kx'(t) * e^{-(K/B)t}$; $T[f](t) = 1 * \left[f'(t) + \frac{K}{B}f(t)\right] = \int_0^t \left[f'(z) + \frac{K}{B}f(z)\right] dz$;
   Yes, they are mutually inverse.

8. From $v' = i'R + \frac{1}{C}i$ deduce that

$$T[i] = 1 * \left[Ri'(t) + \frac{1}{C}i(t)\right] ; K[v] = \frac{1}{R}v'(t) * e^{-t/RC}$$

   Yes, they are mutually inverse.

Part 11

10. (a) $L[e^t] = 12e^t$ (d) $L[te^{-t}] = 0$
   (e) Yes, by (b) and (e) both e^{-t} and te^{-t} are solutions.

Part 15

12. (b) $x^{(4)} + \frac{1}{3}x^{(3)} + x' + \frac{1}{3}x = \frac{1}{3}t - 4$

Part 19

14. Linear; $a_1 = 1$, $a_0 = 3$
16. Linear; $a_2(t) = e^t$, $a_1 = 0$, $a_0 = 0$
18. Not linear; test homogeneity, with $y(t) = t$, $C = 5$
20. Linear; $a_2 = 1$, $a_1(t) = e^t$, $a_0(t) = \cos 2t$

Section B

Part 10

1. $p(s) = s$; root 0 with multiplicity 1; $y(t) \equiv C$
3. $p(s) = s + 3$; root -3 with multiplicity 1; $y(t) = Ae^{-3t}$
5. $p(s) = s^2 + 4s + 5$; roots: $-2 + j$ with multiplicity 1, $2 - j$ with multiplicity 1
   $y(t) = B_1e^{-2t} \cos t + B_2e^{-2t} \sin t$
7. $p(s) = s^2 + 4s + 3$; roots -1, -3, each with multiplicity 1
   $y(t) = A_1e^{-t} + A_2e^{-3t}$

9. $p(s) = s^3 - 4s^2 + 4s$; roots: 0 with multiplicity 1, 2 with multiplicity 2
$y(t) = A_1 + (A_2 + A_3 t)e^{2t}$

11. $p(s) = s^3 + 3s^2 + 3s + 1$; roots: -1 with multiplicity 3
$y(t) = (A_1 + A_2 t + A_3 t^2)e^{-t}$

13. $y^{(4)} - 8y^{(3)} + 24y'' - 32y' + 16y = 0$
$y(t) = (A_1 + A_2 t + A_3 t^2 + A_4 t^3)e^{2t}$

Part 17

16. e^{-4t}, te^{-4t}

22. $e^{2jt}, te^{2jt}, e^{-2jt}, te^{-2jt}, e^{-3t}$

27. $(D - 8)^5$

31. $(D^2 + 1)^3$

33. $y(t) = (A_1 + tA_2)e^{-t}$

37. $y(t) = A_1 + A_2 e^{(-2+j)t} + A_3 e^{(-2-j)t} = A_1 + e^{-2t}(B_1 \cos t + B_2 \sin t)$

39. $y(t) = A_1 e^{2jt} + A_2 e^{-2jt} + A_3 e^{4jt} + A_4 e^{-4jt}$
$= B_1 \cos 2t + B_2 \sin 2t + B_3 \cos 4t + B_4 \sin 4t$

41. $L_1[L_2[y]] = t^2 y''' + 2ty''$, while $L_2[L_1[y]] = t^2 y'''$; $L_1 L_2 \neq L_2 L_1$

43. (b) $2x''' - 10x'' + 12x' = 5e^{3t} - e^t$

44. $z^{(6)} - z^{(5)} + 3z^{(4)} - 2z^{(3)} - z'' = 64e^{4t}$

19. $e^{-4t}, te^{-4t}, e^{-3t}$

25. $(D + 2)^2$

29. $(D^2 + 4\pi^2)$

35. $y(t) = A_1 + A_2 e^{-2t} + A_3 e^{-4t}$

Section C

Part 5

1. $y(t) = \frac{1}{3} + Ae^{-6t}$

8. $y(t) = -\dfrac{1}{36} + \dfrac{t}{6} + te^{-6t} + Ae^{-6t}$

4. $y(t) = -\dfrac{t}{108} + \dfrac{t^2}{36} - \dfrac{t^3}{18} + A_1 + A_2 e^{-6t}$

10. $y(t) = \frac{1}{2}t^2 e^{-2t} + (A_1 + A_2 t)e^{-2t}$

Part 10

13. $y(t) = -\frac{2}{9} + \frac{2}{3}t + Ae^{-3t}$

17. $y(t) = -\frac{1}{5}e^t + (A_1 + A_2 t)e^{-2t}$

23. $y(t) = -\frac{1}{5}e^t + (-1 - \frac{2}{5}e)e^{-2(t-1)} + (2 + \frac{1}{5}e)te^{-2(t-1)}$

15. $y(t) = \dfrac{6 \cos \pi t + 2\pi \sin \pi t}{9 + \pi^2} + Ae^{-3t}$

25. The polynomial $y_p(t) = -\frac{2}{9} + \frac{2}{3}t$ is the only polynomial which is a solution, because all other solutions are of the form $y(t) = -\frac{2}{9} + \frac{2}{3}t + Ce^{-3t}$ with C nonzero. No polynomial of degree zero can be a solution; no polynomial of degree (exactly) two can be a solution.

Part 6

1. (a) $y(t) = At^{-1/2}$ $t > 0$; $y(t) = 2t^{-1/2}$ $t > 0$

2. (b) $y(t) = A_1(-t)^{1/3} + A_2(-t)^{-1/3}$ $t < 0$; $y(t) = \dfrac{-(-t)^{1/3} + (-t)^{-1/3}}{2}$ $t < 0$

3. (a) $y(t) = A_1 t^{-2} + A_2 t^{-2} \ln t$ $t > 0$
$y(t) = (4 - 8 \ln 2)t^{-2} + 8t^{-2} \ln t$ $t > 0$

4. (b) $y(t) = (-t)^3[B_1 \cos (2 \ln (-t)) + B_2 \sin (2 \ln (-t))]$ $t < 0$
$y(t) = (-t)^3[\cos (2 \ln (-t)) + \frac{3}{2} \sin (2 \ln (-t))]$ $t < 0$

Section E

Part 6

1. (a) $v_0(t) = \cos 3t$, $v_1(t) = \frac{1}{3}\sin 3t$
$y(t) = y(0)\cos 3t + y'(0)\frac{1}{3}\sin 3t$

(b) $v_0(t) = \cos 3(t-2)$, $v_1(t) = \frac{1}{3}\sin 3(t-2)$
$y(t) = y(2)\cos 3(t-2) + y'(2)\frac{1}{3}\sin 3(t-2)$

3. (a) $v_0(t) = e^{-4t} + 4te^{-4t}$, $v_1(t) = te^{-4t}$
$y(t) = y(0)(e^{-4t} + 4te^{-4t}) + y'(0)te^{-4t}$

(b) $v_0(t) = e^{-4(t-5)} + 4(t-5)e^{-4(t-5)}$, $v_1(t) = (t-5)e^{-4(t-5)}$
$y(t) = y(5)[e^{-4(t-5)} + 4(t-5)e^{-4(t-5)}] + y'(5)(t-5)e^{-4(t-5)}$

5. (a) $v_0(t) = \frac{1}{2}t^5 + \frac{1}{2}t^{-5}$, $v_1(t) = \frac{1}{10}t^5 - \frac{1}{10}t^{-5}$
$y(t) = y(1)\frac{1}{2}(t^5 + t^{-5}) + y'(1)\frac{1}{10}(t^5 - t^{-5})$ $t > 0$

(b) $v_0(t) = \frac{1}{2}\left[\left(\frac{-t}{2}\right)^5 + \left(\frac{-t}{2}\right)^{-5}\right]$, $v_1(t) = \frac{1}{10}\left[\left(\frac{-t}{2}\right)^5 - \left(\frac{-t}{2}\right)^{-5}\right]$

$y(t) = y(-2)\frac{1}{2}\left[\left(\frac{-t}{2}\right)^5 + \left(\frac{-t}{2}\right)^{-5}\right] + y'(-2)\frac{1}{10}\left[\left(\frac{-t}{2}\right)^5 - \left(\frac{-t}{2}\right)^{-5}\right]$ $t < 0$

7. (a) $v_0(t) = \cos\left(\frac{1}{2}\ln\frac{t}{3}\right)$, $v_1(t) = 6\sin\left(\frac{1}{2}\ln\frac{t}{3}\right)$ $t > 0$

$y(t) = y(3)\cos\left(\frac{1}{2}\ln\frac{t}{3}\right) + y'(3)6\sin\left(\frac{1}{2}\ln\frac{t}{3}\right)$ $t > 0$

(b) $v_0(t) = \cos\left(\frac{1}{2}\ln\frac{-t}{5}\right)$, $v_1(t) = -10\sin\left(\frac{1}{2}\ln\frac{-t}{5}\right)$ $t < 0$

$y(t) = y(-5)\cos\left(\frac{1}{2}\ln\frac{-t}{5}\right) + y'(-5)(-10)\sin\left(\frac{1}{2}\ln\frac{-t}{5}\right)$ $t < 0$

Part 12

9. $e^t = \frac{1}{2}\cosh t + \frac{1}{6}(3\sinh t)$ **12.** $1 = \frac{1}{2}t^2 - (t+1)^2 + \frac{1}{2}(t+2)^2$

14. If $\sin 2t = A_1\cos 2t + A_2\sin t$, then $A_2\sin t = -A_1\cos 2t + \sin 2t$, the function on the right having period π and the function on the left not having any period smaller than 2π. Therefore the function cannot be so spanned.

17. Linearly independent

19. (a) Linearly independent   (b) Linearly dependent

22. Any one of the three functions $t+1$, $t+7$, or $t-1$ may be removed leaving a linearly independent set.

Section F

Part 8

1. $y(t) = -2te^{2t} + A_1e^{2t} + A_2e^{3t}$ **3.** $y(t) = \frac{2}{3}t^2e^{3t} + A_1te^{3t} + A_2e^{3t} + A_3$

5. $y(t) = t^{-2} + A_1t^{-5} + A_2t^{-1}$ **7.** $y(t) = \frac{3}{4} + Ce^{-2t^2}$

9. $c''(t)t + c'(t) = t^{-3}$; $y(t) = \frac{1}{4} + C_1t^2\ln|t| + C_2t^2$

11. $c''(t) - 2c'(t) = 0$; $y(t) = C_1t + C_2te^{2t}$

Appendix

Part 3

1. (b) $(k + 1)b_{k+1} + 5b_k = 0$ $k \geq 0$; $y(t) = 7 \sum_{k=0}^{\infty} \dfrac{(-5t)^k}{k!}$

(c) $(k + 1)c_{k+1} + 5c_k = 0$ $k \geq 0$; $y(t) = c_0 \sum_{k=0}^{\infty} \dfrac{(-5(t-2))^k}{k!}$

3. (b) $b_1 + 5b_0 = 1$; $2b_2 + 5b_1 = 1$; $(k + 1)b_{k+1} + 5b_k = 0$ $k \geq 2$;

$y(t) = 2 - 9(t - 1) + \dfrac{46}{25} \sum_{k=2}^{\infty} \dfrac{(-5(t-1))^k}{k!}$

Part 8

6. $(k + 2)(k + 1)b_{k+2} + 3(k + 1)b_{k+1} + 2b_k = 0$ $k \geq 0$

$y(t) = 1 - t + \dfrac{t^2}{2} - \dfrac{t^3}{6} + \dfrac{t^4}{24} - \dfrac{t^5}{120} + \cdots = \sum_{k=0}^{\infty} \dfrac{(-t)^k}{k!}$

8. $2b_2 + b_0 = 0$; $(k + 2)(k + 1)b_{k+2} + b_k + b_{k-1} = 0$ $k \geq 1$

$y(t) = 1 - \frac{1}{2}t^2 - \frac{1}{6}t^3 + \frac{1}{24}t^4 + \cdots$

10. $t = 0$: irregular singular point

$t = 1$: regular point

$t = 2$: regular singular point

12. Indicial equation $p^2 - 1 = 0$.

$p = 1$: b_0 is arbitrary, $b_1 = 0$, $(k^2 + 2k)b_k = -b_{k-2}$ $k \geq 2$

$$y(t) = b_0 t \sum_{m=0}^{\infty} \dfrac{(-1)^m (t/2)^{2m}}{m! \, (m + 1)!}$$

$p = -1$: c_0 is arbitrary, $c_1 = 0$, $((k - 1)^2 - 1)c_k + c_{k-2} = 0$ $k \geq 2$

For $k = 2$, the recurrence relation implies $c_0 = 0$, contradicting the arbitrary choice. Thus there is no solution of this form corresponding to $p = -1$.

14. Indicial equation $p^2 - (\frac{1}{2})^2 = 0$.

$p = \frac{1}{2}$: b_0 is arbitrary; $b_k = 0$ $k \geq 1$; $y(t) = b_0 t^{1/2}$

$p = -\frac{1}{2}$: c_0 is arbitrary, c_1 is arbitrary; $c_k = 0$ $k \geq 2$; $y(t) = c_0 t^{-1/2} + b_0 t^{1/2}$

17. u_1 is the solution to Problem 12, with a value taken for b_0, say, $b_0 = 1$. The second solution is

$$t \sum_{m=0}^{\infty} \dfrac{(-1)^m (t/2)^{2m}}{m! \, (m + 1)!} \ln t + \left(-2t^{-1} + \tfrac{3}{32} t - \tfrac{13}{27648} t^2 + \cdots \right)$$

INDEX

INDEX

Page numbers in *italic* indicate definitions and theorems; page numbers in **bold-face** indicate illustrations.